Introduction to RF Stealth

Introduction to RF Stealth

David Lynch, Jr.

© 2004 SciTech Publishing Inc.
5601 N. Hawthorne Way
Raleigh, NC 27613
www.scitechpub.com
sales@scitechpub.com

President and CEO: Dudley R. Kay
Page Composition and Graphics: J. K. Eckert & Company, Inc.
Indexing: Cynthia Coan
Cover Design: WT Design
Photos and Figures: Courtesy of DL Sciences, Inc.
Printer: Victor Graphics—Baltimore, MD

Also distributed by:

William Andrew Inc.—Resellers in the U.S. and Canada
13 Eaton Ave.
Norwich, NY 13815
(800) 932-7045
www.williamandrew.com

The Institution of Electrical Engineers—Europe and Members Worldwide
Michael Faraday House
Six Hills Way, Stevenage, SG1 2AY, UK
+44 (0) 1438-313311
www.ieee.org.uk

SPIE—The International Society for Optical Engineering—Members
P.O. Box 10, Bellingham, WA 98227-0010
(360) 676-3290
www.spie.org

International Standard Book Number 1-891121-21-9
Institution of Electrical Engineers ISBN 0-86341-349-8

10   9   8   7   6   5   4   3   2   1

SciTech books may be purchased at quantity discounts for educational, training, or sales promotional use. For information, contact the publisher at (919) 866-1501 or sales@scitechpub.com

This textbook was reviewed and approved for Public Release, Distribution Unlimited, by Aeronautical Systems Center Public Affairs, Wright Patterson Air Force Base, Ohio, Case number ASC-03-2833 on 30 October 2003.

Dedication

George "Knobby" Walsh
My first mentor

Contents

Preface

WHY THE BOOK?

This book is based on material developed in the mid-1970s for stealth weapon systems. The material remained classified for 20 years or more. All of the material in the text is from unclassified sources that have become available in the last 5 to 10 years. Also, much of this material is in the public domain and first appeared in un-copyrighted sources. It appears in many places, although I created much of it over the last 30 to 40 years. A few photos, tables, and figures in this intellectual property were made at Hughes Aircraft Company and first appeared in public documents that were not copyrighted. These photos, tables, and figures were acquired by Raytheon Company in the merger of Hughes and Raytheon in December 1997 and are identified as Raytheon photos, tables, or figures. The book is based on notes that the author and others prepared for Stealth Radar and Data Link courses, which the author taught between 1985 and 1996. There are several references to formerly classified documents. The author may have the only copies in existence, because the government usually destroys them when they become unclassified. The "author" is the editor and compiler of the material in this book. The ideas contained here are the products of an array of very bright people, and it is my privilege to summarize their work. Every attempt has been made to give credit where it is due, but the early originators were not as careful as they should have been in citing references, and some readers will undoubtedly find subjects in the text for which references aren't properly cited. The author apologizes in advance for these errors, but 25 years in the black makes them difficult to correct. In addition, try as one will, there will always be mistakes that escape into the text; please notify me of those you find so that they may be corrected.

The author had the privilege of pioneering modern stealth techniques beginning in 1975. Because of the necessary requirements of security, this material is not widely known. As a result of this limited availability, people keep "reinventing the wheel," keep "going down blind alleys" already explored, and keep making nonsensical claims. The book's purpose is to provide a new generation of designers with a firm basis for new developments and to allow buyers of stealth technology to sort the charlatans from serious engineers. Much stealth technology is retrospectively obvious and mathematically simple but, before it is explained, it is not so easy to see. This introductory book presents first principles in a simple and approximate way. The author has tried to avoid repeating material easily found in other texts except where it is essential to understand the limitations of both stealth and counterpart threat sensors. Some may find this book trivial and too simple even for an introductory text. Again, please accept apologies; the simplicity is because the author did not include anything he couldn't understand himself. Besides, if the book seems trivial, you're already beyond the point of need, anyway. Mathematicians may be bothered by some of the approximations, but these approximations have proven to be adequate for real applications. The text is oriented toward undergraduate seniors and graduate engineers with some prior background in radar, communications, and basic physics. It approaches each topic from a system engineering perspective. This is a labor of love, and it represents years of labor both teaching and writing.

ORGANIZATION OF THE BOOK

This book covers two major topics: *low observables* and *low probability of intercept (LO and LPI) of radars and data links*, collectively sometimes called *stealth*. In most sections, both are covered, because the signatures often interact. Each chapter has examples, student exercises, references, and counterpart appendices that describe the associated software on CDROM. Most of the analysis has been verified by experiment or computer simulation by the "Famous Names of the Radar/Stealth World." Much of the pertinent analysis in the form of computer programs is on the CDROM, which is described in Appendix A. Chapter 1 provides an introduction and history of RF/microwave LPI/LO techniques and some basic LPI/LO equations. Chapter 2 covers interceptability parameters and analysis. Chapter 3 covers current and future intercept receivers and some of their limitations. Chapter 4 surveys exploitation of both the natural and the threat environment and gives examples of one of the "great thoughts" of LPI/LO design, electronic order of battle (EOB) exploitation. Chapter 5 deals with LPIS waveforms and pulse compression. It covers another "great thought" of LPIS: Hudson-Larson complementary code pulse compression. Chapter 6 introduces some hardware techniques associated with LO/LPIS: low sidelobe/cross section antenna and radome design. It includes another of the "great thoughts" of LPI design: separable antenna illumination functions. Chapter 7 describes typical LPIS low-level RF and signal processing, which is often unique relative to conventional radar and data link processing.

David Lynch, Jr.
davidlynchjr@ieee.org
2004

Acknowledgments

The mid-1970s found the happy coincidence of very bright, hard working leaders, scientists, and engineers in both the DoD and industry working on the stealth problem. The author was probably the least of these great people and lucky to be along for the ride. The early government leaders included Jack Twigg, Bill Elsner, Skip Hickey, Ron Longbrake, Pete Worch, Ken Perko, Bruce James, Nick Willis, Bobby Bond, Harlan Jones, and (a little later) Ken Staten, Joe Ralston, Bob Swarts, Jerry Baber, Dave Englund, Don Merkl, Allen Koester, Hans Mark, Tom Swartz, Dick Scofield, Ken Dyson, John Sumerlot, John Griffen, Keith Glenn, Allen Atkins, Jim Tegnelia, John Entzminger, Tony Diana, Carl O'Berry, Jim Evatt, George Heilmeir, Allen Wiechman, and many others. The early industry leaders were Ben Rich, Ed Martin, Alan Brown, Norm Nelson, Dick Sherrer, Denys Overholtzer, John Cashen, Fred Oshiro, Irv Waaland, Dave Green, Eddie Phillips, Sam Thaler, Jim Uphold, Jack Pearson, Bob Hanson, Gene Gregory, Fred Rupp, Pete Bogdanovic, Steve Panaretos, Gary Graham, Mike O'Sullivan, and many others. We made lots of mistakes, as happens in every new technology but, fortunately, there were enough creative people that we worked our way through each failure to success. Like all new technologies, you can't give it away at the beginning, and much of the very early effort wasn't classified, because it had never worked yet. Thanks also go to Dr. Brian Kent and his USAF coworkers for a careful assessment of the security compliance of the book.

1

Introduction to Stealth Systems

1.1 INTRODUCTION

Guessing and knowing are two completely different things. The objective of stealth is to keep the adversary guessing until it is too late. Over the past few years, stealth platforms, especially aircraft, have come into public consciousness. However, stealth research work was conducted in earnest beginning in the mid-1970s and was spearheaded by the Defense Advanced Research Projects Agency (DARPA) in both U.S. Air Force and Navy programs. Most of those programs are still shrouded in secrecy, but a few, especially the earliest, are now declassified, and the basic notions of Stealth technology can be described. For reasons of classification, nothing can be said about deployed aircraft such as the F-117A and the B-2, pictured in Figs. 1.1 and 1.2, respectively, but they grew out of those early programs. Several generations of technology have now passed, as embodied by such aircraft, but the basic stealth concepts remain the same.

The Stealth Challenge

• Survive and prosper in the future environment of improved sensors, dense countermeasures, antiradiation weapons, and emitter locators.

• Become invulnerable or invisible.

Figure 1.1   F-117A Stealth attack fighter (photo by author).

Figure 1.2 B-2 stealth bomber (photo by D. Whelan).

The Stealth Approach

- Force the threats to use active sensors sparingly by employing antiradiation missiles and electronic countermeasures.

- Decrease predictability and increase "randomness" to force the threats to increase complexity and cost of intercept receivers, surveillance, fire control, and missiles.

- Reduce active and passive signatures and increase "hiding" to make weapon systems less visible.

- Use tactics that combine with the order of battle as well as the natural and man-made environment to enhance the effect of the reduced observables.

- Use prior knowledge and off-board sensor cueing to minimize on-board active and passive exposure.

Stealth is not one item but an assemblage of techniques, which makes a system harder to find and attack. Stealth radar and data link design involves the reduction of active and passive signatures. *Active signature* is defined as all the observable emissions from a stealth platform: acoustic, chemical (soot and contrails), communications, radar, IFF, IR, laser, and UV. *Passive signature* is defined as all the observables on a stealth platform that require external illumination: magnetic and gravitational anomalies; reflection of sunlight and cold outer space; reflection of acoustic, radar, and laser illumination; and reflection of ambient RF (sometimes called *splash track*). Active radar and data link signature reduction requires the use of techniques that minimize radiated power density at possible intercept receiver locations. Active signature reduction also depends on the implementation of tactics that reduce exposure time during emission.

The active signature reduction methods are commonly called *low probability of intercept (LPI)* techniques and are illustrated in Fig. 1.3. Passive signature reduction techniques are often called *low observables (LO)*. They require the development of radome, antenna cavity, and antenna designs as interactive elements of a common subsystem that yields low in-band and out-of-band *radar cross section (RCS)*. Additionally, passive radar signatures are reduced in-band by employing special antenna design techniques that minimize retroreflective echoes. *Low probability of intercept system (LPIS)* design is an engineering problem with a larger set of optimization constraints and hence is no different from every other modern design challenge. The stealth designer must always create designs in which complete knowledge of the design isn't much help to the threat.

Conventional antenna designs have sidelobe and mainlobe patterns that do not differ fundamentally from that shown in Fig. 1.4(a). There are a few close-in sidelobes

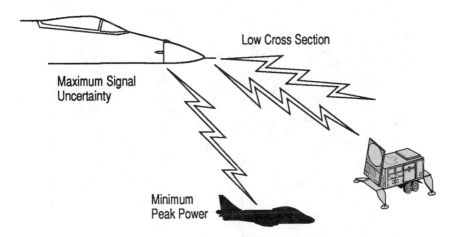

Figure 1.3 LO/LPIS objective.

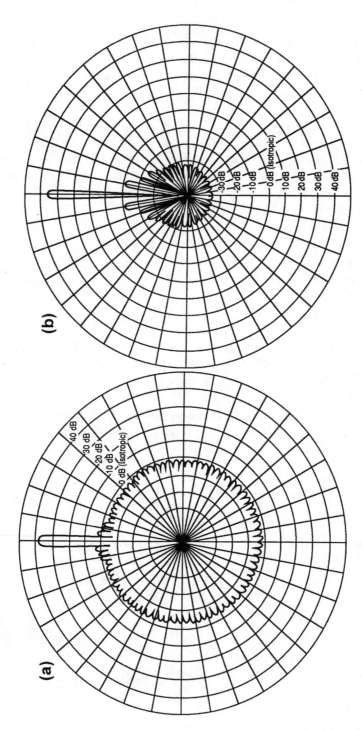

Figure 1.4 (a) Conventional and (b) LO/LI antenna patterns.

and, of course, the mainlobe, which are above isotropic. The remainder of the side-lobes average 3 to 6 dB below isotropic for a conventional antenna. On the other hand, an LO/LPI antenna has sidelobes that are −10 to −30 dB below isotropic and may average more than 20 dB below isotropic. These low sidelobes are realized at the expense of mainlobe gain and full utilization of the total aperture area. An idealized LO/LPI antenna pattern is shown in Fig. 1.4(b).

Similarly, conventional aircraft RCS may be noise-like, but it is generally well above 5 m^2 in most directions, as shown in Fig. 1.5(a). In most cases, very little effort in conventional platform design was devoted to reduction of the platform RCS. Some of this lack of effort was due to the belief that low-RCS vehicles would have undesirable aerodynamic, hydrodynamic, or functional shapes. It is now known that this is not the case; it is lack of planform alignment in the direction of the threat that results in many RCS spikes. A typical stealth aircraft signature strategy might be as shown in Fig. 1.5(b), where five main spikes contain most of the RCS signature.

Figure 1.6 suggests that spikes from each edge result in a signature like that shown on the right in Fig. 1.5. The strategy for a stealth aircraft signature is not fundamentally different from that for an LO/LPI antenna. The idea is to design a platform in such a way that there are only a few RCS spikes in carefully controlled directions. The vast majority of angle space is occupied by RCS that is substantially below that of a conventional platform. This can be thought of as similar to the radar signal ambiguity function, where the total volume under the curve must be preserved. Others have pointed out that the average RCS of a smooth body (see Table 1.5 for eggs and spheres) is on the order of 1/4 the total surface area. As a result, concentrating all of the reflections in a few directions can reduce the RCS in all other directions. RCS enhancement can be designed so that those spikes are not tactically useful to threat sensors, because the geometry is poor relative to either the threat radar horizon or the threat platform velocity vector. The first "great thought" of stealth is planform alignment or spike alignment of all of the major scatterers on a platform. Most RCS reduction comes from shaping. Radar-absorbing materials (RAMs) are applied only in areas where there are special problems, and they have very little to do with the average RCS. Furthermore, mismatches between RAM and free space create a first scatter that must *not* be reflected in a tactically useful direction.

1.1.1 Balanced Design

Active and passive signatures must be balanced relative to a threat. This requires that cross-section reduction and management of the active emissions be controlled in a combined and concerted way. First, the active emission of an LPI system (LPIS) must be controlled in each operating mode, using only the power that is necessary for that mode. In addition, the power must be programmed while tracking a target or communicating by data link so that, as a target or link destination opens or closes in range, the power is adjusted accordingly. This is the first "great thought" of LPIS tech-

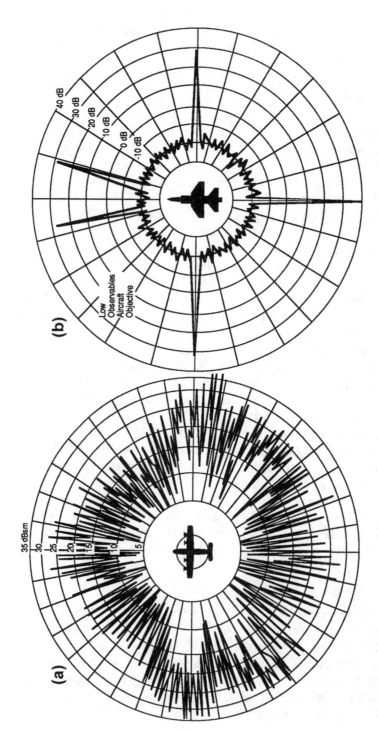

Figure 1.5 (a) Conventional and (b) stealth aircraft RCS signatures.

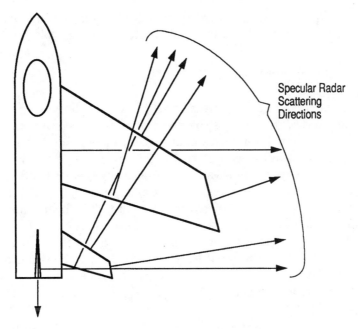

Figure 1.6 Lack of planform alignment results in many RCS spikes.

nology. Next, emitter on-times must be limited. The emitter active signature when the emitter is not transmitting still exists, but it is orders of magnitude lower than when it is on. By using a radar or data link only when necessary and by making maximal use of *a priori* information, the off-time of the emitter can be 60 percent or more, and the system can still perform very successful, useful missions. This is the second "great thought" of LPI design.

1.1.2 "Great Thoughts" of Stealth/LO Technology"

Active emissions control must be coupled with platform cross-section reduction. A number of techniques are used for radar cross-section (RCS) reduction, and they are presented in Chapter 6 of this book. The other "great thoughts," in addition to planform alignment, are

- Facets and shape; i.e., selecting shapes that reflect very low sidelobes in the direction of illumination/threat

- Edge treatment by convolution; i.e. using a mathematical function such as a Gaussian convolved on the junction of two surfaces to create a "blend" that has very low RCS sidelobes

- Impedance control; i.e. ensuring there are no discontinuities in the platform surface impedance, edge treatment by impedance matching of the platform skin to free space
- Exploitation of the environment

Unfortunately, stealth techniques are more expensive both to design and maintain. As a result, a stealth designer must strive to balance all the signatures of the system. The detection or threat range for each signature component should be similar. For example, the infrared threat range should be roughly the same as the radar or intercept threat ranges. Each threat system has a characteristic engagement balloon as shown in Fig. 1.7. The balloons will not be the same shape, in general, but in a modern stealth system, they should be close, as depicted in the figure. This means, for example, that the intercept range for a threat antiradiation missile should be roughly comparable to the range at which a threat radar guided missile would be effective against the platform radar cross section. Each signature should be balanced to the corresponding threat—UV, visible, IR, radar, RF, acoustic, magnetic, and so on. As previously mentioned, the stealth designer must always create designs in which complete knowledge of the LPIS doesn't help much. These constraints are discussed in detail in Chapters 2 and 4. Future and response threats are analyzed in Chapter 3, including massive noncoherent integration (sometimes called *radiometry*). More details are provided in Section 1.5 and Chapter 6.

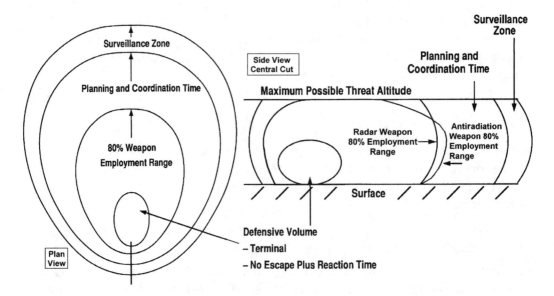

Figure 1.7 Balanced engagement balloons.

With respect to a radar or data link in an LO platform, the first element is a shaped radome. The radome usually will be bandpass, i.e., only passing those frequencies in the normal emitter band of operation. Outside of that band, the radome should be lossy and properly reflective. With appropriate radome shaping, out-of-band RCS can be quite low over a wide range of angles. The radome also can be switchable, which means that, even in-band, the radome may be closed when the emitter is not operating. Under these circumstances, the radome may take on cross-sectional properties of a lossy reflective, properly shaped surface (see Chapter 6).

However, when the emitter is on, the radome must be open and relatively transparent, which can cause the antenna to present a very large RCS in both its surface normal and its pointing direction. Usually, the antenna face must be canted relative to the beam pointing direction. In addition, the antenna should be dynamically stowed when it is off so that it presents the lowest possible RCS to the radar horizon or threat direction. Inside the compartment that contains the antenna, radiation-absorbing material (RAM) should be applied so energy that does enter the radome is well terminated and not retroreflected. Any other internal surfaces inside a radome antenna cavity (e.g., gimbals, mounting plates, auxiliary horns, and so forth) that might be visible at some viewing or multibounce angles should be faceted with their surface normals aligned along the platform planform.

1.1.3 RCS and Power Management Summary

The essential elements of RCS and power management are summarized below. More details are provided in Chapters 2, 5, and 6.

- LPIS power controlled by mode, program in track (Chapters 2 and 5)
- LPIS on-times limited to 40 percent or less (Chapter 2)
- Shaped low-RCS radome (Chapter 6)
- Radome frequency selective (Chapter 6)
- Radome reflective during off-time (Chapter 6)
- Antenna face canted with respect to beam pointing angle (Chapter 6)
- Antenna dynamically stowed during off-time (Chapter 6)
- Antenna cavity lined with radiation absorbent material (RAM) (Chapter 6)

1.2 INTRODUCTION TO LOW-PROBABILITY-OF-INTERCEPT SYSTEMS

A number of "simple" thoughts are robust counters to passive threats to emitters. Some of these have already been introduced in the preceding section. Some will require more background to appreciate. All of these simple but great thoughts will be described in more detail in the chapters that follow. There are many stealth ideas, but the "great thoughts" have been demonstrated to work best in the real environment of modern civil and military electronic systems.[19]

1.2.1 Great Thoughts of LPI Systems Design

The essential elements of active signature control (LPI) are summarized below. More details are provided in Chapters 2, 4, 5, 6, and 7.

- Maximum bandwidth (Chapters 2, 5, and 7)
- Power management (Chapter 2 and 5)
- Emission time control (Chapter 2)
- Creation and exploitation of gain mismatches (Chapter 2)
- Separable antenna illumination functions (Chapter 6)
- Hudson-Larson complementary code pulse compression (Chapter 5)
- Electronic order of battle exploitation (Chapter 4)

LPI counters the many passive threats to emitters such as radars and data links. Table 1.1 summarizes some of those threats with respect to threat type, type of intercept receiver, threat objective, and whether the threat operates primarily in the sidelobes or in the mainlobe of the emitter antenna. The threats range from antiradiation missiles (ARM), direction of arrival (DOA) systems, radar warning receivers (RWRs), electronic countermeasures (ECM) equipment, and electronic intelligence (ELINT) systems. All of these threats are designed to degrade the performance of the emitter in

Table 1.1   Passive Threats to Radars and Data Links

Threat type	Intercept receiver function	Threat objective	Operation through antenna	
			Sidelobes	Mainlobe
ARM	Missile guidance	Hard kill	✔	
DOA	Emitter location	Attack or avoidance	✔	
RWR	Pilot warning	Maneuvers		✔
ECM	Jammer turn-on, set-on, pointing	Degraded detection and track	✔	✔
ELINT	Electronic order of battle (EOB)	Intelligence, battle planning, operations control, ECM design	✔	✔

its mission either immediately or over a span of time and ultimately to destroy the usefulness of the weapons system.

All of these passive threat systems depend on some type of intercept receiver. The intercept receiver performs three basic functions: detection, sorting, and classification. The intercept receiver detection function depends on single-pulse peak-power detection at the outset, and little or no processing gain is obtained for first detection. This is primarily due to uncertainty about the angle of arrival, actual operating frequency, and waveform of emitters, which may include LPI radars or data links. This initial large signal mismatch is one of the principal advantages of an LPIS. The third "great thought" of LPI design is creation and exploitation of gain mismatches.

The second intercept receiver function, sorting, is achieved by a combination of hardware and software. Sorting is designed to operate in a dense signal environment in which there are many intercept signals—perhaps all in the same band, all arriving from different directions. The task of sorting is the separation of individual emitters from this environment so they may be recognized or classified.

The third intercept receiver function is classification. It identifies or recognizes emitter type, or even the specific emitter, and from that it determines the weapon system in which the emitter is carried. Once the weapon system is identified, suitable countermeasures can be employed. The waveforms selected also must make countermeasures difficult to employ.

LPI systems must be designed to degrade all three intercept receiver functions. In the course of the next six chapters, many techniques will be discussed that work on each of these three properties. Obviously, if an emitter is never detected, it cannot be sorted or classified, but mission performance usually suffers too much to depend on it never being detected. There will inevitably be detections, but selected waveform properties, such as a large duty ratio over significant bandwidths, enhance the chance of embedment of multiple emitter signals, making sorting very difficult. Sometimes, in spite of careful waveform selection and bad luck, an LPI emitter will be sorted, and the waveform must also be made hard to classify by spoofing or other agility features.

1.2.2 Detection and Intercept Probability

Two probabilities are associated with detection and intercept by a threat receiver. *Detection probability* is the probability that an intercept receiver will detect an emitter, given that the intercept receiver antenna, RF front end, and processor are directed at the emitter while the emitter is transmitting toward the intercept receiver. *Intercept probability* is the product of the detection probability and the emitter on-time divided by the intercept receiver scan time. The intercept receiver requires a certain amount of time to cover both the frequency range in which the emitter may transmit and the angle space over which it surveys. Because an LPI system is not on or pointing in the same direction all the time, these scan-on-scan probabilities reduce the intercept probability below that which would occur based on power and antenna gain alone.

Detection probability is defined as the detection probability given that the intercept receiver looks at the emitter while emitter is transmitting toward intercept receiver. *Intercept probability* is defined as

$$\text{Intercept probability} = \frac{\text{Detection probability} \times \text{Emitter on-time}}{\text{Intercept receiver scan time}} \qquad (1.1)$$

Obviously, not all systems scan, in which case the emitter on-time/interceptor scan time ratio is 1. It usually turns out that, even if the RF or antenna doesn't scan, the signal/computer processing has a scan or frame or refractory time. Many intercept systems that form simultaneous multiple antenna beams or RF filters have such poor sidelobes (−25 dB) that false alarms overwhelm their sensitivity, and stealth targets are thresholded out. They would have been better off scanning if stealth targets were the only threat.

Such basic observations lead to an obvious set of elements for an LPIS design. To defeat detection, an LPIS must minimize required effective radiated power, effective intercept receiver sensitivity, and required on-time. To minimize effective radiated power, the transmitter power must be managed to the minimum necessary for a given range and mode or function. In addition, low sidelobe antennas must be used for minimum sidelobe effective radiated power. Minimizing required effective radiated power implies that the shortest operating range, based on usable weapon system mission parameters, must be employed at all times, and the largest possible antenna aperture or gain and maximum transmitter duty ratio is required. If receive losses are high, more transmit power must be emitted; therefore, an LPIS requires minimal receive losses and the lowest possible noise figure. Furthermore, because losses are important, maximum coherent integration must be used to save processing losses. To minimize realized intercept receiver performance, the maximum use of *a priori* knowledge and passive sensing is required whenever possible. The weapon system should not emit to determine information that could be obtained by some other method. This, in turn, implies maximal use of prebriefed missions and accurate navigation systems on the weapon system platform. Lastly, LPIS must minimize required on-time and utilize maximum extrapolation between measurements. The least detectable LPIS is one that's off.

1.2.3 Reduced Detectability—Effective Radiated Peak Power

The definition of *effective radiated peak power (ERPP)* is the root mean square power in the transmitted pulse through the antenna mainlobe, sidelobes, and radome into free space. There are two thoughts here: the exact power that can be detected and where to reference it. It is the free space power immediately outside the radome and vehicle skin that is detectable. On a stealth vehicle, one can think of the antenna as the feed and the platform as the actual radiator. Minimizing antenna sidelobes may minimize the sidelobe ERPP—*if* the radome and antenna installations are carefully done. Minimizing receive losses minimizes the power that must be propagated in free space. Because total signal energy determines detection performance, average power over the observation period is what counts for the LPIS—not peak power. The essential elements of ERPP control are summarized below. (More details are provided in Chapters 2, 3, 4, 5, and 7.)

- Signal uncertainties and a broad threat spectrum force intercept receivers to depend on single-pulse peak power detection (Chapters 2, 3, and 4).
- LPI systems require low peak power and high average power (Chapter 2).
- LPI systems are ERPP limited and therefore require very low receive losses and energy management (Chapters 2, 5, and 7).

Figure 1.8 shows one of the simplest kinds of gain mismatch exploitation in LPI, single-pulse peak power detection. Because an interceptor must deal with a wide

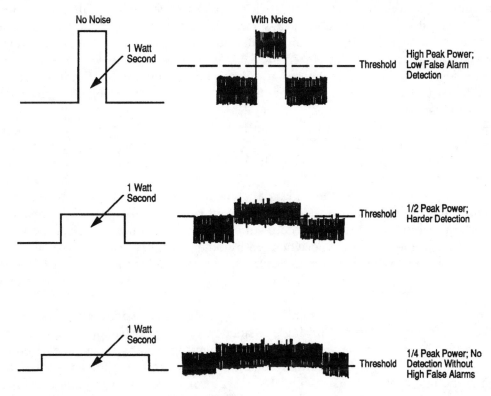

Figure 1.8 Single-pulse peak power detection (SP3D).

range of threats, its predetection bandwidth is set by considerations other than the transmitted pulse width of an LPIS; therefore, noise power in the interceptor's bandwidth is independent of the LPIS transmitted waveform. That being the case, the appropriate LPI strategy is to go to the lowest peak and highest average power waveform obtainable for a given mode of operation. This is shown in simple pictorial form in Fig. 1.8. Either the interceptor must deal with large numbers of false alarms, or the probability of detection must be very low. This mismatch can be expanded if an LPIS intentionally uses a wide range of pulse or chip powers and pulsewidths. Increased sensitivity without selectivity in a dense signal environment just leads to false alarms and no recognitions.[19,24,25]

Minimizing effective radiated peak power requires transmitter power management. Most current radars and data links always transmit the maximum peak power available. They typically use the same power for a 1-mile communication or dogfight mode as for a 100-mile detection or data-sharing mode. In fact, many current radars and data links have no power management capability whatsoever. Power management doctrine dictates that minimum power is transmitted at all times, and the minimum acceptable rms *peak-power-signal-to-mean-noise ratio (SNR)* is used under every condition. Therefore, for each mode, such as air-to-air search, air-to-air track or low bandwidth data handling, and for each selected range, such as 5 or 50 miles, transmitter power is adjusted to the minimum value that will give acceptable performance. For example, a typical target may require 5,000 W peak for 80-mile detection in an air-to-air search mode. On the other hand, to track that same target at 5 miles requires only 75 mW peak. The intercept range is then cut by a factor of $1/256$ of the maximum under this condition.[19]

Every designer who is first introduced to LPIS proposes large noncoherent integration to counter a stealthy emitter. It is the easiest interceptor performance parameter to counter. Section 2.2 deals with the constraints necessary to create time, frequency, and antenna gain mismatches. Those constraints, coupled with the ambient spectra present on a battlefield as analyzed in Sections 4.4 through 4.6, overwhelm high-sensitivity but unselective intercept receivers. As long as there is a 10 percent probability of at least one conventional emitter in the same resolution cell, long coherent integration is overwhelmed by other/false targets. Even some of the most modern intercept equipments can't use the sensitivity they already have because of processing, false alarms, and resolution limitations. The statements made in 1975 are still true in 2000.

Nowadays, almost every radar and data-link mode uses coherent transmission and processing, but it is anything but single-frequency, as shown in Chapter 5. Waveforms are completely geometry dependent, so even the emitter won't know what waveform will be transmitted until the control frame immediately preceding the emission. How then is the interceptor to match filter the waveform unless every possibility is processed? If only noncoherent integration is used, then the LPIS emission only needs to be significantly lower than all the other emitters in that interceptor resolution cell. This is discussed in Section 3.2. These facts force initial detection on the basis of SP3D.

1.2.4   Reduced Detectability—Maximum Signal Uncertainty

An LPIS must also defeat sorting and classification. This, in turn, requires maximum signal uncertainty at all times. Waveform parameters, such as pulse repetition frequency (PRF), chip rate, encryption code, frequency, and on-time, must be randomized so that the interceptor cannot predict how and when the next emission will occur. The waveform location inside the operating band is selected on a pseudorandom basis for each coherent array. The waveform itself is geometry dependent[1]; some examples are given in Chapters 5 and 7. The PRF is "FMed" and adaptively changed to account for maneuvers during a coherent array. The next PRF from a compatible set is chosen on a pseudorandom basis. Other techniques that are commonly used are mimicking another system or spoofing the interceptor. This is achieved either by transmitting an easily detectable mimic waveform that is interleaved with the LPI waveform and that mimics a benign or an adversary weapon system, or by using a cooperative jammer that transmits a low-level spoofing signal, masking the LPIS signal. Simultaneous multibeam antennas are used to more rapidly search the required volume, but they must be implemented in such a way that the same frequency does not visit the same region of space very often. The essential elements of maximum signal uncertainty are summarized below. More details are provided in Chapters 2, 4, 5, 6, and 7.

- Direction of arrival—low sidelobes, multibeams, movement between emissions

- Frequency—large instantaneous bandwidth

- Time of arrival—multiple PRFs, low peak power, time spread, multibeams

1.2.5   LPI Performance Example

As an example of what can be accomplished using stealth notions to achieve the same weapon system mission, consider a high-peak-power fighter radar such as those found in the French Mirage (Cyrano Series)[7] and a similarly performing LPIS. Table 1.2 summarizes the improvements that have been demonstrated with stealth principles. The corresponding intercept sensitivities for the table are given in Section 1.3.3. The threats are typical of 5000 to 10,000 currently deployed worldwide (Chapter 4).

    LPI and electronic counter-countermeasures (ECCM) features go hand in hand. Both LPI and ECCM designs require antennas that have narrow main beams and very low sidelobes. They both require transmitters that have frequency agility, PRF agility, and large instantaneous bandwidths. Furthermore, they both require processors with large time bandwidth products and integration times.

Table 1.2   LPIS Performance Comparison

A/A lookdown example (range in nautical miles)	Target detect	Threat RHAW	Threat ELINT	Threat ARM
Typical fighter radar	20	187	1181	29.7
LPI radar	20	4.6	10.4	0.26

System parameters: 5 W/beam, 9 beams, 320 MHz bandwidth, –55 dB rms sidelobes, LPI waveform for both radar and missile

LPI technology also augments ECCM. The combination of LPI operation and low platform RCS reduces detection by electronic support measures (ESM) equipment because neither the active emissions nor cueing from threat sensors can be used by the ESM systems to exploit an LPI/LO platform. Low antenna sidelobes and adaptive null steering also negate sidelobe jammers. The impact of very low antenna sidelobes coupled with modern adaptive null steering provides many orders of magnitude improvement against sidelobe jamming threats. An LPIS must operate over a large band, or the jammer will place all its power in the narrower band. The jammer can't know what part of the wider band to jam, and so some of the jam power is wasted. Narrowband systems are extremely easy to jam with swept noise. LPIS requires frequency agility for maximum signal uncertainty, but frequency agility controlled by a sniff feature defeats mainlobe continuous wave (CW) and spot noise jammers as well. LPIS with high duty ratios usually requires high pulse or data compression ratios. High compression ratios, in turn, enable signal-to-jam processing gain improvements of an order of magnitude under most circumstances. The net effect of these LPI technologies is to greatly decrease LPIS susceptibilities to some types of countermeasures.

LPI capabilities are not free by any means. First, maximum antenna gain and very low sidelobes have many undesirable effects. The aperture is always bigger than the platform designer wants. A narrow beam limits dwell time in volume search scan, and multiple simultaneous beams may be required to cover a volume adequately. Low sidelobes increase required manufacturing accuracy and always compromise gain. Not surprisingly, these requirements are conflicting and thus carry some weapon-system-level costs.

Second, multiple agile simultaneous antenna beams inevitably result in higher levels of complexity. Each beam should be on a different frequency to limit sidelobe-radiated power density in any direction. The transmission of multiple simultaneous frequencies through a typical modern travelling wave tube (TWT) based transmitter is no problem, but the need for a separate receiver and processor channels for each frequency greatly increases complexity and cost.

Third, high duty ratios often result in diminishing returns, especially in large range blind zones or eclipsing losses that complicate mode design or may require multiple waveforms to obtain the complete desired coverage. Furthermore, high duty ratio usually requires significant amounts of compression. Compression usually has range sidelobes, which may greatly limit performance, so it is not always desirable to have large compression ratios.

Fourth, a large coherent integration time may conflict with mode and target type. For example, in a synthetic aperture radar mode, integration time is determined by geometry, not by the convenience of the mode designer. In air-to-air modes, large coherent integration time is usually limited by potential target geometrical acceleration. In the presence of large accelerations, large coherent integration times may require multiple simultaneous integration's, each using a different target acceleration hypothesis. This greatly increases the complexity of the processing.

Fifth, wide instantaneous bandwidth inevitably results in more costly hardware. Such things as analog-to-digital (A/D) converters, processor memory, and signal processor arithmetic throughput are usually substantially more expensive as they get faster. More instantaneous bandwidth increases both signal processor required speed and memory capacity requirements. For example, a 500-MHz bandwidth gives 1-ft range bins and, to cover a 20-mile search range, requires 120,000 range bins per channel. This is a lot of processing just to improve SNR. In addition, small range bins will not necessarily improve performance if the target extent is greater than 1 ft, which is typical. Similarly small chip widths for data compression require massive "defruiting" due to multipath.

1.3 A LITTLE HISTORY OF STEALTH SYSTEMS

RF stealth is not new. The British were proposing RCS reduction techniques in 1941.[16] The first complete modern demonstration of an LO platform was the DARPA/USAF/ Lockheed Have Blue aircraft. So much has been written about Have Blue that no discussion is necessary here.[10]

The first modern demonstration of LPI techniques was the Navy/Westinghouse Sneaker program. The first complete LPI system program was the LPIR started in the mid-1970s by DARPA, the U.S. Air Force and Navy, with Hughes Aircraft Company as prime contractor. That program resulted in the conclusive demonstration that LPI techniques could be employed and still provide a complete airborne radar weapon system. The performance, modes, and accomplishments are summarized in this section. Flight tests of the first airborne LPIR system were completed in 1979 and 1980. The system was tested against an AN/ALR-62-equipped F-111 from the 57th Test Wing at McClellan Air Force Base. The ALR-62 is a radar homing and warning (RHAW) receiver that has full 360° coverage over a wide range of frequencies. It has a potential equivalent sensitivity of –65 dBm. The ALR-62 used in these tests was especially programmed to detect and identify LPIR waveforms. All available parameters of the LPIR system were provided to the programmers of the RHAW receiver so that an adequate algorithm could be incorporated to detect an LPI system.[11,23–25]

1.3.1 LPIR Program Accomplishments

- First complete airborne LPIR system
- System tested against RHAW (ALR-62) equipped F-111F from 57th Test Wing, Mc-Clellan AFB
- Air-to-air modes demonstrated with 40 km detection, acquisition, and track to crossover without intercept
- Ground-to-air roofhouse demonstration in January 1979, air-to-air demonstration in November 1979
- Air-to-ground modes (ground moving target detection and tracking) demonstrated in April and May 1980, successful LPI detection and track at maximum range of 20 km

The second program to incorporate LPI features was the DARPA/USAF Pave Mover program, which began flight test shortly after the conclusion of the LPIR program. It continued to demonstrate a wider set of LPI modes and features, including alternative LPI strategies, longer range operation, LPI data links, more air-to-ground modes, adaptive ECM nulling, weapon delivery, and the LPI benefits of electronic scanning. Subsequently, there were many programs that have applied LO/LPI features; most are still classified.

1.3.2 LPI Modes Demonstrated through Test

Listed in the table below are the stealthy radar and data link modes that have been demonstrated in unclassified or declassified test programs. For example, air-to-air medium PRF modes were demonstrated with 40-km detection, acquisition, and tracking (shown in Fig. 1.9). Power management was employed with typical maximum

Air–to–air	Air–to–ground
Search	Real beam ground map
Track	Ground moving target indication
Track while scan	Ground target track moving and stationary
Air combat	Doppler beam sharpening
Missile data link	Terrain following
Stealth weapon delivery	Synthetic aperture
ECCM	All weather weapon delivery
	Air-to-ground station data links

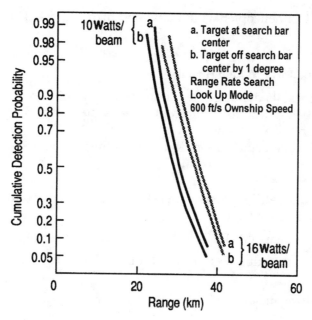

Figure 1.9 Performance of MPRF demonstration radar.

power per frequency of 3 W. Tracking lock-on near maximum range was consistently achieved, and track was maintained all the way to crossover/gimbal limit without any intercepts occurring. In January 1979, there were ground-to-air roofhouse demonstrations and, in November 1979, air-to-air demonstrations. The following year, air-to-ground modes were tested. Both ground moving target detection (GMTD) and ground moving target tracking (GMTT) were demonstrated in April and May 1980, with successful LPI detection and track at the maximum design range of 20 km. The surface vehicles tracked in these tests carried a responsive jammer with an intercept receiver that could be used to detect the LPI signal. In addition, these tests were conducted in an ECM environment and, in most cases, the LPIR was not only able to detect, acquire, and track both air-to-air and air-to-ground targets, but also to defeat the countermeasures.

The program was very successful, and the LPI system met its program objectives. The only time this radar was ever detected by any intercept receiver or ECM gear was when radiated power was intentionally augmented to 100 W per frequency and fixed so that the radar could be detected for intercept receiver performance verification.

Even to this day, there are skeptics about achievable stealth system performance, but this is mostly from those who are uninformed. Especially silly are claims about some Eastern European intercept systems. The stealth weapon systems that have been shot down owe that primarily to poor tactics, as is discussed in Chapter 4. Any system is susceptible to bad tactics or the "golden BB;" just remember that an aborigine shot down a UN helicopter with a bow and arrow and killed the UN secretary general.

1.3.3 Summary of LPIR Program Results

The performance results demonstrated more than 20 years ago in the LPIR program are summarized below. The total intercept threat system sensitivity, P_I/G_I, given below is the intercept receiver sensitivity at the threshold device referred to the front end divided by the interceptor antenna gain including all radome and cabling losses.

- *Radar Performance*

  Air-to-air: 5-m^2 target, lookdown mode, search, acquisition, track, missile guidance—detection range 40 km

  Air-to-ground: RB map, GMTI/T (trucks), DBS (0.3°)—detection range 80 km

  Weapon delivery targeting accuracy: 0.46 milliradian CEP, stationary and moving targets

  ECCM: Like F-16 against DLQ-3B jammer

- *Flight Test*

  One year, 400 passes against military and civilian targets

  LPIR detection ranges: 22–35 km, Track acquisition ranges: 14–22 km

  Track ranges, 0–40 km; power, 3 W peak, 0.75 W average

  RWR: Specially modified ALR-62 in F-111F, the best RHAW the U.S. had at that time, 18 passes, Jan. 11–12, 1979. ALR-62 detected LPIR only when the test profile intentionally allowed it using 100 W peak, 14 km typical intercept range, consistent with ALR-62 specification performance

- *Intercept Threat*

  RWR mainlobe: $R_I \leq 10$ km, $P_I/G_I = -60$ dBm, 1-GHz bandwidth

  ELINT sidelobes: $R_I \leq 20$ km, $P_I/G_I = -120$ dBm, 5-MHz bandwidth

  ARM sidelobes: $R_I \leq 5$ km, $P_I/G_I = -75$ dbm, 30-MHz bandwidth

These demonstrations give rise to an obvious question: how does one verify and validate that an LO/LPI system is succeeding at its mission? Perhaps the intercept receiver or jamming equipment is broken. Is there any difference between a broken intercept receiver or countermeasures equipment and one that is actually being fooled by an LPIS? There must be a sequence of techniques used to verify that the countermeasures or interceptors are actually working during the course of a LO/LPI test program. Most LPI evaluation is identical to conventional emitter evaluation. LPI is measured by placing an intercept receiver at the target or with a fixed offset with respect to the target. Power management is then switched off so that full power will be transmitted to verify that the intercept receiver is working.

One of the items often requested, but usually virtually impossible to do, is suppression of the pseudorandom dynamic waveform selection that is incorporated into the mode design of every LPIS. It is generally not technically feasible to shut off dynamic waveform selection during LPI verification and validation. Finally, the RCS of an LO/LPIS is usually measured statically on indoor or outdoor ranges, and its performance is implied indirectly when mounted on a LO vehicle. The upper bound on the installed RCS of an LO/LPIS is the vehicle cross section in the emitter band, about ±60° with respect to the antenna normal.

What then are typical LPIS performance parameters? First, antenna sidelobes are typically less than –55 dB rms below the mainlobe over the half space. Second, power management is often conducted over a 70-dB dynamic range. Time bandwidth products of 4×10^6 are typically achieved in LPI systems. Even with modern processing technology, that's a great deal of integration. Typical instantaneous bandwidths are between 0.5 and 1 GHz. To achieve very high time-bandwidth products, fully coherent frequency agility and corresponding stability of 1 in 10^{10} are required. A typical number of channels simultaneously processed is on the order of 6 to 9. Tracking is normally performed with a 3-dB SNR. All these performance parameters are achieved with a full complement of modes.

1.3.4 LPIS Typical Technology

Current unclassified LPIS performance is summarized below.

- Antenna sidelobes of < –55 dB rms over half space relative to mainbeam
- Power management over a 70-dB dynamic range
- Large coherent integration gain—time bandwidth products of at least 4×10^6
- Instantaneous bandwidth of 2 GHz, hopped bandwidths of 6 GHz
- Fully coherent frequency agility/stability of $1:10^{12}$
- Dynamic waveform selection—full mode complement/minimum dwell and minimum power
- Low peak power/high duty waveforms—3-dB *SNR* tracking
- Multibeam, multifrequency search—6 to 9 channels
- Adaptive data link spectral spreading—300:1

Two LPI strategies have been used and will be described later in the text: the minimum-dwell strategy and the minimum-power strategy. These strategies are other examples of the creation and exploitation of gain mismatches. Typical dynamic waveform selection is summarized below.

1.3.5   LPI Maximizes Uncertainty

The current unclassified LPI emitter parameters that are variable are as follows:

- PRF: 1000:1 variation
- Pulse width: 8000:1 variation
- Bandwidth: 100:1 variation
- Data link chip width: 100:1 variation
- Power: 10^7:1 variation
- Dwell time: 10:1 variation
- Frequency: 0.5 to 2 GHz variation
- Number of unique pulse or data compression codes: 10^5 easily obtainable, 10^{20} feasible

1.4   BASIC LPI EQUATIONS

The basic LPI equations are specialized applications of the well known radar equation (1.2) and beacon equation (1.3, 1.4).[2, 23] The power received in the radar equation is inversely proportional to R^4. The power received in the beacon or intercept receiver equation is inversely proportional to R^2. At first glance, it appears that the interceptor has an insurmountable advantage, but the condition that typically occurs is shown in Fig. 1.10. There is a radar or data link operating range, R_{Dmax}, at which the required transmitted peak power crosses over the corresponding intercept receiver required power. This range is usually less than the radar horizon, R_H.

1.4.1   Radar and Beacon Equations

To review the radar and beacon equations, we begin by defining the following parameters:

$$R_D, R_{DL} = \text{radar, data link emitter design range}$$
$$\sigma = \text{target RCS}$$
$$\lambda = \text{emitter wavelength}$$
$$P_R = \text{RMS power required at radar or link receiver}$$
$$G_T = \text{emitter antenna gain}$$
$$G_{RP}, G_{DP}, G_{IP} = \text{radar, data link and interceptor processor net gain respectively}$$
$$L_R, L_{DL} = \text{emitter total path losses}$$
$$L_I = \text{losses between emitter and interceptor}$$
$$P_I = \text{power required at intercept receiver for detection}$$
$$P_i = \text{power received at the intercept detector}$$
$$R_i = \text{intercept receiver to emitter range}$$

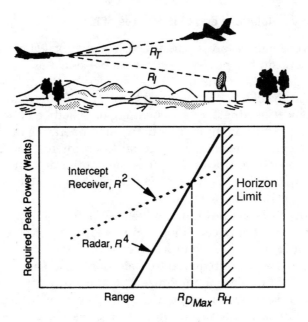

Figure 1.10 Interceptor has R^2 advantage at long range.

$$G_I = \text{intercept receiver antenna gain}$$
$$G_{DL} = \text{data link receive antenna gain}$$
$$G_{TI} = \text{emitter gain in interceptor direction}$$

Then, the radar receiver threshold detector input power is

$$P_R = \frac{P_T \cdot G_T^2 \cdot G_{RP} \cdot \lambda^2 \cdot \sigma \cdot L_R}{(4 \cdot \pi)^3 \cdot R_D^4} \tag{1.2}$$

The data link receiver threshold detector input power is

$$P_R = \frac{P_T \cdot G_T \cdot G_{DL} \cdot G_{DP} \cdot L_{DL} \cdot \lambda^2}{(4 \cdot \pi)^2 \cdot R_{DL}^2} \tag{1.3}$$

The intercept receiver threshold detector input power is

$$P_i = \frac{P_T \cdot G_{TI} \cdot G_I \cdot \lambda^2 \cdot G_{IP} \cdot L_I}{(4 \cdot \pi)^2 \cdot R_i^2} \tag{1.4}$$

1.4.2 Intercept Power Relations and LPIS Figures of Merit

The intercept power relations are summarized in Equations (1.5) through (1.13) below. The required transmitter power, P_T, can be represented as a figure of merit constant, K, times the detection or link range, R_D, divided by the radar antenna gain, G_T [Equation (1.5)]. K is the *effective radiated peak power (ERPP)* normalized by the design range raised to a power of either 2 or 4, depending on whether a one-way or two-way path is involved. K depends on the emitter design and operating mode. The power received at the interceptor threshold, P_i, is equal to an interceptor figure of merit, J, times the emitter transmitter power, P_T, times the emitter antenna gain in the interceptor direction, G_{TI}, divided by the intercept receiver to emitter range squared, R_i^2, as shown in Equation (1.7) (one-way path).

Substitution of the P_T Equation (1.5) into the P_i Equation (1.9) provides a form that contains ratios of radar-to-intercept range and radar-to-interceptor gain [Equation (1.11)]. For an intercept to occur, the power received at the interceptor threshold, P_i, must equal or exceed the power required for detection, P_I, as shown in Equation (1.8). The object for every LPI mode is to minimize K. Radar processing gains, G_{RP}, are defined in Section 1.6.1. Duty factor, integration time, and receiver bandwidth are specifically called out. Similarly interceptor processing gains, G_{IP}, are described in Section 3.3. These gains are obviously signal-to-noise ratio (SNR) gains at the detection threshold and are dealt with more completely in Section 7.2.5.

Let the radar equation for required transmitted power for a given target detection range, R_D, be written as

$$P_T = \frac{K_R \cdot R_D^4}{G_T} \quad \text{(Required radiated peak power for } R_D) \qquad (1.5)$$

where K_R is a radar figure of merit, which depends on the mode and mission and is defined as

$$K_R = \frac{(4 \cdot \pi)^3 \cdot P_R}{\lambda^2 \cdot G_T \cdot G_{RP} \cdot \sigma \cdot L_R} \Rightarrow 10^{-16} \text{ W/m}^4 \text{ typical value} \qquad (1.6)$$

Similarly, for a data link, the required transmitted power for a given range, R_{DL}, can be written as

$$P_T = \frac{K_{DL} \cdot R_{DL}^2}{G_T} \quad \text{(Required power for } R_{DL}) \qquad (1.7)$$

where K_{DL} is a data link figure of merit, which depends on the mode and mission and is defined as

$$K_{DL} = \frac{(4 \cdot \pi)^2 \cdot P_R}{\lambda^2 \cdot G_{DL} \cdot G_{DP} \cdot L_{DL}} \Rightarrow 10^{-8} \text{ W/m}^2 \text{ typical value} \tag{1.8}$$

Let the beacon equation for a given transmitted power and emitter-to-interceptor range, R_i, be written as

$$P_i = J_I \frac{P_T \cdot G_{TI}}{R_i^2} \text{ (Power received by intercept receiver at range} R_i) \tag{1.9}$$

where J_I is an interceptor figure of merit, which also depends on the mode and mission of the interceptor and is defined as

$$J_I = \frac{G_I \cdot G_{IP} \cdot \lambda^2 \cdot L_I}{(4 \cdot \pi)^2} \Rightarrow 2.10^{-3} \text{ m}^2 \text{ typical value} \tag{1.10}$$

Substituting Equation (1.5) into (1.9) yields

$$P_i = J_I \cdot K_R \cdot \frac{R_D^4 \cdot G_{TI}}{R_i^2 \cdot G_T} \tag{1.11}$$

And similarly substituting Equation (1.7) into (1.9) yields

$$P_i = J_I \cdot K_{DL} \cdot \frac{R_{DL}^2 \cdot G_{TI}}{R_i^2 \cdot G_T} \tag{1.12}$$

For example, let $P_i = P_I$ and $G_{TI} = G_T$ (i.e., the interceptor is in mainlobe); then, the intercept range for specified R_D and P_I is

$$R_I = \sqrt{J_I \cdot K_R \frac{R_D^4}{P_I}} = J_I^{1/2} \cdot K_R^{1/2} \frac{R_D^2}{P_I^{1/2}} \tag{1.13}$$

The LPIS concept is to minimize K_R or K_{DL} for each operating mode.

1.4.3 Detection Range versus Intercept Range Equations

As previously mentioned and used below, the radar, the data link, and the interceptor have figures of merit, shown in Equations (1.6), (1.8), and (1.10), respectively. More insight into how LPI radar is successful against an interceptor can be gained by comparing equations for radar detection range versus intercept range. The radar range equation for a given required receive power, P_R, is

$$R_D^4 = \frac{P_T \cdot G_{RP} \cdot G_T^2 \cdot \lambda^2 \cdot \sigma \cdot L_R}{(4 \cdot \pi)^3 \cdot P_R} = \frac{P_T \cdot G_T}{K_R} \qquad (1.14)$$

And, similarly, the data link (beacon) range equation for a given required receive power, P_R, is

$$R_{DL}^2 = \frac{P_T \cdot G_T \cdot G_{DL} \cdot G_{DP} \cdot L_{DL} \cdot \lambda^2}{(4 \cdot \pi)^2 \cdot P_R} = \frac{P_T \cdot G_T}{K_{DL}} \qquad (1.15)$$

The interceptor (beacon) range equation for a given *required* receive power, P_I, is

$$R_I^2 = \frac{P_T \cdot G_{TI} \cdot G_I \cdot G_{IP} \cdot \lambda^2 \cdot L_I}{(4 \cdot \pi)^2 \cdot P_I} = \frac{P_T \cdot G_{TI} \cdot J_I}{P_I} \qquad (1.16)$$

The ratio of R_D^4 to R_I^2 [Equations (1.14) and (1.16)] is given in Equation (1.17). By grouping like elements, it can be seen that best performance for the radar relative to the interceptor occurs when each ratio in Equation (1.17) is maximized. Maximizing each ratio means maximum radar antenna gain relative to radar-to-interceptor gain, smaller radar detection power relative to required interceptor power, and greater radar processing gain relative to interceptor processing gain.

$$\frac{R_D^4}{R_I^2} = \left(\frac{G_T^2}{G_{TI} \cdot G_I} \cdot \frac{P_I}{P_R} \cdot \frac{G_{RP}}{G_{IP}} \cdot \frac{L_R}{L_I} \right) \cdot \frac{\sigma}{4 \cdot \pi} \qquad (1.17)$$

If the maximum range where LPI is maintained is defined as the range at which the radar detection range just equals the intercept range, as in Equation (1.18), then R_{Dmax}^2 is equal to Equation (1.19). Because R_I^2 is proportional to R_D^2, then once R_D is less than R_{Dmax}, the intercept range decreases faster than the detection range, as in Equation (1.21). Thus, to find the maximum LPI radar range, set R_D equal to R_I, which defines R_{Dmax} as follows:

$$R_D = R_I \equiv R_{Dmax} \quad \text{Maximum LPI Range} \qquad (1.18)$$

Substituting Equation (1.18) into (1.17) and grouping like parameters yields

$$R_{Dmax}^2 = \left(\frac{G_T^2}{G_{TI} \cdot G_I} \cdot \frac{P_I}{P_R} \cdot \frac{G_{RP}}{G_{IP}} \cdot \frac{L_R}{L_I} \right) \cdot \frac{\sigma}{4 \cdot \pi} \tag{1.19}$$

Furthermore, because both R_i^2 and R_D^4 are proportional to transmitted power, P_T,

$$R_i^2 \propto P_T \propto R_D^4 \tag{1.20}$$

And normalizing R_i and R_D to R_{Dmax} yields

$$\frac{R_i}{R_{Dmax}} = \left(\frac{R_D}{R_{Dmax}} \right)^2 \tag{1.21}$$

Therefore, there is no intercept if $R_D < R_{Dmax}$. The intercept probability gets very small inside R_{Dmax}. As can be seen in Chapter 2, the intercept probability never quite goes to 0, but it does become adequately small; e.g., the $P_D = 3 \cdot 10^{-3}$ to $3 \cdot 10^{-5}$ at $1/2$ of R_{Dmax}. Similarly, taking the ratio of Equation (1.14) to (1.15) yields

$$\frac{R_{DL}^2}{R_I^2} = \left(\frac{G_T \cdot G_{DL}}{G_{TI} \cdot G_I} \cdot \frac{P_I}{P_R} \cdot \frac{G_{DP}}{G_{IP}} \cdot \frac{L_{DL}}{L_I} \right) \tag{1.22}$$

There are two cases for LPI data links associated with radars. The first is for a data link between close station keeping platforms or between a platform and a missile so that

$$R_{Dmax} \geq R_{DL} \tag{1.23}$$

The right side of Equation (1.22) is almost always greater than 1 but, by Equation (1.23), the left side of Equation (1.22) will be less than or equal to 1, so there will never be an intercept. The second case occurs when an LPI radar sensor must transmit data to a remote processing station so that

$$R_I \geq R_{DL} \geq R_{Dmax} \tag{1.24}$$

For this case, R_{DLmax} is dependent on the specific parameters and geometry. For example, suppose the processing station and the interceptor have the same antenna gain, required received power is about the same, data link bandwidth compression is 100, the interceptor is in the sidelobes, and data link losses are double interceptor losses. Then,

$$R_{DLmax} = R_I \cdot \sqrt{3 \cdot 10^5 \cdot 1 \cdot 1 \cdot 100 \cdot 0.5} = R_I \cdot 3.9 \cdot 10^3 \tag{1.25}$$

If the interceptor is in the mainlobe, R_{DLmax} can easily be very close to R_{Dmax}. Comparison of three different fighter radars with similar missions dramatizes what can be done both in terms of figure of merit and mainlobe intercept range against a typical RWR. Improvements of two orders of magnitude are achievable.

Another example based on the LPIR system, a typical RWR, and Equation (1.19) is shown in Table 1.3. The table compares the various performance elements of the radar and the interceptor and provides an explanation of the differences. Note that the maximum safe range can be a very large number and thus is operationally useful in many real scenarios. One important factor in this example is the adjustment of transmitter power based on the interceptor's known or assumed sensitivity.

Table 1.3   Mainlobe/RWR Intercept Ranges

Weapon/sensor system	$K(W/km^{4,2})$	$K^{0.5}$ (REL)	$TYP \cdot RWR \; R_I \; (km)$
F4E/APQ-120*	23.7*	59.5	595*
F-16/APG-66*	10.9*	40.3	403*
LPIR (HPRF mode)	0.0067	1	10
LPI missile data link	$1.6 \cdot 10^{-6}$	0.015	$3.75 \cdot 10^{-3}$

Air-to-air lookdown, transmitter power adjusted for 40 km detection range.
*Estimated from unclassified sources.

An alternative formulation for R_{Dmax} is given in Equation (1.28). It substitutes aperture areas for gains and required $SNRs$, detection bandwidths, and noise figures for required detection powers. Equation (1.19) can be rewritten in terms of antenna areas, signal-to-noise ratios, bandwidths, and noise figures. Because

$$P_I = SNR_I \cdot B_I \cdot NF_I \cdot k_B \cdot T \text{ and } P_R = SNR_R \cdot B_R \cdot k_B \cdot T \tag{1.26}$$

and

$$G_T = \frac{4 \cdot \pi \cdot A_{eT}}{\lambda^2} \text{ and } G_I = \frac{4 \cdot \pi \cdot A_{eI}}{\lambda^2} \text{ and } G_{TI} = \frac{4 \cdot \pi \cdot A_{eTI}}{\lambda^2} \tag{1.27}$$

Substituting into Equation (1.17) yields

$$R_{Dmax}^2 = \frac{A_{eT}^2}{A_{eTI} \cdot A_{eI}} \cdot \frac{SNR_1 \cdot B_I \cdot NF_I}{SNR_R \cdot B_R \cdot NF_R} \cdot \frac{G_{RP}}{G_{IP}} \cdot \frac{L_R}{L_I} \cdot \frac{\sigma}{4 \cdot \pi} \tag{1.28}$$

Similarly, substituting into the corresponding data link Equation (1.21) yields

$$R_{DLmax} = R_I \cdot \frac{A_{eT} \cdot A_{eD}}{A_{eTI} \cdot A_{eI}} \cdot \frac{SNR_1 \cdot B_I \cdot NF_I}{SNR_{DL} \cdot B_{DL} \cdot NF_{DL}} \cdot \frac{G_{DP}}{G_{IP}} \cdot \frac{L_{DL}}{L_I} \cdot \frac{\sigma}{4 \cdot \pi} \qquad (1.29)$$

where

NF_R, NF_{DL}, NF_I = noise figure for the radar, data link, and interceptor, respectively

k_B = Boltzmann's constant

T = equivalent receiver temperature

A_{eT}, A_{eD}, A_{eI} = antenna effective area for the radar, data link, and interceptor, respectively

A_{eTI} = emitter antenna effective area in interceptor direction

SNR_R, SNR_{DL}, SNR_I = required peak signal to rms noise power ratio for the radar, data link, and interceptor, respectively

B_R, B_{DL}, B_I = noise bandwidth for the radar, data link, and interceptor, respectively

L_R, L_{DL}, L_I = loss fraction for the radar, data link, and interceptor, respectively

The effective antenna area for the radar usually will be higher than the interceptor for operational reasons. The intercept receiver must search all the resolution cells the radar or data link scans. In addition, it must search more angle and signal space because, in the beginning, it can't know where the LPIS emitter is. More interceptor bandwidth causes poorer low-noise amplifier impedance match, and most noise figure increases are caused by mismatch loss (a microwave circuit limitation). Allowable radar SNR can be lower because the radar coverage is smaller. The radar bandwidth is smaller because the waveform is more completely known, and the radar noise figure is lower because the bandwidth is smaller.

A white paper prepared for DARPA by Mitre Corporation on August 22, 1978, concluded that it is extremely difficult to hide a radar or data link from a decent intercept receiver. The theoretical analysis was correct, but it assumed a radar state of the art that was far behind the times, especially in terms of achievable time-bandwidth product, antenna gain, and noise figure advantage over an intercept receiver. It also assumed a substantially better interceptor signal processor than that in the radar and that the interceptor has *a priori* knowledge of the radar direction and waveform, which is impossible in a dense emitter environment, as shown in Chapters 3 and 4. A more realistic example is given in Table 1.4 for equally good radars and interceptors. Mitre, Lincoln Laboratories, and NRL each were funded to develop and test interceptors and ECM for the LPIR and Pave Mover Programs. These systems were so unsuccessful that one could conclude that the state of the art of ECM and intercept systems is considerably behind counterpart radars and data links. Even though it appears that

ECM and intercept systems lag the state of the art, any deployed LPIS will steadily become more vulnerable over time as threat systems improve through the application of more modern RF and signal processing technologies. Any LPIS will require steady upgrades to maintain its stealth advantage.

Table 1.4   LPIR—How It's Done Example

Inverse square beats inverse fourth power, but...				
Parameter	Radar (dB)	Interceptor (dB)	Advantage ratio (dB)	Comment
Detection power *SNR*	13.5	16	2.5	Processor loading
Antenna gain	37	0	37	Direction unknown
Receiver bandwidth (dB Hz)	16.5	90	73.5	Frequency unknown
Noise figure	2	10	8	Bandwidth matching
Integration gain	9	10	−1	Better intercept SNR
Total system losses	−9.1	−8	−1.1	Radar complexity
Duty ratio	−3.5	0	−3.5	Central band filter
RCS per steradian (dB km^2)	−64	0	−64	5 m^2 nature
R_{Dmax}^2 (dB km^2)			51.4	
It's all in the numbers!			*R_{Dmax} = 372 km*	

Example applications of Equations (1.28) and (1.29) for various trade-off parameters are shown in Figs. 1.11 and 1.12. Those parameters, which are held fixed in the trade-offs, are summarized in Table 1.5. It is instructive to look at mainlobe intercept versus radar predetection bandwidth with fixed postdetection integration, as shown in Fig. 1.11. This is really a trade-off between radar data rate and maximum safe detection range. The update period varies from a low of 28 ms with maximum safe radar range (R_{Dmax}) of 12 km to a high of 4 min, 40 sec with maximum safe range of 1131 km. Figure 1.11 shows the dramatic increase in maximum safe operating range as a function of decreasing detection bandwidth (longer coherent integration time). Long coherent integration time requires complex signal processing, because changing Doppler and geometric acceleration require a different filter for each target extent and path. The antenna gain has been set at a value typical of many deployed radar systems (see Chapter 4). The intercept receiver for this example is very simple. Nonetheless, there are more than 10,000 intercept receivers currently deployed worldwide similar to the one used in the examples of Tables 1.4 and 1.5 and Figs. 1.11 and 1.12

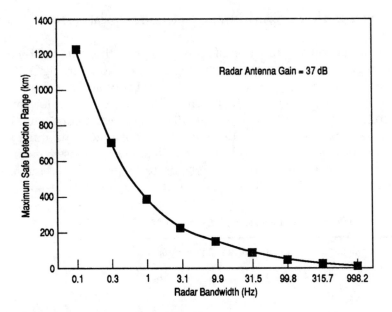

Figure 1.11   Mainlobe intercept versus radar predetection.

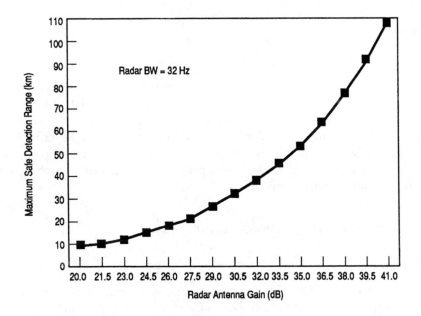

Figure 1.12   Mainlobe intercept versus radar antenna gain.

(also see Chapter 4). The performance parameters are typical of modern fighter radars and RWRs.

A second comparison as a function of antenna gain with fixed (32-Hz) predetection bandwidth and all other parameters, as in Fig. 1.11, is shown in Fig. 1.12. The detection filter bandwidth in Fig. 1.12 is set to a value that is typical and allows simple signal processing (not too stealthy). The antenna gain varies from a low of 20 dB with R_{Dmax} = 9 km to a high of 41 dB with R_{Dmax} = 108 km. Large antenna gains are required to obtain operationally useful safe detection range. The antenna must still be constrained to superior sidelobes. The obvious conclusion from these curves is to choose the largest gain consistent with good sidelobe performance and the smallest predetection bandwidth consistent with threat response time for an LPIS design.

Table 1.5   Mainlobe Intercept Parameters, Figs. 1.11 and 1.12

Emitter parameter	Emitter value	Interceptor value
Noise figure	3 dB	6 dB
Wavelength	0.03 m	0.03 m
Processor gain (postdetection)	28	3
Required SNR	20	40
Antenna gain	See figures	10
Bandwidth	See figures	1 GHz
Target RCS	1 m^2	N/A

The appendices describe the contents of the accompanying CDROM. The CDROM contains an Excel® spreadsheet in .XLS format for Figs. 1.11 and 1.12 for student manipulation. Each chapter has a similar counterpart appendix. Excel®, Mathcad®, and BASIC programs for many of the figures throughout the text are described in the appendices. The CDROM contains the computational details for these figures and many others that do not appear in the text. These programs are meant to be the jumping off point for the book user's own analysis.

1.5 INTRODUCTION TO RADAR CROSS SECTION (RCS)

1.5.1 Mathematical Basis

Whether an observation of an object is optical, infrared, or radar, it is governed by the physics of electromagnetic radiation. James Clerk Maxwell, in a surprisingly modern and prescient book written around 1850, laid out a set of equations describing electromagnetic theory. These equations have stood the test of time quite well, and most microwave engineers use them daily in one form or another. Equations (1.30) below give one form of Maxwell's equations.[3–6,11–13,18]

$$\nabla \times \mathbf{H} = \mathbf{J} + \frac{\partial \mathbf{D}}{\partial t} \quad \nabla \times \mathbf{E} = -\frac{\partial \mathbf{B}}{\partial t} \quad \nabla \bullet \mathbf{D} = \rho \quad \nabla \bullet \mathbf{B} = 0 \tag{1.30}$$

where

  $\mathbf{H}$ = vector magnetic field intensity
  $\mathbf{J}$ = vector current density
  $\mathbf{D}$ = vector electric flux density
  t = time
  $\mathbf{E}$ = vector electric field intensity
  $\mathbf{B}$ = vector magnetic flux density
  ρ = electric charge density

Fortunately, most of us don't have to solve this set of differential equations for every problem, because very smart people have spent the last 150 years solving these equations one way or another for most practical problems. Maxwell's equations are solved as summarized below.

- *Exact solutions.* Separation of variables, orthogonal coordinate systems, boundary value problems

- *Integral forms.* Stratton-Chu integrals, vector Green's functions

- *Approximate techniques.* Geometrical optics (GO), microwave optics (MO), physical optics (PO), geometrical theory of diffraction (GTD), physical theory of diffraction (PTD), uniform theory of diffraction (UTD), method of moments (MM), finite difference time domain (FDTD), finite difference frequency domain (FDFD), conjugate gradient fast Fourier transform (CG-FFT)

- *Hybrids.* Combinations of several of the above techniques

A related set of equations that are easier to visualize are the wave equations. If the fields are in homogeneous regions with no current or charge sources, the wave equations are given in Equation (1.31).

$$\nabla^2 \mathbf{E} - \gamma_c \cdot \mu \cdot \frac{\partial \mathbf{E}}{\partial t} - \varepsilon \cdot \mu \frac{\partial^2 \mathbf{E}}{\partial t^2} = 0$$

$$\nabla^2 \mathbf{H} - \gamma_c \cdot \mu \cdot \frac{\partial \mathbf{H}}{\partial t} - \varepsilon \cdot \mu \frac{\partial^2 \mathbf{H}}{\partial t^2} = 0 \tag{1.31}$$

where

γ_c = the conductivity of the homogeneous medium
μ = the magnetic permeability of the homogeneous medium
ε = the electric permittivity of the homogeneous medium

Note that, in free space or dry air, γ_c is close to 0, and, on a conductor, γ_c is very large (for copper, $\gamma_c = 5.8 \times 10^7$ Siemens/meter) so that the center terms in Equations (1.30) either drop out or are dominant. If the fields are assumed made up of $E(x, y, z)$ $\exp(j\omega t)$ and $H(x, y, z)$ $\exp(j\omega t)$ components, then Equations (1.30) simplify to Equations (1.32).

$$\nabla^2 \mathbf{E} + k^2 \cdot \mathbf{E} = 0 \quad \nabla^2 \mathbf{H} + k^2 \cdot \mathbf{H} = 0 \tag{1.32}$$

where

$k^2 = \omega^2 \varepsilon\mu - j\gamma_c\omega$
$\varepsilon = \varepsilon_r\varepsilon_0$
$\mu = \mu_r\mu_0$
$\mu_r,\ \varepsilon_r$ = the permeability and permittivity relative to free space, respectively

If the fields are in free space, then $\gamma_c = 0$, $\mu_r = 1$, and $\varepsilon_r = 1$, and the equation for k is the familiar form given in Equation (1.33).

$$k^2 = \omega^2 \cdot \varepsilon_0 \cdot \mu_0 = \omega^2/c^2 = (2 \cdot \pi/\lambda)^2 \tag{1.33}$$

where the permittivity of free space is $\varepsilon_0 = 8.85 \times 10^{-12}$ F/m, the permeability of free space is $\mu_0 = 4\pi \times 10^{-7}$ H/m, and the velocity of light in free space is c = 2.9979 $\times$ 10^8 m/sec. One way of thinking about Maxwell's equations and the counterpart wave equations is that they are conservation of energy equations in space and time.

The basic definition of *radar cross section (RCS)* is the ratio of the scattered power to the incident power in the direction of an observer at infinity. Equation (1.34) shows the fundamental RCS equation in terms of the electric or magnetic fields.

$$\sigma = (4 \cdot \pi)\lim_{R \to \infty} R^2 \frac{|\mathbf{E_s}|^2}{|\mathbf{E_i}|^2} = (4 \cdot \pi)\lim_{R \to \infty} R^2 \frac{|\mathbf{H_s}|^2}{|\mathbf{H_i}|^2} \tag{1.34}$$

where

R = range to the observer
$\mathbf{E_i}, \mathbf{H_i}$ = the vector incident electric and magnetic fields
$\mathbf{E_s}, \mathbf{H_s}$ = the vector scattered fields

1.5.2 RCS Phenomenology

As a propagating wavefront meets a boundary or discontinuity in the medium, some energy must be reflected, some is transmitted, and some is reradiated so as to satisfy the physics described by Equations (1.30) through (1.32). Figure 1.13 summarizes the five major sources of reradiation that can result in observable RCS. The first, and the one with which everyone is familiar, is specular or mirror-like reflection or scattering. Although the reflection has a lobe structure, the dominant reflection is at the complement of the incidence angle of the illumination relative to the surface normal. Almost everyone has experienced the mirror-like glint from the sun on a window. A second source of reradiation is diffraction in which a tip scatters spherically or an edge scatters in a specular torus (donut) about the complement to the edge normal. A third source of scattering is travelling waves, which arises when the illuminating wave couples into a surface at a shallow angle. This creates currents propagating in the surface. The effect is most pronounced in long, thin bodies such as wires, egg (prolate ellipsoid) shapes, and ogives. As these surface currents encounter shape discontinuities, material changes (different μ or ε), or the end of the body. Then, for the physics to be satisfied, some reradiation usually occurs. A fourth source of reradi-

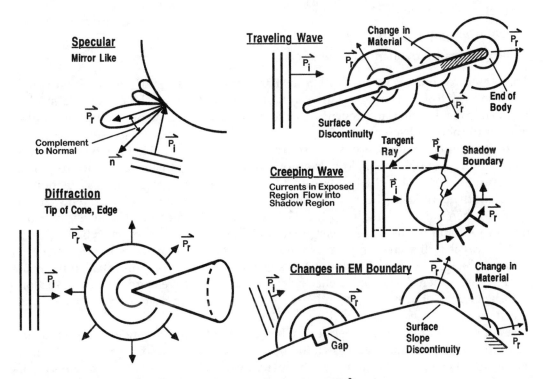

Figure 1.13 Sources of electromagnetic scattering (adapted from Fuhs[3]).

ation or scattering is creeping waves. The illuminating wave couples into the surface creating currents propagating in the surface that flow from the region of illumination to shadowed regions. These currents can flow all the way around a body and then interact with the specular reflection to create a resonance effect. The effect is most pronounced in objects that are good conductors and bodies of revolution (cylinders, spheres, eggs, ogives). The last category includes any coupled wave currents that encounter a change in the electromagnetic boundary conditions such as gaps, surface curvature changes, material changes, conductivity changes, and even smashed bugs.

So what does this all mean in terms of real objects? Almost every single radar or ladar observable depends on the laser or radar cross section of the object creating that effect. The overall RCS determines detection performance, but the unique features of that RCS determine target recognition performance. The recognition of a target is based on its signature which, in turn, is made up of RCS from individual sources. A list of typical RCS signature sources is given below.

• Jet engine, propeller, or vibration modulation

• Roll, pitch, yaw motions about the object center

• Polarization conversion

• Resolution of individual scatterers

• Nonlinear frequency interactions

• Near-resonance frequency response

These features may have small RCS in many cases, thus forcing recognition using these sources to short range relative to first target detection. The object of a stealth platform must be, first, to prevent detection but, next, to prevent recognition on the basis of the above observables.

Some examples of the less obvious sources of RCS are helpful in understanding resonance and edge diffraction. Two examples of the effects of near-resonance response are given in Figs. 1.14 and 1.15. Figure 1.14 shows the RCS of a sphere as a function of wavelength relative to the radius. There are three regions of RCS behavior for a sphere. When the wavelength is large relative to the circumference, the RCS is proportional to λ^{-4} and is called the *Rayleigh region*. When the circumference is between 0.5 and 10 wavelengths, creeping waves are important and constructive interference can enhance the RCS by up to a factor of 4 as suggested in Fig. 1.14. This is called the *resonance region*. When the circumference is large relative to wavelength, the cross section is independent of wavelength and proportional to the projected area, πa^2. This is called the *optical region*.

Similar to the resonance of a sphere, when the incident electric field is perpendicular to the axis of a cylinder as shown in Fig. 1.15, a cylinder exhibits a resonance effect as a result of creeping waves. When the incident electric field is parallel to the axis of the cylinder, the effect of creeping waves is very small. In the Rayleigh region, a cylin-

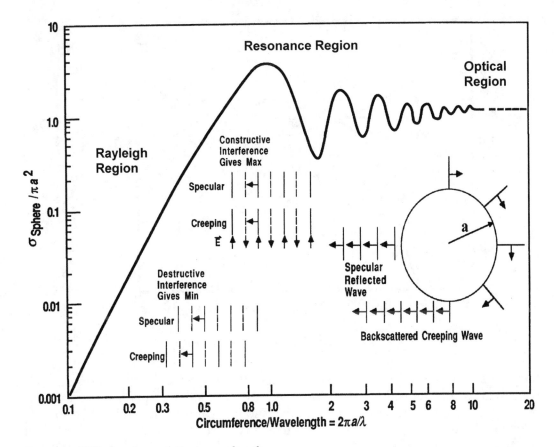

Figure 1.14 RCS of a sphere relative to wavelength.

der exhibits the same λ^{-4} RCS dependence for perpendicular polarization as a sphere, but, for parallel polarization, the RCS exhibits $\lambda^{0.5}$ dependence! This last property is one of the bases for fiber chaff radar countermeasures. The optical region is entered as the wavelength gets smaller relative to the dimensions; then, the RCS is proportional to the circumference divided by λ times the length squared independent of polarization (to a first order!). As the radius of a cylinder gets very small relative to length and wavelength, the parallel polarization RCS approaches L^2/π, and the perpendicular polarization RCS approaches 0.

An example of edge diffraction is shown in Fig. 1.16. These high-resolution images of the RCS of a small drone show that nose-on scattering comes primarily from the inlet edges and inlet interior components, the junctions of major structural features, and the tips of the wings and tail. The physical outline of the drone has been added to the contour plot to assist in visualization of the sources of scattering. Note that the trailing tips have a significantly higher cross section than the leading tips because of trav-

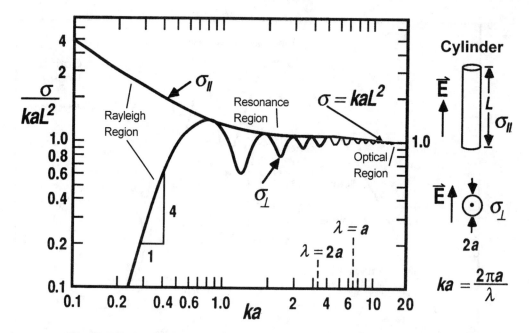

Figure 1.15   RCS of a cylinder relative to wavelength (adapted from Barton[11]).

elling wave scattering. The images have a resolution of 5/8th of an inch using a 10-GHz bandwidth stepped FM waveform and inverse synthetic aperture radar (ISAR) processing. The high-resolution image allows identification of the individual sources of the RCS signature of the object under test. As can be seen from the amplitude plot in the lower left of the figure, the cross section of each source may be quite small, but the summation of all of them may yield a significant RCS. Clearly, low-RCS vehicles require some type of edge treatment. Some examples are provided in Fig. 1.19 and Chapter 6. Obviously, at other angles, a different set of RCS sources, such as the wing leading edges or body, may dominate the signature.

1.5.3   Estimating RCS

Figure 1.16 would suggest that RCS is made up of discrete independent sources that could be summed to provide overall estimates of a vehicle signature (this is not really true, but it can provide a reasonable estimate under many circumstances). A first-order approximation of RCS for a platform or complex object can be obtained by summing the major specular RCS components as given in Equation (1.35).

$$\sigma \approx \left| \sum_{m=1}^{M} \sqrt{\sigma_m} \cdot \exp(j \cdot 4 \cdot \pi \cdot r_m / \lambda) \right|^2 \tag{1.35}$$

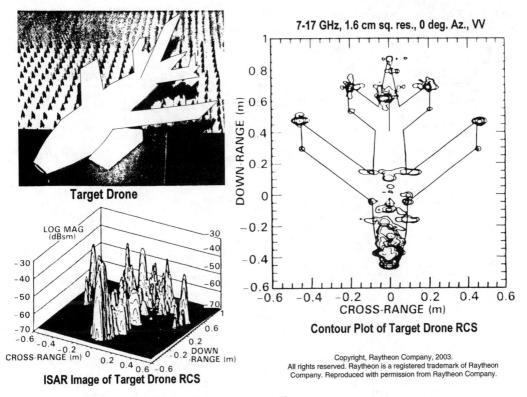

Target Drone

ISAR Image of Target Drone RCS

7-17 GHz, 1.6 cm sq. res., 0 deg. Az., VV

Contour Plot of Target Drone RCS

Figure 1.16 Example sources of scattering (from Raytheon[15]).

where

σ_m = RCS of the individual simple scatterer

r_m = distance to the observer (or some local reference plane)

This formula, which can be used to obtain a second-order approximation to a complex object RCS, can be simplified by dropping the phase term and forming a noncoherent sum for first-order approximations. The idea, which is not new, is to break the object into simple components whose RCS performance is well understood. Then, the specular/major RCS components in each angular region are summed with all the other components in that region to obtain an estimate of RCS. Table 1.6 summarizes the geometrical optics (GO) approximations for the specular components of simple RCS shapes that can be used for a first-order estimate. These components consist of reflections normal to the surface (broadside), principal diffraction from edges and tips at their normals, travelling wave reflections, and reflections from cavities or holes. The formulas in Table 1.6 assume that the components are perfect electrical conductors (sometimes called *PECs* in textbooks) at the illumination wavelength. Obviously,

Table 1.6 First Order Specular RCS Components[12]

Scattering source	Type of scatter	Approx. RCS equation $(m^2) \bullet \Gamma^2$	Approx. beam width (°)	Approx. peak angle (°)
Flat plate	Broadside	$4\pi A_e^2/\lambda^2$	$(57.3\lambda/L_{eff})^2$	$\theta = 0$
	Front edge	L_{eff}^2/π	$(57.3\lambda/L_{eff}) \times 360°$	$\phi = 0$
	Back edge	$< L_{eff}^2/\pi$	$(57.3\lambda/L_{eff})^2$	$\phi = 0, \theta = 49(\lambda/L_{eff})^{0.5}$
	Tips	$\lambda^2\tan^4\alpha/(16\pi)$	omni	all ϕ and θ
Cylinder	Broadside	$2\pi a L_{eff}^2/\lambda$	$(57.3\lambda/L_{eff}) \times 360°$	$\theta = 0, \phi = 0$
	Edges	$a_e\lambda/(2\pi)$	omni	all ϕ and θ
	Trailing rim edge	$\pi\lambda a^2/L_{eff}$	$(57.3\lambda/2a)^2$	$\theta = 0, \phi = \pm90$
	End	$4\pi^3 a_e^4/\lambda^2$	$(57.3\lambda/2a_e)^2$	$\theta = 0, \phi = \pm90$
Cone	Broadside base	$4\pi^3 a_e^4/\lambda^2$	$(57.3\lambda/2a_e)^2$	$\theta = 180, \phi = $ any
	Broadside cone	$8a_e^2/(9\sin^2\alpha \cdot \cos\alpha)$	$(57.3\lambda/2a_e) \times 360°$	$\theta = \alpha, \phi = $ any
	Trailing rim edge	$\pi\lambda a \cdot \tan\alpha/2$	$(57.3\lambda/2a)^2$	$\theta = 0, \phi = $ any
	Tip	$\lambda^2\tan^4\alpha/(16\pi)$	omni	all ϕ and θ
Ogive	Broadside @ surface normal	$\pi r_1^2(1-\cos\alpha/\sin\theta)$	$(90° - \alpha) \times 360°$	$\theta = (90°\pm\alpha), \phi = $ any
	Trailing rim edge	$\pi\lambda a^2/L_{eff}$	$(57.3\lambda/2a)^2$	$\theta = 49(\lambda/L_{eff})^{0.5}, \phi = $ any
	Tip	$\lambda^2\tan^4\alpha/(16\pi)$	omni	all ϕ and θ
Egg (spheroid)	Broadside @ surface normal	$\pi r_1 r_2$	omni	all ϕ and θ
Untreated cavity	Hole	$2A_e\cos^2\theta$	$60°^2$ typical	$\theta = 0, \phi = $ any
	Edges	L_{eff}^2/π	$(57.3\lambda/L_{eff}) \times 360°$	$\phi = 0, \theta = $ edge normal
Dihedral	Broadside	$8\pi a_I^2 b_I^2/\lambda^2$	$90°^2$	$\theta = \pm45°, \phi = $ most
	Edges	L_{eff}^2/π	$(57.3\lambda/L_{eff}) \times 360°$	$\theta, \phi = $ edge normal
Wire	Broadside	L_{eff}^2/π	$(57.3\lambda/L_{eff}) \times 360°$	$\theta, \phi = $ edge normal
Wire loop	Broadside	πa^2	$(57.3\lambda/2a)^2$	$\theta = 0, \phi = $ any

Note: λ is the wavelength, L_{eff} is the effective length, a is the radius, a_e is the effective radius, a_I and b_I are the side lengths, α is the cone or edge half angle, A_e is the effective area, and ϕ and θ are the usual spherical coordinate angles assuming that z is normal to the surface or edge.

real objects aren't perfect conductors, and so the scattered fields and RCS are usually somewhat smaller (usually not nearly enough without special effort for stealth platforms). The complex nature of scattering objects can be modeled on a first-order basis by applying a complex reflection factor, Γ^2, to each RCS component.

Geometrical optics assumes that each photon or ray is like a billiard ball; i.e., the ray is reflected from a surface at the complement of the angle between the incident ray and the surface normal. Thus, only rays that are normal to a surface—the specular— would be retroreflective and give rise to monostatic RCS. Obviously, there is the potential that a complex object could have multiple reflections that, in combination, would be retroreflective. The obvious examples of this are right dihedrals and corner reflectors. Stealth vehicles must avoid such configurations at all costs. Usually, ray tracing is used until the last bounce, and the RCS of this last surface is then applied to estimate the overall RCS (sometimes called *shooting and bouncing rays*).

As an example of the use of Equation (1.35) and Table 1.6, consider the hypothetical aircraft shown in Fig. 1.17. From an RCS point of view, there are seven major components of the hypothetical example aircraft as listed in the figure (marked 1 through 7). These components give rise to 12 sources of scattering, which are alphabetically listed in the figure as (a) through (i). Each of these is characterized in Table 1.6 with a description of the type of the scatter, approximate equation for RCS times a reflection factor, the approximate width of the main reflection, and the approximate angle of the peak reflection. The peak angular directions of these sources are generally indicated in the figure next to the letter designations. For example, ogive tip diffraction and the exhaust duct travelling wave scatter are directly off the aircraft nose, the specular from the canopy exists at a wide range of angles, and the specular from the nose ogive ranges from $90° - \alpha$ to $90°$. To keep the example simple, the ogive and canopy are assumed to be good RF reflectors, and the scattering from inside the canopy is significantly lower (obviously not always true). The lengths and radius of curvatures of the major components are noted in the figure as *L1* through *L7* and *r1* through *r4*. The ogive angle and wing sweep angles are α and β. The areas of the cavities are A_I and A_E. All of the dimensions of this type for a real aircraft can be obtained from Jane's *Aircraft Recognition Guide* or similar books. This makes a first-order approximation straightforward, as shown below.

For the example here, assume dimensions comparable to an F-16. Let the length of the ogive, *L1*, be 2.5 m; the length of the remainder of the body, *L2*, 12 m; the wing leading and trailing edges, *L3*, 4.5 m; the horizontal stabilizer leading and trailing edges, *L4*, 1.25 m; the wing tip edge, *L5*, 1 m; the horizontal stabilizer tip edge, *L6*, 0.5 m; and the inlet perimeter, *L7*, 5 m. Let the ogive long dimension curvature, *r1*, be 5 m; long dimension canopy radius, *r2*, 1 m; the canopy short dimension radius, *r3*, 0.5 m; and the fuselage curvature (modeled as a cylinder, it can worse than this but not better), *r4*, 1.25 m. Assume that the vertical stabilizer is the same size as the horizontal stabilizer. Let the engine inlet and exhaust areas, A_I and A_E, be 1 m^2. Assume a

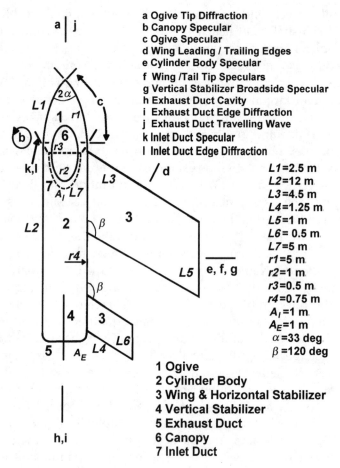

a Ogive Tip Diffraction
b Canopy Specular
c Ogive Specular
d Wing Leading / Trailing Edges
e Cylinder Body Specular
f Wing /Tail Tip Speculars
g Vertical Stabilizer Broadside Specular
h Exhaust Duct Cavity
i Exhaust Duct Edge Diffraction
j Exhaust Duct Travelling Wave
k Inlet Duct Specular
l Inlet Duct Edge Diffraction

L1=2.5 m
L2=12 m
L3=4.5 m
L4=1.25 m
L5=1 m
L6= 0.5 m
L7=5 m
r1=5 m
r2=1 m
r3=0.5 m
r4=0.75 m
A_I =1 m
A_E=1 m
α =33 deg
β =120 deg

1 Ogive
2 Cylinder Body
3 Wing & Horizontal Stabilizer
4 Vertical Stabilizer
5 Exhaust Duct
6 Canopy
7 Inlet Duct

Figure 1.17 Example first-order RCS estimate from specular segments.

30° wing sweep so β will be 120°, and α is a computed angle from the other dimensions that is approximately 33°. The last assumption is the choice of wavelength, λ; let it be 0.06 m, which is typical of SAM threats. With these basic dimensions, an approximation of RCS can be computed as a function of angle in the horizontal plane as summarized in Table 1.7 below. The table has the first-order specular equation for each of the 12 main scattering centers. Once the observer is away from the specular direction, the geometrical optics RCS is 0. Obviously, only perfect scattering bodies have 0 RCS away from the normal. In real life, this can't be the case, of course, so 0 is arbitrarily set to –60 dBsm in the table. As seen in Chapter 6, at most frequencies, manufacturing tolerances dominate the off-specular RCS, and –60 dBsm is a reasonable lower bound—even for more accurate solutions of Maxwell's equations and real objects.

Table 1.7 Example Calculation of First-Order RCS

Figure 1.16 feature	Equation	RCS @ 0° (dBsm)	RCS @ 10° (dBsm)	RCS @ 20° (dBsm)	RCS @ 30° (dBsm)	RCS @ 40° (dBsm)	RCS @ 50° (dBsm)	RCS @ 60° (dBsm)	RCS @ 70° (dBsm)	RCS @ 80° (dBsm)	RCS @ 90° (dBsm)
a	$\lambda^2 \tan^4 \alpha/(16\pi)$	-40	-40	-40	-40	-40	-40	-40	-40	-40	-40
b	$\pi r_3^* r_2$	-7	-5	-2	1	2	3	4	4	5	5
c	$\pi r_1^2(1-\cos \alpha/\sin \theta)$	-60	-60	-60	-60	-60	-60	-10	7	9	10
d	$2(L3^2 + L4^2)/\pi$	-60	-60	-60	11	-60	-60	-60	-60	-60	-60
e	$2\pi r4 \cdot L2^2/\lambda$	-60	-60	-60	-60	-60	-60	-60	-60	-60	-60
f	$(L5^2 + L6^2)/\pi$	-60	-60	-60	-60	-60	-60	-60	-60	-60	-4
g	$4\pi(L4 \cdot L6\sin\beta)^2/\lambda$	-60	-60	-60	-60	-60	-60	-60	-60	-60	18
h	$2A_E$	-60	-60	-60	-60	-60	-60	-60	-60	-60	-60
i	$(2\lambda)^2/\pi$	-23	-23	-23	-23	-23	-23	-23	-23	-23	-23
j	$\pi\lambda(r4)^2/(L1 + L2)$	-21	-60	-60	-60	-60	-60	-60	-60	-60	-60
k	$2A_I \cos^2 \theta$	3	2.9	2.5	1.8	-0.7	-0.8	-3	-6.3	-12	-60
l	$(L7)^2/\pi^{**}$, $(2\lambda)^2/\pi$	9	-23	-23	-23	-23	-23	-23	-23	-23	-23
Total	Noncoherent Sum	10	2.9	3.8	12	3.8	4	5.4	9	10.4	19

*r_3 and r_2 are related to r3 and r2 but are equal only at 90° as shown in Equation (1.36).
**For polarization parallel to the plane of the inlet edge; for polarization perpendicular to the plane of the edge, the mainlobe is almost the same, but the sidelobes drop off faster.

$$r_2 = -\frac{((r2)^2 \cdot \cos^2(w) + (r3)^2 \cdot \sin^2(w))^{1.5}}{r2 \cdot r3} \quad \text{and} \quad r_3 = \frac{-(r2)^2}{[(r3)^2 \cdot \cos^2(w) + (r2)^2 \cdot \sin^2(w)]^{0.5}}$$

(1.36)

where $\theta = \operatorname{atan}\left(\frac{r3}{r2} \cdot \tan(w)\right)$ and (w = an independent variable in spheroid space)

The important focus of the example is the method, not the actual numbers used. Although the example does give the right order of magnitude for such an aircraft, routine or incidental changes can affect the RCS by 3:1, such as canopy and radome RF transparency for any given frequency/wavelength.

Note that minor changes in conductivity and scattering from the inlet would significantly reduce the nose-on RCS (k, l in Fig. 1.17) where the maximum threat usually occurs, provided the radome is reflective at the threat frequency. If the inlet impedance were somehow matched to the impedance of the air around the aircraft, there would be no reflection. (More about this in Chapter 6.)

In the example above, edges are a significant or dominant contributor to RCS. As previously mentioned, two of the great thoughts of LO technology are edge treatment by convolution and by impedance matching. An example of edge treatment by convolution is given in Fig. 1.18, in which two sides of the same vehicle are shown; one side is convolved with a Gaussian to create a "blended" contour, and one side is faceted. The idea is to take the faceted shape, which directs the speculars in safer directions, and convolve a second function on it. The resulting "blended" shape is very special,

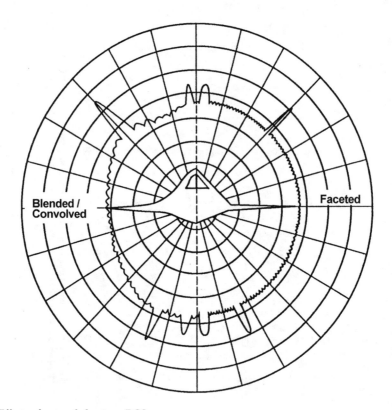

Figure 1.18   Effects of general shape on RCS.

and most traditional aerodynamic lofting does not achieve low-RCS lobes. This "blending" slightly broadens the main specular spike but can dramatically reduce the off-normal RCS lobes. The effect is especially useful when the RCS at the horizontal is dramatically reduced, because that is where the long range threat is. More about the desired convolving functions is given in Chapter 6, but even a rectangular convolving window whose width is only 1/10th of the specular surface width can produce a 10:1 improvement at the horizon.

The second way to reduce the RCS of an edge is by impedance matching. The impedance of free space is approximately 377 Ω. The leading edges of most major RCS contributors can be made to match free space as shown in Fig. 1.19. The most common method is to impregnate some lightweight substrate (e.g., honeycomb) with resistive ink whose resistivity in ohms per square is 377. The material then is protected with a high strength-to-weight ratio RF transparent skin. Trailing edges can also be treated, but the impedance must match the arriving travelling wave propagating in the surface at a much lower impedance, as well as a primary illumination from free space. This usually requires multiple layers with different impedances. In addition, some surfaces may be treated with ferrite-loaded radiation-absorbing material (RAM) to attenuate special illumination directions such as ground bounce. Ferrite RAM is used sparingly, because it is heavy, expensive, and not very rugged. The main advantage to ferrite RAM is that it can be tailored to frequency and angle of arrival. Impedance matching can improve edges and off-normal scattering by 100:1.

Second-order estimates of vehicle RCS can be generated by using the physical optics (PO) approximations to simple shapes and then summing these using Equation (1.35) with their corresponding relative phases based on their location on the vehicle. It is important to continually remind oneself that none of these equations gives an exact estimate of RCS. Even so-called exact solutions are based on models that do not

Clean Shape with Leading Edge Ogival and Trailing Edge Wedge Shaped

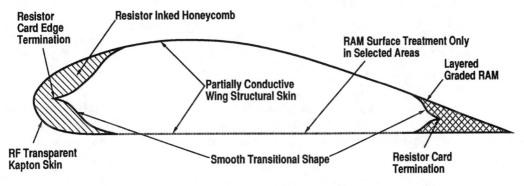

Figure 1.19 Typical RCS edge treatments (adapted from Bhattacharyya[4] and Adam[6]).

account for everything on a real object. The approximations can get better and better, to within 1 to 3 dB, but are never a substitute for measurement.

Figure 1.20 summarizes PO and GO for some simple shapes. The approximations in the figure assume that the wavelength is small relative to the dimensions of the object and that the observation distance is very large relative to those same dimensions. The edges mentioned for many of the objects can be estimated using the edge term or the wire loop equation. Irregular loops can be approximated with the circular loop if the enclosed area is the same.

Most dimensions, angles, and variables are defined in Fig. 1.20, with the following exceptions: the sinc(x) function, which is sin(x)/x, and sinc(x,y), which is (sin(x)/x) · (sin(y)/y); the variable U, which is $(\pi/\lambda) \cdot a \cdot \sin\theta \cdot \cos\phi$; and V, which is $(\pi/\lambda) \cdot b \cdot \sin\theta \cdot \sin\phi$. Although these formulas are significantly more accurate away from the specular at surface normal, they are still approximations and not a substitute for verification by test. They will help a system architect synthesize a design or estimate the probable performance of an existing design.

1.6 INTRODUCTION TO SIGNATURE BALANCE

One natural question arises: "How much stealth does a system need?" The waggish answer is: "How much money do you have?" Realistically, because stealth is expensive, one must balance all of the signatures to make a best fit for the money available for a weapon system. Hopefully, by the end of this book, the student will have a good idea of how to achieve this balance. As an introduction to the problem, consider the following radar, IR, and intercept threats.

1.6.1 Radar Threat

Consider the threat to a stealth platform as a function of radar cross section (RCS) from the following four classes of radar.

1. Long-Range Early Warning (LREW)

2. Surveillance and Target Acquisition (S&TA)

3. X-Band Fire Control (XFC)

4. Ku-Band Fire Control (KFC)

Some typical parameters for the four classes are given in Table 1.8. These parameters do not represent any specific systems but are typical of deployed systems.[3] The same required *SNR* (100), duty ratio *Duty* (0.05), and dwell time, T_D, (0.01 sec) are assumed for each system. For this simple case, the radar processor gain, G_{RP}, is assumed to be

$$G_{RP} = Duty \cdot T_D \cdot B_R \tag{1.37}$$

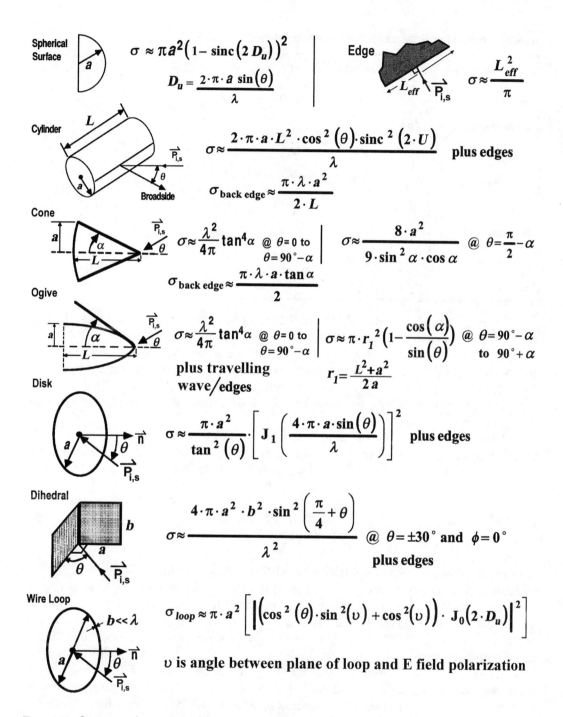

Spherical Surface

$$\sigma \approx \pi a^2 \left(1 - \text{sinc}(2\,D_u)\right)^2$$

$$D_u = \frac{2 \cdot \pi \cdot a\, \sin(\theta)}{\lambda}$$

Edge

$$\sigma \approx \frac{L_{eff}^2}{\pi}$$

Cylinder

$$\sigma \approx \frac{2 \cdot \pi \cdot a \cdot L^2 \cdot \cos^2(\theta) \cdot \text{sinc}^2(2 \cdot U)}{\lambda} \quad \text{plus edges}$$

Broadside

$$\sigma_{back\;edge} \approx \frac{\pi \cdot \lambda \cdot a^2}{2 \cdot L}$$

Cone

$$\sigma \approx \frac{\lambda^2}{4\pi} \tan^4\alpha \quad @\ \theta = 0\ \text{to}\ \theta = 90° - \alpha$$

$$\sigma \approx \frac{8 \cdot a^2}{9 \cdot \sin^2\alpha \cdot \cos\alpha} \quad @\ \theta = \frac{\pi}{2} - \alpha$$

$$\sigma_{back\;edge} \approx \frac{\pi \cdot \lambda \cdot a \cdot \tan\alpha}{2}$$

Ogive

$$\sigma \approx \frac{\lambda^2}{4\pi} \tan^4\alpha \quad @\ \theta = 0\ \text{to}\ \theta = 90° - \alpha$$

plus travelling wave/edges

$$\sigma \approx \pi \cdot r_1^2 \left(1 - \frac{\cos(\alpha)}{\sin(\theta)}\right) \quad @\ \theta = 90° - \alpha\ \text{to}\ 90° + \alpha$$

$$r_1 = \frac{L^2 + a^2}{2\,a}$$

Disk

$$\sigma \approx \frac{\pi \cdot a^2}{\tan^2(\theta)} \cdot \left[J_1\left(\frac{4 \cdot \pi \cdot a \cdot \sin(\theta)}{\lambda}\right) \right]^2 \quad \text{plus edges}$$

Dihedral

$$\sigma \approx \frac{4 \cdot \pi \cdot a^2 \cdot b^2 \cdot \sin^2\left(\frac{\pi}{4} + \theta\right)}{\lambda^2} \quad @\ \theta = \pm 30°\ \text{and}\ \phi = 0°$$

plus edges

Wire Loop

$$\sigma_{loop} \approx \pi \cdot a^2 \left[\left| \left(\cos^2(\theta) \cdot \sin^2(\upsilon) + \cos^2(\upsilon)\right) \cdot J_0(2 \cdot D_u) \right|^2 \right]$$

υ is angle between plane of loop and E field polarization

Figure 1.20 Summary of second-order RCS approximations.

Table 1.8   Typical Radar Threat Performance Parameters

Parameter	LREW	S&TA	XFC	KFC
Wavelength (m)	1.5	0.33	0.04	0.015
Bandwidth (MHz)	0.1	0.3	1	2
Peak power (kW)	2000	1000	50	25
Antenna gain	250	500	5000	10000
Duty ratio	0.001	0.001	0.01	0.01
Dwell time (sec)	0.3	0.1	0.025	0.01
Noise factor	1.6	2	2.5	4
Losses	2	2	3	4

Later, it will be shown that this is probably optimistic for a stealth target, but for this example it is a good first-order assumption. Target cross sections vary all over the map with aspect and frequency but, typically, as the wavelength increases, the RCS of normal and stealth targets increases. The good news for stealth platforms is that, even though detectability increases with wavelength, LREW systems usually have such poor position accuracy that exploitation of the detection in a dense target environment is limited.

Utilizing the parameters of Table 1.8, detection range versus cross section can be calculated for each type of threat as shown in Fig. 1.19. The particular version of the radar range equation for Fig. 1.21 is given in Equation (1.38).

$$
R_D = \left[\frac{P_T \cdot G_T^2 \cdot \lambda^2 \cdot \sigma \cdot L_R}{(4 \cdot \pi)^3 \cdot SNR_R \cdot k_B \cdot T \cdot B_R \cdot NF_R} \right]^{1/4}
\tag{1.38}
$$

In terms of sheer quantity, surface-based threats are dominant. The operating altitude of a stealth system and the altitude of the threat are important. What is important for LPIS detection is the lower edge of the threat antenna mainbeam, not the phase center. As soon as the lower edge of the main beam subtends the Earth, whether in the near field or far field, the side/scattering lobes go to pot. There are so many other targets with cross sections or emissions greater than an LPIS that the threat radar/interceptor will be overwhelmed with ambiguous/false targets when the lower edge strikes the ground. By the way, both azimuth and elevation multipath as well as civilian interference are serious problems at VHF, which otherwise is the best frequency band to detect stealth platforms.

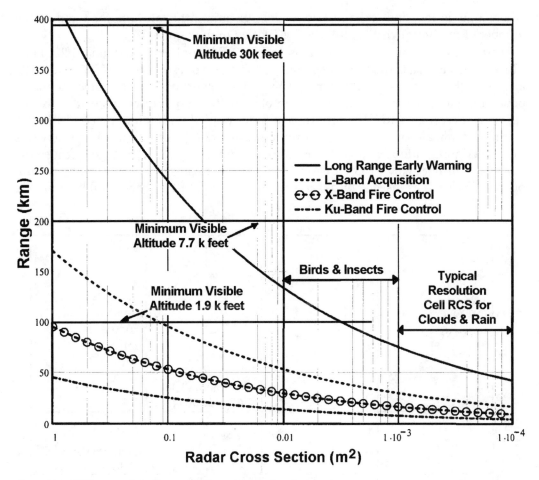

Figure 1.21   RCS versus detection range for various radar classes—long range.

The radar horizon and probability of visibility are analyzed at length in Section 4.3. In simple terms, the radar horizon using a 4/3 Earth model is

$$R_H \approx 1.228 \cdot (\sqrt{h_s} + \sqrt{h_t}) \text{ nautical miles} \qquad (1.39)$$

where the height of the surveillor, h_s, and the height of the target, h_t, are in feet. The minimum visible altitude for a given detection range, R_D, for a threat radar antenna whose lower edge, h_s, is 6 ft above the ground using the 4/3 Earth model is approximately

$$h_v = [(R_D - 5.57)/2.2748]^2 \text{ feet} \qquad (1.40)$$

where h_V is in feet and R_D is in kilometers. Assuming this altitude for the threat, a number of minimum visible altitudes are plotted in the figures.

There are a number of observations to be made from Figs. 1.21 and 1.22. First, LREW, if ground based, can use its range performance only against very high-altitude targets. LREW can easily be countered by flying lower, which, of course, degrades penetrating aircraft range, but there is an obvious trade-off between stealth and fuel efficiency. Second, if LREW is airborne, altitude coupled with interference and clutter rejection dominate performance, not the range equation. Although it is not obvious from the figures, altitude subjects the threat radars to more interference, because the horizon is further away. Third, low-altitude, stealthy targets can be masked by natural phenomena such as insects, birds, and weather. Typical RCS for natural phenomena

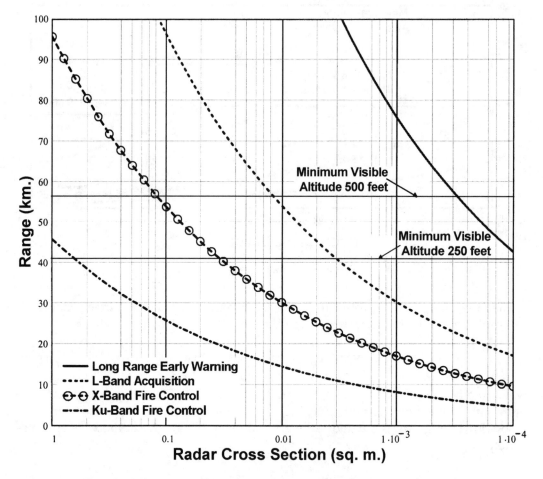

Figure 1.22 RCS versus detection range for various radar classes—short range.

and representative resolution cells is also plotted in the figure. Fourth, the combination of low altitude and modest stealth greatly reduces the effectiveness of surface based fire control. Fifth, airborne fire control will be dominated by clutter rejection for low-altitude, stealthy targets.

In summary, high altitude is a bad place to be for stealthy aircraft unless one can afford very low radar cross section. Although it is not obvious, higher operating altitude for the threat sensors subjects them to more interference, because the horizon is further away, and the probability of a bright discrete is proportional to area. Obviously, outer space is the worst place to be, as a platform is always visible and susceptible to ISAR imaging at inch resolution. In space, stealth must involve deception and spoofing rather than invisibility. Chapter 3 gives some spoofing examples. Chapter 4 contains more details on the interference environment as well as clutter and masking.

1.6.2   Infrared Threat

The infrared (IR) threat depends first on the temperature difference between a stealthy target and its immediate background. The temperature contrast is a function of both the blackbody radiation from a stealthy platform and the reflection of the apparent temperature of the surroundings by the platform. The power flux from blackbody radiation is governed by the Planck radiation law. An approximation to that law is given in Equation (1.41), below.[8,9]

$$L(\lambda, T) = \frac{1.19 \cdot 10^{-16}}{\lambda^5 \cdot [\exp(1.44 \cdot 10^{-2}/(\lambda \cdot T)) - 1]} \text{ watts/m}^2\text{-steradian} \tag{1.41}$$

where λ is in microns and T is the apparent temperature of the vehicle in kelvins. The IR range equation is given in Equation (1.42).[8,9]

$$R_{IRMAX} = \left[e \cdot A_{Teff} \cdot I \cdot L_{band} \cdot A_o \cdot L_o \cdot \frac{D_{avg}}{A_D^{0.5}} \cdot \frac{L_p}{B_{IR}^{0.5} \cdot SNR_{IR}} \right]^{0.5} \text{ cm} \tag{1.42}$$

In addition, using the following approximation to convert from free space to in-atmosphere detection range provides the upper and lower bounds on detection performance depending on surveillance altitude.

$$R_{IR} \approx \frac{\ln(0.85 \cdot ec \cdot R_{IRMAX} + 1)}{0.85 \cdot ec} \tag{1.43}$$

where

R_{IRMAX} = maximum range independent of atmospheric attenuation

e = emissivity

A_{Teff} = effective area (cm^2) of the target in the direction of the IR sensor

I = radiant intensity (W/cm^2 × steradian) and is the integral of $L(\lambda)$ over the band of sensor wavelengths

L_{band} = average atmospheric propagation loss fraction across the band

ec = loss or extinction coefficient per kilometer between the target and the IR sensor across the band

A_o = area of the optical entrance pupil (cm)

L_o = optics loss fraction

D_{avg} = normalized average detectivity—the reciprocal of the noise power (W/cm-Hz$^{0.5}$)

A_D = IR detector area (cm^2)

L_p = processor loss fraction

B_{IR} = detection bandwidth of the sensor

SNR_{IR} = required signal-to-noise ratio for detection

For many aspect angles, the IR signature of a penetrating aircraft is dominated by its aerodynamically heated skin. A stealth vehicle may have its skin treated for low emissivity, but the basic effects are still prominent. Because friction drag coefficient goes down with Mach number, a simple approximation at sea level for conventional aircraft skin heating is given in Equation (1.44).

$$T_S = \left[373.8 + 45.4 \cdot V^2 - \frac{84}{1 + V} \right] \text{kelvins} \tag{1.44}$$

where V is the velocity in Mach number, and atmospheric temperature and pressure conditions are 288 K and 14.7 psi, respectively.

These equations can be combined to estimate the detection performance of an IR sensor against a penetrating stealth aircraft as a function of velocity. For example, consider a typical modern IR sensor and target that have the following parameters: e = 0.5, A_{Teff} = 10^4 in cm^2, L_{atm} = 0.5, A_o = 100 in cm^2, L_o = 0.6, D_{avg} = 5 ×10^{11} in cm-Hz$^{0.5}$/W, A_D = 2 × 10^{-3} in cm^2, L_p = 0.5, B_{IR} = 20 Hz, and SNR_{IR} = 10. The radiant intensity is given by Equation (1.45).

$$I = \int_{\lambda_1}^{\lambda_2} L(\lambda, T_S) \cdot d\lambda \tag{1.45}$$

The upper and lower bounds on the detection range shown in Fig. 1.23 were generated from Equation (1.42) by substituting Equations (1.41), (1.43), (1.44), and (1.45). For Fig. 1.23, the IR detection band was from λ_1 = 1 to λ_2 = 3 μm (details in the appendices and on the disk). The upper bound is the performance that might be obtained from a surveillance satellite roughly over the stealth platform. The lower bound is the performance that might be obtained from a surface-based IR sensor and stealth plat-

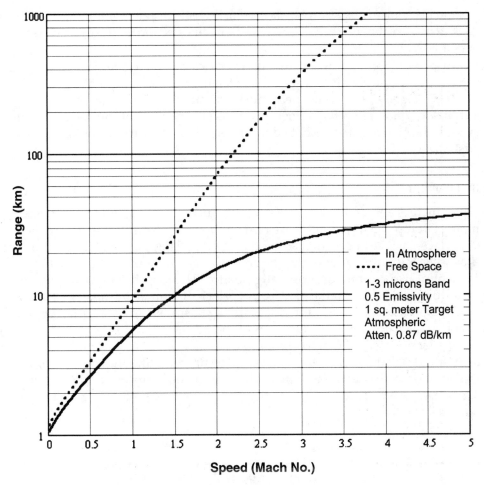

Figure 1.23 Typical IR detection range versus aerodynamic heating.

form at low altitude or near the horizon. Figure 1.23 shows that there is an obvious trade-off between penetrator speed and its IR signature. Suppose an aircraft weapon system has been designed to have an RCS signature that limits threat detections to 10 km; then, a balanced design would also limit the penetration speed to Mach 1 or less, depending on threat surveillance altitude.

Note also that an aircraft attempting to use supersonic speed to flee an incoming missile dramatically increases the IR detection range and targeting accuracy. Similarly, if a system is designed to travel at Mach 3.5, don't spend a penny on stealth, because its IR signature is so large it will be detected at the IR sensor horizon or from space. The skin heating effect is not as dramatic at longer IR wavelengths, because the peak of blackbody radiation shifts down in wavelength with increasing skin temperature. Similar problems exist on surface vehicles as a result of sunlight heating, not speed. Surface treatments help somewhat, but high absorptivity usually results in higher emissivity.

Of course, there are times when an otherwise stealthy vehicle might choose to abandon stealth to flee or transit to a safe area. Stealthy vehicles might also choose to enhance their signature during peacetime to maintain an element of surprise. Unfortunately, once a stealthy vehicle's location is known with precision, tracking is often possible by the means described in Chapter 3. So the trick is not to get caught in the first place.

1.6.3 Intercept Threat

Similarly, there is a trade-off/balance between interceptability and RCS. Consider the example of a ship versus a dual-mode missile shown in Fig. 1.24. Suppose a medium-sized surface ship has been treated so that its RCS is approximately 1000 m^2 from most aspects. Suppose the threat to the ship is a sea-skimming missile with radar as well as antiradiation homing terminal guidance. The ship has a surface search and target acquisition (S&TA) radar that is the principal emitter in the engagement scenario. The ship uses various countermeasures against this type of missile, but the primary counter outside the last mile is an off-board decoy that is launched when the missile is first detected. The decoy has a transponder designed to simulate a ship signature with 10^4 to 10^5 m^2 cross section; i.e., 10 to 100 times the ship signature. The LPI design question is, "What improvements must be made to the existing S&TA so that it is as good as the rest of the ship?" so that the decoy will succeed. Table 1.9 summarizes the relevant parameters.[25]

Table 1.9 Example Threat Parameters

Parameter	Missile active/passive	Decoy	Current S&TA radar
Wavelength (m)	0.063	0.063	0.063
Power (W)	50/N.A.	Up to 2	280k
Sensor gain	10/1	800	30
Antenna gain (dB)	23.5/23.5	1.5	30
Sensitivity (dBm)	–90/–60	–60	–90
Sidelobes (dBi)	–3	Close to omni	–3
RCS (dB sm)	0.1	40	35 in mainlobe –41 in sidelobes
Losses (dB)	–3	–3	–6

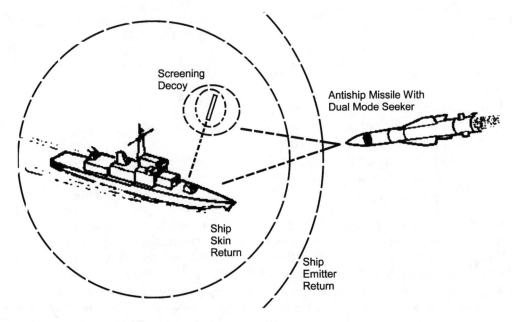

Figure 1.24 Example scenario for intercept threat trade-off.

The decoy equivalent RCS must be above the general ship RCS but also above the RCS of the mainlobe of the S&TA radar, because the radar will illuminate the antiship missile and provide an enhanced RCS to the threat. Outside the mainlobe of the S&TA radar, the antenna RCS will be much smaller, and emissions will dominate.

One could guess that the current radar is detectable in the sidelobes beyond the missile's horizon. To verify the assertion, substitute the above parameters in Equation (1.15) as shown below.

$$R_I^2 = 2.8 \cdot 10^5 \cdot 0.5 \cdot 2.24 \cdot 10^2 \cdot 1 \cdot (0.063)^2 \cdot 0.5/(4 \cdot \pi)^2 \cdot 10^9 = 3.94 \cdot 10^9 \tag{1.46}$$

and then, as expected,

$$R_I = 63 \text{ km} \tag{1.47}$$

This is well beyond the horizon of a sea-skimming missile. To gain an intuition for this scenario, one needs to set Equations (1.1) and (1.3) equal to one another, define the *ERPP*, and solve for it in terms of RCS, σ, as shown in Equation (1.48).

$$ERPP = \frac{P_T \cdot G_T \cdot G_{RP}}{4 \cdot \pi \cdot R_i^2 \cdot G_{IP}} \cdot \sigma \text{ and } ERPP = P_E \cdot G_{EI} \tag{1.48}$$

Thus,

$$\sigma = \frac{4 \cdot \pi \cdot R_i^2 \cdot G_{IP} \cdot P_E \cdot G_{EI}}{P_T \cdot G_T \cdot G_{RP}}$$

where

P_T, G_T, G_{RP} = missile radar power, antenna gain, and processor gain

G_{IP} = missile interceptor processor gain

P_E, G_{EI} = on-ship emitter rms peak power and gain in the missile direction treated as an emitter

σ = true or apparent RCS for the ship, decoy or shipboard emitter

Several assumptions are implicit in the above equation: the missile antenna is used both for the radar and the intercept functions; the wavelengths, losses, and required SNRs are close enough for both functions to be considered the same; and, finally, the decoy is close enough to the ship to consider its range approximately the same.

Figure 1.25 shows the example scenario in terms of the power received at the missile versus range. The on-ship radar and data link emissions should be close to the missile radar skin return from the ship. The on-ship radar emissions must be significantly less than the decoy so that the missile will not switch to passive guidance and be guided to the ship rather than the decoy. Under this constraint, the allowable radar emitter peak power can be calculated for various possible values of emitter sidelobes.

Figure 1.26 shows the allowable emitter power as a function of range for a ship-based emitter with screening decoy. At first, these curves seem counterintuitive, as they show the allowable power going down when the missile is farther away. But the reason is that the power out of the decoy goes down with greater range to the missile, and so it will not screen higher emissions. The conclusion is that, even with state-of-the-art −30 dBi sidelobes, the effective radiated peak emitter power must be substantially reduced through such measures as pulse compression, better processing, longer dwell times, higher PRF, and so on. Alternatively, the decoy must be 10^6 m^2. The problem with a very large decoy signal is that it is obviously not the ship target, and many countermeasures schemes will detect this and begin a search for an alternative return from the real target. Another possible solution to this challenge is the use of some off-platform sensor (satellite, manned aircraft, or UAV) coupled with a lower peak power on the platform sensor for confirmation.

1.7 EXERCISES

1. Calculate the radar figure of merit, K, for a radar antenna gain, G_T, of 40 dB; processing gain of 35 dB; RCS, σ, of 0.1 m^2; a received power of −110 dBm; and a wavelength, λ, of 0.3 m.

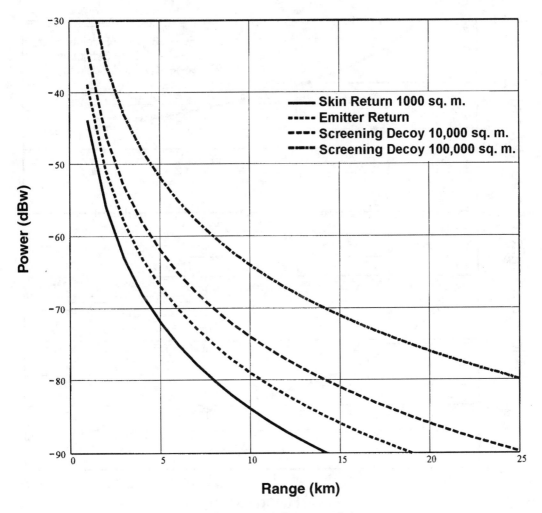

Figure 1.25 Desired available power at threat missile for emitter with decoy.

2. Calculate the interceptor figure of merit, J_I, for an interceptor antenna gain of 20 dB and interceptor processor gain of 40 dB, with wavelength, λ, of 0.06 m.

3. Calculate R_{Dmax} for the parameters of Exercise 1 with the interceptor processing gain, G_{IP}, of 30 dB; interceptor antenna gain of 30 dB; interceptor sensitivity (power) of –70 dBm; and radar gain in the interceptor direction, G_{TI}, of –20 dB.

4. Calculate R_{Dmax} for a radar antenna effective area of 4 m², SNR of 4 dB, bandwidth of 100 kHz, noise figure of 4 dB, processor gain of 60 dB, target RCS of 1 m², and for an interceptor effective area of 10 m², SNR_I of 16 dB, interceptor band-

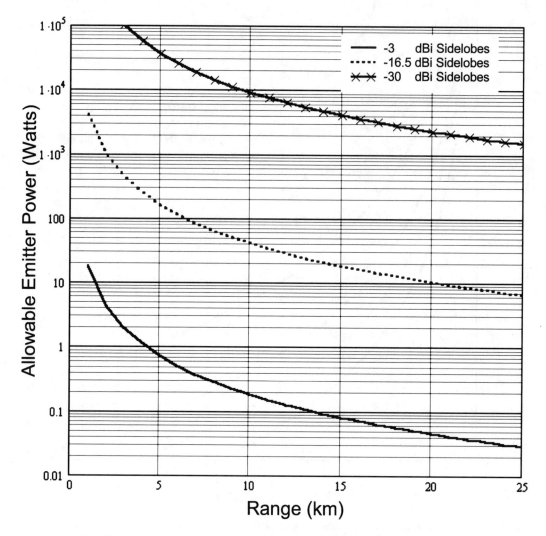

Figure 1.26   Allowable on-ship emitter peak power with screening decoy.

width of 25 MHz, noise figure of 4 dB, and interceptor processing gain of 30 dB, with radar effective area in the interceptor direction 1 cm^2.

5. For the parameters of Fig. 1.11, calculate the maximum safe range, R_{Dmax}, for a radar bandwidth of 10 Hz and interceptor bandwidth of 25 MHz.

6. For the parameters of Fig. 1.12, calculate the maximum safe range for radar antenna gain of 32 dB, interceptor bandwidth of 25 MHz, interceptor processing gain of 25 dB, and interceptor antenna gain of 15 dB.

7. For the hypothetical aircraft of Fig. 1.16, calculate the RCS with angle from 0 to 180° in 10° steps for a wavelength of 0.03 and 0.3 m.

1.8 REFERENCES

1. Lynch, D., "Radar Systems for Strike/Fighter Aircraft," *AOC Third Radar/EW Conference Proceedings*, February 12–13, 1997.
2. Berkowitz, R. S., Ed., *Modern Radar*, John Wiley and Sons, 1965, pp. 6–27.
3. Fuhs, A., "Radar Cross Section Lectures," *NPGS*, 1984.
4. Bhattacharyya, A. and Sengupta, D., *Radar Cross Section Analysis & Control*, Artech House, Boston, 1991, pp. 64, 148.
5. Jenn, D., *Radar and Laser Cross Section Engineering*, AIAA, Washington, DC, 1995, p. 81.
6. Adam, J., "How To Design An 'Invisible Aircraft'," *IEEE Spectrum*, April 1988, pp. 26–30.
7. Blake, B., Ed., *Jane's Radar and Electronic Warfare Systems 1990–1991*, Jane's Information Group, 1990, pp. 28–81, 224–225.
8. Schlessinger, M., *Infrared Technology Fundamentals*, Marcel Dekker, New York, 1995, pp. 6–91, 245–325.
9. Macfadzean, R., *Surface Based Air Defense System Analysis*, Macfadzean, 2000, pp. 22–30.
10. Sweetman, W., *Lockheed Stealth*, MBI Publishing, 2001, pp. 21–432.
11. Barton, D., *Radar System Analysis*, Artech House, 1976, pp. 66–74.
12. Crispin, J. and Maffett, A., "Radar Cross-Section Estimation for Simple Shapes," *IEEE Proceedings*, August 1965, pp. 833–848.
13. Ross, R., "RCS of Rectangular Flat Plates as a Function of Aspect Angle," *IEEE Trans. on Antennas and Propagation*, May 1966, pp. 329–335.
14. Heilenday, F., *Principles of Air Defense and Air Vehicle Penetration*, Mercury Press, 1988, pp. 8-1 through 8-18.
15. A few photos, tables, or figures in this intellectual property were made at Hughes Aircraft Company and first appeared in public documents that were not copyrighted. These photo, tables, or figures were acquired by Raytheon Company in the merger of Hughes and Raytheon in December 1997 and are identified as Raytheon photos, tables, or figures. All are published with permission.
16. Watt, W., "Camouflaging of Aircraft at Centimetre Wavelengths," *Telecommunications Research Establishment Report*, August 1941.
17. Adamy, D., "EW101 Tutorial," September 1998, http://www.jedonline.com/jed/html/new/sep98/ew101.html.
18. Stadmore, H., "Radar Cross Section Fundamentals for the Aircraft Designer," *AIAA Aircraft Systems and Technology Meeting*, August 20–22, 1979, A79.47898.
19. Stimson, G., *Introduction to Airborne Radar*, 2nd ed., Scitech Publishing, 1988, pp. 525–528.
20. Dorr, R., *Desert Storm Air War*, Motorbooks International, 1991.
21. Sweetman, W. and J. Goodall, *Lockheed 117A*, Motorbooks International, 1990.
22. Sweetman, W., *Lockheed Stealth*, Motorbooks International, 2001.
23. Skolnik, M.I., *Introduction to Radar Systems*, 2nd ed., McGraw-Hill, 1980, pp. 15–61.
24. Schleher, D., *Introduction to Electronic Warfare*, Artech House, 1986, pp. 171–177.
25. Schleher, D., *Electronic Warfare in the Information Age*, Artech House, 1999, pp. 147–192, 201–257, 498–531, 539–578.
26. McNamara, D., C. Pistorius, and J. Malherbe, *Introduction to the Uniform Theory of Diffraction*, Artech House, 1990.

2

Interceptability Parameters and Analysis

LPI radar or data link performance is a complex function of many variables. Three main elements of LPIS performance affect a stealthy system's success.

1. The actual design features incorporated into the emitting system specifically for LPI.

2. The ESM strategy and corresponding implementation utilized by threat forces. Hostile forces' ESM strategy and implementation are very important factors affecting any LPI design. Design strategy begins with an analysis of individual threat receiver characteristics against which the LPIS may be deployed. The EOB deployment and location strategy for intercept receivers must also be considered so as to create the most successful design features. Additionally, the issue of whether or not individual interceptors are netted has a strong influence on the strategies an LPIS might use.

3. The geometry between the area of regard (AOR) and threat receivers.

These design and performance considerations are summarized in Fig. 2.1. LPI design constraint features, some netting constraints, and target or emitter to threat receiver geometry are covered in this chapter. Chapter 3 covers individual threat receiver characteristics. Chapter 4 covers deployment and location strategies.

2.1 INTERCEPTABILITY PARAMETERS

There are seven main interceptability parameters or constraints.

1. LPI system mainlobe power at the intercept receiver

2. LPI system sidelobe power level at the intercept receiver

3. Area of mainlobe and sidelobes on the ground or at a certain threat altitude

4. Time of AOR illumination for mapping, tracking, or targeting

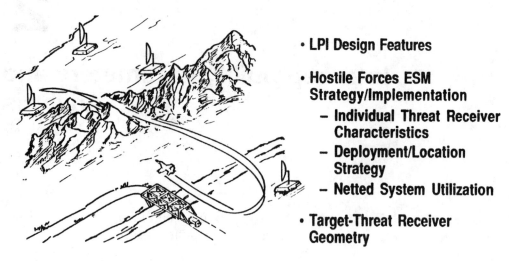

• LPI Design Features

**• Hostile Forces ESM
Strategy/Implementation**

**– Individual Threat Receiver
Characteristics**

**– Deployment/Location
Strategy**

– Netted System Utilization

**• Target-Threat Receiver
Geometry**

Figure 2.1 LPI performance is a complex function of many variables.

5. Intercept receiver density and search time

6. Intercept receiver detection response

7. Power management strategy

Each parameter is described in some detail in the chapter ahead. As will be seen, engagement scenarios must be assumed so as to properly perform LPI evaluation, and more details are described in Chapter 4.

2.1.1   Interceptability Footprints

The first three interceptability parameters are depicted in a simple model shown in Fig. 2.2. LPIS surface (or constant altitude) power contours can be computed to determine interceptability using a model that assumes that intercept receivers are distributed on some regular grid on a surface plane perpendicular to the down vector beneath the emitter or at a constant altitude such as on an orbiting surveillance aircraft. The power intercepted at each of these grid coordinates can be calculated based on the emitter geometry. This consists of repetitive calculation of the beacon Equation (1.3). The emitter if airborne has some flight path, altitude, and corresponding ground range from the ground track. The emitter, if surface based, has some altitude difference with the interceptor and some surface range to a point under the interceptor. Based on those known parameters, the emitter line of sight (LOS) is calculated for each grid coordinate in the geometry. The emitter power captured by intercept receivers located at each of these grid intersections then can be calculated. If one assumes that each of the interceptors has a specific sensitivity and that its antenna coinciden-

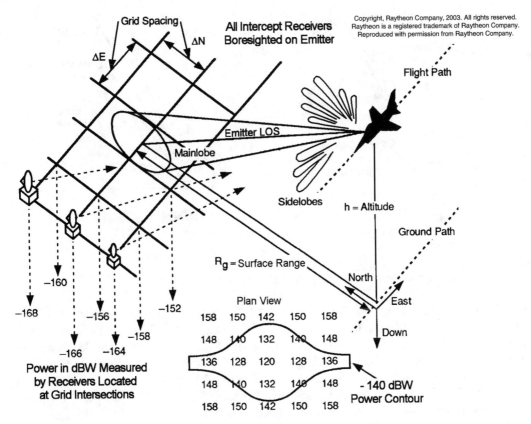

Figure 2.2   Ground power contours computed to determine interceptability.[13]

tally is pointed at the emitter, one can calculate the area containing all those intercept receivers in which the emitter received signal is larger than the minimum threshold. Corresponding power contours can be interpolated from the grid points (as shown in Fig. 2.2). The details of the calculation process are given in Section 2.4.[12,14]

In addition, there are far sidelobes in the immediate vicinity of the emitter, and their area will be larger because of the emitter's proximity to the ground (see Fig. 2.3). At some range, the emitter horizon occurs and, although there is a small amount of refraction that causes the emitter to see a little over the geometrical horizon, there is a horizon cutoff that occurs rather abruptly; beyond that range, the emitter antenna mainlobe no longer intercepts the Earth's surface. In general, the low-altitude, shallow-grazing-angle case creates the largest area in which the emitter can be detected and thus is the most dangerous from an LPI point of view.[14]

Of course, actual footprints created by emitters on a surface vary with surface shape, grazing angle, antenna shape, and antenna sidelobe weighting. Three example

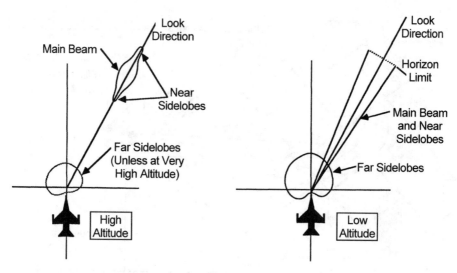

Figure 2.3 Platform altitude affects antenna "footprint."

footprints on a flat Earth are shown in Fig. 2.4. In this particular case, three antennas inscribed inside a 37-in circle (typical large fighter antenna diameter) are compared. For this example, the surface range to the center of the mainbeam is 80 km, and the depression angle is 8°. The first antenna is a circular disk with Sonine weighting.[1] It has substantially more gain; thus, one's intuition would be that it would have the best intercept footprint for a given detection performance. The second example is a square inscribed in the 37-in circle with 33-in sides and 40-dB Taylor weighting that is separable in two dimensions.[2] The third example is a diamond-shaped antenna, again inscribed in the 37-in circle, with 40-dB separable Taylor weighting. The actual footprint area for each of these cases has been calculated and is shown below each footprint. What is surprising is that the circular amplitude weighting gives rise to near boresight sidelobes that are above an interceptor threshold; hence, the actual footprint area on the ground is 3234 km².

In the case of the inscribed square, even though it has lower gain and, hence, a broader mainbeam, the enhanced sidelobe performance gives rise to less than half the footprint area—a little over 1600 km². Lastly, the 33-in diamond with a Taylor separable weighting gives rise to slightly less than 1600 km² in area. By the way, it is shown later, in Chapter 6, that separable illumination functions are easier to manufacture, and so actual sidelobes are usually better implemented this way. Thus, it can be seen that the amplitude weighting, especially separable amplitude weighting functions, and actual antenna shape have a significant impact on the intercept footprint. Antenna footprints and their calculation are discussed in more detail later in this chapter. The problem is primarily one of geometry and approximations to simplify calculations.

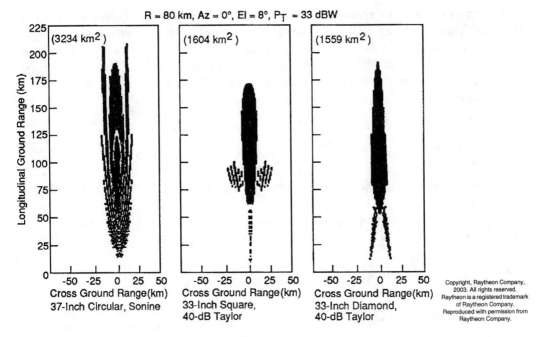

R = 80 km, Az = 0°, EI = 8°, P_T = 33 dBW

Figure 2.4 Footprint varies with antenna shape.[13]

2.1.2 Interceptor Time Response

In addition to footprint area, intercept probability is a function of several other variables. The density of intercept receivers per unit area obviously is an important element; so are the illumination time of the interceptor by the emitter, or *time-on-target* (T_{OT}), and the *intercept receiver search time* (T_I).

The intercept probability problem can be modeled as a sequence of repeated independent trials with two constant-probability, possible outcomes for each trial—intercept and no intercept. This is called a Bernoulli trial, and a great body of literature and analysis exists on problems of this type.[3] Unordered Bernoulli trials give rise to the binomial distribution as shown in Equation (2.1).

$$P(m;n,p) = \begin{pmatrix} n \\ k \end{pmatrix} \cdot p^m \cdot (1-p)^{n-m} \qquad (2.1)$$

where

m = the number of successes required (i.e., intercepts)
n = the number of trials (i.e., interceptor illuminations)
p = the probability of success per trial (i.e., intercept probability per illumination)

One can think of many reasons why such a model isn't exactly correct, such as netting, multipath, distance changes, waveform changes, and so forth but, practically speaking, the model provides excellent correlation with actual experiment. In most cases of interest, the probability of intercept per illumination is small, and the number of illuminations is large, which allows the use of the Poisson approximation to the binomial distribution,

$$\mathcal{P}(k;\ell) = \frac{\ell^k}{k!} \cdot \varepsilon^{-\ell} \qquad (2.2)$$

where $\ell = n \cdot p$. For $\ell < 0.3$, which is usually the case for an LPIS, the approximation is extremely good.

The probability of intercept is the probability that one or more intercepts have occurred or, said another way, the probability that *no* intercepts have *not* occurred, i.e.,

$$\mathcal{P}_i = 1 - \mathcal{P}(0;\ell) = 1 - \varepsilon^{-\ell} \qquad (2.3)$$

It can be seen that ℓ is approximately

$$\ell = A_F \cdot D_I \cdot \frac{\min(T_{OT}, T_I)}{T_I} \qquad (2.4)$$

where
A_F = antenna beam "footprint" in km^2
D_I = density of intercept receivers per km^2

The probability that the emitter beam illuminates an intercept receiver location is approximately the product of the radar footprint and the density of intercept receivers per unit area, which is shown in Equation (2.5).

$$A_F \cdot D_I \qquad (2.5)$$

The probability that an intercept receiver interrogates the emitter location when the emitter is active corresponds to the minimum of either T_{OT} or T_I, normalized by T_I. Usually, T_{OT} is much smaller than T_I, so this probability is approximately T_{OT} divided by T_I, as shown in Equation (2.6). Therefore, the probability that the intercept receiver interrogates emitter location when the emitter is active is approximately

$$\frac{\min(T_{OT}, T_I)}{T_I} \approx \frac{T_{OT}}{T_I} \text{ assuming: } T_{OT} \leq T_I \qquad (2.6)$$

where

T_I = interceptor search time

T_{OT} = illumination time by emitter of interceptor

Several other probabilities come into play, but they usually are so close to 1 that they are often not included in the probability of intercept calculation. These others are the probability that the intercept receiver is tuned to the emitter frequency during the emitter T_{OT} and the probability that the intercept receiver detects the emitter signal when it's above the threshold. These two elements then give rise to a probability of intercept equation of the form shown in Equation (2.7). Substituting and including other important probabilities yields

$$\mathcal{P}_i = \left\{ 1 - \exp\left[-\left(A_F \cdot D_I \frac{\min T_{OT}, T_I}{T_I} \right) \right] \right\} \cdot \mathcal{P}_F \cdot \mathcal{P}_D$$

$$\mathcal{P}_i \approx A_F \cdot D_I \cdot \mathcal{P}_F \cdot \mathcal{P}_D \cdot \frac{T_{OT}}{T_I} \tag{2.7}$$

where

$\mathcal{P}_F$ = probability that interceptor is tuned to emitter frequency

$\mathcal{P}_D$ = probability that emitter is detected assuming illumination and proper tuning

Later in this chapter, this equation will be revisited.

2.1.3   Receiver Sensitivity versus Intercept Probability

An intercept receiver may be made more sensitive by improving the sensitivity through the use of low noise components and by minimizing losses. Further increases in sensitivity may be achieved by the use of narrower bandwidth and/or higher antenna gain, which may require frequency and/or spatial scanning. Because frequency scanning implies that the total RF band of interest is not monitored continuously, and spatial scanning implies that the total (solid) angular sector of interest is not monitored continuously, these two approaches involve a trade-off between system sensitivity and intercept probability.

Intercept probability is especially important in any situation in which a time-varying emitter is to be detected. Time variations may result from emitter platform motion, antenna scanning, waveform on-time, and frequency variations. Given that the emitter presents (in some sense) the intercept receiver with an opportunity to detect it, the intercept probability is the likelihood that the intercept receiver system is both pointed in the right direction and tuned to the right frequency when that opportunity occurs as given previously in Equation (2.7).

For example, suppose that an air-to-ground LPIS scans across the location of a ground-based intercept receiver so that the LPIS beam remains pointed toward the receiver for a time, T_{OT}, referred to as the *time on target*. Also suppose that $\mathcal{P}_D = 1$. During the time on target, the receiver will be able to "listen" to a total of N_L combinations of beam positions and frequency channels; then,

$$N_L = T_{OT}/t_L \tag{2.8}$$

where t_L = the time that the receiver listens in each beam position and frequency channel.

Suppose that the receiver scans a total of N_B beam positions and N_{Freq} frequency channels, one of which corresponds to the LPIS position and frequency. Then, the intercept probability, the probability that an intercept occurs during the dwell time, T_{OT}, assuming that the probability of detection is 1 ($\mathcal{P}_D = 1$) is given by

$$\mathcal{P}_I \approx \left\{ \begin{array}{cc} N_L/(N_B \cdot N_{Freq}) & \text{for } N_L \leq N_B \cdot N_{Freq} \\ 1 & \text{for } N_L > N_B \cdot N_{Freq} \end{array} \right\} \tag{2.9}$$

Recall that T_I denotes the intercept receiver total search time, the time that it takes for the receiver to scan through its set of beam positions and frequency channels. Then,

$$T_I = N_B \cdot N_{Freq} \cdot t_L \tag{2.10}$$

And thus,

$$\mathcal{P}_I \approx \left\{ \begin{array}{cc} T_{OT}/T_I & \text{for } T_{OT} \leq T_I \\ 1 & \text{for } T_{OT} > T_I \end{array} \right\} \text{for}(\mathcal{P}_D) = 1 \tag{2.11}$$

These somewhat obvious equations are valid in the case where the intercept receiver scans its set of beam positions and frequency channels uniformly. In cases where nonuniform scans are performed, the intercept probability depends on the joint probability of the scan space, frequency, and time density, and on the position emission density of the LPIS relative to the intercept receiver.

The listening time, t_L, is considered to be constrained by the waveform of the emitters of interest to the receiver. In general, t_L should be long enough to permit several emitter pulses to be received so as to implement the threat deinterleaving and identification process. Note, however, that if an emitter is sufficiently unique in its pulse characteristics (e.g., pulse width), then a single pulse may suffice for its identification, in which case t_L need only be as long as the longest pulse repetition interval. If t_L is shorter than the pulse repetition interval, then the intercept probability will often be reduced.

An additional effect, which may reduce the intercept probability, arises in cases where the emitter changes frequency during the dwell time by an amount that is larger than the bandwidth of the intercept receiver frequency channels. The receiver may guard against this possibility by using channels of sufficiently large bandwidth, at a sacrifice in sensitivity or through use of multiple simultaneous frequency channels (which may also degrade sensitivity). In this example, assume that the interceptor design is such that degradation of the intercept probability caused by emitter frequency agility or long pulse repetition interval does not occur.

For a given listening time, t_L, number of beam positions, N_B, and number of frequency channels, N_{Freq}, the receiver may decrease the search time, T_I, through the use of parallelism; that is, by listening to some number, N_S, of beam positions and frequency channels simultaneously. The search time is then reduced to

$$T_I = \frac{N_B \cdot N_{Freq} \cdot t_L}{N_S} \tag{2.12}$$

The number of beam positions and frequency channels scanned during the time on target is increased to

$$N_L = \frac{T_{OT} \cdot N_S}{t_L} \tag{2.13}$$

When these expressions are substituted into Equation (2.9), the intercept probability is still given by Equation (2.11).

Although parallelism in beam and frequency channel formation are both feasible, the latter approach is far more economical, as only the last receiver stages need be duplicated in the hardware. The term *n-channelization* will be used to refer to this latter form of parallelism without concern for the particular hardware realization (e.g., filter banks, transform, and so forth) employed. It is assumed that channelization is used in the receiver to decrease the frame time and avoid potential problems associated with emitter frequency uncertainty or agility while achieving the sensitivity associated with narrow channel bandwidth.

To relate the intercept probability to the receiver system sensitivity, the sensitivity will be expressed in terms of the parameters N_B and N_{Freq}. First, however, a brief discussion of receiver sensitivity is necessary (more details are given in Chapter 3).

To achieve the desired detection probability and false alarm rate, a certain rms *signal-to-noise (SNR)* power ratio is required. The noise power, referenced to the receiver input, is given by

$$n_i = k_B \cdot T_0 \cdot NF_i \cdot B_i \tag{2.14}$$

where
  k_B = Boltzmann's constant
  T_0 = 290-Kelvins reference temperature
  NF_i = receiver noise figure
  B_i = intercept receiver channel noise bandwidth

The same notation is used for the interceptor parameters subscripts as in Chapter 1; i.e., the lower-case i represents the actual interceptor value, and the upper-case I represents the required value. The required intercept power is simply the minimum power at the receiver threshold detector for which the desired detection probability and false alarm rate may be achieved. [Modifications of Equation (2.14), which are necessary under certain conditions, are discussed further below.] The intercept receiver required detection power is then defined as

$$P_I = SNR_I \cdot n_I \cdot \frac{G_{IPn}}{G_{IP}} = k_B \cdot T_0 \cdot NF_I \cdot B_I \cdot SNR_I \cdot \frac{G_{IPn}}{G_{IP}} \tag{2.15}$$

where G_{IPn} =the intercept processor noise gain and all other variables are as defined previously. Note that G_{IP}/G_{IPn} is equivalent to the internal system losses plus signal gain.[15–17]

In defining the sensitivity of the total intercept receiver system, including the antenna and other components preceding the receiver itself, the antenna gain and component losses must be taken into account. The precise meaning of the definition of SNR_I is more obvious if the ratio of the required detection power to the interceptor noise power is formed using Equation (1.3) and (2.15) as follows:

$$SNR_I = \frac{P_I \cdot G_{IP}}{n_I \cdot G_{IPn}} = \frac{P_T \cdot G_{TI} \cdot G_I \cdot G_{IP} \cdot \lambda^2 \cdot L_I}{k_B \cdot T_0 \cdot NF_I \cdot B_I \cdot G_{IPn} \cdot (4 \cdot \pi \cdot R_I)^2} \tag{2.16}$$

where, as before, P_T is the emitter radiated power, and G_{TI} is the emitter gain in the direction of the intercept receiver for an emitter at range, R_I, and where L_I represents any path losses between the emitter and receiver.

Referring to Equation (2.15), define the interceptor system sensitivity, S_I, as the ratio of the required detection power to the interceptor antenna gain as follows:

$$S_I = \frac{P_I}{G_I} = \frac{SNR_I \cdot \boldsymbol{n}_I \cdot G_{IP\boldsymbol{n}}}{G_I \cdot G_{IP}}$$

$$= \frac{k_B \cdot T_0 \cdot NF_I \cdot B_I \cdot G_{IP\boldsymbol{n}} \cdot SNR_I}{G_I \cdot G_{IP}} \tag{2.17}$$

The system sensitivity is expressed as the power received by a hypothetical isotropic, lossless receiver at the position of the actual receiver such that the corresponding power density is just sufficient to result in the required signal-to-noise ratio in the actual receiver.

Equation (2.16) can now be related to the intercept probability. Suppose that the receiver must monitor a solid angle, Ω, and a total bandwidth, B. If the receiver has a circular beam with beamwidth, θ_B, and beams are hexagonally scanned over the solid angle Ω, the required number of beam positions is given by

$$N_B = \frac{\Omega}{6.928 \cdot c_{SP\theta}^2 \cdot (1 - \cos(\theta_B/2))} \cong \frac{\Omega}{0.866 \cdot c_{SP\theta}^2 \cdot \theta_B^2} \tag{2.18}$$

where $c_{SP\theta}$ is the beam spacing factor, the ratio of center-to-center spacing to the beamwidth.

For a typical beam pattern, the beamwidth and gain are related to the antenna diameter D by

$$\theta_B \cong 1.2 \cdot \lambda/D \text{ and } G_I \cong 6 \cdot D^2/\lambda^2 \tag{2.19}$$

Substituting one into the other gives the well known approximation,

$$G_I = 8.64/\theta_B^2 \tag{2.20}$$

Finally, combining Equations (2.18) and (2.20) yields

$$N_B = \frac{\Omega \cdot G_I}{7.48 \cdot c_{SP\theta}^2} \tag{2.21}$$

This result also is obtained for an elliptical beam pattern if it is assumed that the hexagonal scan is simply stretched in proportion to the ellipse axes. The number of frequency channels required is assumed to be given by

$$N_{Freq} = \frac{B}{c_{SPF} \cdot B_I} \tag{2.22}$$

For example, if $B = 1$ GHz, $c_{SPF} = 0.79$, and $B_I = 6.3$ MHz, then $N_{FREQ} = 200$. And when Equations (2.22) and (2.21) are substituted in (2.17), it yields

$$S_I = \frac{k_B \cdot T_0 \cdot NF_I \cdot B \cdot \Omega \cdot SNR_I \cdot G_{IP\mathbf{n}}}{7.48 \cdot c_{SP\theta}^2 \cdot c_{SPF} \cdot N_{Freq} \cdot N_B \cdot G_{IP}} \qquad (2.23)$$

And Equations (2.9) and (2.13) may be combined to give Equation (2.24) (for $\mathbf{P}_I < 1$),

$$N_B \cdot N_{Freq} \approx \frac{T_{OT} \cdot N_S}{t_L \cdot \mathbf{P}_I} \qquad (2.24)$$

This finally yields Equation (2.25) below, relating the required system sensitivity to the intercept probability.

$$S_I \approx \frac{k_B \cdot T_0 \cdot NF_I \cdot B \cdot \Omega \cdot G_{IP\mathbf{n}} \cdot SNR_I \cdot t_L \cdot \mathbf{P}_I}{7.48 \cdot c_{SP\theta}^2 \cdot T_{OT} \cdot N_S \cdot c_{SPF} \cdot G_{IP}} \qquad (2.25)$$

For example, let $k_B = 1.38 \times 10^{-23}$, $T_0 = 290$ K, $NF_I = 4$ dB, $B = 1$ GHz, $\Omega = 0.666\pi$, $G_{IP}/G_{IP\mathbf{n}} = 3$ dB, $c_{SP\theta} = 0.79$, $c_{SPF} = 0.79$, $t_L = 10^{-2}$ sec, $T_{OT} = 10^{-2}$ sec, $N_S = 200$, and $\mathbf{P}_I = 0.1$; then, for a one-per-minute false alarm rate, a probability of detection of 90 percent, and a nonfluctuating emitter, the SNR_I must be 15.5 dB. Similarly, for a much more realistic Swerling case 1 emitter, the SNR_I must be 23.5 dB.

$$S_I = -103.2 \text{ dBm, nonfluctuating; } S_I = -95.2 \text{ dBm, fluctuating}$$

2.1.4   Power Management

The next important parameter of interceptability is power management. In general, the strategy is to transmit the minimum power for each mode, target time, and range scale. However, there are still many alternative strategies and parameters. First, as the emitter range equation readily predicts, each target type requires a different power-range profile: radar point targets require R^4, ground mapping requires R^3 (area), and weather detection requires R^2 (volume). Next, it's important to utilize the minimum acceptable SNR in all cases. For radar ground map, it has been determined from experiments run in the early 1970s that the necessary average SNR per pixel is approximately 3 to 4 dB (USAF FLAMR program[8-11]). It was determined in the LPIR program that for acceptable tracking performance, SNRs of approximately 3 dB were required at the tracker input. For detection, of course, it's been well known since the 1940s that the required SNR is typically 13 dB for 50 percent detection probability with low false alarm rate.

The following is a list of the primary power management parameters:

1. Transmit minimum power for each mode, target type, and range scale—for example, point targets, R^4; map, R^3; weather, R^2.

2. Utilize minimum acceptable SNR—for example, ground map SNR, 3 to 4 dB; tracking SNR, 3 dB; Detection SNR, 13 dB.

3. Program power in track as a function of range: R^3 or R^4 are typical.

4. Program according to hand control in acquisition, i.e., range proportional to acquisition gate.

5. Operator override for ECM or emergency.

6. Array-to-array power control.

A power management strategy also requires that the power transmitted be programmed in track as a function of the tracking range. In the DARPA/USAF LPIR program, experiments were conducted using R^3 and R^4 power profiles as a function range. In addition, attempts were made to program radar power proportional to receive SNR in track. These latter experiments were rather unsuccessful, primarily because of the interaction between the tracking loop filter phase shift and target scintillation. These parameters are close enough in cutoff frequency that it is virtually impossible to maintain both adequate target dynamics response and a stable track with closed-loop power management. For the acquisition phase, the transition between detection and track, the power is usually programmed according to the location in range of the acquisition gate. Usually, an operator override is provided in the power management function to allow higher radiated powers in the presence of ECM not automatically detected or for other emergency reasons.

One other advantage to power management is that an average power received estimate does not give any information about the emitter's range from the interceptor. For example, the aspect, knowledge, and dampness of the terrain in *terrain following (TF)* have a larger influence on radiated power than the range equation.

In the LPIR program, it was discovered that coherent-array-to-coherent-array power control was the most successful and the most stable. That means that the power measured in one coherent array is used to set the transmitted power for the next coherent array in what is essentially an open loop, feed-forward fashion. This is similar to some very high-performance, high-power consumer stereo music amplifiers. (It should be noted that careful stability analysis is still required for each case.)

The actual application of power management can most easily be seen by the use of graphs. Plotted in Fig. 2.5 are the powers required by various types of radar as a function of range. In this figure, the ordinate is the equivalent isotropic intercept receiver sensitivity, P_I, divided by G_I. One can see (Fig. 2.5, point 2), that the power at an inter-

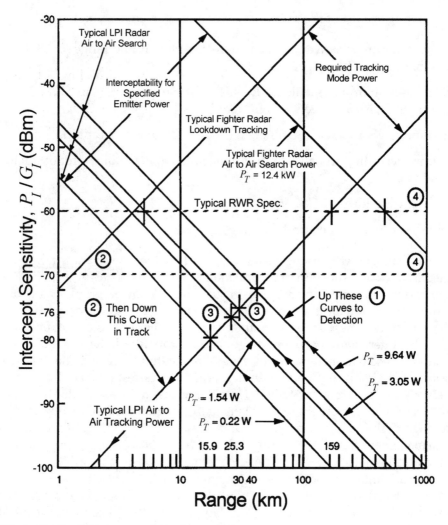

Figure 2.5 Application of power management.

cept receiver for a typical fighter radar with constant power in a lookdown mode steadily increases as a function of range by R^4. Similarly, an LPIR with a high PRF mode also increases as R^4. One also can see (Fig. 2.5, point 3) that the power received at an intercept receiver steadily increases as the required LPIS detection range gets greater for the four example cases: 15.9, 25.3, 30.0, and 40.0 km. Also plotted on this graph are contours of constant emitter power (Fig. 2.5, point 1) received at an intercept receiver as a function of range between the emitter and intercept receiver: the LPIS radiated powers range from a little less than 0.25 W up to a little less than 10 W. Additionally, a typical fighter radar peak power, a little over 12 kW, is also plotted.

Several example RWR sensitivities (Fig. 2.5, point 4) are shown as dashed lines; they, of course, are flat. As will be shown in a later chapter, this sensitivity is usually not limited by thermal noise but by emitter density in the LPIS operating band.

One can determine the maximum acceptable range for a given intercept receiver sensitivity by graphical means. Intersections between fixed power levels radiated from the radar with the actual power level required by a particular radar track mode suggest a strategy in which a fixed power is radiated until an RWR-carrying platform is detected and then programmed in track. This, in turn, limits the maximum range that's usable against that threat. At the same time, it is the largest power that ever need be transmitted for that operating range. If that power is chosen to be less than the sensitivity of the intercept receiver, once the radar goes into track, the transmit power can be decreased steadily as the target closes. This way, the maximum power that will ever be received at the intercept receiver is at the moment of detection.

One can see that, for a typical high-peak-power, low-duty-ratio fighter radar, the intercept receiver sensitivity has to be −50 dBm or above for no intercept to occur. With an LPIR, however, it's possible to detect targets at ranges of 40 km while maintaining power at the intercept receiver below −70 dBm. The graphical procedure is to move up constant power curve 1 until it intersects a required track mode power curve. That point of intersection is the lowest maximum required operating power for that indicated range. If that power is below the intercept receiver sensitivity, once track is achieved, moving down the track curve 2 is all that's needed. This is the simplest way to implement power management. Therefore, it's possible to trade off maximum range versus intercept receiver sensitivity. Low-sensitivity intercept receivers allow maximum ranges well in excess of 100 km. High-sensitivity receivers may force an LPIS to operate at 50 km or below.

2.2 INTERCEPTABILITY ANALYSIS

2.2.1 Intercept Receiver Sensitivity

As was mentioned before, LPI is highly dependent on scenario and the sensitivity of the intercept receiver. Although much more is said in Chapters 3 and 4 about intercept receiver sensitivities, to proceed with an analysis, some assumptions must be made about intercept receiver sensitivities. Representative intercept receiver sensitivities are listed in Table 2.1.

The table shows five categories, essentially organized by their use; these categories don't represent any specific intercept receiver but are typical of the actual intercept receivers currently deployed. Simple warning receivers have reasonably low sensitivities and wideband omnidirectional antennas. Their limitations are usually the computational horsepower in the intercept receiver processor. Improved warning receivers typically are characterized by higher sensitivities and most often have narrowband filtering with omnidirectional antennas. Again, they are limited primarily

Table 2.1   Intercept Receiver Sensitivities

Use	Sensitivity[*]	Parameters	Limitations
Simple warning	Low (–40 to –50 dBm)	Wideband Omni-angle coverage	Signal overload Processor limited
Improved warning	Moderate (to –60 dBm)	Narrowband[†] Omni-angle coverage	Signal overload Processor limited
ESM/ECM Direction finding	High (–80 dBm)	Narrowband[†] Directional angle coverage	Scan-on-scan Processor limited
ECM/ELINT Direction finding/location	Higher (to –100 dBm)	Narrowband[†] Very directional angle coverage	Scan-on-scan Long scan time
Future	Highest (–120 dBm)	Elastic	Massive number crunching

[*]Net including antenna gain
[†]Improved detection/recognition

by signal processing and sorting overload. ESM and intercept receivers, which are associated with ECM direction finding, have higher sensitivities and usually have narrowband filtering and directional antennas. Usually, their limitations are that the directional antennas limit scan rate, and hence there is a scan-on-scan probability that can be exploited. ECM or ELINT receivers do both direction finding and location as well as recognition. They have the highest sensitivity and usually have narrowband detection filters and highly directional antennas. They usually have long scan times, and scan-on-scan intercept probability is their main weakness. When ELINT system antennas are pointed at an emitter, however, they have very high probabilities of detection and intercept. Typical intercept performance versus interceptor sensitivity in the emitter antenna mainlobe is shown in Fig. 2.6.

A simple LPI performance calculation of the power received at an interceptor collocated at a target, assuming a unit gain intercept receiver antenna [using Equation (1.2)], is shown below in Table 2.2, using Equation (2.26).

$$S_i = \frac{P_i}{G_I} = \frac{P_T \cdot G_T \cdot G_{TS} \cdot \lambda^2 \cdot L_I}{(4 \cdot \pi)^2 \cdot R_i^2} \tag{2.26}$$

In this particular example, a peak power per transmitted frequency of 100 W is chosen. The mainlobe gain of the radar antenna is chosen to be 35 dB, and the relative radar antenna sidelobe attenuation is chosen to be –55 dB or 20 dB below isotropic. These are typical for an LPIS (of course, sidelobes are 0 dB in the mainlobe). A wavelength at X-band and one-way loss of 3 dB are assumed. An intercept receiver range of 100 km has arbitrarily been selected. One can see that the power received in the mainlobe is –100 dBW or –70 dBm, and the power received through the sidelobes is

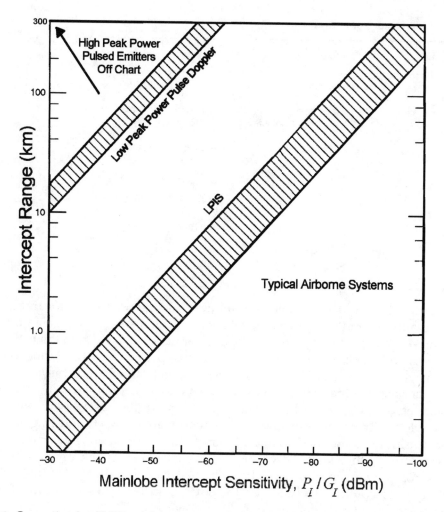

Figure 2.6   Conventional and LPIS mainlobe intercept range versus sensitivity.

–155 dBW or –125 dBm. By referencing Table 2.1, it can be seen that both simple warn-ing receivers and improved warning receivers could detect such a radar in its main-lobe but could not detect it in its sidelobes. Furthermore, even the more advanced ESM and ELINT receivers, depending on the gain of the antenna, may not be able to detect an LPIS in its sidelobes.

2.2.2   Sidelobe Intercept Range

Sidelobe intercept range can be understood graphically with Fig. 2.7. In the figure, in-tercept range in kilometers is plotted along the ordinate for a range of intercept re-

Table 2.2   Interceptor Power Received

Parameter		Explanation	Mainlobe	Sidelobes
P_T		Peak power/frequency 100W	20 dBW	Same
G_{TI}	$= G_T$ $+ G_{TS}$	Antenna mainlobe gain Antenna relative sidelobe level	35 dB 0 dB	Same −55 dB
λ^2		Wavelength 0.0316 m	−30 dB	Same
L_I		One-way loss (incl. weather)	−3 dB	Same
R_I^2		$(100\ km \cdot 4\pi)^2$	−122 dB	Same
S_I		Received power mainlobe	−100 dBW	
S_I		Received power sidelobes		−155 dBW

ceiver sensitivities in decibels below a milliwatt along the abscissa for three typical classes of airborne radar: high-peak-power pulsed radars such as those in the F-111, B-1, and F-16; low-peak-power pulse Doppler radars such as those in the F-15 and F/A-18; and, finally, LPIS. There is a range of possible performances as a function of the actual power transmitted from these radars as well as whether they incorporate power management; many of the most modern radars already have power management for reasons other than LPI. Also depicted along the abscissa are several currently deployed intercept receivers, represented by an alphabetical listing A through H. The sensitivity is the total, including both interceptor antenna gain and receiver processor-limiting threshold. As will be seen in a later chapter, sensitivity is limited as much by computing and pulse density as by thermal noise. Because of this fact, high sensitivity interceptors usually are on the surface, in large transport aircraft, or on large spacecraft, and their density/mobility is low.

2.2.3   Interceptor Detection Probabilities

The graph in Fig. 2.7 assumes that the detection threshold is actually infinitely non-linear and absolute. In reality, detection probability versus SNR is represented by a probability density function in which the threshold value provides a given probability of detection that is not usually 1, and an associated probability of false alarm that is usually a very small number. Even at very low operating signal levels, an LPIS is detectable; it just turns out that the probability of detection can be a very, very low number. Normally, when detection probability versus *SNR* curves are plotted, they are plotted for high SNR values.[4] Very low probabilities of detection are either squeezed onto the bottom of the graph or not plotted at all. An approximation to the

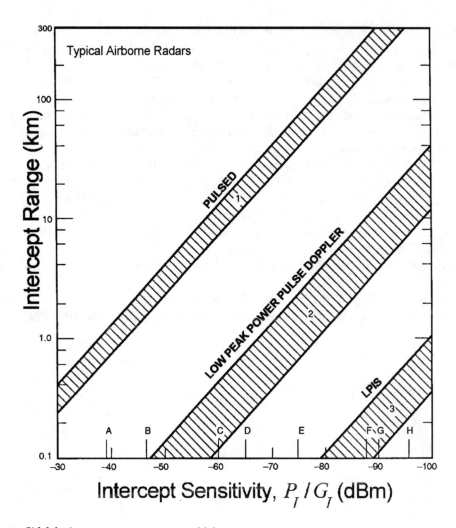

Figure 2.7 Sidelobe intercept range versus sensitivity.

detection performance for a wider range of rms signal power to noise power ratios using probability of false alarm as a design parameter was described by Rice in the mid-1940s. Rice's equation for a linear detector with a steady target in noise is given in Equation (2.27).[5]

$$\mathcal{P}_D = 2 \cdot \int_{\sqrt{-\ln(\mathcal{P}_{FA})}}^{\infty} y \cdot I_0(2 \cdot y \cdot \sqrt{SNR}) \cdot \exp[-(y^2 + SNR)]dy \qquad (2.27)$$

Figures 2.8 and 2.9 show the condition that exists for the typical LPIS as mentioned in Chapter 1. It shows the logarithm of the probability of detection of a steady signal plus noise based on single-pulse peak power detection in a linear detector.

Normally, detection thresholds are set as shown in Fig. 2.8, with a 90 percent probability of detection (P_D) and a false alarm probability (P_{FA}) of perhaps 10^{-8}. What is the probability of detection for an LPIR that may be operating at as much as 10 dB below the intercept receiver detection threshold? The answer is not zero, but it can be a very small number. If one examines a probability of false alarm of 10^{-8} and assumes that a threshold has been set at approximately 14 dB (the usual value for Gaussian noise), one can see in Fig. 2.9 that the probability of detecting an LPIS that's 10 dB below the threshold is 8×10^{-8}. A small number, but still a very long way from zero. If an LPIR emitted 300,000 pulses per second (which it might do in a high PRF mode), one detection would occur in 42 sec of observation. It's possible for long observation times to result in detection under certain circumstances. Falling raster displays similar to underwater acoustic sonograms can be used for long observation LPIS detection.

What saves a typical LPIS, however, is that, during that same amount of observation time, the number of false alarm opportunities is usually 10 to 100 times greater,

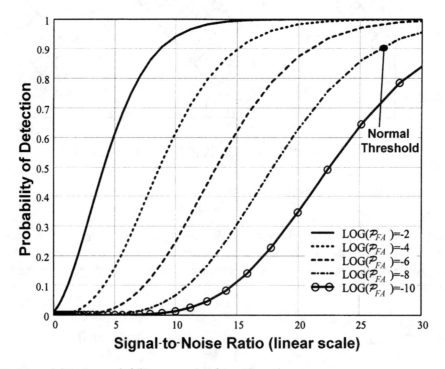

Figure 2.8 Normal detection probability versus signal-to-noise ratio.

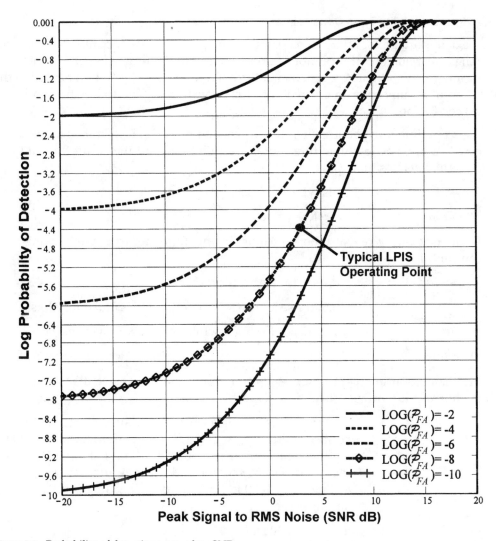

Figure 2.9 Probability of detection versus low SNR.

even though the probability of false alarm per measurement is 10^{-8}. The conclusion drawn from this is that the intercept performance is actually scenario dependent and greatly depends on the search space over which the intercept receiver must operate. The potential for other emitters to mask an LPIS is quite high. Some potential masking sources will be discussed in Chapters 3 and 4. There is as much protection in being below the intercept receiver threshold as there is from the number of opportunities for false alarms that occur as the intercept receiver searches its observation space. Computational details are given in the appendices.

2.2.4 Interceptability Time Constraints

The next important topic is interceptability time constraints. Because an intercept does the threat no good if it cannot respond, control of the time when intercepts may occur can greatly mitigate the risk if an intercept does occur. This notion has led to the evolution of a number of interceptability time constraints. The following is a list of typical constraints:

1. Intercept time must be less than threat reaction time, T_R; i.e., emitter time on target, T_{OT}, must be less than interceptor reaction time, and the emitter dwell time, T_D, is always less than T_{OT}.

    $T_{OT} < T_R$ T_R is typically 5 sec.

    Therefore, $T_D < T_{OT} < 5$ sec.

2. Frequency agility revisit time, T_{FA}, must be much greater than the tracking time constant, T_{TC}.

    $T_{FA} >> T_{TC}$ = 0.3 sec for electronic trackers

                         = 0.5 to 1 sec for missile aerodynamics and autopilots

                         = 1 to 10 sec for track-while-scan (TWS) systems

    Therefore, $T_{FA} \approx 2.5$ to 100 sec

3. Multilateration intercept time, T_{MI}, separation must be greater than one-half of the horizon-to-horizon travel time, T_{HT}.

    $T_{MI} > (0.5 \cdot T_{HT})$ Example 1: Assuming a surveillance aircraft interceptor and surface emitter, V_{AC} = 700 ft/sec, aircraft to horizon at 20 kft is 245 mi, then $0.5 \cdot T_{HT}$ = 30 min, 48 sec. Example 2: Assuming a surface interceptor and a penetrating aircraft, V_{AC} = 850 ft/sec at 1000 ft, horizon to A/C 45 mi, then $0.5 \cdot T_{HT}$ = 4 min, 38 sec.

    Therefore, $T_{MI} > 5$ to 30 min.

4. Time between mainlobe intercepts, T_{MLI}, (single interceptor) must be less than a missile fly-to-time, T_M, plus the interceptor reaction time.

    $T_{MLI} > T_M + T_R$ Example 3: Assuming an emitter-to-missile-launcher distance of 12 mi, missile velocity of 1200 ft/sec, then T_M = 53 sec.

    Therefore, $T_{MLI} > 53 + 5 = 58$ sec.

As shown, there are four main types of interceptability time constraints. The first constraint is that the intercept time must be less than the reaction time; that is, the emitter dwell time (T_D) must be less than the interceptor reaction time (T_R). T_R for man-in-the-loop systems is typically 5 sec or greater; hence, the emitter dwell time must be 5 sec or less.

The second time constraint is the frequency hopping or agility revisit time (T_{FA}). T_{FA} must be much greater than threat tracking time constants (T_{TC}). Tracking time constants are typically 0.3 sec for electronic trackers, 0.5 to 1 sec for missiles limited by aerodynamics, and 1 to 10 sec for TWS systems. This, in turn, means that the typical frequency-hopping or agility revisit time should be greater than 2.5 to 100 sec, depending on the threat. This time is a great deal longer than the frequency agility revisit times that most frequency-hopping systems exhibit today. If an LPIS revisits the same frequency in less than or equal to a tracking time constant, then it doesn't make any difference how great a frequency excursion was made in between those measurements. The tracker essentially can get an update once per time constant, which is all that is required for accurate tracking and the guidance of a threat to the LPI system.

Multilateration is the almost simultaneous intercept of the emitter signal at multiple sites so that the emitter can be located in two or three spacial dimensions. The third important time constraint is the multilateration intercept time (T_{MI}). Multilateration intercept time separation must be greater then half the horizon-to-horizon travel time (T_{HT}) of either the observer or the LPIS platform. For example, a surveillance aircraft with an intercept receiver and a surface radar emitter might have a relative velocity of 700 ft/sec and an aircraft-to-horizon distance of approximately 245 mi at an altitude of 30,000 ft. It follows, then, that half the horizon-to-horizon travel time is approximately 30 min and 48 sec. Similarly, a surface interceptor tracking a penetrating aircraft with an LPIS on it, travelling at 850 ft/sec at 1000 ft, has a horizon-to-aircraft distance of 45 mi; it follows that half the horizon-to-horizon travel time is approximately 4 min and 38 sec. Furthermore, because multilateration is primarily an antenna sidelobe problem, there is a trade-off between the antenna complexity and total operating duty cycle. In the worst case, the duty cycle must be less than T_D/T_{HT}; e.g., 5 sec/1848 sec, or 0.0027. A data link typically operates a much larger percentage of the time and is susceptible to radiometric intercept threats (see Chapter 3). The emitter must travel a minimum of several null-to-null radiometer beamwidths between emissions. The conclusion is that multilateration intercepts must occur less frequently than 5 to 30 min, depending on relative altitude and velocity.

The fourth constraint is the time between mainlobe intercepts for a single intercept threat. The time between mainlobe intercepts must be greater than a threat missile fly-to time, T_M, plus the threat reaction time, T_R. For example, if the distance from an emitter to a missile-launching interceptor is 12 mi, and the missile velocity is 1200 ft/sec, then T_M will be approximately 53 sec. T_M added to a typical T_R of 5 sec means that the time between mainlobe intercepts (T_{MLI}) must exceed 58 sec. Typically, it turns out that these time constraints are very difficult to meet simultaneously while still allowing performance of the weapon system mission. Nonetheless, one or more of these threats often occur during a mission, and the LPI system must meet several or all of these constraints simultaneously.

2.2.5 Interceptability Frequency Constraints

Another important parameter constraint is the LPIS emission frequency and band-width. Although emission in an unexpected band may contain the element of sur-prise, one must assume that the threat will discover this frequency usage plan by spying, exploitation of captured assets, or intercept. Usually, operation in a band with many other masking emissions or natural protection is preferable. Furthermore, the band of operation should be wide enough to prevent long term integration to pull up a signal otherwise deeply buried in noise for detection. Narrowband center frequen-cies or leakage harmonics must be avoided. Master clock countdowns to selected PRFs, pulse compression chip widths, frequency hop sequences, pulse compression codes, frequency channel centers, and uniformity of total band occupancy must be adequate to prevent exploitation of the whole strategy of operation based on a single or small number of widely spaced intercepts. Although frequency hopping is useful, the best strategy from an LPI point of view is to occupy the entire operating band all during the emission time so that the lowest peak power is observable per unit time *and* per unit frequency. In other words, a stealth system should not allow long-term integration in time, frequency, or angle. The threat must be forced to surveil the larg-est region of frequency, time, and space that creates the largest probability of false alarms and the lowest probability of detection. A good rule of thumb is to maintain less than 3 dB ripple over the operating band.

2.2.6 Antenna Gain Mismatch

The next important analysis topic for an LPI system is the exploitation of antenna gain mismatch between the LPIS and the interceptor. Basically, there are two alterna-tive LPI strategies: the minimum power strategy and the minimum dwell strategy. The minimum power strategy requires that an LPIR transmit at minimum power at all times and use maximum signal integration. The intent is always to stay below the interceptor threshold. This strategy works with low-sensitivity interceptors that allow use of reasonable power to perform the mission. The minimum dwell strategy as-sumes that, when transmitting, the radar will be detected and, hence, the strategy is to keep the exposure time as short as possible. This means that the radar transmitter should be on for a time as short as possible, or the transmitter should be in its high-power mode for the minimum time possible. In addition, it requires the use of mini-mum power compatible with the short integration time. The minimum dwell strategy works when the interceptor is very sensitive but also has a long time volume scan so that one gets scan-on-scan advantage by having a minimum dwell. Usually, highly sensitive interceptors do have long scan times, because their sensitivity is achieved with higher-gain antennas.[6,12]

An example of the minimum power strategy is shown in Fig. 2.10. The graph shows probability of intercept as a function of elevation angle for contours of con-

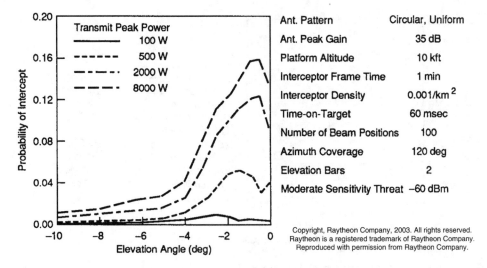

Ant. Pattern	Circular, Uniform
Ant. Peak Gain	35 dB
Platform Altitude	10 kft
Interceptor Frame Time	1 min
Interceptor Density	$0.001/km^2$
Time-on-Target	60 msec
Number of Beam Positions	100
Azimuth Coverage	120 deg
Elevation Bars	2
Moderate Sensitivity Threat	−60 dBm

Figure 2.10   Example of minimum power strategy.[12]

stant transmit peak power. Other radar performance parameters also tabulated in the figure are antenna pattern and gain, platform altitude, interceptor density and frame time, time-on-target, number of beam positions, azimuth coverage, and number of elevation bars. Four different transmit peak powers are shown in the graph: 8000, 2000, 500, and 100 W. What can be seen from the graph is that, for depression angles of 4° and greater, the footprint on the ground is so small that power has little effect on the interceptability. Probability of intercept is limited only by the probability that there is an interceptor in the above threshold footprint. For shallow grazing angles, where the radar is illuminating the limb of the Earth, and at higher powers, the intercept range is limited only by the horizon. The probability of intercept rises rapidly with decreasing grazing angle and increasing power. Note, however, that for very low peak power, illuminating a large horizon area is not a great threat, because the power at the horizon is quite low. One conclusion is that, if shallow grazing angles are required, low peak radar powers must be used for typical interceptor performance parameters.

An example of the minimum dwell strategy is given in Fig. 2.11. The graph shows probability of intercept as a function of peak power and time-on-target, because peak power can be traded for time-on-target in many situation geometries. The performance factors are similar to those in Fig. 2.10 with a 2° grazing angle (close to the worst grazing angle in that figure). What can be seen is that, if the probability of intercept is high, one can actually reduce this probability by transmitting more power. This is counterintuitive but is still the case, because the scan-on-scan probability is reduced. As can be seen, a 3:1 decrease in probability of intercept can be achieved by transmitting more power for less time in the same geometry as in Fig. 2.10.

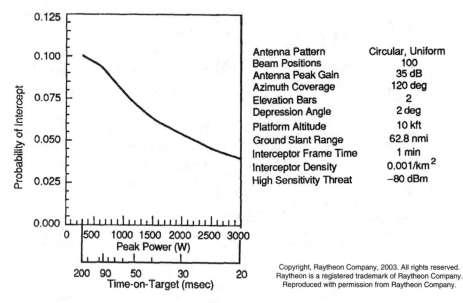

Figure 2.11 Example of minimum dwell strategy.[12]

Throughout this discussion, scan-on-scan probabilities have been mentioned. A typical ground map mainlobe probability of intercept is computed in Table 2.3. The probability of intercept is given in Equation (2.28) [a slightly simplified form of Equation (2.7), which assumes that the values of $\mathcal{P}_D$ and $\mathcal{P}_F$ are each 1]. Recall that, if the time on target is more than the intercept time, then the probability of intercept is only dependent on the area of the emitter footprint and the density of interceptors.

Table 2.3 Simple Probability of Intercept Calculation with Typical Ground Map Parameters

Parameter	Symbol	Value	Units	DB units
Radar time-on-target	T_{OT}	0.06	sec	−12.22
Surface interceptor frame time	T_I	60	sec	−17.78
Footprint area	A_F	160	km^2	22.04
Surface interceptor density	D_I	0.001	km^{-2}	−30.0
Probability of intercept	$\mathcal{P}_i$	0.00016		−37.96

$$\mathcal{P}_i = 1 - \exp\left\{-\min\left(\frac{T_{OT}}{T_I},1\right) \cdot A_F \cdot D_I\right\} \qquad (2.28)$$

Here, one can see that the ratio of the time-on-target to the interceptor frame time is crucial in the probability of intercept. In addition, the area of the footprint and the ground interceptor density, which really determine the number of opportunities, are also important factors. The net result in the example shown is that the probability of intercept is substantially less than 1 percent, but one can see that, if the radar time-on-target went from 60 msec to 1 sec, the probability of intercept would grow to substantially larger than 2 percent. The time-on-target is a function of the antenna gain as well as the mode of operation necessary for weapon system performance.

2.2.7 Cumulative Probability of Intercept

The next important concept is the cumulative probability of intercept, i.e., not just the intercept that occurs from a single opportunity, but the accumulation of the probability on each opportunity times the number of opportunities. Equation (2.29) shows that the cumulative probability of intercept is 1 minus the product of the individual probabilities of no intercept.

$$\mathcal{P}_{icum} = 1 - \prod_{IBP=1}^{NBP} (1 - \mathcal{P}_{iIBP}) \tag{2.29}$$

where

$\mathcal{P}_{icum}$ = probability of an interception over the entire scan volume
IBP = interceptable beam position
NBP = total number of beam positions in the scan volume
$\mathcal{P}_{iIBP}$ = probability of an interception on the ith beam position

If each beam position has the same probability of intercept, and using the parameters of Table 2.3, then

$$\mathcal{P}_{icum} = 1 - 1 + -(1.60 \cdot 10^{-4})^{100} = 1.587 \cdot 10^{-2} \tag{2.30}$$

Thus, the probability of intercept in each beam position magnified by the total number of beam positions can result in an accumulative number that is quite high, even if the individual probability of intercept is low. That means that not only must the time-on-target be optimized, but the total number of beam positions in the scan volume should be minimized as well.

A better understanding of the foregoing constraints and requirements can be obtained by working through an extended example, a typical ground map radar system with general parameters, as shown in Table 2.4. The table example is for a radar surveilling the ground from an altitude of 10,000 ft. Performance parameters might be typical of the type of radar found in a modern high-performance aircraft. Time-on-targets are typically 60 msec, and the number of beam positions over the surveilled area is 100; in this particular case, the ground slant range is a little less than 12 mi, and the radar footprint area and interceptor density are roughly the equivalent of those used in the examples in Figs. 2.10 and 2.11. The probability of intercept calculated per beam position is a small number: $1.6 \cdot 10^{-4}$. The cumulative probability of intercept is approximately $1.59 \cdot 10^{-2}$; i.e., near 2 percent, a dangerously high number. The mean time between intercepts can be calculated at 6.3 min, and that divided by the radar duty allows intercepts that are 10.5 min apart.

Table 2.4   Ground Map Probability of Intercept Parameters

Radar operating duty = 0.6	Single look detection probability = 0.679
Platform altitude = 10 kft	Number of beam positions = 100
Transmit peak power = 20 W	Center ground slant range = 11.5 nmi
Dwell time = 20 msec	Interceptor frame time = 1 min
Time on target = 60 msec	Footprint area = 160 km²
Frame time = 6 sec	Interceptor density = 0.001 per km²
Looks per ground patch = 1	Single look intercept probability = 0.00016
Antenna azimuth center = 20° relative to velocity vector	Cum. intercept probability = 0.016
Antenna elevation center = –8°	Mean time between intercepts (MTBI) = frame time/cum $\mathcal{P}_i$ = 6/0.016 =6.3 min.
MTBI/radar operating duty = 6.3/0.6 =10.5 min	

Figure 2.12 shows a constant intercept power footprint for the example of Table 2.4. It assumes a high intercept receiver sensitivity but a very low intercept receiver scan frequency of 1/min. As can be seen from the plot, the footprint area on the ground of the LPI system is made up of three main pieces: the mainlobe, cardinal sidelobes, and far sidelobes. The mainlobe often intercepts the limb of the Earth and will be truncated in range depending on the interceptor assumed altitude. The cardinal sidelobes are very large amplitude narrow beam sidelobes, typically along a single surface, and are the result of the method used to synthesize the antenna pattern. The random far sidelobes are the result of manufacturing tolerances and, although low, are immediately below the aircraft and still have significant area above the intercept receiver threshold. Usually, some care is taken to ensure that the cardinal sidelobes are not at the horizon for any significant amount of time by a combination of aircraft maneuvers, antenna design, and antenna installation and mounting. Mainlobe or cardinal sidelobes on the horizon for any period of time approaching the threat scan time are dangerously bad.

The main saving grace for far sidelobes is that the detection range is usually so short that there is no reaction time available to the threat. Further "revenge weapons" that try to engage an opening stealth target usually don't have enough detection range or maneuver energy for success.

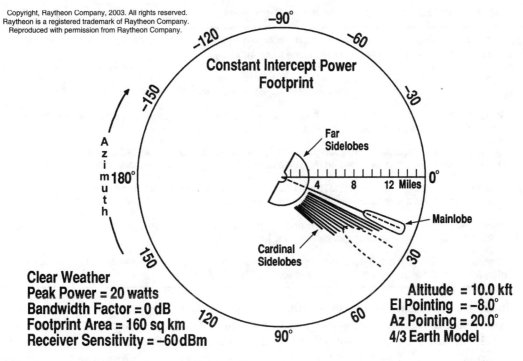

Figure 2.12 Example intercept footprint plan view.[13]

2.3 EXAMPLE MODE INTERCEPTABILITY

2.3.1 Example Radar Mode Interceptability Calculations

An improved understanding of the concepts presented in the previous sections can be achieved by working through a consistent set of emitter modes against a specific intercept threat. Table 2.5 describes five modes that might be employed in a single radar. Radar modes similar to these were used in the DARPA/USAF Pave Mover program in the early 1980s. Two possible mismatch strategies are calculated: minimum dwell and minimum power. The radar operating frequency is X-band with 40 dB antenna gain and 3 dB noise figure. The selected operating range is 50 km, which gives rise to 3 dB of external losses and an approximate 11.2 km^2 mainlobe footprint at an altitude of 35,000 ft. All other operating parameters are mode dependent and are given in Table 2.5. The assumed required SNR is 13 dB for hard targets and 4 dB for ground maps, with the exception of *spotlight moving target detection (SMTD)* minimum dwell, where a double threshold scheme can be used to lower the threshold 3 dB. Pulse compression uses Barker or compound Barker codes for the low-resolution, wide-coverage modes (thus, wide Doppler) and complementary

Table 2.5   LPI Radar Modes and Performance Parameters

Common parameter: wavelength 0.0298 m, antenna gain 40 dB, noise figure 3 dB, Operating altitude 35,000 ft, range 50 km, external losses 3 dB, mainlobe footprint 11.2 km^2

Wide area moving target detection (WAHTD): windshield wiper MTI for large numbers of moving vehicles, ±60° scan, 5 simultaneous transmitted frequencies, 25-m resolution, 150 beam positions

Wide area hard target detection (WAHTD): ground map underlay for wide area MTI, single frequency, ±60° scan, 300-m resolution, 150 beam positions

Synthetic aperture radar mapping spotlight (SMS): SAR map on a small spotlighted area, 2 × 2 mi/3.3 × 3 km, 5 simultaneous frequencies, 3-m range resolution, 2 to 6 beam positions, depending on antenna

Spotlight moving target detection (SMTD): MTI for a small area with many moving targets, 2 × 2 mi/3.3 × 3 km, 3-m resolution, 5 simultaneous frequencies, beams as above

Moving target tracking (MTT): tracking moving and hard targets in a small area, 3-m resolution, frequency, PRF, filter placement based on target parameters, 1 beam position

Missile data link (MDL): guidance commands to missile, 1 beam position on missile, dwell time/message time 50 msec, 5 frequencies, 100:1 spread spectrum, 2000 data bits

codes for the high-resolution, small-coverage (narrow Doppler) modes. Minimum power modes require longer integration and, therefore, have greater losses. The bandwidth in the table is the reciprocal of the uncompressed pulsewidth. The PRFs are ambiguous and are chosen to be the highest value consistent with multiple time around echo rejection, duty ratio, and Doppler ambiguity resolution. In actual practice, there would be several PRFs and pulsewidths whose average would be the entry in the table. The assumed antenna is electronically scanned in a pseudorandom pattern to cover the area of interest.

The first step is the calculation of the required radar power for each of the 10 cases in Tables 2.4 and 2.5. The applicable radar equation for the transmitter power is given in Equation (2.31).

$$P_T = \frac{(4 \cdot \pi)^3 \cdot R^4 \cdot k_B \cdot T_0 \cdot B \cdot NF \cdot SNR}{G_T^2 \cdot G_{RP} \cdot \lambda^2 \cdot \sigma \cdot PRF \cdot T_D \cdot N_{Freq} \cdot L_E \cdot L_P \cdot L_F} \tag{2.31}$$

The next step is the calculation of *probability of intercept*, $\mathbfcal{P}_i$, for each case using Equation (2.32).

$$\mathbfcal{P}_i \cong MF \cdot (2 \cdot P_i/P_I)^{C_o} \cdot D_I \cdot T_{OT}/T_I \tag{2.32}$$

where
  MF = mainlobe footprint area (3 dB)
  P_i = power received at intercept receiver
  P_I = power required at intercept receiver for detection

D_I = density of intercept receivers

G_{RP} = radar processor gain

T_D = radar dwell time

T_I = interceptor scan time

L_p = receiver/processor losses

L_E = external losses (atmosphere, radome)

L_F = noncoherent integration losses due to multifrequency

N_{Freq} = number of separate simultaneous frequency channels

PRF = pulse repetition frequency

B = 1/uncompressed pulsewidth

C_O = footprint/sensitivity scaling factor—0.2 and 0.477 typical for unweighted rectangular and circular apertures, respectively

Equation (2.32) assumes that the LPIS pulse compression is weighted such that

Pulse compression ratio (PCR) ≈ pulse compression output SNR improvement

The decibel form of the radar equation for these ten cases is given in Tables 2.7 and 2.8. Note that, with suitable radar parameter selection, the power per transmitted frequency can be substantially less than 1 W. Normally, radars require multiple frequency "looks" at a target to improve scintillation and ECM performance. However, by transmitting these frequencies all at once, the benefits of spread spectrum and multiple frequency looks can be obtained in the shortest dwell time or at the lowest power per frequency. Depending on the intercept threat, the frequency spacing can and should range from something slightly greater than the reciprocal of the pulse compression chip width to 100 times that value.

The second step is the calculation of the single beam position probability of intercept. If the individual probability of intercept is very much smaller than 1 ($\mathcal{P}_i \ll 1$), then Equation (2.7) can be approximated as shown in Equation (2.32). Equation (2.32) also contains an approximation to the radar footprint area above the intercept receiver threshold as shown in Equation (2.33) below.

$$A_F \cdot \mathcal{P}_F \cdot \mathcal{P}_D \approx MF \cdot (2 \cdot P_i/P_I)^{C_o} \tag{2.33}$$

This approximate curve fit will be discussed in the next section. With random electronic scanning $T_{OT} = T_D$, if sequential scanning is used then $T_{OT} \approx (2 \cdot P_i/P_I)^{C_o/2} \cdot T_D$ can be used.

The intercept threat chosen for this extended example is the "Future" threat of Table 2.1. A detailed description of such a high-performance intercept threat is given in Table 2.6. This threat is elastic in that its parameters are selected to optimize its performance against a given LPIS. In other words, it is a response threat subsequent to the deployment of that system. This is based on the model developed by Gary Gordon of

Table 2.6   High-Performance Intercept Threat (Response)

Parameter	Value
Bandwidth instantaneous coverage	1 GHz
Antenna gain, fixed once selected	0 to 40 dB variable
Predetection bandwidth	5 MHz
Number of channels	200
Dwell time/beam position	10 msec
Noise figure	4 dB
Receiver/processor loss	3 dB
Threshold/noise ratio	17 dB
Scan coverage	200° az. $\times$ 54° el.
Interceptor spacing	0.001/km^2
Interceptor threshold	–113 dBW or 5×10^{-12} W

RDA.[6] (Additional description of the model will be given in Chapter 3.) The high-performance threat has a variable-gain antenna whose beamwidth is chosen to have the best match to the radar operating strategy. The optimal strategy for the interceptor is to match its scan time to the radar dwell time so that, if the radar ever illuminates the interceptor, there will be a detection opportunity. A narrower beamwidth degrades detection probability as a result of scan-on-scan, and a broader beamwidth (less gain) reduces system sensitivity and spatial selectivity. The response threat, once designed, is relatively fixed in performance. For example, once the antenna size is selected, gain can't change much. There might be multiple scan rates but usually at the expense of either coverage or sensitivity. The LPIS has somewhat similar limitations except that it gets to select and match the waveform and engagement geometry.

The threat is also assumed to have a high-dynamic-range broadband channelized receiver, a dwell time long enough to receive a few radar pulses, a threshold high enough for reasonable false alarms, very high sensitivity, and *a priori* knowledge of the general direction of the LPIS (i.e., not in outer space, coming from the FEBA direction, and so on). The spacing between interceptors is 1 every 1000 km^2, which is roughly equivalent to an interceptor every 32 km in all directions. Although this threat is perfectly feasible technically, it is "ten feet tall," and no currently deployed interceptor system has this combination of parameters and distribution density.

For the modes analyzed here, the response threat interceptor scan time should be 5 sec, and the corresponding gain is approximately 32 dB.

Table 2.7 Minimum Dwell Examples for Multiple Modes

Minimum dwell	Wide area moving target detection		Wide area hard target detection		Synthetic aperture map spotlight		Spotlight moving target detection		Moving target tracking	
Parameter	Value	Value, dB	Value	Value, dB	Value	Value, dB	Value	Value, dB	Value	Value, dB
Numerator										
$(4 \cdot \pi)^3$ (steradians^3)	1.98E+03	33.0	1.98E+03	33.0	1.98E+03	33.0	1.98E+03	33.0	1.98E+03	33.0
Noise constant (W/Hz)	4.00E-21	-204.0	4.00E-21	-204.0	4.00E-21	-204.0	4.00E-21	-204.0	4.00E-21	-204.0
Bandwidth (kHz)	4.00E+01	46.0	4.00E+01	46.0	2.50E+01	44.0	2.50E+01	44.0	2.50E+01	44.0
Noise figure	2.00E+00	3.0	2.00E+00	3.0	2.00E+00	3.0	2.00E+00	3.0	2.00E+00	3.0
SNR	2.00E+01	13.0	2.51E+00	4.0	2.51E+00	4.0	10.2E+01	10.1	3.24E+00	5.1
Range^4 (km^4)	5.00E+01	188.0	5.00E+01	188.0	5.00E+01	188.0	5.00E+01	188.0	5.00E+01	188.0
Denominator										
Wavelength^2 (m^2)	8.90E-04	30.5	8.90E-04	30.5	8.90E-04	30.5	8.90E-04	30.5	8.90E-04	30.5
Antenna gain^2	1.00E+04	-80.0	1.00E+04	-80.0	1.00E+04	-80.0	1.00E+04	-80.0	1.00E+04	-80.0
External losses	2.00E+00	3.0	2.00E+00	3.0	2.00E+00	3.0	2.00E+00	3.0	2.00E+00	3.0
Target RCS (m^2)	1.00E+01	-10.0	2.50E+02	-24.0	1.00E+01	-10.0	1.00E+01	-10.0	1.00E+01	-10.0
Rcvr./proc. losses	2.57E+00	4.1	2.57E+00	4.1	2.57E+00	4.1	2.57E+00	4.1	2.57E+00	4.1
Dwell time (sec)	2.00E-01	7.0	5.00E-02	13.0	1.25E+00	-1.0	2.00E-01	7.0	2.00E-01	7.0
PRF (kHz)	1.00E+01	-40.0	1.00E+01	-40.0	6.00E+00	-37.8	6.00E+00	-37.8	6.00E+00	-37.8
Transmit power (W)	6.14E-02	-6.4	4.58E-03	-23.4	2.38E-03	-23.2	1.22E-01	-9.2	3.05E-02	-15.2
	5 FREQ, L=1dB	-6.00	1 FREQ		5 FREQ, L=1.6dB	-5.40	5 FREQ, L=1dB	-6.00	1 FREQ.	
Power/freq. (W)	5.71E-02	-12.4	4.58E-03	-23.4	1.38E-03	-28.6	3.05E-02	-15.2	7.67E-03	-21.2

Table 2.8   Minimum Power Examples for Multiple Modes

Minimum power Parameter	Wide area moving target detection		Missile data link		Synthetic aperture map spotlight		Spotlight moving target detection		Moving target tracking	
	Value	Value dB	Value	Value dB	Value	Value dB	Value	Value dB	Value	Value dB
Numerator										
$(4\pi)^3$ (steradians^3)	1.98E+03	33.0	1.58E+02	22.0	1.98E+03	33.0	1.98E+03	33.0	1.98E+03	33.0
Noise constant (W/Hz)	4.00E-21	-204.0	4.00E-21	-204.0	4.00E-21	-204.0	4.00E-21	-204.0	4.00E-21	-204.0
Bandwidth (kHz)	4.00E+01	46.0	4.00E+01	46.0	2.50E+01	44.0	2.50E+01	44.0	2.50E+01	44.0
Noise figure	2.00E+00	3.0	2.00E+00	3.0	2.00E+00	3.0	2.00E+00	3.0	2.00E+00	3.0
SNR	2.00E+01	13.0	5.01E+01	17.0	2.51E+00	4.0	2.00E+01	13.0	3.24E+00	5.1
Range^4 (km^4)	5.00E+01	188.0	5.00E+01	94.0	5.00E+01	188.0	5.00E+01	188.0	5.00E+01	188.0
Denominator										
Wavelength^2 (m^2)	8.90E-04	30.5	8.90E-04	30.5	8.90E-04	30.5	8.90E-04	30.5	8.90E-04	30.5
Antenna gain^2	1.00E+04	-80.0	5.62E+02	-55.0	1.00E+04	-80.0	1.00E+04	-80.0	1.00E+04	-80.0
External losses	2.00E+00	3.0	2.00E+00	3.0	2.00E+00	3.0	2.00E+00	3.0	2.00E+00	3.0
Target RCS (m^2)	1.00E+01	-10.0	N/A	0.0	1.00E+01	-10.0	1.00E+01	-10.0	1.00E+01	-10.0
Rcvr./proc. losses	3.47E+00	5.4	2.57E+00	4.1	5.13E+00	7.1	5.50E+00	7.4	2.57E+00	4.1
Dwell time (sec)	1.00E+00	0.0	5.00E-02	13.0	5.00E+00	-7.0	5.00E+00	-7.0	5.00E+00	-7.0
PRF (kHz)	1.00E+01	-40.0	1.00E+01	-40.0	6.00E+00	-37.8	6.00E+00	-37.8	6.00E+00	-37.8
Transmit power (W)	6.14E-02	-12.1	2.29E-07	-66.4	2.38E-03	-26.2	2.03E-02	-16.9	1.54E-03	-28.1
	5 FREQ, L=1dB	-6.00	5 FREQ, L=1dB	-6.00	5 FREQ, L=1.6dB	-5.40	5 FREQ, L=1dB	-6.00	1 FREQ	
Power/freq. (W)	1.54E-02	-18.1	5.75E-08	-72.4	6.87E-04	-31.6	5.09E-03	-22.9	1.54E-03	-28.1

With those parameters, $\boldsymbol{P}_i$ can be calculated as shown in Tables 2.9 and 2.10. Note that if the interceptor matches to 5 sec emitter dwell, then it is mismatched for the shorter dwells, and vice versa. Thus, *a priori* knowledge of the threat allows choice of the most mismatched waveform for a given mode.

The third step is calculation of the cumulative probability of intercept, $\boldsymbol{P}_{icum}$, using Equation (2.29), the number of beam positions, *NBP*, from Table 2.8, and the probability of intercept in the *I*th beam position, $\boldsymbol{P}_{iIBP}$, from Equation (2.29), as tabulated in Tables 2.8 and 2.10. The number of beam positions arises from the desired coverage in each mode, the beam overlap required to minimize SNR variations (scalloping) and the beam shape. Clearly, there is a trade-off between cumulative probability of intercept and acceptable SNR and coverage. For example, the *wide area moving target detection (WAMTD)* modes shown in Table 2.8 aren't stealthy with $\boldsymbol{P}_{icum}$ values of 0.9 and 0.58. To make these modes stealthy, coverage in angle or range must be decreased, and scalloping must be increased. Also note that the minimum power strategy is less stealthy for most of the example modes. That said, however, many other scenarios and threats would result in the opposite result. Table 2.10 summarizes the calculations but also shows the effect of reducing the density of intercept receivers by a factor of four as well as reducing the interceptor sensitivity by a factor of ten. Both of these reductions are likely in real scenarios.

Returning to Equation (1.3), the power at the interceptor can be calculated for each mode transmit power, which, along with interceptor parameters from Table 2.6, can be inserted into Equation (2.29) to estimate the intercept probability in Table 2.9.

$$P_i = \frac{P_T \cdot G_{TI} \cdot G_I \cdot \lambda^2 \cdot G_{IP} \cdot L_I}{(4 \cdot \pi)^2 \cdot R_i^2}$$

$$= \frac{P_T \cdot 10^4 \cdot 1585 \cdot (0.03)^2 \cdot 0.5 \cdot 0.5}{(4 \cdot \pi \cdot 50000)^2}$$

$$= P_T \cdot 9.033 \cdot 10^{-9} \tag{2.34}$$

Substituting Equation (2.34) and the other fixed parameters into Equation (2.32), assuming an unweighted circular antenna, yields

$$\boldsymbol{P}_i \cong MF \cdot (2 \cdot P_i/P_I)^{C_o} \cdot D_I \cdot T_{OT}/T_I$$

$$= 11.2 \cdot \left(\frac{2 \cdot P_T \cdot 9.033 \cdot 10^{-9}}{5 \cdot 10^{-12}} \right)^{0.477} \cdot \frac{0.001 \cdot T_{OT}}{5}$$

$$\cong 0.11 \cdot P_T^{0.477} \cdot T_{OT} \tag{2.35}$$

Table 2.9 Example Intercept Calculations for Multiple Modes

Mode/parameter	Minimization strategy	Time on target (sec)	Transmit power (dBW)	Intercept probability	No. of beams	Cumulative intercept probability
Wide area moving target detect	Dwell	0.2	−12.4	0.0057	150	0.5762
Wide area hard target detection	Dwell	0.05	−23.4	0.0004	150	0.0622
Synthetic aperture map spotlight	Dwell	1.25	−28.6	0.006	4	0.0239
Spotlight moving target detect	Dwell	0.2	−15.2	0.0042	4	0.0168
Moving target tracking	Dwell	0.2	−21.2	0.0022	1	0.0022
Wide area moving target detect	Power	1	−18.1	0.0153	150	0.9007
Missile data link	Power	0.05	−72.4	1.97E−06	1	1.97E−06
Synthetic aperture map spotlight	Power	5	−31.6	0.0173	4	0.0675
Spotlight moving target detect	Power	5	−22.9	0.045	4	0.1683
Moving target tracking	Power	5	−28.1	0.0254	1	0.0254

Table 2.10 Intercept Probability Summary

Parameter/mode	Wide area (W/A) moving target detection		W/A hard target detection	SAR mapping spotlight		Spotlight moving target detection		Moving target TWS		Missile data link
Peak power (dBW) (min. pwr., min. dwell)	−18.1	−12.4	−23.4	−31.6	−28.6	−15.2	−22.9	−28.1	−21.2	−72.41
Corresponding dwell time (sec)	1.00	0.20	0.05	5.00	1.25	0.20	5.00	5.00	0.20	0.05
Approx. intercept probability	0.0153	0.0057	0.0004	0.0173	0.0060	0.0450	0.0042	0.0254	0.0022	1.97E−06
Cum intercept probability	0.9007	0.5762	0.0622	0.0675	0.0239	0.1683	0.0168	0.0254	0.0022	1.97E−06
Cum. probability—1/4 density	0.4367	0.1928	0.0159	0.0172	0.006	0.0443	0.0042	0.0064	0.0005	4.92E−07
Cum. probability—0.1 sensitivity	0.5351	0.2485	0.0212	0.0229	0.008	0.0587	0.0056	0.0085	0.0007	6.56E−07

Spreadsheet printouts used to generate Tables 2.7 through 2.10 are included in the CDROM appendices. All the CDROM tables are in Microsoft Excel® .xls file format. The *optimum strategy for interceptor* implies: Total interceptor scan time, T_I = radar maximum dwell time, T_D. Since the radar maximum dwell time = 5 sec; T_I = 5 sec, and so the interceptor optimum gain is G_I = 32 dB, which is implicit in Tables 2.9 and 2.10.

2.3.2   Data Link Mode Interceptability Example

The missile data link in the preceding examples is easily very stealthy, both because of its limited data rate and its limited use. Often, for modes that involve SAR mapping, data link data rates and dwell times are high. There are three advantages a stealthy data link has over stealthy radar emissions: it employs one-way versus two-way transmission, the ground station usually has a high-gain antenna, and the ground station is usually in relatively controlled territory. This implies that the data link intercept threat is primarily a sidelobe threat. Consider the case of the interceptor parameters used in Equation (2.34) in the data link sidelobes and a data link path R_{DL} = 200 km. A SAR mapping data link might have the following parameters: λ = 0.02 m, G_T = 30 dB, G_{DL} = 40 dB, G_{TI} = –20 dBi, G_{DP} = 10, SNR_{DL} = 17 dB, NF_{DL} = 3dB, B_{DL} = 10 MHz, T_{DDL} = 1 sec, L_{DL} = 0.5. Substituting these parameters in Equation (2.36) below yields

$$P_{TDL} = \frac{(4 \cdot \pi)^2 \cdot R_{DL}^2 \cdot k_B \cdot T_0 \cdot B_{DL} \cdot NF_{DL} \cdot SNR_{DL}}{G_T \cdot G_{DL} \cdot G_{DP} \cdot L_{DL} \cdot \lambda^2}$$

$$P_{TDL} = 0.32 \text{ mW}$$

(2.36)

Similarly to Equation (2.34), the power at the intercept receiver is

$$P_i = P_{TDL} \cdot 4 \cdot 10^{-15} = 1.28 \cdot 10^{-18} \tag{2.37}$$

And so the probability of intercept is

$$\mathcal{P}_i = \frac{1.28 \cdot 10^{-18}}{5 \cdot 10^{-12}} = 2.57 \cdot 10^{-7} \tag{2.38}$$

Thus, a mapping data link interceptability can be of the same order as a missile data link. The cumulative probability of intercept is roughly 5×10^{-7}.

2.4 FOOTPRINT CALCULATION

2.4.1 "Cookie Cutter" Footprints

Throughout Chapter 2, the shape and area of footprints have been discussed without showing just how to calculate them. Two methods will be shown in this section: the "cookie cutter" method and a more exact method. For many applications, a computationally simple method is more than adequate. Consider the radiation of a typical aperture, as shown in Fig. 2.13. Most of the sidelobes will be far below the interceptor threshold and might reasonably be ignored. Suppose the angle pattern was replaced with a cylinder whose angular diameter was the same as the beamwidth at the 3, 10, or 20 dB point, as shown in Fig. 2.14. This approximation is commonly called the "cookie cutter" antenna pattern. The pattern is an elliptical cone in rectilinear space (see Fig. 2.15). If the antenna pattern is circularly symmetric, the footprint is an ellipse on a planar Earth. Referring back to Fig. 2.2, if a North East Down (NED) coordinate system is defined as shown in the figure, with the radar LOS in the North Down plane for simplicity (and with no loss of generality), the locus of the intersection can be calculated.

Referring to the expanded "cookie cutter" footprint geometry of Fig. 2.16, the beamwidth for an elliptical cone for an angle ϕ about its center line of sight (LOS) is

$$BW = 2 \cdot \operatorname{asin}\left(\frac{U \cdot \lambda}{2 \cdot \pi \sqrt{a^2 \cdot \cos^2 \phi + b^2 \cdot \sin^2 \phi}}\right) \tag{2.39}$$

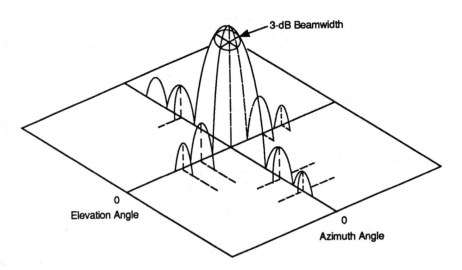

Figure 2.13 Radiation pattern of a typical aperture.

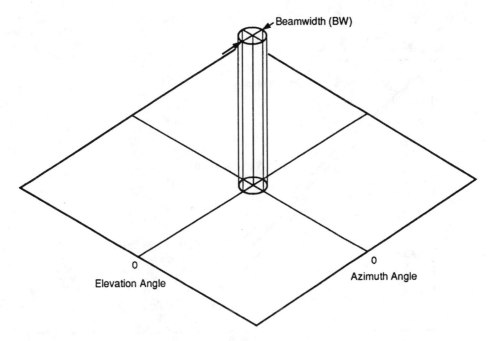

Figure 2.14 "Cookie cutter" antenna pattern approximation.

The projection of the beam into the North Down plane is

$$BW_N = 2 \cdot \operatorname{asin}\left(\frac{U \cdot \lambda \cdot \cos\phi}{2 \cdot \pi\sqrt{a^2 \cdot \cos^2\phi + b^2 \cdot \sin^2\phi}}\right) \tag{2.40}$$

Similarly, the beamwidth in the North East dimension is

$$BW_E = 2 \cdot \operatorname{asin}\left(\frac{U \cdot \lambda \cdot \sin\phi}{2 \cdot \pi\sqrt{a^2 \cdot \cos^2\phi + b^2 \cdot \sin^2\phi}}\right) \tag{2.41}$$

and the cross-range component of the range, R, from Fig. 2.17 is

$$R_C = R_g \cdot \tan\left(\frac{BW_E}{2}\right) \tag{2.42}$$

The ground range component of the range, R_g, from Fig. 2.15 is

$$R_g = h \cdot \tan\left(90° - \varepsilon - \frac{BW_N}{2}\right) \tag{2.43}$$

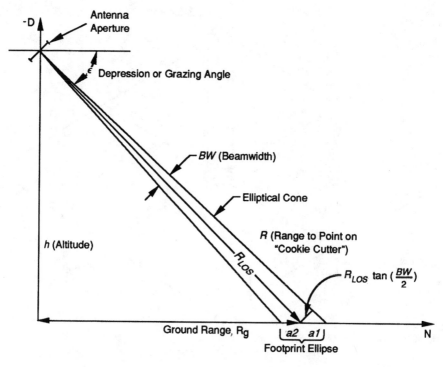

Figure 2.15 "Cookie cutter" footprint geometry.

The maximum beamwidths in the East and North directions are approximately

$$BW_{N\max} = 2 \cdot \mathrm{asin}\!\left(\frac{U \cdot \lambda}{2 \cdot \pi \cdot a}\right) \tag{2.44}$$

and

$$BW_{E\max} = 2 \cdot \mathrm{asin}\!\left(\frac{U \cdot \lambda}{2 \cdot \pi \cdot b}\right) \tag{2.45}$$

where

ϕ = angle of rotation around the radar LOS, measured from the North Down plane

ε = depression or grazing angle of the radar LOS with respect to the North East plane for a flat Earth

λ = wavelength in units compatible with a, b, where a is one-half the aperture height and b is one-half the aperture width

U = beamwidth at some predefined power roll-off point in normalized $\sin\theta$ space

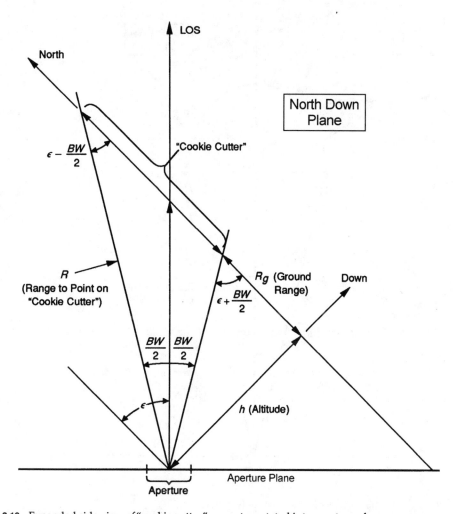

Figure 2.16   Expanded side view of "cookie cutter" geometry rotated into aperture plane.

In Chapter 6, a more generalized definition of U is provided. By substituting Equations (2.44) and (2.45) into Equations (2.42) and (2.43), the maximum and minimum down range, R_{gmax}, and R_{gmin}, and maximum cross range, R_{Cmax}, can be approximated.

$$R_{gmax}, R_{gmin} = h \cdot \tan\left(90° - \varepsilon \pm \frac{BW_{Nmax}}{2}\right) \tag{2.46}$$

The maximum cross range is close to the middle of the down range footprint for small beamwidths and can be approximated as

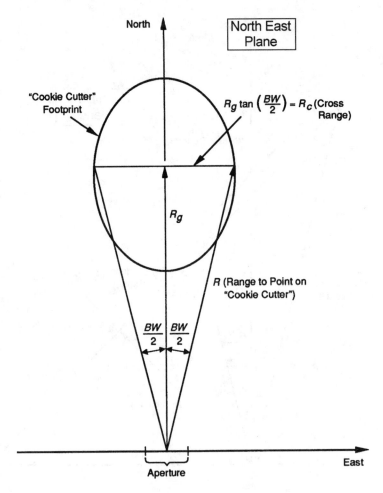

Figure 2.17   Expanded plane view of "cookie cutter" geometry.

$$R_{C\max} = (R_{g\max} + R_{g\min}) \cdot \tan\left(\frac{BW_{E\max}}{2}\right) \tag{2.47}$$

The mainlobe footprint area is approximately

$$MF = \frac{\pi}{4} \cdot R_{C\max} \cdot (R_{g\max} - R_{g\min}) \tag{2.48}$$

Two examples of "cookie cutter" footprints for U at the 10-dB down beamwidth for two different aperture shapes are given in Fig. 2.18 and 2.19. Figure 2.18 is the far-field pattern projected on a flat Earth for a square array that is 38 in on a side with a wave-

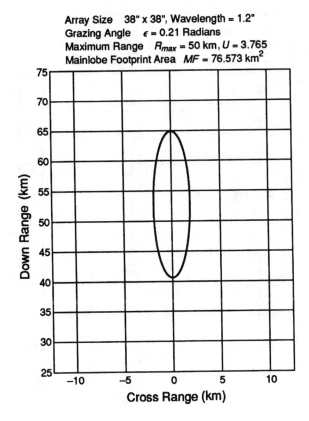

Figure 2.18 "Cookie cutter" footprint square array.

length of 1.2 in. Figure 2.19 is the field pattern projected on a flat Earth for a rectangular array that is 38 in on one side by 77 in on the other side with a wavelength of 1.2 in. The calculation detailed results in Mathcad® are provided in the appendix CDROM. It often turns out that the choice of amplitude weighting in the elevation dimension is different from the azimuth dimension for implementation or mission performance reasons, and so these patterns are unrealistic but still a good first approximation. The ratio of the U values in the two dimensions for the examples is 1.25, i.e., different azimuth and elevation weighting.

With "cookie cutter" models, the assumption is made that the power inside the perimeter is above the interceptor threshold, while outside the perimeter it is not.

A correction can be made to provide a more accurate analysis using Equation (2.32). To derive that equation, some representative values of U need to be found. Consider the cardinal cut pattern of a uniformly illuminated circular aperture (Fig. 2.20). Note that the plot is normalized for the ratio between the aperture diameter and wavelength as well as the warping associated with phase space.

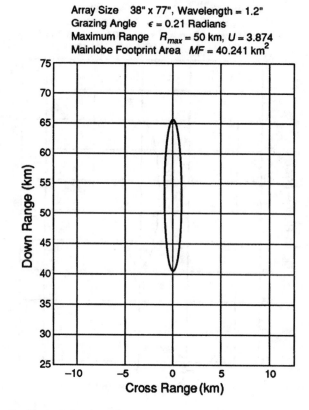

Array Size   38" x 77", Wavelength = 1.2"
Grazing Angle $\epsilon = 0.21$ Radians
Maximum Range $R_{max} = 50$ km, $U = 3.874$
Mainlobe Footprint Area $MF = 40.241$ km^2

Figure 2.19 "Cookie cutter" footprint rectangular array.

The equation for an ideal uniformly illuminated circular aperture antenna is given in Equation (2.49) assuming that the antenna is many wavelengths across. (More details are given in Chapter 6).

$$G(\theta,\phi) \cong \frac{4 \cdot \pi^2 \cdot a^2}{\lambda^2} \cdot \frac{J_1^2(U)}{U^2} \qquad (2.49)$$

where

$$U = \frac{2 \cdot \pi \cdot a}{\lambda} \sin(\theta) \text{ and } a = \text{antenna radius}$$

The 3, 10, and 20 dB beamwidths can be found from Equation (2.49) by finding the arguments that give 0.5, 0.1 and 0.01 gain. Because the wavelength is very small rela-

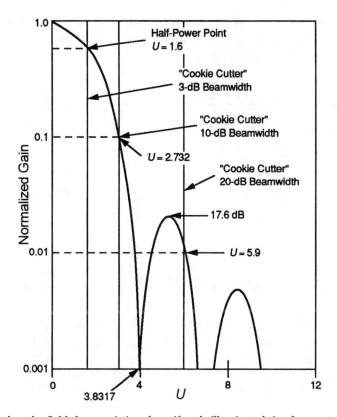

Figure 2.20 Cardinal cut far-field characteristics of a uniformly illuminated circular aperture.

tive to the antenna radius, the arcsine can be approximated by its argument as given in Equations (2.50).

$$3 \text{ dB Beamwidth} = 2 \cdot \text{asin}\left(\frac{1.6 \cdot \lambda}{2 \cdot \pi \cdot a}\right) \cong 1.02 \frac{\lambda}{2 \cdot a}$$

$$10 \text{ dB Beamwidth} = 2 \cdot \text{asin}\left(\frac{2.732 \cdot \lambda}{2 \cdot \pi \cdot a}\right) \cong 1.74 \frac{\lambda}{2 \cdot a}$$

$$20 \text{ dB Beamwidth} = 2 \cdot \text{asin}\left(\frac{5.9 \cdot \lambda}{2 \cdot \pi \cdot a}\right) \cong 3.76 \frac{\lambda}{2 \cdot a}$$

$$(2.50)$$

The value of U that corresponds to the half-power beamwidth is 1.6, and the "cookie cutter" beam would be a cylinder with radius of 1.6 in sin θ space. The 10 dB and 20 dB "cookie cutter" beamwidths are also shown in Fig. 2.20 at $U = 2.732$ and $U = 5.9$.

The equation for an ideal uniformly illuminated rectangular aperture antenna is given in Equation (2.51), assuming that the antenna is many wavelengths across. (More details are given in Chapter 6.)

$$G(\theta, \phi) = \frac{4 \cdot \pi \cdot a \cdot b}{\lambda^2} \cdot \frac{\sin^2(U)}{U^2} \cdot \frac{\sin^2(V)}{V^2} \cdot \frac{[1 + \cos(\theta)]^2}{4} \tag{2.51}$$

or more commonly,

$$G(\theta, \phi) = \frac{4 \cdot \pi \cdot a \cdot b}{\lambda^2} \cdot \frac{\sin^2(U)}{U^2} \cdot \frac{\sin^2(V)}{V^2}$$

where

$$U = \frac{\pi \cdot a}{\lambda} \sin(\theta) \cdot \cos(\phi) \text{ and } V = \frac{\pi \cdot b}{\lambda} \sin(\theta) \cdot \sin(\phi)$$

The sidelobe patterns at $\phi = 0$ and $\pi/2$ are called the *cardinal cuts*. For the cardinal cut pattern of a uniformly illuminated rectangular aperture as shown in Fig. 2.21, the 3, 10, and 20 dB beamwidth "cookie cutter" cylinders have radius $U = 1.392$, 2.319, and 8.423, respectively. As before, the 3, 10, and 20 dB beamwidths are found from Equations (2.51) as given in Equations (2.52).

$$3 \text{ dB Beamwidth} = 2 \cdot \text{asin}\left(\frac{1.39 \cdot \lambda}{\pi \cdot (a, b)}\right) \cong 0.88 \frac{\lambda}{(a, b)}$$

$$10 \text{ dB Beamwidth} = 2 \cdot \text{asin}\left(\frac{2.319 \cdot \lambda}{\pi \cdot (a, b)}\right) \cong 1.48 \frac{\lambda}{(a, b)}$$

$$20 \text{ dB Beamwidth} = 2 \cdot \text{asin}\left(\frac{8.423 \cdot \lambda}{\pi \cdot (a, b)}\right) \cong 5.36 \frac{\lambda}{(a, b)} \tag{2.52}$$

Lastly, a more representative case is shown in Fig. 2.22, a cardinal cut of a rectangular slotted aperture array with Hamming weighting. This is the pattern of a real array of the type used on programs such as the USAF Pave Mover, Joint STARS, Swedish Erieye, and Israeli Phalcon.

The E-field equation for a normalized ideal Hamming weighted illumination, slotted radiator, planar rectangular aperture antenna cardinal cut in the $\phi = 0$ plane is given in Equation (2.53) assuming that the antenna is many wavelengths across. (More details will be given in Chapter 6).

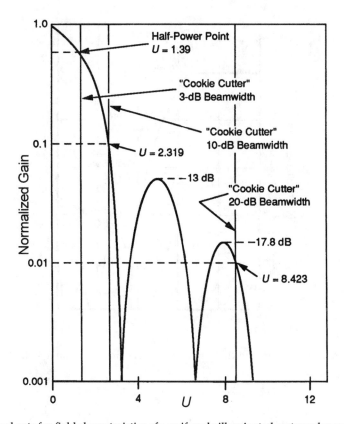

Figure 2.21 Cardinal cuts far-field characteristics of a uniformly illuminated rectangular aperture.

$$E_{norm}(\theta, \phi = 0) = E_e(\theta, \phi = 0) \cdot \begin{bmatrix} \dfrac{0.54 \cdot \sin(\pi \cdot N \cdot d \cdot \sin(\theta)/\lambda)}{N \cdot \sin(\pi \cdot d \cdot \sin(\theta)/\lambda)} + \\[2em] \dfrac{0.23 \cdot \sin\left(\pi \cdot N \cdot d \cdot \sin\left(\theta - \dfrac{\lambda}{N \cdot d}\right)/\lambda\right)}{N \cdot \sin\left(\pi \cdot d \cdot \sin\left(\theta - \dfrac{\lambda}{N \cdot d}\right)/\lambda\right)} + \\[2em] \dfrac{0.23 \cdot \sin\left(\pi \cdot N \cdot d \cdot \sin\left(\theta + \dfrac{\lambda}{N \cdot d}\right)/\lambda\right)}{N \cdot \sin\left(\pi \cdot d \cdot \sin\left(\theta + \dfrac{\lambda}{N \cdot d}\right)/\lambda\right)} \end{bmatrix} \qquad (2.53)$$

where N is the number of radiators in the x dimension and d is the spacing between radiators. The slotted radiator pattern, E_e, and U are defined as given in Equations (2.54).

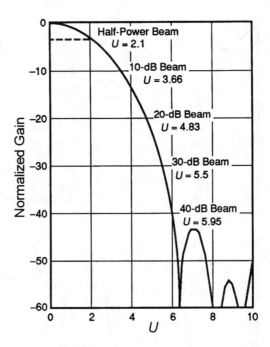

Figure 2.22 Cardinal cut ($\phi = 0$) far-field characteristics of a Hamming weighted, slotted planar rectangular aperture.

$$E_e(\theta, \phi = 0) = 0.5 + 0.5 \cdot \cos(1.33 \cdot \theta)$$

and

$$U = \frac{\pi \cdot N \cdot d}{\lambda} \sin(\theta)$$

(2.54)

As before, the corresponding 3, 10, 20, 30, and 40 dB cylinders have U values of 2.1, 3.66, 4.83, 5.5, and 5.95, respectively. These cylinders are elliptical cones in angle space. As can be seen in the antenna chapter (Chapter 6), this array also has an element pattern, $E_e(\theta, \phi = 0)$, which accounts for mutual coupling. Note also that U values don't increase nearly as fast with angle when an array is weighted (naturally!).

Using this same analysis, an example of array size versus footprint area for alternative synthetic array radar mapping strategies is shown in Table 2.11. Note that the footprint decreases as the aperture increases, but the number of beam positions and the dwell time get larger. Larger beamwidths also overilluminate the area to be mapped. There are multiple trade-offs based on the threat density, response time, and type of threat response.

Table 2.11   Array Size versus Interceptability Trade-Offs Ground Map Example

Parameter	Antenna size				
Array dimensions (inches)	38×38	38×77	38×115	38×154	58×154
BW_{AZ} (deg)	2.3	1.2	0.9	0.6	0.6
BW_{EL} (deg)	3	3	3	3	2.1
Gain (dB)	37	39	41	42	44
RMS sidelobes (dB)	−55	−55	−55	−55	−55
Mainlobe −3 dB footprint[*]					
Azimuth (km)	2	1.05	0.79	0.52	0.52
Elevation (km)	12.2	12.2	12.2	12.2	8.5
−10 dB footprint area (km²)	76.6	40.2	30.3	19.9	13.9
SAR array time (sec)[†]	2.5	2.5	2.5	2.5	2.5
No. arrays to map 10 km²	2	3	4	6	6
Time to map 10 km² (sec)	5	7.5	10	15	15
Overillumination area (km²)	142	110	110	108	72

[*]$h = 35$ kft, $R = 50$ km, look angle = 30° off normal, 60° off velocity vector
[†]$T = \lambda \cdot R/\Delta R \cdot V$, $\lambda = c/10$ GHz, $\Delta R = 3$ m, $V = 400$ kt

The footprints of multibeam antennas can be calculated as though each beam is completely separated and independent. The accuracy of "cookie cutter" footprints can be improved by calculating footprint areas for several different classes of aperture as a function of sensitivity relative to the mainlobe peak using the foregoing analysis. Figure 2.23 shows a plot of relative footprint area versus relative intercept receiver sensitivity using the more accurate footprint model in the next section (see appendices), including weather and a uniform weighting function on circular and rectangular apertures for grazing angles near 6°. Several straight-line approximations are drawn on the graph, which represent the exponent, C_o, of Equation (2.32). Fully weighted apertures have very little growth in footprint area as a function of increased interceptor sensitivity; for example, Hamming weighting roughly corresponds to C_o equal to 0.12. In the absence of detailed knowledge of the aperture and its weighting, the straight-line approximations are useful. An exponent, C_o, of approximately 0.2 splits the difference between unweighted and heavily weighted apertures.

2.4.2   More Accurate Footprints

Footprints that are still more accurate can be calculated by incorporating actual path performance to each interceptor in the calculations. This includes the actual antenna

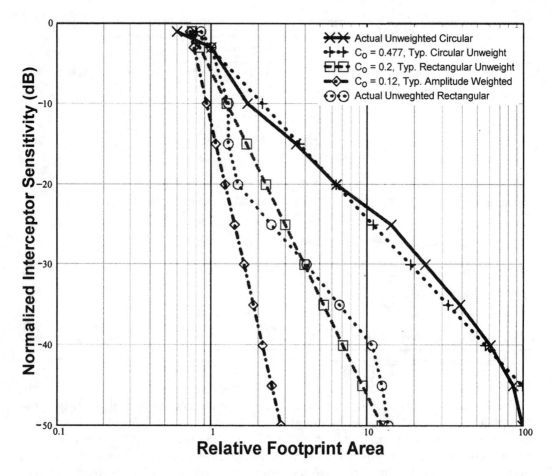

Figure 2.23 Relative receiver sensitivity versus relative footprint area.

pattern along the path to the interceptor, the atmospheric and weather losses, and the beacon equation. Consider the geometry[7,12] of Fig. 2.2, as shown in Fig. 2.24, the range, R, to the interceptor at the l, m grid point is given in Equation (2.55),

$$R(l,m) = \sqrt{h^2 + (R_g + l \cdot \Delta N)^2 + (m \cdot \Delta E)^2}$$ (2.55)

The maximum design range, R_{max}, (or R_{Dmax} if LPI) and the operating altitude, h, provide the LOS range, R_{LOS}, and grazing angle, ε. They are

$$R_{LOS} = \sqrt{h^2 + R_{max}^2}$$ (2.56)

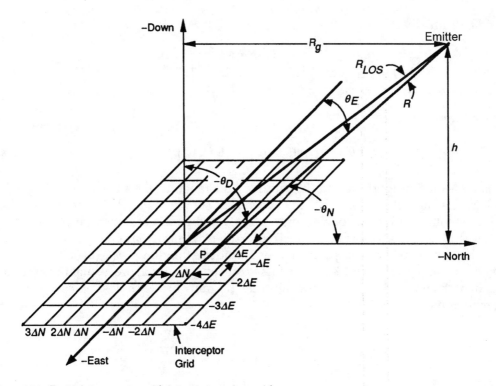

Figure 2.24 Footprint geometry with intercept receiver grid.

$$\varepsilon = \text{atan}\left(\frac{h}{R_{max}}\right) \tag{2.57}$$

The antenna gain pattern is assumed to be pointed along the LOS so that, if the pattern is defined in normalized θ, ϕ space with peak gain, G_0,

$$G_0 \cdot G[U(\theta',\phi)] = G_0 \cdot G[U(\theta-\varepsilon,\phi)] \tag{2.58}$$

The angles θ' and ϕ' about the LOS to a point in the interceptor grid are

$$\theta'(l,m) = \text{acos}\left(\frac{l \cdot \Delta N \cdot \cos(\varepsilon) + R_{LOS}}{R(l,m)}\right) \tag{2.59}$$

and

$$\phi'(l,m) = \text{atan}\left(\frac{m \cdot \Delta E}{l \cdot \Delta N \cdot \sin(\varepsilon)}\right) \tag{2.60}$$

And recalling that

$$U(l,m) = \frac{2 \cdot \pi \cdot \sin(\theta') \cdot \sqrt{a^2 \cos(\phi') + b^2 \cdot \sin^2(\phi')}}{\lambda} \tag{2.61}$$

Substituting,

$$G_0 \cdot G(l,m) = G_0 \cdot G(U(l,m)) \tag{2.62}$$

The losses are proportional to the altitude and range to the interceptor. The atmospheric loss model consists of three elements, as shown in Fig. 2.25. The clear air loss is exponential, with the exponent proportional to the ratio of the range, $R(l,m)$, to the altitude, h, and loss constant K_{CA}. The losses resulting from cloud cover are path length dependent beginning at the upper altitude of the clouds, h_{CL} (usually the freezing line) with loss constant K_{CL}. The losses caused by rainfall are also path length dependent beginning at the upper altitude of rain, h_r, with loss constant K_r. In combination,

$$L(l,m) = 10^{\left(R(l,m) \cdot \left[\frac{K_{CA} + K_{CL} \cdot h_{CL} + K_r \cdot h_r}{h} \right] \right)} \tag{2.63}$$

Combining the above equations, the power received at the intercept receiver is

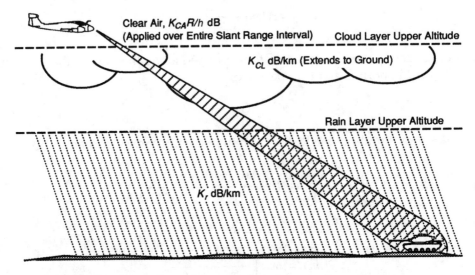

Figure 2.25 Atmospheric model used for more accurate footprint generation.

$$P_i = \frac{P_T \cdot G_0 \cdot G(l,m) \cdot L(l,m) \cdot G_I \cdot \lambda^2}{(4 \cdot \pi \cdot R(l,m))^2} \tag{2.64}$$

The intercept detections occur if P_i is greater than P_I. Using the Heavyside step function, Φ, the interceptors above the threshold are

$$A(l,m) = \Phi\{P_i(l,m) - P_I\} \tag{2.65}$$

The intercept footprint area is the sum of all $A(l, m)$ times the area of a single grid cell.

$$A_F = \Delta E \cdot \Delta N \cdot \sum_m \sum_l A(l,m) \tag{2.66}$$

The appendices contain more details of the foregoing analysis, including the calculation of Figs. 2.24, 2.26, and 2.27. The appendices contain enough details for the student to calculate footprints with different antenna patterns, interceptor densities, or interceptor sensitivities as well as almost any emitter location in the atmosphere. The model is a flat, smooth Earth model and has limitations, which are discussed in Chapter 4. The Mathcad® files are enclosed on the accompanying appendix CDROM.

An example of the foregoing analysis is given in Figs. 2.26 and 2.27 (see Table 2.12). This example uses an unweighted circular aperture of 1 m in diameter, a 6° grazing angle, maximum range of 100 km, transmitter power of 100 W, intercept receiver sensitivity of –80 dBm, intercept receiver spacing of 3 km, interceptor antenna gain of 10 dB, K_{CA} of 2 dB, K_{CL} of 0.1 dB/km and cloud height 6 km, K_r of 0.5 dB/km and rain height 3 km, and wavelength of 0.03 m (X-band). Figure 2.27 shows four cuts through the mainlobe. Figure 2.26 shows the footprint amplitude above the interceptor threshold and has area of 909 km^2.

Table 2.12   Leading Parameters for Higher Accuracy Footprint Example

Transmit power, P_T, W	100	Cloud attenuation, K_{CL}, dB/km	0.1
Antenna shape	1 m, circular	Cloud height, km	6
Aperture weighting	Uniform	Rain attenuation, K_r, dB/km	0.5
Wavelength, λ, m	0.03 (X-band)	Rain height, km	3
Maximum operating range, km	100	Interceptor sensitivity, dBm	–80
Altitude, h, km	10.45	Interceptor antenna gain, dB	10
Clear air attenuation, K_{CL}, dB	2	Interceptor density, D_I, /km^2	0.1

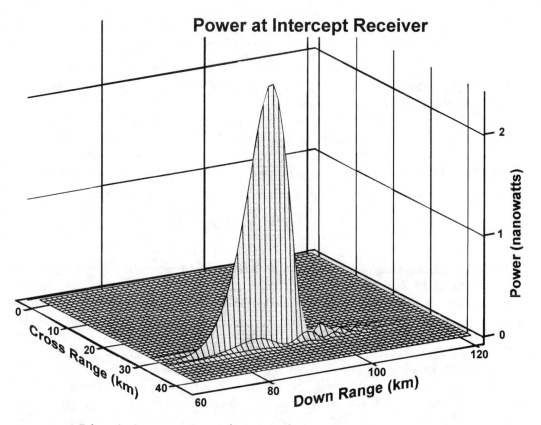

Figure 2.26 3-D footprint in power, φ, θ space for 1-m circular aperture.

2.5 EXERCISES

1. Calculate the required transmitter powers for minimum power and minimum dwell and, from this, the probability of intercept using the high-performance threat optimized for dwell, for the SMS mode, 9 frequencies, $\sigma = 1$ m^2, range 75 km, 38- by 154-in antenna, and all other parameters as in the chapter examples.

2. Calculate the false alarm rate and probability of detection by an interceptor with a 16-dB SNR threshold and an LPIS SNR of 0 dB, for a 300-kHz PRF, a 30-msec dwell, interceptor bandwidth of 5 MHz, and interceptor search time of 5 sec.

3. Calculate the horizon-to-horizon travel time for an aircraft altitude of 200 ft and Mach 0.85 velocity.

4. Calculate the cumulative probability of intercept for the example of Figure 2.12 for T_{OT} of 40 msec.

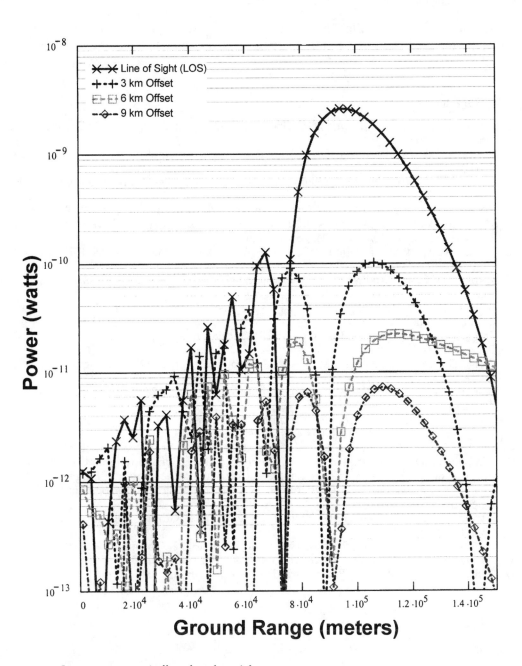

Figure 2.27 Intercept power—4 offsets from boresight.

5. Calculate the cumulative probability of intercept for the example of Figure 2.11 for an elevation of –6° and P_T of 2000 W.

6. Calculate a "cookie cutter" footprint for an array size of 18 by 36 in with $\lambda = 0.5$ in using the 20-dB point on a circular aperture at a range of 75 km and grazing angle of 6°.

2.6 REFERENCES

1. Ramsey, J.F., "Lambda Functions Describe Antenna/Diffraction Patterns," *Microwaves*, June 1967, pp. 69–106.
2. Elliot, D., Ed., *Handbook of Digital Signal Processing*, Academic Press, 1987, pp. 253–286, 975–985.
3. Feller, W., *An Introduction to Probability Theory and Its Applications I*, John Wiley & Sons, 1957, pp. 135–198.
4. Skolnik, M.I., *Introduction to Radar Systems*, 2nd ed., McGraw-Hill, 1980, pp. 28, 470–474.
5. Rice, S.O., "Mathematical Analysis of Random Noise," *Bell System Technical Journal*, Vol. 23, 1944, pp. 282–332 and Vol. 24, 1945, pp. 46–156.
6. Gordon, G.A., "Low Probability of Intercept Radar Evaluation," *RDA-TR*-173500-Ml, December 1978.
7. Sokolnikoff, I., and Redheffer, R., *Mathematics of Physics and Modern Engineering*, 1st ed., McGraw-Hill, 1958, p. 360.
8. Pearson, J.O., "FLAMR Signal to Noise Experiments," *Hughes Report* P74-501, December 1974, Declassified 12/31/1987.
9. Craig, D., and Hershberger, M., "Operator Performance Research," *Hughes Report* P74-504, December 1974.
10. Craig, D., and Hershberger, M., "FLAMR Operator Target/OAP Recognition Study," *Hughes Report* P75-300, 1975.
11. Pearson, J. et al., "SAR Data Precision Study," *Hughes Report* P75-459, December, 1975, Declassified 12/31/1987.
12. Hormel, E., and Hill, D., "Antenna Footprint Simulation and Probability of Intercept Calculations," *Hughes IDC*, 2311.11/807, October 1987.
13. A few photos, tables, or figures in this intellectual property were made at Hughes Aircraft Company and first appeared in public documents that were not copyrighted. These photos, tables, or figures were acquired by Raytheon Company in the merger of Hughes and Raytheon in December 1997 and are identified as Raytheon photos, tables, or figures. All such appear by permission.
14. Clapp, "A Theoretical and Experimental Study of Radar Ground Return," *MIT Radiation Laboratory Report* 1024, 1946.
15. Skolnik, M., Ed., *Radar Handbook*, McGraw-Hill, 1970, pp. 19–7 through 19–12, 38–4.
16. Berkowitz, R. S., Ed., *Modern Radar*, John Wiley and Sons, 1965, p. 11.
17. Nathanson, F., *Radar Design Principles* 2nd ed., McGraw-Hill, p. 64.

3

Intercept Receivers

3.1 SURVEY OF CURRENT AND FUTURE INTERCEPT RECEIVERS

An LPIS must be designed to counter many different current intercept receivers. Any LPIS design also must anticipate future intercept threat trends. There are several trends in both airborne warning receivers and ground-based receivers that will continue well into the future. First, Chapter 3 provides an overview of the general operating techniques for six separate types of intercept receiver implementations. Second, intercept receiver performance limitations are discussed. Third, possible future threats are analyzed. Finally, two LO/LPI countermeasures for intercept receivers are suggested.

There are six general kinds of intercept receiver implementation (see Table 3.1). The first and simplest class is wideband channelized crystal video receivers (CVRs) with RF preamplifiers. The second class, which is widely used, is instantaneous frequency measuring (IFM), usually employing preamplifiers and wideband channelization. The third class is digitally controlled scanning superheterodyne/homodyne receivers, which are characterized by narrowband filters that are swept over the frequency range of interest. Fourth are completely channelized high-dynamic-range intercept receivers that are characterized by wide frequency coverage but broken into reasonably small filter bins, implemented with multiple discrete filters. These provide very high dynamic range. Fifth are transform receivers (microscan, Bragg cell, or compressive), which in essence form a filter bank from a frequency-dispersive or optical device. Last are hybrids of the above types, which allow cueing of high-resolution, high-dynamic-range analysis receivers. The six types are described in more detail in Section 3.2.

Table 3.1   Intercept Receiver Typical Performance

Parameter/type	CVR	IFM	Superhet	Channelized	Transform	Cueing
Instantaneous bandwidth	Excellent	Excellent	Poor	Good	Good	Excellent
Simultaneous signals	Poor	Poor	Poor	Good	Good	Good
Frequency resolution	Poor	Good	Excellent	Good	Good	Excellent
Dynamic range	Good	Fair	Excellent	Very good	Fair	Excellent
Usable sensitivity	Poor	Fair	Excellent	Good	Fair	Excellent

The performance trends in radar warning receivers are primarily improvements to obtain higher dynamic range and higher frequency resolution, not in the direction of improved sensitivity. As shown in this chapter, usable sensitivity is always limited by the environment and not by Boltzmann's constant. The second area where improvements will occur is signal processing. In most cases today, the signal processor in an intercept receiver is considerably less powerful than its counterpart in radars or data links. Processors also will be dynamically programmed to threat frequency bands based on netted situational awareness. The rate of the developments in signal processing is so great relative to the rest of the technology that dramatic improvements in signal processing can be expected over the coming decades. What this will require of stealth systems is spreading the emissions over wider and wider operating bands and creating more complex operating and spoofing waveforms.

3.2   Receiver Types (Similar to Schleher)[3]

From the point of view of detection performance, intercept receivers can be classed on the basis of how well matched they are to the emitter waveform. Figure 3.1 shows four intercept receiver response conditions. The crystal video receiver is very mismatched and has the poorest detection performance. For a mildly channelized re-

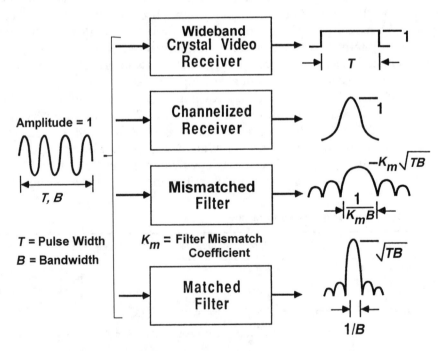

Figure 3.1   Intercept receiver relative emitter signal mismatch.

ceiver, such as a crystal video or IFM receiver with a few preselection channels, peak output amplitude is roughly comparable to the input amplitude, and its performance is primarily limited by the noise bandwidth of the individual channels (Chapter 1). The crystal video receiver has no compressive gain, for all practical purposes. Obviously, there is a continuum of mismatched filters where the mismatch may be mild—say a mismatch factor of between 1 and 10. For such filters, which are characteristic of superhets, instantaneous frequency measurement, and channelized receivers, the amplitude response is K_m times the root of the time bandwidth product. A transform receiver might have such a response. Subsequent to full matched filtering, the detection magnitude is the square root of the time-bandwidth product. On the high end of receivers, the cued receiver may be perfectly matched to the target emitter pulse and its detected energy-to-noise ratio is close to the performance of the emitter receiver. Another important limitation of all intercept receivers is their instantaneous dynamic range, which is limited by filter sidelobes and filter transient response.[29] The dynamic range limit is very important in dense signal environments. (Section 3.2 was first used in the Phase 1 LPIR review in 1977 but parallels Schleher Chapter 2 very closely.)[3,28]

3.2.1   Crystal Video Receiver

As mentioned, there are six basic types of intercept receivers. First and most common are crystal video receivers (CVRs). These are the types widely used in radar automobile speed trap detectors available in most auto parts stores. Crystal video receivers have broad channelization (e.g., X, K_u, IR, and UV) and may have preamplifiers. CVRs are very mismatched in the optimal detection sense, but they were designed for high peak power and short ranges for which they are quite adequate. Usually, CVRs have too much sensitivity and create many false alarms unless thresholds are set very high (–35 to 40 dBm).

The crystal video receiver is characterized by broad instantaneous bandwidth and low sensitivity. Its sensitivity may be between the levels of –35 to –55 dBm without a preamp and down to –85 dBm with an RF preamplifier. Front-end bandwidths are typically 200 MHz up to 4 GHz. A typical crystal video receiver block diagram is shown in Fig. 3.2. It consists of one or several omnidirectional antennas, which are frequency multiplexed into a small number of RF amplifier channels. Subsequent to this channelization, which typically might be 1 GHz in bandwidth, the output is detected and compressed in a log compressive video amplifier. Subsequent amplitude compressed outputs in each of the bands are processed with respect to pulse train shape, angle of arrival, and general frequency of operation.

The advantages of such a system are that the design is simple by military standards, relatively inexpensive, not spoofed by complex waveforms, and, in a low-density signal environment, every bit as good as the more elaborate receivers. The principal disadvantage is lack of selectivity, which is a severe problem in dense sig-

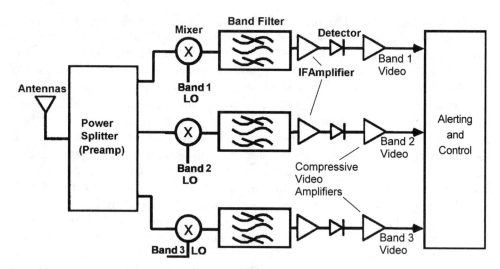

Figure 3.2 Crystal video receiver concept.

nal environments. Its sorting capability is limited, because frequency resolution and angle of arrival accuracy is usually poor. In addition, sensitivity is usually limited by RF preamplifier bandwidth. The sensitivity limitation, however, is usually not significant, because the poor selectivity prevents the handling of very many signals so that noise limited thresholds can never be approached. Table 3.2 gives typical crystal video receiver performance parameters.

3.2.2 Instantaneous Frequency Measurement

The next most common types are all channelized receivers. One form of channelized receiver is the instantaneous frequency measurement (IFM) receiver. In this scheme, a delay line is used as a frequency measuring reference and, hence, has the effect of channelizing the frequency spectrum to an accuracy that is on the order of the reciprocal of the delay line time length. The instantaneous frequency measurement intercept receiver (IFM) trades additional complexity for improved frequency accuracy and more usable sensitivity. It can have the same bandwidth coverage as the crystal video receiver and similar sensitivities. The basic block diagram for a typical instantaneous frequency measuring system is shown in Fig. 3.3.

The IFM receiver principal can be understood from Fig. 3.4. An input signal of amplitude, A, and instantaneous frequency, ω, is split and applied to both inputs of a delay line phase detector. One signal is time delayed by an amount, τ, relative to the other signal. When the signals are mixed inphase and quadrature, they create four square law outputs as shown in Fig. 3.4.

The inphase component is proportional to the square of the amplitude and cosine of the input frequency, ω, times delay, τ. Similarly, the quadrature component is pro-

Table 3.2   Typical CVR Performance

Parameter	Value
Frequency coverage	4 to 32 GHz
Bands	3
RF bandwidth	Octave
Video bandwidth	20 Mhz
Sensitivity	−35 to −55 dBm
Instantaneous dynamic range at frequency separation	45 dB @ 85 MHz
Pulse traffic capacity	<100 k pulses/sec
Frequency resolution	Octave
Adjacent pulse recovery time	300 nsec
Coverage scan rate	1 sec
Advantages	Simple, broad instant bandwidth, inexpensive, easy to use if sensitivity low
Disadvantages	Limited sensitivity, limited to low-density signal environments, very limited sorting capability

portional to the square of the amplitude and sine of the instantaneous input frequency times the delay, τ. In the simplest form of IFM receiver, these two signals, the inphase and quadrature components, are applied to the x and y axes of a linearly deflected cathode ray tube. The angular location, θ, of the output indication on the cathode ray tube corresponds to the input frequency, and the radius from the center of the cathode ray tube is proportional to two times the input amplitude squared. In es-

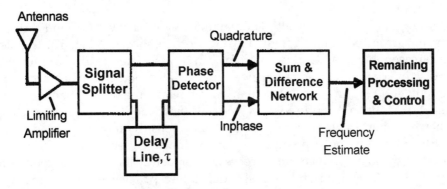

Figure 3.3   Instantaneous frequency measurement (IFM) concept.

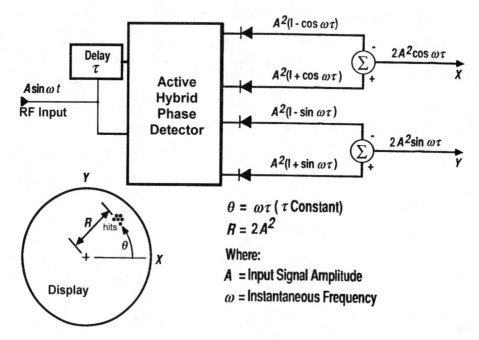

Figure 3.4 IFM processing.

sence, the delay line is being used as a wavelength or time reference. The original IFMs were amazingly simple, as the delay line was usually microstrip. For high signal-to-noise ratios, they are surprisingly good against most wide and narrow band waveforms.

Intensity is usually thresholded to prevent writing the entire display with noise. The signal above that threshold is usually log compressed by a pre-emphasis circuit so that there is a correspondence with a range reticle on the display (the value, R, in Fig. 3.4 is *not* range).

Modern digital IFMs have multiple delay lines with binary ratios and thresholded outputs that are encoded into a frequency tag word as shown in Fig. 3.5. Transient response is reasonably good, and pulse trains can be associated readily as long as there is no pulse overlap. Major advantages of an IFM are good frequency accuracy and moderate to good sorting capability in an interference-free or low-signal-density environment. IFMs are not confused by wide-bandwidth modulations such as chirp and frequency hopping. As was mentioned previously, sensitivities can be good—as much as –85 dBm with a low-noise preamp. In general, however, sensitivities are limited not by noise but by signal density and processing capacity. The IFM is a good compromise between channelized and crystal video intercept receivers for many applications.

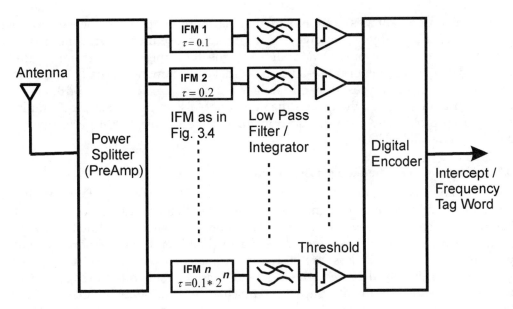

Figure 3.5   Digital multioctave IFM.

Table 3.3   Typical IFM Performance

Parameter	Value
Frequency coverage	1 to 18 GHz
Bands	5
RF bandwidth	Octave
Video bandwidth	20 Mhz
Sensitivity	–65 to –85 dBm
Instantaneous dynamic range at frequency separation	65 dB @ 275 MHz
Pulse traffic capacity	<100 k pulses/sec
Adjacent pulse recovery time	50 nsec
Frequency accuracy	0.001 of band center
Frequency resolution	Octave
Coverage scan rate	1 sec
Advantages	Good frequency resolution, good sorting in low-density signal environments, handles wideband modulation
Disadvantages	Confused by overlapping pulses, more complex and expensive than CVR

The major disadvantages of an IFM are that it is more complex and, hence, more expensive than a crystal video receiver. In addition, because there often is no high dynamic range frequency preselection ahead of the IFM, these systems are typically confused by overlapping or pulse-on-pulse signal environments. Usually, in the presence of overlapping pulses, dynamic ranges are never greater than 25 dB, and the more overlapping pulses, the greater the number of false targets. A simple example of the overlap problem is analyzed later in this chapter. Typical IFM performance is summarized in Table 3.3.

3.2.3 Scanning Superheterodyne Receivers

Two common channelization schemes are used today. The first is a digitally controlled scanning superheterodyne design. The scanning superhet has a single or small number of narrowband filters whose frequency locations are swept according to some preprogrammed strategy. These are more narrowband and have improved noise performance but, nonetheless, are still very mismatched with respect to the intercepted signal. Usually, the sweep rate is low in dense threat bands and high in bands in which there are few threats. Superhets are characterized by reasonable sensitivity but, because they are time multiplexed, they suffer from the scan-on-scan limitations discussed in Chapter 2.

The scanning superheterodyne receiver in Fig. 3.6 consists of a swept preselection filter: e.g., swept YIG resonator, a swept local oscillator, a narrowband intermediate frequency filter detector, a logarithmic amplitude compressor, and a signal encoder. All subunits are controlled by computer as to frequency, sweep rate, and threshold by a threat table. The scanning superheterodyne is very similar to a modern instrument-style spectrum analyzer. A preselection filter is required to reduce the effects of harmonic distortion, and preselector bandwidth is determined by local oscillator harmonic spacing—the higher the local oscillator frequency, the wider the preselector bandwidth can be. The controlled frequency synthesizer successively locates the nar-

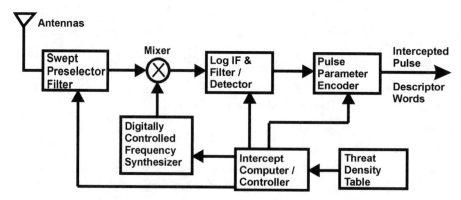

Figure 3.6   Scanning superheterodyne receiver.

rowband IF filter-detector across the band of interest by means of the mixer output signal (usually the first difference term). The bandwidth of the IF filter is selected to be approximately matched to the narrowest pulse to be detected. Subsequent to filtering, the pulses are detected and amplitude compressed. Amplitude compression is required to compress the typical 10^{12} dynamic range down to perhaps 100:1. The encoder output usually includes the pulse amplitude, pulse width, time of arrival, frequency, and special parameters—for example, CW or FM waveforms.

The dynamic range for superheterodynes is extremely high and often is limited by the threat table to a value related to lethal range. For example, the ALR-56 threshold is programmed by frequency band based on the threat effective engagement range that reduces pulse traffic in the receiver. Table 3.4 shows typical scanning superheterodyne performance.

Table 3.4   Typical Superhet Performance

Parameter	Value
Frequency coverage	0.5 to 18 GHz
Bands	6
RF bandwidth	500 MHz
Video bandwidth	10 Mhz
Sensitivity	–85 dBm
Instantaneous dynamic range at frequency separation	70 dB @ 1 GHz
Pulse traffic capacity	<1 M pulses/sec
Adjacent pulse recovery time	10 µsec
Frequency accuracy	0.001 of band center
Frequency resolution	10 MHz
Coverage scan rate	1 sec
Advantages	Very good sorting capability, high usable sensitivity, good in dense signal environments, very good frequency resolution
Disadvantages	Long scan times, narrow bandwidth limits, detection of wideband modulation, low probability of intercept for transient emitters

3.2.4   Channelized Receivers

A fourth type of receiver is the fully channelized receiver, in which multiple simultaneous frequency bands roughly matched to the target emitter spectrum are completely processed and detected. This type of processor is very complex and has been used primarily in ground-based and transport aircraft ELINT receivers up to the current time. New airborne systems, however, exploiting modern miniaturization techniques, will increasingly use fully channelized receivers.

The ultimate in both complexity and performance is the channelized intercept receiver. It consists of an antenna or set of antennas, multiplexers, and filters. RF channels are frequency multiplexed to an intermediate frequency where multiple local oscillators place 50 to 200 MHz segments of the received band into narrow filters. The outputs of these individual filters are down converted again, folded in frequency by resampling to save hardware, detected, and video filtered. Those sampled outputs are encoded in amplitude, phase, pulse width, and so on. A typical channelized system is shown in Fig. 3.7.

The major advantage of a channelized receiver is that sensitivity is usually limited only by thermal noise, because sorting problems are minimized. Channelized receivers have excellent frequency resolution, which not only provides good sorting capability but also minimizes interference in a dense signal environment. The disadvantages of such systems primarily revolve around the complexity and expense of a channelized receiver. Other limitations are the spectral purity of the multiplicity of local oscillators and signal ambiguity in a dense emitter environment caused by folding. Although a less serious problem, the narrow bandwidth may also result in some desensitization or false targets when chirp, frequency-hopping, or phase-coded signals are utilized by the target emitters (see Section 3.1.7).

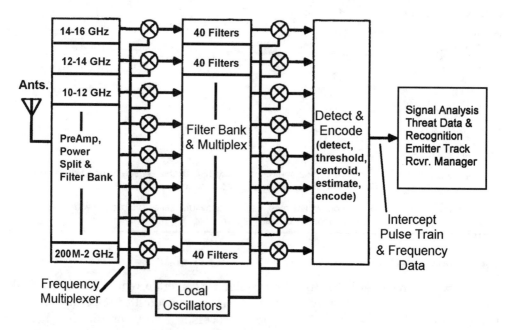

Figure 3.7 Channelized intercept receiver.

A simple example will dramatize the advantage that a channelized receiver has over the instantaneous frequency measuring receiver. Consider two signals, one with amplitude A_1 at frequency F_1, and the second with amplitude A_2 at frequency F_2 as shown in Fig. 3.8. When both of the signals overlap in an IFM, the vector difference between the two signals is what is actually measured. If the two signals have equal amplitude, then the frequency measurement is approximately the mean of the two frequencies. If one of the signals is very large relative to the other, then the difference factor is dominated by the larger, and the smaller is for all practical purposes lost. This gives rise to dynamic range limitations of approximately 20 dB or less for signals that overlap. If the signals are sufficiently separated in time, then the dynamic range can be quite large. Dynamic ranges of as much as 70 dB can be achieved if a swept preselection filter-amplifier is placed in front of an IFM. This, of course, defeats the main reason for having an IFM in the first place, and that is broadband coverage in a simple way. If one considers the same example for a channelized system, also shown in Fig. 3.8, then the two signals fall in separate filters, and the dynamic range is limited only by the filter sidelobe ratios, which can be 40 dB for adjacent filters and as much as 80 dB for distant filters. No frequency averaging or centroiding occurs to a first order when multiple signals overlap.

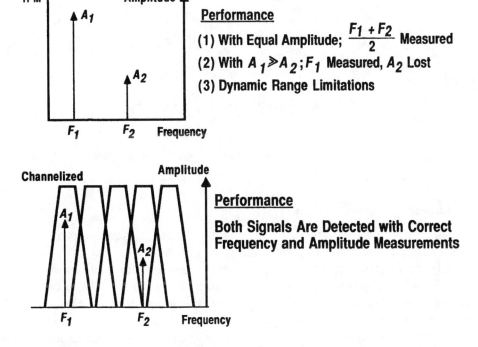

Figure 3.8 IFM versus channelized frequency measurement.

The typical leading parameters for a channelized receiver are shown in Table 3.5. Although they may not look much different from some of the other receivers, the reduced processing load allows dramatically improved usable sensitivity and much better emitter classification.

Table 3.5 Typical Channelized Receiver Performance

Parameter	Value
Frequency coverage	0.2 to 16 GHz
Bands	8
RF bandwidth	2 GHz
Video bandwidth	10 Mhz
Sensitivity	–85 dBm
Instantaneous dynamic range versus frequency separation	0 dB @ 10 MHz 50 dB @ 30 MHz 70 dB @ 200 MHz
Pulse traffic capacity	1 M pulses/sec, processor limited
Adjacent pulse recovery time	200 nsec
Frequency accuracy	3 MHz
Frequency resolution	10 MHz
Coverage scan rate	1 sec
Advantages	Sensitivity limited only by thermal and ambient noise, excellent frequency resolution, excellent sorting capability in dense signal environment
Disadvantages	Complex, expensive, narrow bandwidth limits, detection of wideband modulation

3.2.5 Transform Intercept Receivers

The fifth general type of intercept receiver is the transform receiver, which approximates a Fourier transform. They can be implemented in a variety of ways such as Bragg cells, microscan, and other compressive devices. Their scan times are designed to be on the order of the pulse lengths from the target emitter class. In essence, the microscan again is a channelized receiver. It suffers the same difficulties that an IFM receiver suffers, which is a limitation in instantaneous dynamic range. Instantaneous dynamic range limitations are a serious problem in dense emitter environment. Multiple signals simultaneously received cause harmonic distortion in the inevitable nonlinearities that exist in all receivers. These harmonics cause the signals to be misclassified in frequency, direction, amplitude, number of pulses present, and so on. Furthermore, nonlinearities cause small signal suppression and, hence, in an environment in which LPI systems exist with conventional systems, the conventional emitter often completely masks the LPI systems.

The microscan compressive receiver is sometimes called a *chirp transform receiver*. It consists of an antenna set, frequency translation to a convenient band, and multiplication by a swept local oscillator so that any constant frequency signal will be chirped. The chirp output then goes through the chirp matched filter or dispersive delay line. The output is amplitude weighted to improve chirp filter sidelobes, and the signal is encoded with respect to amplitude, frequency, and pulse width. The chirp sweep is usually repeated at a period equivalent to the smallest expected emitter transmitted pulsewidth. Such a system is shown in Fig. 3.9.

The advantages of such a scheme are those obtained for any chirp transform system. The chirp transform has good frequency resolution, a wide instantaneous frequency coverage, and sensitivity that is comparable to that of a channelized receiver. The major disadvantages are that they are complex and expensive. The subsequent signal processing is usually difficult, because the output bandwidth from the transformer is comparable to the total frequency coverage. In addition, dynamic range is limited by chirp linearity and sidelobes from the chirp filter. It does not seem to matter whether these filters are implemented digitally or by analog means, such as dispersive surface acoustic wave delay lines. The current state of the art appears to achieve filter near sidelobes that are on the order of –30 dB down, and so the dynamic range is not substantially better than that in a dense signal environment.

The idea of a microscan compressive receiver is shown in Fig. 3.10. The waveforms at the numbered points are shown below the block diagram in Fig. 3.10. The chirp local oscillator is multiplied by the intercepted signal so that each constant frequency input is converted into a chirp with the same frequency-time slope, but with an offset

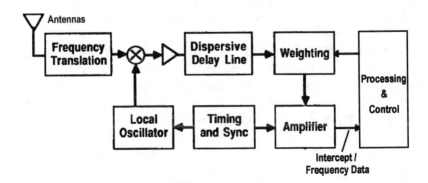

Figure 3.9 Microscan transform receiver.

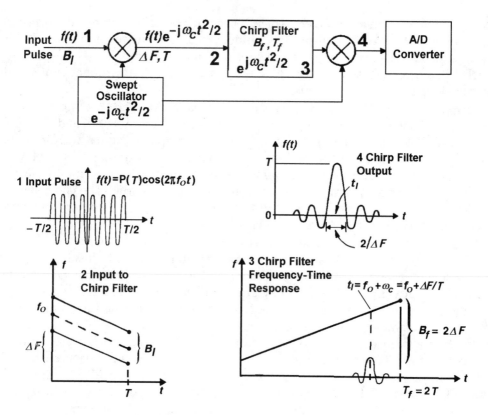

Figure 3.10 Microscan processing concept (similar to Schleher[3]).

frequency proportional to the original intercepted input signal. The multiplied output is imposed on a chirp filter or dispersive delay line with impulse response matched to the chirp slope. The approximately linear correspondence between frequency and time in the dispersive device results in the original constant frequency components being translated to pulses with time displacements proportional to the frequency of the original intercepted inputs.

The configuration shown can be analyzed from the diagram of Fig. 3.10 by referring to the figure point by point. Start by considering the input waveform, $f(t)$, modulated by a unity window function whose duration is the same as the premultiplying chirp (point 1). For a signal that is almost constant with frequency over the pulse duration, T, the signal can be represented as a CW signal modulated by a rectangular window function that is T seconds long. The incoming signal, $f(t)$, then is multiplied by the chirp waveform, $S(t) = \exp(-j\omega_c t^2/2)$, where ω_c is the slope in radians per second for the linear-FM swept local oscillator signal (point 2). This signal is applied to a dispersive delay line or surface acoustic wave (SAW) filter whose slope is matched to the premultiplying chirp but with opposite sign. The impulse response of this filter is given by the function $\exp(j\omega_c t^2/2)$ (point 3). The output of the chirp filter is the convolution of the multiplying waveform, $f(t) \exp(-j\omega_c t^2/2)$ with the impulse response of the chirp filter.

The Fourier transform of the product of two time functions is found by convolving the individual function Fourier transforms. Thus, the compressive receiver output is an approximate Fourier transform of the input signal. The output, $g(t)$, is a video waveform whose amplitude is the shape of the Fourier spectrum of $f(t)$ except for edge effects with scale determined by the chirp slope, ω_c, and referenced in time to the premultiplying chirp waveform. As shown in Fig. 3.10, the resulting time output is multiplied by a phase term $\exp(j\omega_c t^2/2)$, which can be eliminated through post multiplying by the waveform $\exp(-j\omega_c t^2/2)$ to obtain the Fourier transform time waveform, $g(t)$, which is proportional to the Fourier transform of the input signal as an approximately linear function of time referenced to the premultiplying chirp waveform and proportional to the chirp slope (point 4). The resulting approximate sinc(x)-shaped video waveform has a peak at t_l equal to $f_o + \Delta F/T$, where ΔF is the bandwidth of the premultiplying chirp with an output pulse width on the order of $1/\Delta F$ at peak normalized amplitude of approximately 0.64. The output of the compressive filter is the time distribution proportional to the Fourier transform of the CW signal, which is an impulse, and the sinc(x) transform of the rectangular window function. The time position of the video waveform within the overall sweep width gives the center frequency of the emitter relative to the total band, B_I. The advantage to the configuration shown is that the premultiplying chirp can be used to chirp modulate the signal *and* translate its frequency to a range where a dispersive filter can be implemented more easily.

As described later in this chapter, both emitter frequency and angle of arrival are sorting parameters. A Hilbert transforming A/D converter is used at the output of the post-multiplier to preserve phase. The configuration that allows direction finding uses two compressive receiver channels whose outputs are applied to a phase detector (similar to that shown in Section 3.3.4.2). Each compressive receiver is connected to a spatially separated antenna connected in an interferometer configuration. The phase difference between the signals is related to the angle of the instant phase front of the emitter allowing angle arrival (AOA) to be calculated. The phase information is retained in the compressive receiver before detection due to its wide sense linear nature. Employing a wideband phase detector, and comparing the two compressive receiver channels before detection, allows the angle of arrival to be measured, while the frequency can be measured using an envelope detector on either channel. This system must maintain a tight phase match between both channels. A match of a few degrees is typical.

To implement the compressive receiver algorithm depicted in Fig. 3.10 with practical chirps, finite limits must be imposed on time and frequency extent. For example, assume that a 1-GHz analysis bandwidth is required and that the shortest pulse to be channelized is 0.3 μsec. One way to design the compressive receiver would be to use a premultiplying sweep that covered the 1-GHz operating band in 0.3 μsec, or a chirp rate of 3.3 GHz/μsec. Because the input signal can be located anywhere in the coverage band, the frequency range at the chirp filter input must span a little over twice the bandwidth, or 2 GHz in the example. Because the magnitude of the chirp rate of the impulse response of the filter is the same as the multiplying chirp, the filter delay must be twice the period of the multiplying chirp, or 0.6 μsec. Also, the ability to separate emitters of equal amplitude is on the order of $1/T$, or about 3 MHz for the previous example.

The configuration shown has been used in ELINT receivers that require the detection of many short pulse emitters. However, it also has several disadvantages. First, the bandwidth of the chirp filter must be twice the bandwidth of the analysis band, which results in a time bandwidth product that is four times that of the premultiplying chirp. Because the maximum bandwidths of practical SAW filters are approximately 2 GHz, this limits the bandwidth of an individual practical compressive receiver to approximately 1 GHz (obviously, multiple parallel channels can be used).

Second, the first sidelobe of –13 dB and the slow subsequent rolloff, which results from an unweighted rectangular window, severely limits the resolution of the emitters. The solution is to use a window function such as the Hamming window function (see Chapter 6), which provides a first sidelobe of –42.8 dB but spreads the mainlobe by a factor of 1.5. One solution to amplitude weighting is to use a SAW dispersive filter with the appropriate weighting to generate the premultiplying chirp by impulse excitation and a linear multiplier instead of a hard limited mixer.

A third limitation is the necessity of making the convolving filter's dispersive delay twice as long as the premultiplying chirp time duration, and successive premultiplying chirps must be separated by $2T$. This causes each compressive receiver channel to operate at a 50 percent duty ratio, and hence a 50 percent probability of intercept if no provisions are made. Thus, interlaced multiplexed chirp channels often are used.

Another type of transform receiver is based on electro-optical techniques. It is usually called a *Bragg cell receiver*. Such a configuration is shown in Fig. 3.11. The time waveform input modulates a collimated light beam, which then goes through a lens

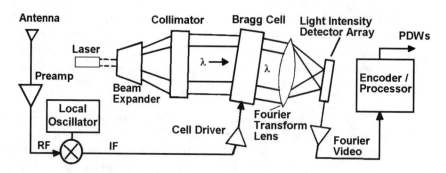

Figure 3.11 Brag cell transform receiver.

acting as a spatial Fourier transformer of the input spatial modulation (similar to early synthetic aperture radar film processors). The lens output falls on a detector array, which is then sequentially readout providing a time waveform proportional to the Fourier transform of the input waveform. The bandwidth of such systems can be quite large, but they suffer from the same problems that doomed early SAR film processors: i.e., extremely poor dynamic range that overwhelms performance in a dense target environment. Near sidelobe dynamic range is typically no better than 20 to 30 dB. Angle measurement is also usually very limited. Low-cost, very broadband coverage, limited only by the front end antennas and preamps, is the Bragg cell principal advantage. The remaining processing after digitization is essentially the same as all the other channelized receivers that are discussed in later sections.

Laser or monochromatically filtered light power is usually surprisingly high as a result of low overall optical efficiency. Optical modulating cells of the general electro/ magneto acoustic type have been in use since the early 1930s televisions, and some of the most efficient are the most dangerous. (One uses nitrobenzene with a few thousand volt modulation!) Image orthocons have also been used in related configurations to provide transform receivers dating back to the 1960s. The potential for this class of receiver grows as the worldwide telecommunications net expansion in fiber optics drives the underlying device technologies.

The typical transform receiver leading parameters are tabulated in Table 3.6. The principal feature of this class of receiver is virtually unlimited pulse traffic performance as available processing power grows.

3.2.6 Hybrid or Cueing Receivers

The *hybrid* or *cueing receiver* is a combination of two or more of the previous receiver types. The first receiver type is used to provide alerting, coarse angle of arrival, and coarse frequency estimates that are used to cue a second receiver with higher resolution and dynamic range for tracking, sorting, and classification. The coarse receiver can be crystal video, IFM, channelized, or compressive. Wideband delay lines provide enough signal storage so that the coarse receiver can command the high-resolution receiver to the correct frequency band and angular quadrant for high-resolution measurement. A typical block diagram is shown in Fig. 3.12. The hybrid receiver is usually quite complex but, with modern miniaturized microwave and IF components, it is finding its way into aircraft as well as ground installations. It is capable of wide coverage while still providing high enough resolution to simplify processing, which allows higher sensitivity.

More details of a typical single channel in Fig. 3.12 are shown in Fig. 3.13. An input RF pulse is simultaneously applied to a wideband channelizer or filter bank and a wideband delay line. The filter bank might consist of 50 filters of 20 MHz each for 1 GHz total coverage. The filter bank outputs are detected and centroided. The resultant goes to cueing logic, which commands one or more fast-slewing local oscillators

Table 3.6   Typical Transform Receiver Performance

Parameter	Value
Frequency coverage	1 to 10 GHz
Bands	4
RF bandwidth	1 GHz
Video bandwidth	2 GHz
Sensitivity	−75 dBm
Instantaneous dynamic range versus frequency separation	0 dB @ 3 MHz 30 dB @ 30 MHz 50 dB @ 300 MHz
Pulse traffic capacity	150 M pulses/sec, processor limited
Adjacent pulse recovery time	200 nsec
Frequency accuracy	3 MHz
Frequency resolution	3 MHz
Coverage scan rate	0.3 μsec
Advantages	Good frequency resolution, wide instantaneous bandwidth, good sensitivity
Disadvantages	Complex, expensive, poor dynamic range limits dense, signal environment performance, difficult signal processing

(FSLOs). Note the receiver timing in Fig. 3.13 shows the centroided channelizer detection output delayed relative to the input pulse. The FSLO requires some additional time to slew and settle to the commanded frequency with adequate spectral purity to allow fine measurements. All of that time must be less than the wideband delay. The slower the impulse response of the coarse channelizer and FSLO, the more the difficulty in providing a high-fidelity wideband delay. There are inevitable compromises in delay line, filter skirt selectivity, and FSLO noise performance that limit achievable sensitivity. Hybrid systems are better by 10 to 20 dB in many scenarios, but not as much better as one might expect on the surface. When the delayed pulse is applied to the parameter measurement unit, additional filtering and circuitry estimates fine frequency (FF), signal phase (φ), amplitude (AMP), time of arrival (TOA), and pulse width (PW).

One major advantage of this configuration is that the higher resolution analysis band can more easily separate overlapping pulse trains and similar signals, because the probability of many signals over a narrower angle and frequency space is greatly reduced. Reduced signal overlap allows higher sensitivity to be used, because the probability of processor overload is reduced. The hybrid receiver approach is not

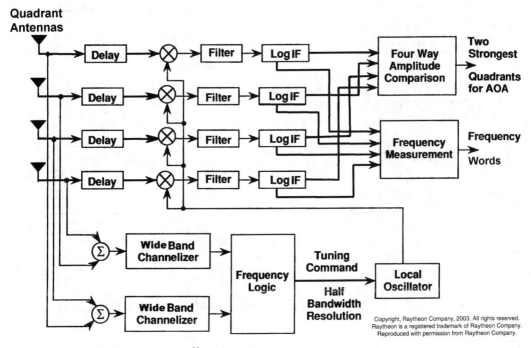

Figure 3.12 Hybrid or cueing receiver.[30]

new, but the availability of very broadband delay lines has made them much more practical. For example, delay lines based on amplifying optical fibers allow gigahertz bandwidths for microseconds of high-fidelity delay in very small, rugged packages with development and improvement driven by the internet juggernaut. There are many variations on this theme, but the stealth problem doesn't change much, because initial detection and alerting depends on the coarse receiver performance/sensitivity.

The disadvantages of the hybrid class of receiver (see Table 3.7) are complexity, cost, and sensitivity to overload if all the signals come from the same quadrant and frequency bands (which they often do in wartime).

3.2.7 Intercept Receiver Processing

Most modern interceptors use similar processing for the pulse descriptor words that they receive from the front end. Figure 3.14 shows most of the elements of intercept receiver processing. The processing falls into the following three categories:

1. Interface and control of the receiver assets

2. Periodic calibration and test

3. Emitter classification

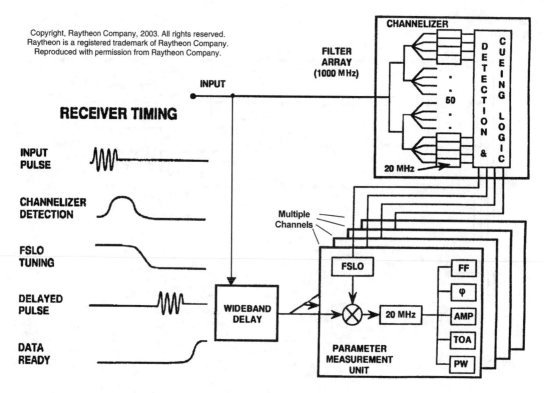

Figure 3.13 Typical hybrid receiver single channel.[30]

Pulse descriptor word processing consists of angle of arrival (AOA), hit counting, and emitter RF frequency prebinning as well as PRI correlation and pulse train deinterleaving. High duty ratio and high/medium PRF pulse doppler excision are required to prevent processing overload. Emitter classification is performed by emitter analysis in conjunction with the predefined threat data files and the currently tracked, active emitter files. Emitter analysis parameters are RF, PRI, PRI type, pulse width, presence of chirp, presence of biphase modulation, master clock period, RF agility, scan type, polarization type, scan period, and mainlobe beamwidth. Each of these must be measured or inferred so as to classify an emitter.[31]

The number of unique emitters that must be classified is typically between 2000 and 3000. Depending on the altitude of the intercept receiver, the number of emitters visible will range from a few tens to 2000, and the total pulse traffic can be up to a few tens of million pulses per second. As the number (N) of visible emitters grows, the sorting problem grows at least proportionately by $N \log_2 N$ and often as a power of N. The potential of processor overload usually results in the detection thresholds being set significantly higher than the best achievable by the front end hardware. In addi-

Table 3.7   Typical Hybrid Receiver Performance

Parameter	Value
Frequency coverage	0.2–35 GHz
Bands	18
RF bandwidth	2 GHz
Video bandwidth	10 MHz
Sensitivity	–85 dBm
Instantaneous dynamic range versus frequency separation	0 dB @ 10 MHz 50 dB @ 30 MHz 70 dB @ 200 MHz
Pulse traffic capacity	>1 M pulses/sec, processor limited
Adjacent pulse recovery time	200 nsec
Frequency accuracy	3 MHz
Frequency resolution	10 MHz
Coverage scan rate	1 sec
Advantages	Best dynamic range on a small emitter sub-set, good in dense signal environment, good for broadband emitters
Disadvantages	Complex, expensive, longer search time

tion, receiver self-noise resulting from local oscillator (LO) AM, FM, and "birdies," power supply ripple, and microphonics often will consume processor resources un-less thresholds are set significantly above thermal noise.

The concept for the first part of the processing, AOA, and RF prebinning, is shown in Fig. 3.15. Pulse descriptor words are associated by angle of arrival, RF frequency, and the number of hits per unit time. An old versus new comparison is made to asso-ciate the new measurements with prior intercepts, e.g., one dwell to the next. New emitters are declared, and old emitters are continuously tracked. If the number of hits per unit time is higher than a threshold, the intercept is declared as a pulse Doppler emitter and is excised from the PRI deinterleave process.

The concept for PRI correlation and deinterleave is shown in Fig. 3.16. An AOA and RF bin may contain multiple interleaved pulse trains, which must be separated so as to measure the PRF of each emitter. Starting with an arbitrary pulse, the proces-sor forms the hypothesis that there is only one almost regular pulse train, takes the difference between that pulse and the next, and uses that interval to test the differ-ence between the second pulse and the third pulse. Should that interval fail to find

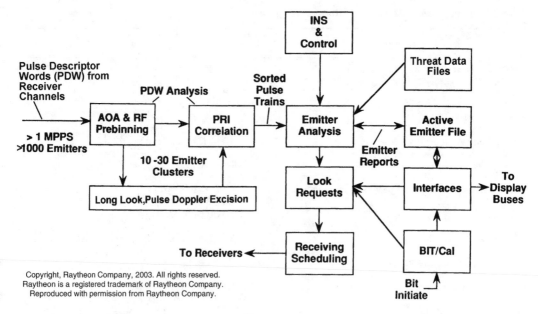

Figure 3.14   Typical intercept receiver processing.[30]

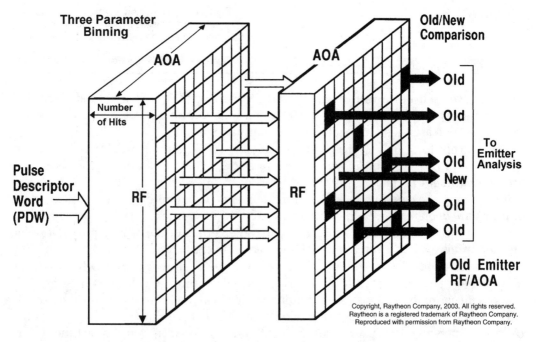

Figure 3.15   Angle of arrival, RF prebinning, pulse doppler excision.[30]

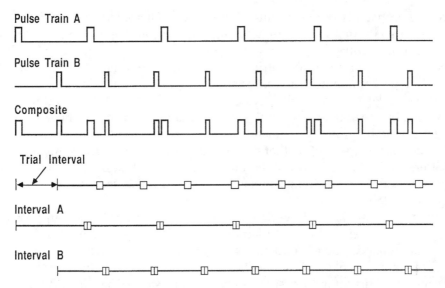

Figure 3.16   PRI correlation/deinterleave.

coincidence, which it does in the example, then a second hypothesis is formed using the time interval between the first and third pulses, and that interval is tested for coincidence on the rest of the pulse train using a tolerance window consistent with the time of arrival measurement accuracy. For the example shown, the hypothesis is correct and is declared as Interval A in Fig. 3.16. All of the pulse descriptor words coincident with Interval A are tagged. The interval between the second and fourth pulse is used to form a new hypothesis, and that interval is tested for coincidence. In the example shown, the test is successful and Interval B is declared. For the example, all of the pulses are accounted for with just two pulse trains, and the deinterleave is complete. Should many pulses still be unaccounted for, the process is continued. There are obvious details, such as missing pulses, noise spikes, and pulse train coincidence that must be accounted for in an operational PRI correlation algorithm.

These PRI intervals become part of the parameter set used to classify the emitter in the emitter analysis processing that follows. The emitter analysis consists of taking each emitter vector whose dimensions are all the measured parameters and mapping it into the multidimensional classification space of all the emitters in the threat data files. A probability of the emitter classification is calculated. If the correlation is not high enough, then an unknown is declared, and additional interceptor time is allocated, or the emitter is handed off to another sensor.

Even in the absence of measurement errors, emitters cannot always be classified unambiguously. For example, Fig. 3.17 shows one way that ideal classification probabilities are calculated. The figure shows an example of the overlap in a two-dimensional measurement space with dimensions of RF and PRI between two hypothetical emitter entries. The probability of overall unambiguous classification is the fraction of the total parameter space that is unambiguous for a given parameter family *(Pe)* times the number of emitter entries in that family *(Ne)* summed over all parameter families normalized by the total number of entries. For the example of Fig. 3.17, there are 10 parameter families containing 161 emitter entries that have ambiguities and 811 entries that have none. *Pe* for each family is calculated by taking the total unambiguous area and dividing by the total area for that parameter family. For the example, the total unambiguous probability is approximately 0.96. There are obvious weaknesses with this method such as the assumption that the measurement space is uniformly occupied by the emitters, and so on. Superimposed on this ideal calculation are the additional ambiguities created by measurement inaccuracies, self-noise, external noise, sidelobes and signal overlap.

An example of one kind of overlap problem is given in Fig. 3.18. Suppose a high-duty-ratio LPIS waveform (B) and a larger amplitude conventional pulse train (A) occupy the same band. There will be a high probability that A will be embedded in B with the operating level set by A. The beat frequency PRI will give rise to a third declared pulse train resulting from small signal suppression effects. In addition, the beat RF frequency or its harmonics may result in a third RF frequency being declared. There are several foreign intercept systems that respond exactly as described in Fig. 3.18. The only way around this embedment problem is for the intercept system to have a very large linear dynamic range or much more processing power to sort out the spurious signals. This is achievable but expensive.

The effect of pulse-on-pulse (POP) interference is shown in phasor diagram of Fig. 3.19. While part of the pulses overlap, the larger signal has the smaller vector superimposed on it, which causes an apparent phase and amplitude ripple. Depending on the relative amplitudes, the ripple can be large enough to cause a false AOA measurement or, if hard limiting is used, false intercepts. The typical LPIS signal by de-

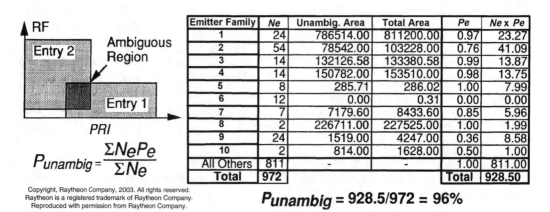

Emitter Family	Ne	Unambig. Area	Total Area	Pe	Ne x Pe
1	24	786514.00	811200.00	0.97	23.27
2	54	78542.00	103228.00	0.76	41.09
3	14	132126.58	133380.58	0.99	13.87
4	14	150782.00	153510.00	0.98	13.75
5	8	285.71	286.02	1.00	7.99
6	12	0.00	0.31	0.00	0.00
7	7	7179.60	8433.60	0.85	5.96
8	2	226711.00	227525.00	1.00	1.99
9	24	1519.00	4247.00	0.36	8.58
10	2	814.00	1628.00	0.50	1.00
All Others	811	-	-	1.00	811.00
Total	972			Total	928.50

$$P_{unambig} = \frac{\Sigma Ne Pe}{\Sigma Ne}$$

$$P_{unambig} = 928.5/972 = 96\%$$

Figure 3.17 Example classification probability calculation.[30]

Gain Controlled/Amplitude Limited Pulse Trains A > B

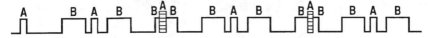

- Beat Frequency PRI Declared as 3rd Pulse Train
- Beat Frequency RF Declared as 3rd RF Frequency
- Non Linearities May Cause Additional Spurious Signals

∴ LPI Forces Very Large Dynamic Ranges & Higher Processing Load

Figure 3.18 LPI waveform embedment.

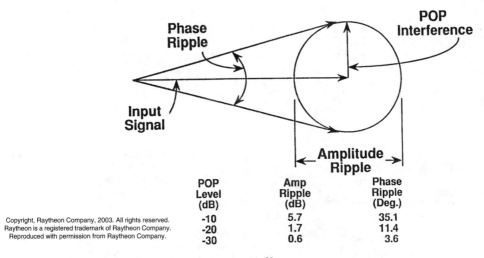

POP Level (dB)	Amp Ripple (dB)	Phase Ripple (Deg.)
-10	5.7	35.1
-20	1.7	11.4
-30	0.6	3.6

Figure 3.19 Pulse on pulse amplitude and phase ripple.[30]

sign will be –10 to –20 dB below most intercepts and will normally have a much larger duty ratio than most other signals in that band.

A measure of the probability that there will be no overlap is given in Fig. 3.20, assuming that the total pulse traffic is a stationary random process with parameter ζ. The parameter, ζ, is the product of the total pulse traffic times the sum of the average intercepted pulsewidth plus the reset or pulse recovery time, T_{RST}. For example, with pulse traffic averaging 1 million pulses per second (MPPS), average pulse width of 1.5 μsec and reset time of 0.5 μsec, ζ has a value of 2. Even with narrow pulses, the probability of overlap is very high with the current per band pulse traffic of 1 MPPS in populated regions. Higher levels of channelizing can help this problem but doesn't eliminate it because of the "rabbit ear" effect shown in Fig. 3.21. For example, consider two pulses separated in frequency and differing in amplitude by 60 dB presented to an

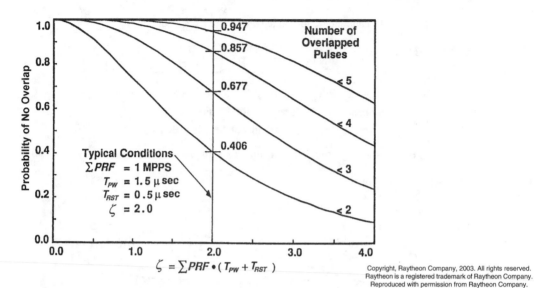

Figure 3.20   Probability of no pulse overlap example.[30]

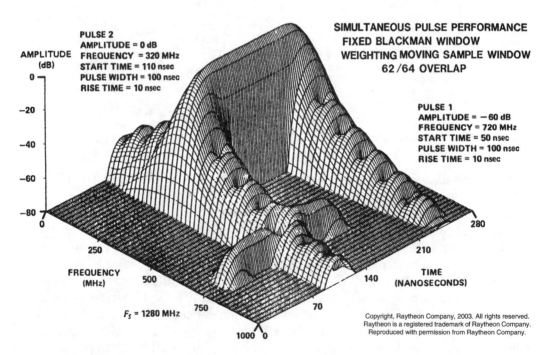

Figure 3.21   Channelizer pulse overlap example.[30]

intercept receiver channel bank. The figure shows an amplitude-frequency-time plot of a high-performance digital finite impulse response filterbank over a 1.28 GHz sampling band. The weighting window selected for these filters has a very good transient response. The two pulses are 400 MHz apart, but the rise and fall time on the larger pulse produces a transient that rings filters across the band, masking the smaller pulse. Note that the leading edge rabbit ear of the large pulse cuts right through the middle of the smaller pulse and actually causes destructive interference at its center. Most modern channelizer filterbanks have gating which eliminates the "rabbit ears" that would otherwise create many false intercepts. Fortunately for the stealth emitter, that gating also eliminates or corrupts its intercepted signal (the smaller one) and makes recognition very difficult.

Another important aspect of intercept receiver processing is the software and timing. The processor executive usually has functions that are timing driven and repeat on a regular cycle. It also has functions that are slower, have irregular timing, and are interrupt driven, such as motion sensing, blanking from other sensors, operator commands, refresh requests from the display subsystem, calibration, built-in test, diagnostics, and so forth. Most of these events can wait for short periods.

The real time control and processing collapses if not serviced during a narrow timing window. The real-time tasks are shown in Fig. 3.22. The real-time tasks are executed with a timing regimen as shown in Fig. 3.23. A master clock generates regular interrupts that schedule the specific set of bands, time intervals, antenna beams, front end gains, and so forth. Those scheduled events are executed in the front-end hardware during the next timing interval. When the gathering phase for a specific collection event is complete, those pulse descriptor words are transferred to signal analysis where pulse deinterleaving, fine frequency, AOA, PRI, and so on are estimated. Those results, when complete, are passed to the emitter manager for identification, correlation, and tracking. As will be seen later, these latter processes get progressively slower

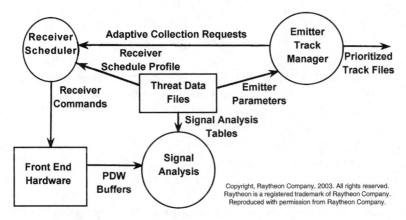

Figure 3.22 Interceptor processor key software functions.[30]

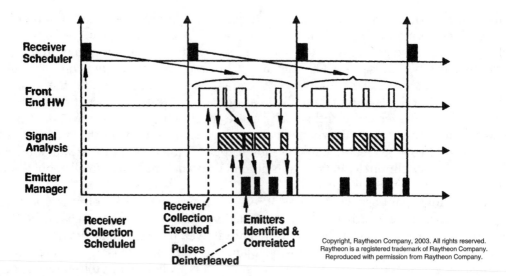

Figure 3.23   Simplified processor software time line.[30]

if pulse density gets higher. It is often to the stealth emitter's advantage to spoof by transmitting large numbers of pulses resembling the adversary's systems.

Figure 3.23 does not show the real issue with respect to processor loading. The time required for signal analysis is strongly related to the number of pulse trains in each frequency band and angle cell. Because dwell times of both radars and data links are typically measured in milliseconds, the number of pulses processed per millisecond is a key design parameter. The processing of intercepted signals is dominated by sorting functions. The best available sorting algorithms have execution times that are proportional to $N \cdot \log_2(N)$, where N is the number of items to be sorted. Usually, the processing time is proportional to $c_c \cdot N \cdot \log_2(N) + c_{oh}$ of the total number of pulses per millisecond (ΣPRF) in each band/cell. Although some operations are dependent on the number of emitters (not the number of pulses), the above relationship provides a reasonable approximation to the required processing time. The coefficient, c_c, is hardware and software state-of-the-art dependent and currently is about 225 processor operations per pulse. In addition, there is some fixed overhead, c_{oh}, associated system interfaces, self-test, and other housekeeping, which is currently about $2.9 \cdot 10^6$ operations per second. (More about this in Chapter 4.)

Referring to the processor executive of Fig. 3.22 and the time line of Fig. 3.23, typical software subprograms in an intercept receiver processor are given in Table 3.8. The main subprograms, their complexity in equivalent assembly language orders, and their execution times are as shown in the table for a 36 million operations per second (MOP) processor designed for real-time embedded applications (an Intel 80960, often used in laser printers).

Table 3.8   Subprogram Execution Time

Subprogram	Equivalent assembly language orders	Execution time (msec)
Executive and waiting	100,000	40
Receiver scheduling	5,000	4
Interface servicing	3,000	2
Display update	1,000	4
Environment snap shot	30,000	10
Signal analysis tables	1,000	2
PRF deinterleave	5,000	$2.6\ KN + 5$
Generate pulse descriptor	30,000	$0.7\ KN + 6$
Emitter track files	5,000	$0.5\ KN + 4$
Emitter ID and correlation	50,000	$2.4\ KN + 82$
Totals	230,000	$6.2\ KN + 82$

In Table 3.8, $KN = N \cdot \log_2(N)$, the best available sorting time for a random data set. If the number of pulse trains to be sorted is 1, then the execution time for a 36-MOP processor is 82 msec. If the number of pulse trains of 1 kHz PRF to be sorted in 1 msec is 30, then KN is 147.21, and total execution time is a little under 995 msec. If the number of pulse trains to be sorted is 250 of 1 kHz each, then KN is 1991.45, and total execution time is a little under 12.43 sec using the values of Table 3.8. If the required update time is once per second, then required processor throughput is a strong function of front end selectivity as shown in Table 3.9. Because deployed processors always seem to require more than ideal amounts of processing, the throughput required is very optimistic.

Table 3.9   Example Processor Throughput Trade-Offs

Receiver capability AOA and freq. res. (18 kft AGL, –60 dBm)	No. pulse trains in band, 90% probability	Required processing speed (MOPS)
3° and 10 MHz	1	2.9
22.5° and 50 MHz	30	36
90° and 1 GHz	250	~450

3.3 INTERCEPTOR MEASUREMENT ACCURACY

In the case of all the previously described receivers, the physics that we all know still applies. The accuracy of measurement, and hence ultimately the success of classification and countermeasures, depends on the of signal to noise and interference ratio. The characterization of the measurement performance of intercept receivers is discussed in the sections that follow. Swerling[8] and others have calculated the Cramer-Rao lower bounds for regular, unbiased estimation cases for all signal-to-noise ratios (SNRs) that, with minor definition differences, can be applied to intercept receivers. These lower bound results for the case of white noise and intercepted signal matched filter are provided without proof, which is available in many radar texts including the references. Because of filter mismatching as shown in Fig. 3.1 and hardware limitations, these lower bounds are never achieved (are usually off by at least 2:1) but are useful in assessing LPIS versus interceptor performance.

3.3.1 Frequency Measurement

The measurement of RF frequency is the first parameter in most emitter analyses. The frequency discriminant is obtained by comparing the signal to a reference such as a delay line (IFM), master oscillator, multiple overlapping filters (superhet), or a filter-bank (channelized). All of these methods depend on the characterization and accuracy of the hardware as well as the SNR. Historically, the SNRs were so high that only hardware limitations limited performance. With the advent of stealth radars and data links, the SNR limits interceptor performance. In fact it is a design objective of LPIS to maximize signal uncertainty, thus limiting classification and tracking.

The rms error lower bound for a frequency estimate, assuming a rectangular intercepted pulse with a matched filter in the presence of white noise, is given in Equation (3.1) below.

$$\sigma_f \geq \frac{\sqrt{3}}{\pi \cdot T_{rb} \cdot \sqrt{2 \cdot SNR}} \text{ Hz} \tag{3.1}$$

where T_{rb} is the pulse width and SNR is the signal-to-noise ratio. For example, for the noise-limited performance only, let $T_{rb} = 1$ μsec and SNR = 0.1. Then,

$$\sigma_f = 1.233 \text{ MHz}$$

This is clearly adequate to sort fixed-frequency pulse emitters on the basis of frequency. Suppose the emitter used a pulse compression waveform with an equivalent pulse width $T_{rb} = 10$ nsec, and the interceptor was again matched to the emitter spectrum. Also assume the same SNR and operating frequency as in the previous example. Then,

$$\sigma_f = 123 \text{ MHz}$$

This is not necessarily adequate to sort emitters, let alone classify them depending on the receiver type as was shown in earlier sections.

Furthermore, a more realistic lower bound for rms frequency measurement includes filter mismatch and multipath scintillation. The rms error is given in Equation (3.2) below.[9]

$$\sigma_f \geq \frac{c_m \cdot B}{\sqrt{2 \cdot SNR}} \text{ Hz} \tag{3.2}$$

where c_m is the multipath scintillation coefficient, typically 2, and B is the detection bandwidth of the intercept receiver. For example, let $B = 10$ MHz, and $SNR = 0.1$, $c_m = 2$. Then,

$$\sigma_f = 31.6 \text{ MHz}$$

This is greater than the detection bandwidth; therefore, multiple 10 MHz would have detections, and some centroiding would be required. Also note that an SNR of 1 would give the same result. This is more than adequate to begin tracking an emitter but may not be adequate to classify most emitters, depending on the operating band.

Similarly, the rms error lower bound for phase measurement is given in Equation (3.3) below.[10]

$$\sigma_\varphi \geq \frac{c_\varphi}{\sqrt{2 \cdot SNR}} \text{ radians} \tag{3.3}$$

where c_φ is a phase detector coefficient, usually between $2^{0.5}$ to 2.

3.3.2 Pulse Amplitude and Width Measurement

The rms error lower bound for an amplitude estimate, assuming a rectangular intercepted pulse with a matched filter in the presence of white noise, is given in Equation (3.4) below.

$$\sigma_A \geq \frac{A}{\sqrt{2 \cdot SNR}} \tag{3.4}$$

where A = the amplitude to be measured.

For example, let $A = 1$ and $SNR = 0.1$. Then,

$$\sigma_A = 2.24, \text{ or a 7-dB error}$$

This is clearly not adequate to allow the use of amplitude only to estimate range (as some systems do). This still might be adequate to cue a higher resolution sensor to this emitter.

The rms error lower bound for a pulse width estimate, assuming a rectangular intercepted pulse with a matched filter in the presence of white noise, is given in Equation (3.5) below.

$$\sigma_{PW} \geq \frac{T_{rb} \cdot \sqrt{3}}{\pi \cdot \sqrt{SNR}} \text{ Hz} \qquad (3.5)$$

A more realistic estimate of the pulse width error, taking into account bandwidth mismatch and multipath, is given in Equation (3.6) below.

$$\sigma_{PW} \geq \frac{c_m}{B \cdot \sqrt{2 \cdot SNR}} \qquad (3.6)$$

For example, let B = 10 MHz, c_m = 2, and SNR = 1. Then,

$$\sigma_{PW} = 141 \text{ nsec}$$

This is probably more than adequate for classification sorting.

3.3.3   Time of Arrival and PRI Measurement

The rms error lower bound for a time-of-arrival estimate, assuming a rectangular intercepted pulse with a matched filter in the presence of white noise, is given in Equation (3.7) below. This equation also assumes that the measuring timebase or clock is orders of magnitude more accurate than the measured interval (often but not always true!).

$$\sigma_{TOA} \geq \frac{T_{rb} \cdot \sqrt{3}}{\pi \cdot \sqrt{2 \cdot SNR}} \text{ Hz} \qquad (3.7)$$

A more realistic estimate of the pulse width error, taking into account bandwidth mismatch and multipath, is given in Equation (3.8) below.

$$\sigma_{TOA} \geq \frac{c_m}{2 \cdot B \cdot \sqrt{SNR}} \qquad (3.8)$$

For example, let B = 10 MHz, c_m = 2, and SNR = 1. Then,

$$\sigma_{TOA} = 100 \text{ nsec}$$

This is more than adequate in almost all cases. *SNR* must be 0.001 before classification is degraded enough to prevent use of this parameter. Similarly, the rms error for PRI interval measurement is given in Equation (3.9) below.

$$\sigma_{PRI} \geq \frac{c_m}{2 \cdot B \cdot \sqrt{SNR \cdot n}} \tag{3.9}$$

where n is the number of PRI intervals used to make the estimate.

3.3.4 Angle of Arrival Measurement

The biggest exploitable weakness for stealth emitters in intercept receivers is *angle of arrival (AOA)* measurement. This weakness also cascades into range estimation. As shown in the earlier measurement sections, there is a signal-to-noise-limited lower bound on measurement accuracy. If one assumes the best case, which is a matched filter in the presence of white noise, then the lower bound rms error will be as given in Equation (3.10) below.[4]

$$\sigma_\theta \geq \frac{\lambda \cdot \sqrt{3}}{\pi \cdot \cos\theta \cdot \ell \cdot \sqrt{2 \cdot SNR}} \tag{3.10}$$

where ℓ = the aperture length, and θ is the AOA of the intercepted signal.

In most cases, the noise-limited measurement of AOA will not be within 2:1 of this lower bound. For example, for the noise-limited performance only, let $\ell = 6$ in, $\lambda = 8$ in, $\theta = 45°$, and $SNR = 1$. Then,

$$\sigma_\theta = 0.437 \text{ radians or } 25°$$

An angle error of ±25°probably defeats sorting, prevents range estimation, and greatly enhances successful spoofing in a dense signal environment.

3.3.4.1 Amplitude Angle of Arrival. Angle of arrival (AOA) measurement is usually done either by amplitude or phase comparison. Most radar warning intercept receivers (RWRs) use amplitude comparison because it is simple. The concept usually uses broadband cavity-backed spiral antennas in each quadrant whose pattern is approximately Gaussian, i.e., parabolic in dB. The gain pattern in one plane for such an antenna is shown in Fig. 3.24. Antennas of this type have gain approximately as shown over bandwidths of almost 20 GHz. The antennas typically are installed on the intercept platform so that their patterns overlap as shown in Fig. 3.25.[32]

The signal in each quadrant after filtering goes through a logarithmic amplifier, and the difference between any two quadrants is multiplied by an angle scale factor and fitted to a correction curve to estimate angle of arrival. This system is simple and,

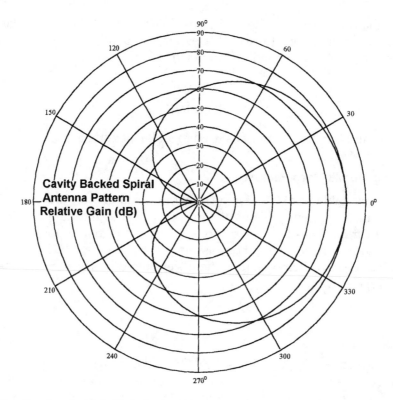

Figure 3.24 Broadband cavity-backed spiral gain pattern.

when the signal-to-noise ratio is very high, as it often is for conventional radars and data links, the angle accuracy can be surprisingly good.

Frequency must be separately estimated before angle estimation to allow accurate detection. Figure 3.26 shows three idealized quadrant antenna patterns; it can be seen that the differences can be used to uniquely determine the AOA if the signal-to-noise ratio is high. The difference between the log-compressed output voltages in any two channels is approximately linear and therefore makes an excellent discriminant if the signal-to-noise ratio is high enough.

Assuming 90° antenna spacing, the emitter angle is estimated from Equations (3.11) and (3.12).

$$\hat{\theta}_E = \frac{(20 \cdot \log(Q_i) - (20 \cdot \log Q_{i+1}))}{k_m} \tag{3.11}$$

$$Q_i = E_0 \cdot \exp\left[-k^2 \cdot \left(\theta_E - i \cdot \frac{\pi}{2}\right)^2\right] \tag{3.12}$$

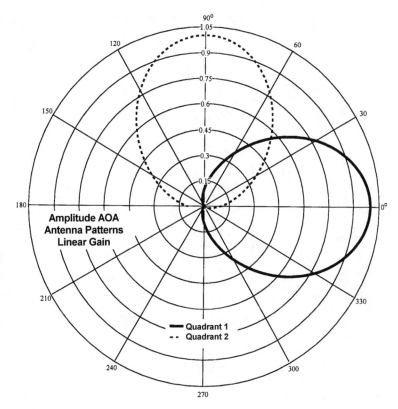

Figure 3.25 Amplitude AOA antenna pattern placement.

where

Q_i = pattern of the antenna in the ith quadrant

$k_m = -8.68 \cdot k^2 \cdot \pi$

E_0 = peak output voltage of the quadrant antenna and amplifiers before log compression

$k = 5/(3 \cdot \theta_B)$

θ_E = true direction of the emitter

$\theta_B = 1/4$ power antenna beamwidth

Suppose the AOA was 0°; then, the signals in quadrants 1, 2, and 3 of Fig. 3.26 would be –5 dB, –5 dB, and –40 dB relative. The difference between the antenna channel 1 and 2 log amplifiers would be 0 and would be designated as 0° AOA. The other channel differences would be larger than 20 dB, thresholded, and discarded. Because the log compression process is very nonlinear (but monotonic), there is a "capture effect" for large signals, which provides excellent high-SNR performance. The nonlin-

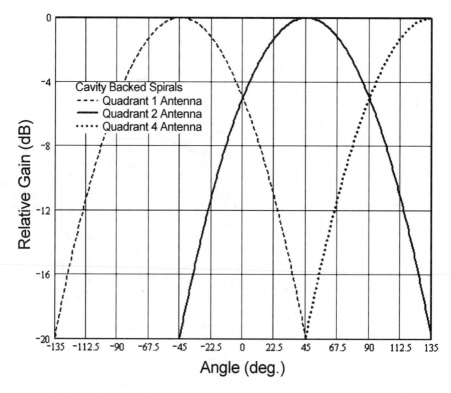

Figure 3.26 Angle of arrival measurement by amplitude comparison.

ear nature of the processing makes it difficult to make simple estimates for low SNR performance. As will be shown shortly, low signal-to-noise ratio performance has both a large random component and a large bias component. If the SNR is low, this scheme still provides general direction but with very poor accuracy (±45°), which the stealth emitter can exploit.

Beginning with Equations (3.11) and (3.12) and a simple statistical model based on the notion of "circular error probable" measurement error probability density, rough estimates of the angle discriminant output density as a function of *SNR* have been made. The model details are given in the appendices in .mcd format files on the software disk. The behavior of the angle discriminant can easily be seen from Fig. 3.27. As the signal-to-noise ratio gets smaller, the standard deviation of the discriminant gets progressively larger, and the mean shifts from the true angle of arrival (32 V in the example of Fig. 3.27), creating not only a larger variation but a bias as well. The bias cannot be removed by additional postdetection filtering. Depending on the angle to the emitter, the bias can be as much as 45°, which allows the stealth platform to spoof, as there is no angle discrimination. The details of this noise calculation also are given in the Mathcad® files in .mcd format on the appendix disk.

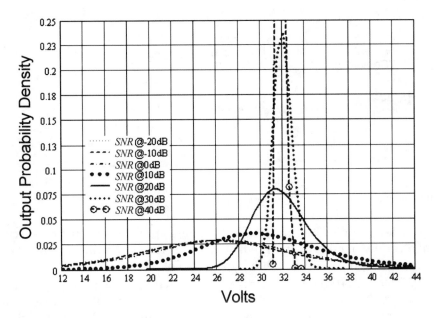

Figure 3.27 Angle discriminant spread with signal-to-noise ratio.

Obviously, deviations from an ideal pattern, ideal logarithmic conversion, component temperature, and time drift and amplifier linearity will result in AOA errors. A ±1-dB error in amplitude difference can result in an error of up to ±4.5°. A similar problem, although somewhat less severe, exists for phase comparison AOA. To compensate for this deficiency, periodic calibration is performed in most of these systems (the period is usually seconds to minutes). A calibration table is then used to correct the measurements in each frequency band. One example configuration is shown in Figure 3.28. To improve accuracy, each quadrant has a calibration source that can generate multiple frequencies over the operating band. The source is switched in periodically during operation to calibrate the RF, log IF amplifiers, and analog signal processing as shown in Fig. 3.28. The antennas, which are passive and have repeatable temperature characteristics, are measured before installation. A second calibration source is used to measure the amplitude performance of the log IF and remaining processing.

The output from these calibration signals is used to create an amplitude offset versus frequency calibration map. Antennas of the type described here have some polarization variations that are known at the time of installation. However, unless there is a way to estimate arrival polarization, these cannot be corrected in the AOA estimate. Periodic calibration used in an experimental system with the block diagram of Fig. 3.28 is given in Table 3.10. The amplitude calibration table plus a predetermined first-order polynomial antenna correction as shown in the table can produce AOA ac-

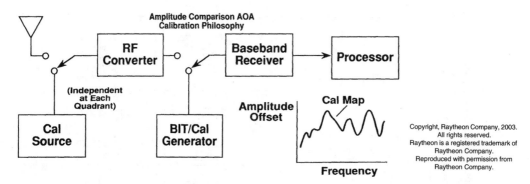

Figure 3.28   Amplitude comparison AOA calibration.[30]

curacies of a few degrees for high SNRs. The intercept receiver must also receive signals from a variety of polarizations, which has the effect of adding another 2 to 2.5° of angle error as shown in the last column in the table.

Figures 3.29 and 3.30 show total error versus signal-to-noise ratio for first-order approximations to the nonlinear signal and nonlinear noise outputs from the typical amplitude AOA processing described above (Figs. 3.26 through 3.28 and Table 3.9) for two different emitter arrival angles. These plots assume that, at the input to the log IF, the inphase and quadrature noise is a zero mean Gaussian random variable resulting from the front-end noise temperature, that the signal from the emitter is steady and nonfluctuating, and that the log detection is a perfect calculation of the logarithm of the true magnitude of the input signal. Of course, none of these assumptions is really true, but they are true enough to provide realistic assessments of performance, and degradation from nonideal functions only makes the realized accuracy worse for the interceptor system. Actual systems are usually within a few decibels or 0.1 radian of the predictions below. Although the high-SNR performance appears quite good, if there is pulse embedment or low signal-to-noise ratio, as previously mentioned, AOA accuracy can be an order of magnitude worse than this.

Also plotted in these figures is the simple approximation given in Equation (3.13). As can be seen in the figures, there is both good news and bad news for the interceptor; at low SNRs, random errors are slightly better than would be predicted by the general theory whereas, at high SNRs, errors are slightly worse. The real killer is that, at low SNRs, bias errors are dramatically worse—even at the best arrival angles where the emitter signal is equal in two quadrants. When the emitter is centered in one quadrant, the bias error can be as large as 45°. An approximate mean bias for all angles of arrival as a function of SNR is given in Equation (3.14). Also note that, even at 10 dB, SNR angle error is quite high.

Theoretically, the bias can be estimated if the angle to the emitter can be changed in a regular way and the rms noise can be estimated accurately. In the presence of scintillation, multipath and emitter frequency changes, bias estimation is practically impos-

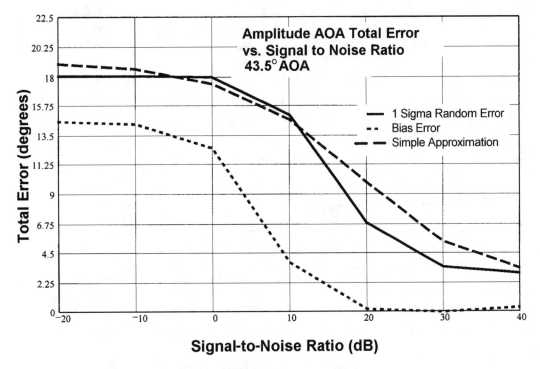

Figure 3.29   Amplitude AOA error for an emitter at 43.5°.

sible. Most systems solve this problem by thresholding emitter SNRs near 20 dB, which naturally works in the stealth emitter's favor. Because some form of rate-of-change of angle is used to estimate range, denying angle denies range. Educated guesses about angle and range can be made, but the threat doesn't need an intercept system for that. The obvious strategy for any stealth emitter in the presence of amplitude AOA threats is to maintain low SNR at the interceptor so that angle discrimination is poor. The emitter or other support assets can couple this with spoofing signals to completely confuse the interceptor and overload its processor.

In summary, the random component of amplitude AOA error can be approximated by Equation (3.13), which is plotted in Figs. 3.27 and 3.28. (See the appendices for more detail.)

$$
\sigma_\theta = \sqrt{\left[\frac{3 \cdot \theta_B}{5 \cdot \pi \cdot (1 + \sqrt{SNR})}\right]^2 + \left[\frac{\sigma_{Amatch}}{k_m}\right]^2}
\tag{3.13}
$$

where, for the parameters of Figs. 3.24 through 3.30 and Table 3.9, $\theta_B = 0.55 \cdot \pi$, σ_{Amatch} is the amplitude match in dB, and $k_m = -25.371$ dB per radian. The mean bias that can-

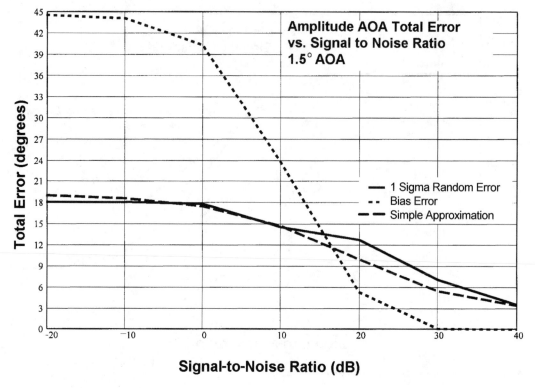

Figure 3.30 Amplitude AOA error for an emitter at 1.5°.

not be removed by any postdetection processing can be approximated by Equation (3.14).

$$\theta_{Bias} = \frac{\theta_B}{3 \cdot (1 + 0.5 \cdot \sqrt{SNR})} \tag{3.14}$$

3.3.4.2 Phase Angle of Arrival. Phase comparison AOA depends on measuring the time of arrival difference of the same signal at two or more antennas by comparing the relative phases from each antenna in a phase discriminator as shown in Fig. 3.31. An incoming wavefront from a distant point source will arrive as almost a plane wave. For a wavefront arriving at an angle of θ relative to the normal to the antenna baseline, the distance, r, is $\ell \cdot \sin \theta$ which can be related to time by dividing by the velocity of light, c. For a single emitter frequency, the phase difference between the two antennas is the product of the distance and the angular frequency divided by the speed of light as shown in Equations (3.15).

Table 3.10 Example AOA Error Resulting from Polarization Mismatch and Pattern Fitting versus Frequency

Frequency	Polynomial coefficients		RMS error due to pattern (°)	RMS error including polarization (°)
	A_0 (°)	A_1 (°/dB)		
2.0	3.297	4.172	1.79	3.09
3.0	3.183	3.306	1.27	2.68
4.0	3.121	2.848	1.01	2.40
5.0	3.091	2.581	0.88	2.22
6.0	3.085	2.419	0.83	2.16
7.0	3.102	2.322	0.84	2.17
8.0	3.144	2.266	0.88	2.24
9.0	3.214	2.236	0.95	2.33
10.0	3.316	2.220	1.02	2.43
11.0	3.453	2.209	1.08	2.51
12.0	3.627	2.193	1.13	2.57
13.0	3.837	2.164	1.15	2.60
14.0	4.075	2.113	1.17	2.59
15.0	4.333	2.036	1.17	2.58
16.0	4.594	1.932	1.19	2.59
17.0	4.843	1.802	1.26	2.66
18.0	5.063	1.653	1.37	2.86
Average RMS error			1.14	2.52

$$\theta = \operatorname{asin}\left(\frac{r}{\ell}\right) \text{ and } \varphi = \frac{2 \cdot \pi \cdot f \cdot \ell}{c} \sin(\theta)$$

Because: $c = f_0 \cdot \lambda_0$, then, $\varphi = 2 \cdot \pi \cdot \left(\frac{f}{f_0}\right) \cdot \left(\frac{\ell}{\lambda_0}\right) \sin(\theta)$

$$(3.15)$$

The frequency of the arriving signal must be measured to correctly determine the angle of arrival. Phase AOA is clearly a narrowband technique, and filters before and

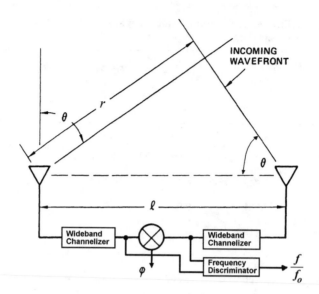

Figure 3.31 Angle of arrival measurement by phase comparison.

after the phase detector are required. There is usually a frequency discriminator function in parallel with the phase detector to provide the other variable (f/f_0) required for angle of arrival estimation. Also note that noise through an arcsine function has a nonlinear distortion that gets progressively worse as emitter AOA approaches endfire, causing some bias errors at low SNRs. These phase AOA bias errors are not nearly as severe as for amplitude AOA.

Then, solving Equation (3.15) for θ provides phase comparison AOA. For an emitter at θ_E, the result is given in Equation (3.16) below.

$$\theta_E = \text{asin}\left(\frac{\lambda_0}{2 \cdot \pi \cdot \ell} \cdot \frac{f_0}{f} \cdot \varphi\right) \tag{3.16}$$

where

   λ_0 = wavelength at band center
   f_0/f = the reciprocal of the output from a frequency discriminator centered at band
         center, f_0
   ℓ = spacing between phase measuring antennas
   φ = the measured phase

The overall phase AOA error has three components: the emitter multichannel signal-to-noise ratio, the emitter frequency measurement error, and the phase measurement error, which is given in Equation (3.17).[1] The frequency and phase measurements, in turn, have signal-to-noise dependence. The frequency measurement is the least sensitive to SNR, and hardware limitations can easily be orders of magnitude lower than phase or antenna location errors. One can think of many other sources of error, but these are the dominant contributors.

$$\sigma_\theta = \sqrt{\frac{\lambda_0^2}{2 \cdot \pi^2 \cdot \ell^2 \cdot SNR} + \frac{\sigma_f^2 \cdot \sin^2\theta_E}{f_0^2} + \frac{\lambda_0^2 \cdot \sigma_\varphi^2}{2 \cdot \pi^2 \cdot \ell^2}} \qquad (3.17)$$

where σ_f is the rms frequency error as in Equation (3.2) and σ_φ is the rms phase error as in Equation (3.3), but it includes $\sigma_{\varphi match}$, the rms hardware phase error.

Another form of this equation, substituting Equations (3.2) and (3.3), is given in Equation (3.18) below.

$$\sigma_\theta = \sqrt{\frac{\lambda_0^2}{2 \cdot SNR} + \left[\frac{1 + 0.5 \cdot c_\varphi^2}{\pi^2 \cdot \ell^2} + \frac{c_m^2 \cdot B^2 \cdot \sin^2\theta_E}{c^2} \right] + \frac{\lambda_0^2 \cdot \sigma_{\varphi match}^2}{2 \cdot \pi^2 \cdot \ell^2}} \qquad (3.18)$$

where

c_φ = the phase detector coefficient with value typically $2^{0.5}$ to 2
c_m = a coefficient to account for scintillation
c = velocity of light
$\sigma_{\varphi match}$ = the phase match in radians
B = the bandwidth in which the frequency of the emitter is estimated

For example, let ℓ = 6 in, λ_0 = 8 in, B = 20 MHz, θ_E = 45°, c_m = 2, c_φ = 2, SNR = 1, and $\sigma_{\varphi match}$ = 0.1. Then,

$$\sigma_\theta = 0.52 \text{ radians or } 29.8°$$

This seems no better than amplitude AOA. But suppose that, by using multiple antennas with different spacing long baseline, ambiguities could be resolved. For example, let ℓ = 60 in, keeping everything else the same.

$$\sigma_\theta = 0.11 \text{ radians or } 6°$$

The above example shows why phase AOA with multiple antennas is preferred for high-performance intercept systems. Such a long-baseline system requires significant additional complexity; e.g., 8 to 16 antennas and RF amplifiers per quadrant, dozens of RF switches and phase detectors, not to mention orders of magnitude increases in processor loading. Such systems are restricted to large surface or transport aircraft installations.

Just as for amplitude comparison AOA, phase deviations from an ideal antenna pattern, ideal RF and IF conversion, VSWR induced phase pulling resulting from mismatch, component temperature/time drift, frequency discriminator errors, and amplifier and phase detector linearity will result in AOA errors. To improve accuracy, each antenna has a calibration source that can generate multiple frequencies over the operating band. The source is switched in periodically during operation to calibrate the RF and IF amplifiers, phase detectors, discriminators, and analog signal processing as shown in Fig. 3.32. A second calibration source is used to measure the performance of the IF and remaining processing. The output from these cal signals is used to create a phase offset versus frequency calibration map. The antennas that are passive and have repeatable temperature characteristics are measured before installation. Usually, there are more antennas in a phase comparison scheme (versus amplitude) to resolve phase ambiguities over a wide frequency range. After phase calibration, the typical error residue budget for a phase comparison AOA scheme is also shown in Fig. 3.32. For high signal-to-noise ratios, residual phase error of 6° is achievable.

An electrical phase error of 6° translates into a small AOA error, depending on antenna separation, as shown in Fig. 3.33. Once the separation ratio is near 1, more than two antennas are required to resolve ambiguities. Figure 3.33 assumes that the emitter AOA is 45°, but it can be shown that emitter AOA has very little effect until the angle is near the antenna array endfire. Figure 3.33 also assumes a very high SNR (>40 dB) and ignores some hardware realizability issues at the extremes; not withstanding all of that, it is possible to achieve very high AOA accuracy under some circumstances.

When the signal-to-noise ratio is low or measurement bandwidth is large, then phase AOA is better than amplitude AOA but still quite limited. Limited AOA performance is one of the major advantages of stealth radar and data links in a hostile environment. This observation is shown in Figs. 3.34 and 3.35. Notice the dramatic

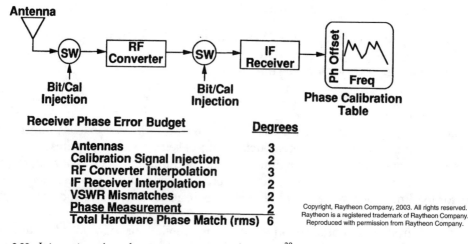

Receiver Phase Error Budget	Degrees
Antennas	3
Calibration Signal Injection	2
RF Converter Interpolation	3
IF Receiver Interpolation	2
VSWR Mismatches	2
Phase Measurement	2
Total Hardware Phase Match (rms)	6

Figure 3.32   Intercept receiver phase measurement performance.[30]

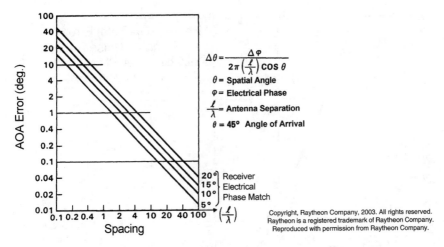

Figure 3.33 AOA error as a function of antenna phase match and separation.[30]

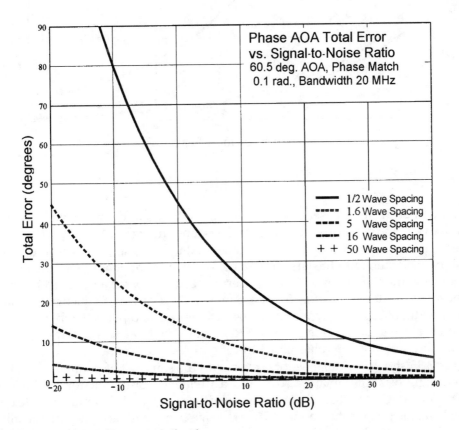

Figure 3.34 Phase AOA total error, narrowband.

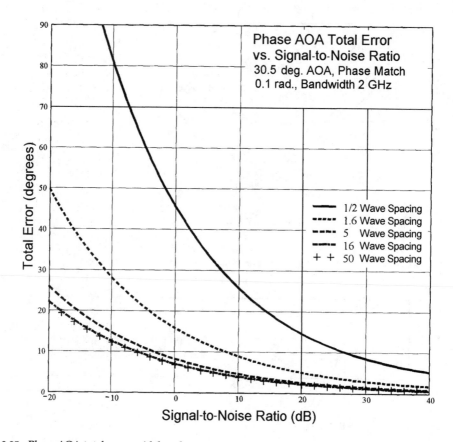

Figure 3.35 Phase AOA total error, wideband.

degradation of the long-baseline accuracy for low SNRs in the wideband case of Fig 3.35 relative to the performance of the long baselines in Fig. 3.34.

The figures each show a set of curves generated by Equation (3.17) as a function of signal-to-noise ratio with the largest spacing for an array of phase AOA antennas as a parameter. Figure 3.34 shows total angle error versus SNR for a phase AOA array at X-band with emitter angle of 60.5°, phase match of 0.1 radian, frequency discriminator bandwidth of 20 MHz, degradation coefficients of 2, and perfect ambiguity resolution (not always possible at low SNRs). Note that, at large array sizes and narrowband detection, SNR has very little effect on angle accuracy.

Figure 3.35 shows a similar set of curves with the same set of parameters but at an emitter angle of approximately 30.5° and a frequency discriminator bandwidth of 2 GHz. Note that, if the discriminator bandwidth is large and the emitter angle off the array normal is greater than about 45°, array size helps with accuracy but is still significantly degraded by low SNRs. For example, the angle error is 1/3 worse at 0 dB

SNR and 75.5° than it is at 30.5°, as shown in Fig. 3.35. This fact is one of the reasons why stealth systems must have large instantaneous bandwidth.

Note also that there is a constraint on minimum usable signal-to-noise ratio that still allows resolution of ambiguities for a multielement phase AOA array of the type analyzed in Figs. 3.34 and 3.35. Equation (3.19) gives the lower bound SNR for 95 percent confidence ambiguity resolution for a given element spacing in a multielement array.

$$SNR_i \geq \left(\frac{\sqrt{3 \cdot \ell_i}}{\pi \cdot \ell_{i-1}} \right)^2 \tag{3.19}$$

where ℓ_i is current phase AOA element spacing, and ℓ_{i-1} is the next smaller element spacing. For the figures above with spacing ratios of 3.16, the minimum SNR is

$$SNR = 3.04$$

This is probably unrealistically low for most system implementations, and double this value usually is required for 5 percent error ratio because of larger than Gaussian tails in the noise density.

3.3.5   Range Estimation

Once an intercept system has detected an emitter and estimated an angle to it, range can be estimated by multiple observations commonly called *triangulation*. The basic idea is to solve the trigonometric "angle-side-angle" (ASA) equation to make a single estimate of range; i.e., the distance to the third vertex of a triangle. The "side" in question is either the displacement or motion of the interceptor platform between measurements for an almost stationary emitter or a baseline between two or more almost simultaneous measurements for almost stationary interceptors. Such a geometry is shown in Fig. 3.36. The almost stationary single interceptor range estimation problem has no general solution,[13] but some cases can be approximated. The ASA equation that estimates the range to the emitter is given in Equation (3.20).

$$R_E = \frac{D_I \cdot \sin\theta_1 \cdot \sin\theta_2}{\sin(\theta_2 - \theta_1) \cdot \sin\left(\dfrac{\theta_2 + \theta_1}{2}\right)} \tag{3.20}$$

where
D_I = the distance the interceptor has traveled between measurements or the baseline between interceptors
θ_1 = the first angle or first interceptor measurement in a common coordinate system
θ_2 = the second angle or interceptor measurement
R_E = the range to the emitter from the baseline along the angular bisector between θ_1 and θ_2

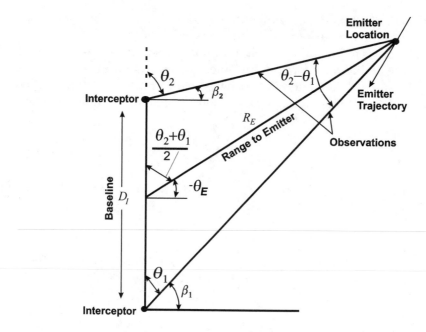

Figure 3.36 Passive location, stationary observers.

One can observe several important characteristics of range estimation by inspection of Equation (3.20). Typically, the baseline is much shorter than the range to the emitter, and small errors in the measured angles are magnified. In addition, the mean error is biased toward longer range. Therefore, very high angular accuracy is required, even to make a poor estimate of range. Usually, a single set of measurements that yields an emitter range estimate must be integrated over multiple observations to obtain a "true" current range and to reduce the effects of scintillation, multipath, and thermal noise. Obviously, the integration must take into account any relative velocity, acceleration, jerk, and so on between the emitter and the interceptor(s). This typically is done in a filter in Earth stabilized cartesian coordinates (NED) so that error propagation is lower. This is because either the emitter or the interceptor (or both) is stationary or is attempting to travel straight and level.

Some amount of relative motion is beneficial to range measurement but, under most geometries, even when relative velocities are 400 knots or greater, the geometry takes several minutes to change significantly for the benefit of the interceptor. Measurement convergence times are typically 5 min or more for low-SNR intercepts. Using Equation (3.20) and a very simple statistical model that assumes a uniform angle error probability proportional to the arctangent of the reciprocal of the square root of the SNR with a mean equal to the true angle to the emitter, rough estimates of range error as a function of SNR have been made. The arctangent of the square root of SNR divided into a constant of proportionality is a simple approximation for Equation (3.20). Figure 3.37 is one model output and shows raw range error, range bias, and smoothed range error as a function of SNR for an AOA of 45°, angle error for phase-AOA with 5 λ element spacing, and 100.85 mi emitter range with a 160 sec smoothing window.

Note that the estimated range mean range doesn't approach the true range until the SNR is 50 dB. The reason the range bias appears to get smaller at very low SNRs is that the strategy used in this model eliminates range or angle estimates that are not physically realizable. As the angle deviations get very large, so many measurements are eliminated that the bias goes down. Note, however, that the rms range error continues to increase, so there is no net improvement.

When the interceptor platform is moving relatively fast and the emitter is almost stationary over the observation time, the interceptor essentially must fly out the baseline, D_I, to achieve reasonable accuracy. Figure 3.38 shows the time required to achieve 10-mi accuracy on an emitter at 100-mi range for various initial angles of ar-

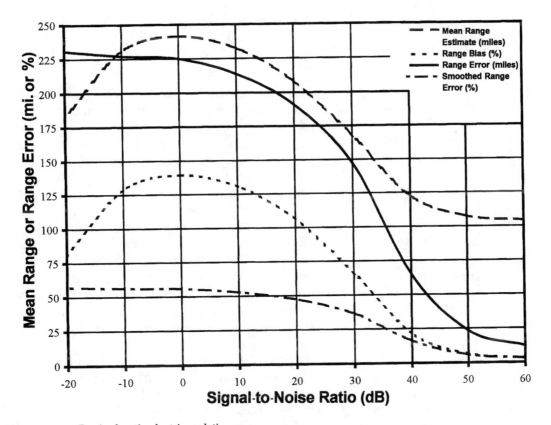

Figure 3.37   Passive location by triangulation.

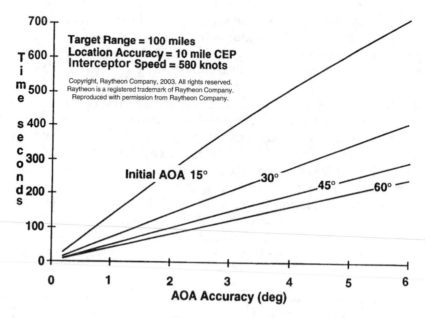

Figure 3.38 Passive location convergence time.[30]

rival and raw AOA accuracy with an interceptor platform speed of 580 kt. For example, for an initial AOA (θ_E) of 15°, a sliding window smoothing function of 36 independent samples, an SNR of 36 dB, and a corresponding rms angle accuracy of 1.5° assuming a 5-wavelength spacing (similar to Fig. 3.35), the platform must fly out a baseline of 36 mi to achieve 10-mi range accuracy at 100 mi (see appendices). This requires 194 sec, as shown in the figure. For a stealthy emitter and a more likely accuracy of 6°, the convergence time can be more than 10 min, but such emitters might only radiate for a few tens of seconds in that period. Hence, range denial can easily be achieved by short aperiodic emissions accompanied by an unpredictable maneuver. Unpredictability is always an essential element of any stealth strategy, not only with pseudorandom emission times but also random maneuvers of the platform.

Another ranging scheme usable from aircraft is phase rate of change, which combines phase AOA with a widely spaced set of antennas. This is used primarily at higher frequencies. The two antennas might be 100 to 500 wavelengths apart and must have an SNR that is high enough to resolve ambiguities. SNRs typically need to be 40 dB (see Fig. 3.39).

The concept is to compare the interceptor platform position change with the measured phase change on the emitter signal. The equation to be solved to find range is shown in Equation (3.22). For a given ownship velocity, assumed to be in the x direction (without loss of generality), and an almost stationary emitter, the angle and phase rate of change will be as given in Equations (3.21).

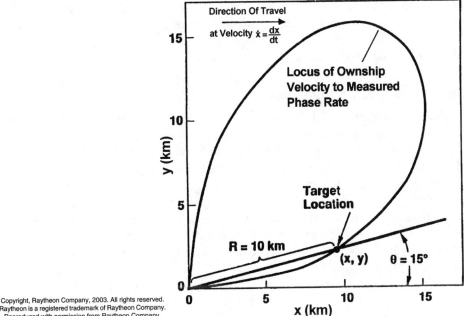

Figure 3.39 Passive ranging by phase rate of change.[30]

$$\frac{d\theta}{dt} = -\frac{\sin(\theta)}{R} \cdot x' \text{ and } \varphi' = 2 \cdot \pi \cdot \left(\frac{f}{f_0}\right) \cdot \left(\frac{\ell}{\lambda_0}\right)\cos(\theta) \cdot \frac{d\theta}{dt}$$

where: $x' = \dfrac{dx}{dt}$ and $\varphi' = \dfrac{d\varphi}{dt}$

(3.21)

Solving for range yields Equation (3.22).

$$R = 2 \cdot \pi \cdot \left(\frac{f}{f_0}\right) \cdot \left(\frac{\ell}{\lambda_0}\right)\cos(\theta)\sin(\theta)\frac{x'}{\varphi'}$$

(3.22)

Because the rate of change of phase is an experimental derivative, it magnifies the errors in a comparable manner to the cases previously analyzed; i.e., SNRs must be greater than 50 dB. The principal feature is that convergence times can be short if SNR is very high.

Many other multilateration schemes have been tried, such as those in the ALSS and PLSS systems, which depend on time of arrival rather than AOA. They all suffered from different versions of the same problem described above. Noise and multipath does them in unless the SNR is very high. These limitations, plus many others, prevented these systems from being deployed, even after many years of government-sponsored development work. A stealth system needs to reduce SNR only moderately at the interceptor to deny range.

3.4 INTERCEPT RECEIVER THREAT TRENDS

Intercept receivers are designed to counter the forecast electronic warfare environment. Current forecasts anticipate 10,000 modern fire control radars and associated data links. These, in turn, may generate visible pulse densities of 3×10^7 per second. This will drive future trends more than any single threat type.

Intercept receiver threats fall into two main categories. The first threat class is the mainlobe threat receiver whose sensitivity is adequate to detect an emitter in its antenna mainlobe but not in its sidelobes. The typical airborne radar warning receiver (RWR) is of this type. The angle arrival is usually estimated to at least a quadrant and perhaps to angle resolutions of 10°. Classification is usually on the basis of gross frequency, PRF, and pulse width.

The mainlobe intercept threat enhancement trends to be expected are as follows:

1. Channelized crystal video receivers with preamplifiers and enhanced processing

2. Instantaneous frequency measurement sets with swept preselection filters and enhanced digital processing

3. Digitally controlled scanning superhets with programmability of a small number of frequency analyzing assets that can be placed at many different parts of the spectrum

4. Wider deployment of compressive receiver systems

Furthermore, with the advent of digital radio, direct RF-to-digital receivers will be deployed by 2010.

All of these, however, will be oriented primarily toward improvements in dynamic range and frequency resolution, not sensitivity. Future improvements will be primarily in signal processors. And, for the foreseeable future, usable sensitivity will always be limited by signal pulse densities in the warning receiver environment. To limit processor complexity, processors probably will be programmed to threat frequency bands dynamically and will cover the smallest amount of frequency space possible. LPI systems then, must spread over progressively wider bands so as to force intercept receivers to utilize more of their assets searching space that potentially could be occupied but probably isn't.

The second class of threat is one in which sensitivity is high enough to detect an LPI emitter through its sidelobes. Of course, such a threat can also detect the radar or data link in the mainlobe. A significant issue is that sidelobe detections must be assumed to be essentially continuous, and that will allow the interceptor substantially longer times to classify and track the emitter. A typical sidelobe threat is shown in Table 3.11. The receiver type might be any of the ones previously discussed, but transform or channelized are the most likely possibilities. The receiver sensitivities may be on the order of –85 to –90 dBm. Instantaneous dynamic ranges might be 30 to 70 dB. Such a system usually has a high-gain antenna associated with it, but dwell time might be 10 msec per beam position or greater. Instantaneous RF bandwidth might be 2 GHz across the band with a 5 MHz instantaneous detection bandwidth. Frequency scan times might be very short, perhaps a fraction of a microsecond. Processing will probably be quite elaborate, but the same problems still exist; i.e., sensitivities are usually limited by processing rather than thermal noise.

Table 3.11   Future Threat Receiver Forecast

Parameter	Value
Frequency coverage	0.2–35 GHz
Bands	18
RF bandwidth	2 GHz
Video bandwidth	5 MHz
Sensitivity	Up to –85 dBm if usable
Instantaneous dynamic range versus frequency separation	0 dB @ 5 MHz 50 dB @ 15 MHz 70 dB @ 100 MHz
Pulse traffic capacity	>10 M pulses/sec Still processor limited
Adjacent pulse recovery time	200 nsec
Frequency accuracy	3 MHz
Frequency resolution	5 MHz
Angular resolution	0.03 steradian
Coverage scan rate	Multibeam, 1 sec for 0.1 steradian
Other features	Netting, ultra-low sidelobe antennas
Advantages	Best dynamic range on a small emitter subset, good in dense signal environment, good for broadband emitters
Disadvantages	Complex, expensive, requires very high-performance antennas or adaptive arrays

The future trends in sidelobe threats, which usually will be on transport aircraft or surface based receivers, primarily revolve around improved processing. All of the current techniques have adequate sensitivity: IFM, scanning superhet, transform, channelized, and hybrid. These high-sensitivity systems usually have high-gain antennas to limit pulse traffic with narrow antenna beams. The gain of the antenna will be limited by scan time or search volume constraints, not by hardware limitations. Improvements in these systems, both in signal processing and in recognition, will follow the computer state of the art. As computational horsepower gets better, these systems inevitably will have more sensitivity. Interceptor sensitivity seems to be getting better at the rate of about 1 dB per year. However, radars and data links, in general, are getting better at the same rate, because they are exploiting the same processing advances, and so the current relative performance advantage may be constant a long way into the future.

Programming for anticipated threats has really improved usable sensitivities. Greatly improved performance of surface-based or large platform-based intercept receivers could be obtained if they were netted. The effective tactical use of either ground or large airborne stations netted for intercept to provide missile guidance and control requires effective command, control, and communications links at high levels of accuracy and large information bandwidth, which are not currently achievable. Developmental systems have had limited success at best. There may be fundamental limitations in netted systems associated with measurement and correlation latency that ultimately prevent targeting-quality detection and tracking.

To anticipate the effective trends both intercept receiver sensitivity and improved radar sensitivity, a plot of sidelobe detection range in kilometers as a function of intercept receiver sensitivity for radars of various generations is shown in Fig. 3.40. Type 1 radars were designed in the 1960s and are conventional high-peak-power pulse radars. Type 2 radars were designed in the 1970s with high- and medium-PRF waveforms and some power management. Radars of type 3 are typical of LPI radars of the 1980s and 1990s. Plotted along the lower axis in Fig. 3.40 are intercept receiver sensitivities including both antenna gain and noise limited thresholds in decibels below a milliwatt (dBm). The latest of the threat systems lie between −60 and −95 dBm, and the general trend is to higher levels of sensitivity. As shown in the next section and Chapter 4, the future threat must have extremely low sidelobe antennas to exploit its potential sensitivity.

3.4.1 Typical Response Threats—Elastic Threat (after Gordon)[6]

To define a family of intercept receivers that would constitute a stressing threat to an LPIS, one can choose high-performance values for the key receiver parameters, allowing the receiver antenna gain to be a variable. Suppose that the receiver designer may choose the gain to achieve a desired balance between sensitivity and intercept probability. Imagine the receiver design to incorporate more than one antenna and/or feed

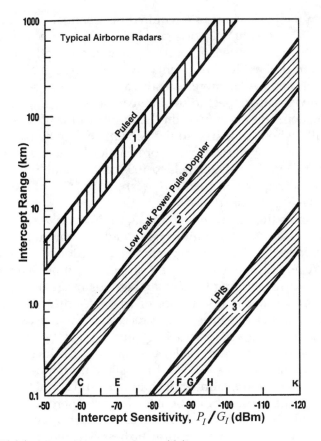

Figure 3.40 Future sidelobe intercept range versus sensitivity.

mechanism so that the gain would be selectable corresponding to a set of specific LPI systems. The interceptor having selected an antenna gain is then constrained to it for the search time. In any case, one should consider the range of possible values in defining the threat.

For example, assume that the total bandwidth B is 1 GHz and that the number of channels N_S is 200. Thus, the noise bandwidth of each channel might be about 6.3 MHz. The 1-GHz total bandwidth supposes that the interceptor is tuned to the emitter general operating band with a 1-GHz uncertainty. A larger number of channels might be assumed. However, once the channel bandwidth becomes smaller than the emitter bandwidth, the sensitivity no longer increases. Thus, a 6.3-MHz channel bandwidth is reasonable. (The effect of broader emitter bandwidth is referred to further below.)

The number of channels and emitter waveform have an effect on the required SNR. For example, if the total system false alarm rate is to be one per minute and there are

200 channels, then the false alarm rate required per channel would be 8.3×10^{-5} per second. With a 6.3 MHz channel noise bandwidth, the false alarm probability required for a rate of 1 per minute is 1.3×10^{-11}. Then, for a nonfluctuating emitter and 90 percent detection probability, approximately a 15.5 dB signal-to-noise ratio is required (very optimistic!).[4] A far more realistic case for an emitter whose geometry is changing very slowly with respect to the interceptor is Swerling 1, which requires a 23.5 dB SNR. The dependence of SNR on the number of channels is rather modest, however, especially when compared to the benefits of channelization. For example, assuming that the channel bandwidth is approximately matched to the emitter waveform, decreasing the number of channels to 100 (i.e., by doubling the channel bandwidth) decreases the required signal-to-noise ratio by much less than 1 dB, thus decreasing the sensitivity by nearly 3 dB.

For example, assume that the listening time, t_L, is 10 msec. As noted above, shorter listening times are possible if the emitter has a high minimum PRF or if few pulses suffice for emitter identification. The minimum must be longer than the time for stabilization of the RF environment at each new angular position. If a radar emitter PRF were 1000 per sec, corresponding to an unambiguous range of 150 km, then the receiver would be able to detect a sequence of 10 pulses during a 10-msec listening time, which may be sufficient for required processing. Finally, assume nominal values for the beam and filter spacing factors, $c_{SP\theta}$ and c_{SPF}, of 0.79 each (giving about 1-dB shape loss), a 4-dB noise figure, NF_I, $G_{IP}/G_{IP\eta} = 3$ dB in processing gain over noise, Boltzmann's constant of $k_B = 1.38 \times 10^{-23}$, standard temperature of $T_0 = 290$ K, bandwidth of $B = 1$ GHz, and a number of simultaneous channels $N_S = 200$. Then, for a 1/min false alarm rate, a probability of detection of 90 percent, and nonfluctuating emitter, the SNR_I must be 15.5 dB. With these values substituted into Equation (2.25), Equation (3.23) results. This is a very optimistic best sensitivity.

$$S_I = 1.86 \cdot 10^{-14} \cdot \frac{\Omega \cdot \mathcal{P}_I}{T_{OT}} \tag{3.23}$$

Or, using Equation (2.11),

$$S_I = 1.86 \cdot 10^{-14} \cdot \frac{\Omega}{T_I} \tag{3.24}$$

The latter expression is perhaps more relevant to the intercept receiver designer for whom the emitter dwell time may not be known. As will be seen below, however, the first expression is of interest to the LPIS designer, to whom the receiver frame time may not be known. The results of Equation (3.24) are shown graphically in Fig. 3.41 for a range of frame time values and for three receiver scan sectors, each of which defines a family of receivers. The largest scan sector shown is a hemisphere ($\Omega = 2\pi$ steradians). The intermediate scan sector shown is an all azimuth scan from

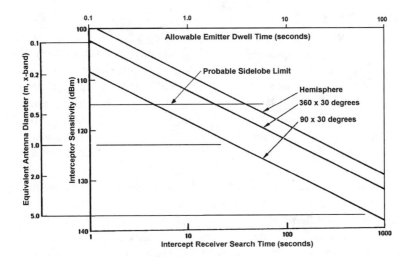

Figure 3.41 Example elastic threat specifications.

the horizon up to 30° elevation. This sector is similar to scans typically performed by air defense search radars. Finally, as a more common battlefield case, consider a similar scan that is limited to a quadrant in azimuth directed toward the FEBA. A second scale on the figure indicates the antenna diameter at X-band required to achieve a given system sensitivity. Also shown on the same graph is the probable maximum sensitivity assuming –65 dB rms sidelobes, which is currently beyond the state of the art.

It is evident that rather high system sensitivities may be achieved at the expense of long scan times and large antenna diameters. Typical long-range search radar frame times are approximately 10 sec, and such radar sites often contain intercept systems. Scan times an order of magnitude larger may yield sensitivities approaching –130 dBm. This assumes that the emitter is moving very slowly, which is very unlikely. Sensitivities this high require a 2-m antenna diameter with unrealistically low (–80 dB rms!) sidelobes. Interceptor antenna and filter sidelobes ultimately limit sensitivities well above –130 dBm, as shown in examples in Chapter 4.

3.4.2 Corresponding Specification of LPIS Emissions

Having defined a family of elastic threat receivers as discussed above, one may use these results to specify limits on the effective radiated power (ERP) and dwell time of stealth systems, so that the LPIS vulnerability to any receiver in the family is limited. The reasoning used proceeds as follows. For a given emitter/receiver geometry and emitter characteristics (i.e., power, beam pattern, and scan rate), the dwell time must be short enough to limit the receiver intercept probability to a low value. That is, the time on target, T_{OT}, must be short compared to the frame time of the intercept receiver (in the family), which is just sensitive enough to detect the emitter. To make these notions more precise, refer to Equation (3.23) in which the time on target and intercept probability appear explicitly. This equation also is depicted graphically in Fig. 3.41 using the top axis, which shows the achievable receiver system sensitivity as a function of the emitter time on target or dwell time, assuming that the intercept probability is 10 percent. In the absence of a comprehensive deployment model for the intercept receivers and operational definition of the LPIS, any value of intercept probability assumed would be somewhat arbitrary, but it is almost never higher than 10 percent.

Suppose that one chooses to use the family of receivers corresponding to the bottom line in the figure (i.e., the most stressing of the three cases shown) to derive an LPIS specification. If the beam pattern of the LPIR were constant over some solid angle (i.e., a "cookie-cutter" pattern), the specification would be particularly simple. First, for a geometry of interest, one would compute the power at the receiver represented by Equation (2.26). Recall that this "reference" power is that which would be received by an idealized isotropic, lossless receiver, and which (as follows from the definition of the system sensitivity) may be directly compared to the actual system sensitivity of the intercept receiver. Note that if the emitter bandwidth is wider than the receiver channel bandwidth, then the power should be reduced to the value within a single receiver band. Having calculated the reference power, the LPIS beamwidth and scan rate can be used to determine the dwell time. Then, the reference power and dwell time define a point relative to the threat receiver family, as depicted in Fig. 3.42. If the point is below the line, then the LPIS beats the threat receiver family. That is, none of the threat receivers is sensitive enough to detect the emitter while achieving more than 10 percent intercept probability.

When one considers that an actual LPIS has a nonideal beam pattern, the situation is somewhat more complex. Then, Equation (2.26) defines the reference power for a given receiver position within the beam, as represented by the value of G_I used. For a given scan rate (and emitter/receiver geometry), the time on target is a function of the reference power. At one extreme of this function, the peak LPIS gain dwells on the receiver for zero time. At the other extreme, the far sidelobes dwell on the receiver essentially continuously (as long as the emitter is on).

The reference power/dwell time function for a hypothetical LPIS might appear as is suggested in Fig. 3.42. As before, if the LPIS characteristics lie below the threat line, then the LPIS beats the receiver family. At some point, the sidelobes will always be above the threat line, unless the maximum dwell time is limited by turning off the radar periodically. The maximum receiver sensitivity is limited, as discussed above, by maximum frame time, antenna diameter, and sidelobe considerations. The sensitivity limit corresponding to a 2-m interceptor antenna diameter is shown in the Fig. 3.42. Sidelobes, which lie below the maximum sensitivity, would not be detectable.

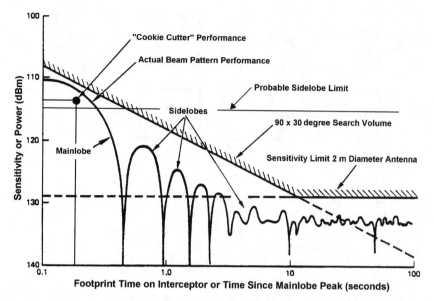

Figure 3.42 LPIS limiting requirements against the elastic threat.

3.4.3 Typical Response Threats—Radiometric

The following analysis is somewhat speculative, given that the threat is yet to be invented. One threat repeatedly postulated is the radiometric threat. This is the radio astronomy approach to intercept. Whereas the radio astronomer can pick a "quiet" site, the time of day for observation, and long integration times, and spend months analyzing the data, the battle area intercept receiver cannot. A radiometer is essentially a crystal video receiver (CVR) with a large antenna and very large postdetection integration. The idea is to treat the emitter as a spatial hot spot and to detect and track hot spots independent of waveform. Microwave radiometric systems have been around at least since the 1940s. During the author's tenure of 25 years at Hughes Aircraft, many microwave radiometric systems for imaging, missile guidance, zero visibility aircraft landing, terrain following, and so forth were developed and tested, but none was deployed. Their principal weaknesses are their susceptibility to interference, clutter, and low false alarm threshold setting. Not withstanding all of the above, there are certain circumstances in which a radiometer will provide superior sensitivity.

The basic radiometric intercept receiver is shown in Fig. 3.43. It consists of a high-gain antenna and an RF front end matched in bandwidth to the emitter total spectrum or to a single pulse compression chip width, B_I. This is followed by a detector and a multibeam sliding window noise (temperature) estimator. The output of the detector

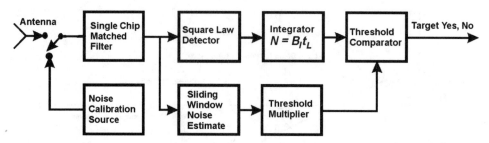

Figure 3.43 Basic radiometric intercept receiver.

goes to an integrator matched to the look or dwell time, t_L, of the antenna beam; i.e., a time bandwidth product of N, which typically has a value between 10^3 and 10^7. The noise estimate is multiplied by a variable threshold multiplier designed to keep the false alarms at an acceptably low value of perhaps 1 per minute, i.e., 10^{-8} to 10^{-12} per trial. The output of the integrator is then compared to the threshold to determine if a target is present.

One proposal has been to use the radiometer for tracking only, but detection and acquisition is always required to invoke track. Unfortunately for the interceptor, the only proposed configuration that has adequate theoretical detection sensitivity for many emitters is the radiometer. As will be shown shortly, microwave radiometers are limited to sensitivities orders of magnitude poorer than their theoretical performance by external noise and the counterpart threshold required to achieve low false alarms.

The difficulty with radiometer threshold setting is presented in Fig. 3.44. The figure shows a possible output from the integrator labeled as current beam output. As inte-

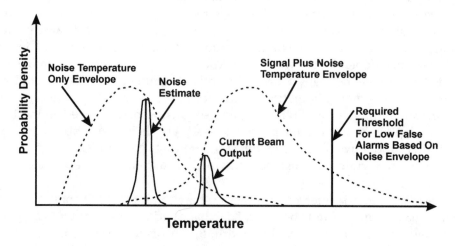

Figure 3.44 Radiometer threshold setting problem.

gration is increased, the normalized temperature variance will get progressively smaller, and it will converge on the temperature line shown. Similarly, the noise estimate formed from angle bins around the current beam also converges on a specific temperature with more integration. If that noise estimate were a true estimate of the noise in the current beam (which cannot be known from the interceptor measurement alone!), then clearly a threshold could be set that would provide detection of an emitter equivalent temperature to the one shown in the figure. Unfortunately, both the noise estimate and the current beam output lie inside the envelope of the opposite hypothesis. Table 3.12 shows the buildup of the possible sources of equivalent noise temperature over a search volume for a ground-based radiometric interceptor. From that table, an envelope for the noise can be estimated. Some components can be compensated as a function of beam pointing angle, but many are time varying and hence not easily compensated. The net result is that the threshold must be set at a higher level than the current beam variance would imply. Figure 3.44 shows the required placement of the threshold to achieve a low false alarm probability (10^{-8} to 10^{-10}), and it is generally independent of the signal. The high threshold, in turn, translates into higher required input SNRs for high probability of detection (0.9).

Recall that the equivalent intercept temperature is related to the intercepted power by Equation (3.25).[14]

$$TE_{int} = P_i/(k_B \cdot B_I) \qquad (3.25)$$

Recall also that the system noise temperature is composed of the available input temperature to an ideal antenna plus the sum of the equivalent temperatures of each cascaded component of the receiver system, divided by the gain between the ideal antenna input and that component, as shown in Equation (3.26).

$$TE_S = TE_a + \sum_{i=I}^{m} TE_i/G_i \qquad (3.26)$$

where
TE_a = the ideal antenna available noise temperature input
TE_i = the equivalent temperature of the ith component
G_i = the gain from the ideal antenna to the ith component

The antenna noise temperature, TE_a, is made up of a number of contributors, including galactic noise, sun noise, the ground, the atmosphere, clouds, clutter, the physical temperature of the antenna and radome, urban microwave emissions, interference from other radars, and solar glint. The input temperature contribution in the antenna sidelobes is semiconstant (hours). The mainlobe contribution is quite variable beam position to beam position, but each contribution has a probability of occurance. For example, clouds can cause an equivalent temperature change of up to 100 K,

Table 3.12 Antenna Available Temperature and Probability

Elevation (°)	Sidelobes, temperature, K					Mainlobe, temperature, K				Temperature, K, vs. probability				
	$0.1*TE_{gr}$ $P=0.9$	$0.1*TE_{phy}$ $P=0.9$	$10*TE_{sint}$ $P=0.8$	TE_{urb} $P=0.8$	TE_{mint} $P=0.01$	TE_{clut} $P=0.8$	TE_{atm} $P=0.9$	TE_{glint} $P=0.7$	TE_{sun} $P=0.001$	P_{cum} $=0.1$	P_{cum} $=0.2$	P_{cum} $=0.3$	P_{cum} $=0.99$	P_{cum} $=0.999$
0	29	29	290	100	154,872	200	145	420	16,000	203	793	1213	156,085	172,085
5	29	29	290	100	154,872	150	33	420	16,000	91	631	1051	155,923	171,923
10	29	29	290	100	154,872	100	20	–	16,000	78	568	568	155,440	171,440
30	29	29	290	100	154,872	100	9	–	16,000	67	557	557	155,429	171,429

depending on RF frequency.[27] The transition from a clear sky to a rain cloud can happen one beamwidth to the next, and therefore 100 K of the clutter variation can be attributed to weather. Certain mainlobe contributors are the result of man-made objects and therefore are elevation dependent.

The available noise power in the sidelobes is weighted by the sidelobe level relative to an isotropic radiator times the number of sources when there is more than one. Some of these coefficients are arbitrary but based on typical experience and similar values used in other texts. The antenna temperature calculation is shown in Equation (3.27).

$$TE_a = 0.1 \cdot TE_{\text{ground}} + 0.1 \cdot TE_{\text{physical}} + 10 \cdot TE_{\text{sinterferer}} + 10 \cdot 0.1 \cdot TE_{\text{urban}}$$
$$+ TE_{\text{atmosphere}} + TE_{\text{clutter}} + TE_{\text{minterferer}} + TE_{\text{glint}} + TE_{\text{sun}}$$

$$(3.27)$$

The calculation of Equation (3.27) is summarized in Table 3.12. The contribution from the ground temperature times the integrated sidelobes intercepting the ground is about 0.1. The antenna internal losses are about 0.1. It is assumed that there are 10 inband interferers at a 20-km mean range coupling sidelobe to sidelobe as calculated below. It is assumed that there are 10 urban interferers coupling sidelobe to sidelobe with 80 percent probability; these are typically communications transponders generating out-of-band third and fifth harmonic spurious signals. There is a 1 percent probability (the radiometer look time divided by search radar total scan time multiplied by the number visible, e.g., 0.01 sec. look × 10 visible/10 sec search) that the mainlobe of an air defense or airborne intercept radar will illuminate the site of the radiometer. (This probability is very likely to be much higher, as shown in the next chapter). The probability of a sun induced glint is about 70 percent, depending on time of day, and search radar induced glints may be an order of magnitude higher for low beam elevations. The clutter temperatures are from *Threshold Signals*.[15] The atmosphere temperature contribution is from the *Radar Handbook*.[14] The sun equivalent temperatures are from *Modern Radar*.[16] The other values are calculated below.

In the microwave region most man-made and many natural surfaces are mirrorlike and hence a source of glints from the sun and other emitters. Anyone who commutes to work has experienced visual glints daily and they are just as common at microwave. Similarly, the glint temperature can be calculated as shown below.

$$TE_{\text{glint}} = L_{\text{glint}} \cdot TE_{\text{sun}} \cdot \Omega_{\text{sun}} \cdot A_{el} \cdot [A_{\text{eglint}}/(R_{\text{glint}} \cdot \lambda^2)]^2 \qquad (3.28)$$

where TE_{sun} is the apparent temperature of the sun at the operating wavelength (e.g., TE_{sun} ranges between 8000 and 35000 K with a typical value of 16000 K at X-band), Ω_{sun} is the solid angle subtended by the sun at the Earth, or approximately 6.8×10^{-5} steradians, and A_{eglint} is the effective area of the glinting surface.[17] For example, if

L_{glint} is 0.5, TE_{sun} is the typical value, A_{el} is 1 m^2, A_{eglint} is 25 m^2, λ is 0.03 m, and R_{glint} is 1000 m, then the glint temperature is $TE_{glint} = 420$ K. Glint is primarily a close-range problem and is sometimes mitigated with clutter fences. The primary weakness with a clutter fence is that it takes away long-range near-horizon surveillance and detection, which is exactly the region where a high-sensitivity interceptor receiver is needed. If you can see the emitters with the naked eye before you detect them, you don't need the intercept receiver in the first place. High elevation angle target detections usually don't result in successful engagements.

The appendix software CDROM contains additional temperature and radiometer performance analysis.

The temperature of a radar interferer can be calculated as shown in Equation (3.29).

$$TE_{interferer} = \frac{P_T \cdot G_{TI} \cdot G_I \cdot \lambda^2 \cdot L_i}{(4 \cdot \pi \cdot R_i)^2 \cdot k_B \cdot B_I} \tag{3.29}$$

For $P_T = 1000$ W, $G_{TI} = 0.5$, $G_I = 0.1$, $\lambda = 0.03$ m, $L_i = 0.5$, $R_i = 20$ km, and $B_I = 1$ GHz, then a sidelobe-to-sidelobe interferer equivalent temperature, $TE_{sinterferer}$, is about 25.8 K. Similarly, if $G_{TI} = 3000$ and all other parameters are the same, then a mainlobe-to-sidelobe interferer equivalent temperature, $TE_{minterferer}$, is about 154,872 K. Table 3.12 assumes there are ten such emitters visible with mean distance 20 km. As mentioned, a guard channel can be used, but 100 such emitters often will be visible to the radar horizon and, when they are in the sidelobes, they kill mainlobe beam positions.

Table 3.12 summarizes the contributors to the antenna available temperature as well as providing an estimate as to its probability of occurrence. All of these temperatures are arguable in any given situation but, as shown in the next chapter, the ambient noise/temperature is high, and these representative values may be low. One solution to the mainlobe-to-sidelobe interferer problem is to use a guard channel to delete sidelobe interference. But because the beam position is lost if the sidelobe canceller is invoked, the optimal strategy for friendly forces is to use small amounts of sidelobe jamming to deny large regions of space to a system of this type. As will be shown shortly, the jamming power required is trivial.

Returning to the overall system noise, a radiometer will have a certain antenna temperature as a function of elevation with a given probability as provided in Table 3.12. This antenna temperature, coupled with internal noise, yields the overall equivalent noise temperature for a given probability as shown in a simplified form of Equation (3.26) in Equation (3.30) below.

$$TE_S = TE_a + \frac{290 \cdot (NF - 1)}{1 - L_a} + \frac{TE_R}{G_{lna} \cdot (1 - L_a)} \tag{3.30}$$

where

NF = the noise figure

L_a = the antenna and plumbing loss up to the *low-noise amplifier (lna)*

TE_R = the equivalent temperature of the remainder of the receiver system

G_{lna} = the gain of the lna

One typically finds that the gain of the lna is high enough that the remaining receiver temperature has very little effect on the system temperature. For example, the cumulative 90 percent probability antenna temperature for 0° elevation is 1213 K (from Table 3.12), L_a is 0.1 in the table; let NF be 2.5 (4 dB.), G_{lna} be 100, and TE_R be 3000 K. Then,

$$TE_S = 1213 + \frac{290 \cdot (2.5 - 1)}{1 - 0.1} + \frac{3000}{100 \cdot (1 - 0.1)} = 1729.67 \text{ K} \tag{3.31}$$

As previously pointed out, the threshold might have to be 10 to 100 times this equivalent temperature to yield acceptable false alarm performance. The natural question to ask is, "What are typical LPIS equivalent noise temperatures?" Figure 3.45 shows the equivalent available noise temperature for two different radiometer antenna diameters for sidelobe intercepts for an LPI data link and radar as a function of range. A probable high false alarm first threshold in a two-threshold detection scheme is also shown in the figure. Notice that even that threshold doesn't provide tactically useful detection ranges in the sidelobes. The leading performance parameters for the two emitters and two different antennas on the radiometer system are given in Table 3.13. The assumption is that the interceptor does whatever processing is necessary to detect the corresponding emitter temperature. Without care, there will be mainlobe intercepts from time to time, and their equivalent temperatures will be well above the threshold.

The losses shown in the table are used only once in the intercept power equation. The power is the net power into free space. The radar has a duty ratio to allow listening, which the data link and the interceptor do not. But, depending on the location of the interceptor, it may also have a duty cycle associated with nearby equipment blanking. Many sources[9,10] show that the optimal threshold for systems of this type is proportional to the noise power, the number of samples integrated, and the logarithm of the desired false alarm probability as given in Equation (3.32).

$$P_{TH} = c_{TH} \cdot B_I \cdot t_L \cdot \boldsymbol{\mathcal{N}}_I \cdot \ln(1/\boldsymbol{\mathcal{P}}_{FA}) + c_T \tag{3.32}$$

where c_{TH} is a threshold multiplier to take into account the nature of the "tails" of the noise statistics, and c_T is a minimum threshold. All the other variables have been previously defined. If the world was filled with well behaved normal statistics, then c_T

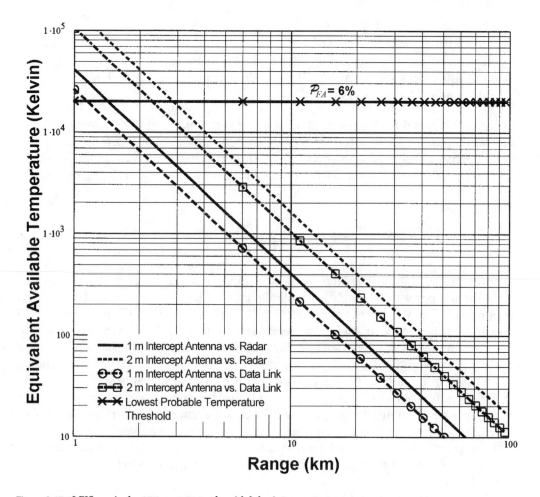

Figure 3.45   LPIS equivalent temperatures for sidelobe intercepts.

would be 0, and c_{TH} would be 2. Unfortunately, because of interferers, "hot clutter," hardware "birdies," and component drift, c_{TH} usually is several hundred higher, and c_T is 1 to 2. For the case of Table 3.12, experimental values of 50 to 500 for c_{TH} provide low false alarms, depending on total integration time. If t_L is 0.01 sec, then TE_{mint} is as shown in the chart, and, using Equation (3.32), a c_{TH} of 46 can be calculated for an elevation of 10° and a $\mathcal{P}_{FA}$ of 0.001. The threshold multiplier must be higher at lower elevations. If t_L is longer, then the probability of a mainlobe-to-sidelobe intercept goes up proportionately, and c_{TH} must be higher. If t_L is about 0.11 sec, then c_{TH} must be about 500.

The power out of the radiometer integrator shown in block diagram of Fig. 3.42 can be calculated as shown in Equation (3.33) below.

Table 3.13   Parameters for Figure 3.45

Parameter	Radar	Data Link	Interceptor	Interceptor
Power (W)	50	8		
Noise figure	2.5	2.5	2.5	2.5
Mainlobe	5000	5000	5000	20,000
Sidelobes	0.0063	0.0063	0.1	0.1
Loss	0.25	0.25	0.25	0.25
Bandwidth (GHz)	1	1	1	1
Duty	0.25	1	1	1
Wavelength (m)	0.03	0.03	0.03	0.03

$$P_{OUT} = \frac{P_T \cdot Duty \cdot G_{TI} \cdot G_I \cdot G_{IP} \cdot L_i \cdot \lambda^2}{(4 \cdot \pi \cdot R_i)^2} + k_B \cdot T_0 \cdot B_I \cdot NF_I \cdot G_{IP}n \qquad (3.33)$$

where *Duty* is the transmitter duty ratio, and all other variables are as previously defined for Equations (2.12) through (2.24). The noise gain grows directly as the integration time bandwidth product, and the signal gain grows approximately as shown in Equation (3.34) below, which takes into account both theoretical low SNR integration losses and arithmetic losses associated with large integration ratios and associated dynamic range limitations. This noncoherent integration signal gain is similar (within 5 dB) but not identical to those calculated in Barton or Peebles, because it takes into account more losses.

$$G_{IP} = 0.85 \cdot (B_I \cdot t_L)^{2.15 - 0.066 \cdot \log(B_I \cdot t_L)} \qquad (3.34)$$

A plot of the output of the radiometer integrator for the radar and intercept parameters of Table 3.13, first column and last column, using Equations (3.32) through (3.34), with an optimistic value of c_{TH} of 50 and false alarm rate of 10^{-8}, is shown in Fig. 3.46. The output power and threshold, in decibels relative to a milliwatt as well as the corresponding SNR, are given as a function of integration time. One property that may seem counterintuitive from Fig. 3.46 is the steady increase in the threshold. Recall from Equation (3.32) that the threshold is set by, among other things, the observation time. For a given false alarm rate, the longer observation time causes a higher probability of a false alarm. Because a radiometer by definition has very limited signal discrimination, any energy source can produce a detection or false alarm.

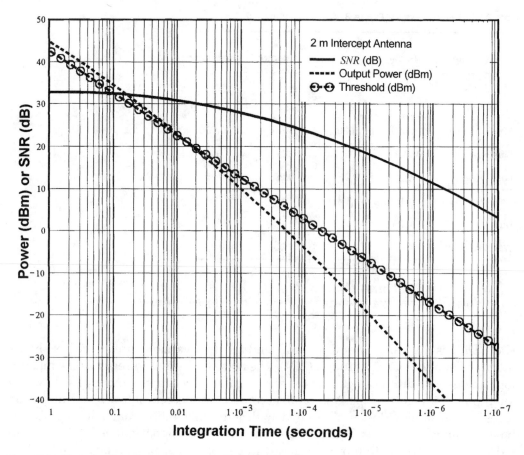

Figure 3.46 Typical radiometer performance with integration time.

Several observations can be made. First, there are very diminishing returns for ultra-long integration times. Second, even with a large antenna at longer range, detections are problematical, as the threshold is going up at almost the same rate as the potential signal. Third, the corresponding search times are also going up at a rate that may make detections tactically useless. The time limitations of any interceptor system with a high gain antenna are shown in Fig. 3.47 below. Utilizing Equation (2.21) from Chapter 2 and the antenna gains given for the interceptor of Table 3.13, it is easy to calculate the required number of beam positions for a given solid angle and multiplying the number of beams by the look time yields the total search time. Figure 3.47 shows the total search time for a typical 120° by 30° sector. Search times over 10 sec make scan to scan correlation progressively more difficult. Search times over 100 sec may be tactically useless and downright dangerous, as that is the typical time of flight of most missiles launched from 20 miles.

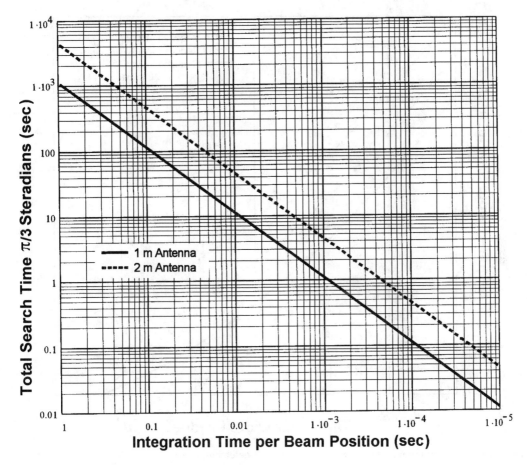

Figure 3.47 Integration time per beam position (sec).

Using an equation of the form of (3.32) and an estimated integration improvement ratio, it is possible to estimate the required input signal-to-noise ratio that will just cross the threshold for a given integration time bandwidth product. The time bandwidth product corresponds to the number of independent samples integrated. For very large integration ratios (10^{10}), the SNR grows as $N^{1.5}$. Because the threshold is related only to the input noise, the number of independent samples to be integrated, the required false alarm rate, and the noise statistics, an estimate can be readily made. If c_{TH} and c_T are as given in the paragraphs above, then the required input SNRs are shown in Fig. 3.48. The low values of c_{TH} (50 and 150) are very optimistic, as the radiometer has a wide open front end sensitive to energy only and a very narrow post detection filter. These systems have a tendency to be triggered by glints from surface vehicles, buildings, and clouds. For example, an industrial smokestack can easily

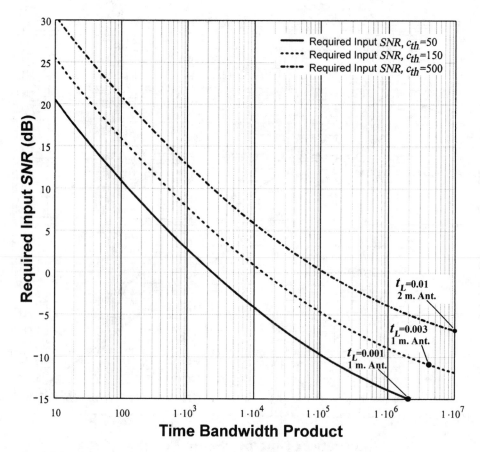

Figure 3.48   Minimum required input SNR for radiometer detection.

have an equivalent temperature of 300°C, which is the same equivalent temperature that the radar used for Fig. 3.45 has at 10 km from the radiometer.

As can be seen from the figure, even an integration of 10 million samples allows only a small negative input SNR for the most likely long dwell time, c_{TH} (500). This is because the "tails" of the ambient noise are not benign. Obviously, if the allowable false alarm rate is higher, the input SNR can be lower, but false alarms go up very fast. For example, decreasing the threshold by just 2.4 dB increases the false alarm rate from approximately 10^{-10} to 10^{-6}. Because RF front ends drift, threshold uncertainties of 2 to 3 dB are common, even with periodic calibration circuits.

One word of caution, however: because a radiometer threat is a response threat and yet to be invented, its performance can be anything someone can dream up, and it could be formidable. There is no counterstealth silver bullet in radiometric intercept receivers, however.

3.4.4 Typical Response Threats—Correlation

Another response threat that has been postulated is the correlation receiver. This analysis is also speculative, as the threat is yet to be invented. Under certain circumstances, it has the potential for high sensitivity detection just as the radiometric receiver has. The correlation intercept receiver is shown in Fig. 3.49.

This receiver consists of a high-gain antenna with monopulse or two high-gain antennas and two RF front ends matched in bandwidth to the emitter total spectrum or to a single pulse compression chip width, B_I. The outputs of the two channels are multiplied together. This "correlated" output is followed by an envelope detector and a multibeam sliding window noise estimator. The output of the detector goes to an integrator matched to the look or dwell time, t_L, of the antenna beam; i.e., a time bandwidth product of N, which typically has a value between 10^3 and 10^7. The noise estimate is multiplied by a variable threshold multiplier designed to keep the false alarms at an acceptably low value of perhaps 1 per minute (i.e. 10^{-8} to 10^{-12}) per trial. The output of the integrator is then compared to the threshold to determine if a target is present.

The motivation for the correlation receiver is that the emitter waveform is used to demodulate itself. If the SNR is high enough, no matter how complex the waveform, it can be integrated. Another claimed advantage is that it has better performance against internal and external noise that is uncorrelated in angle of arrival. The disadvantages are that it has a narrow correlation angle and thus requires longer search times or channels with delay spaced 1/2 chip width over the equivalent total beamwidth. For example, referring to Fig. 3.31 and assuming a binary phase code, if the phase centers of the two antennas (or monopulse channels) are 10 wavelengths apart and the received signal is at bandcenter, f_0, then the correlation has an extinction at 1.4° off the antenna baseline normal. These extinctions occur every 2.8° with a steadily increasing eclipsing loss as the total delay between channels approaches the chip width. To counteract the extinction problem, the two channels must use inphase and quadrature multipliers or a frequency offset. The eclipsing loss can be dealt with only by the 1/2 chip delay channels. If the transmitted frequency is X-band and the chip width is 1 nsec, then total decorrelation occurs at roughly 90°. This latter fact does help in the side-

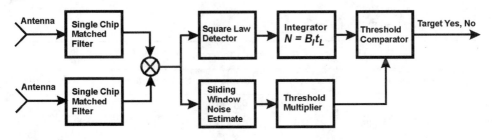

Figure 3.49 Correlation intercept receiver.

lobes, as some sidelobe space will be attenuated. There will be many angles in the sidelobes in which narrowband interferers may fully correlate and be undetectable at the front end. Unfortunately for the correlation receiver, interferers or noise sources from narrow space angles will correlate and successfully limit detection.

Although the noise statistics for the radiometer interceptor are a little different from the correlation interceptor at the input to the integrator, the central limit theorem still applies, and the output statistical performance is virtually identical. The correlation receiver has the same noncoherent integration losses and twice the input noise power, because there are two independent channels. The magnitude of the output of the correlation multiplier has an approximate χ^2 density function with 2 degrees of freedom and a premultiplier of 0.5. Naturally then, the output of the square law detector is χ^2 squared [*not* $1/4 \exp(-x)$], which is then integrated. As is well known, the "tails" of this particular density function have much more area at larger standard deviations than the normal density. Beginning with an input mean of 0 and a variance of 1, and summing $2N$ squared samples, one would expect the approximate output mean to be $2N$ and the approximate output variance to be $8N$. Table 3.14 summarizes a simulation experiment with noise only, noise and a −10 dB SNR emitter, and noise and one mainlobe to sidelobe interferer with the same parameters as in Table 3.12 and with a time bandwidth product of 10,000. The simulation used 4 million random samples and calculated 100 simulated outputs, which were used to estimate the experimental mean and variance (see appendices and software disk for details). Not surprisingly, the outcomes are as expected. The estimated standard deviation for each is given, as well as the threshold multiplier necessary in Equation (3.32), to provide low false alarms (10^{-10}). Note that the SNR for the benign noise case has improved by $N^{0.578}$, as one would expect for noncoherent integration and large N. The SNR improvement is 23.09 dB as calculated from the table using the equation below. The output SNR is still not adequate for normal low false alarm detection.

$$SNR_{out} = 20 \cdot \log\left(\frac{21870 - 19990}{439}\right) = 12.63 \text{ dB and } SNR_{in} = -10.45 \text{ dB} \qquad (3.35)$$

Just as in the radiometer case, it is easily possible to have an interferer that is 100 times the thermal noise power, and the longer the interceptor integrates, the higher

Table 3.14   Output Mean Shift with interferer

Simulation case	Time bandwidth product 10^4			
	Mean output	Variance	STD deviation	c_{TH} for $\mathcal{P}_{FA} = 10^{-10}$
Noise only	19,990	165,100	406	1
Noise plus −10 dB SNR signal	21,870	193,100	439	1
Noise + interferer	$1.02 \cdot 10^8$	$1.133 \cdot 10^{11}$	336,500	477

the probability that more than one interferer will illuminate the interceptor. Neither the correlator interceptor nor the radiometer interceptor can distinguish narrowband from broadband emitters and, hence, cannot easily eliminate interferers. As can be seen, one interferer completely swamps any small signal, and the threshold must be set quite high to avoid these false alarms. In conclusion, Fig. 3.48 applies to the correlation receiver as well as to the radiometric interceptor (for more details see the appendices on disk).

3.5   LPIS versus Interceptor

3.5.1   Screening Jamming

A high-sensitivity intercept receiver is susceptible to sidelobe screening jamming. Suppose a standoff jammer is used that covers the same band as the LPIS. The equation that governs the required screening jamming power is given below [Equation (3.36)].

$$P_{TJ} = \frac{c_{screen} \cdot P_I \cdot (4 \cdot \pi \cdot R_{JI})^2}{G_{TI} \cdot G_I \cdot \lambda^2 \cdot L_{JI}} \qquad (3.36)$$

where
   c_{screen} = the factor of excess power required to provide screening
   R_{JI} = the jammer to interceptor range
   L_{JI} = the loss between the jammer and the interceptor

and all other factors are as previously defined.

For example, suppose that the high-sensitivity interceptor is known as to location either to 5° in azimuth and elevation through some outside intelligence, or it is known as to quadrant by an educated guess so that G_{TI} is either 13.6 or 32.2 dB. Assume that the loss, L_{JI}, is 2 dB, the screening coefficient, c_{screen}, is 4, the interceptor sensitivity, P_I, is −130 dBm., the wavelength, λ, is 0.03 m, and the sidelobe gain G_I is −20 dB. Then, the required jammer power is given in Fig. 3.50 below. For the second case, where the least is known and given a maximum range of 250 km, only 20 W is required. For the first case, where the location is known and given a 250-km standoff range, a little over 0.25 W is required. Older-generation radar and data link systems generate 0.25 W or more spurious emission routinely, even though it is strictly illegal to do so. There can be dozens of emitters visible to the horizon to such a high-sensitivity system.

Far more likely usable sensitivities are limited to P_I = −110 to −115 dBm, in which case the required jammer power for the quadrant scenario approaches 600 to 1000 W at 250 km. Of course, the lower sensitivity allows higher stealth system emissions.

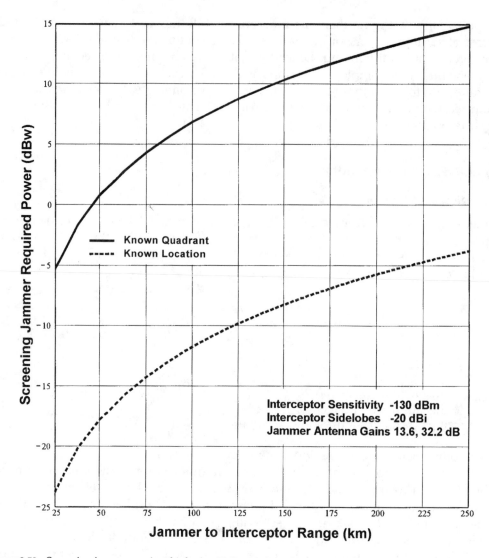

Figure 3.50 Screening jammer against high-sensitivity receiver.

3.5.2 Spoofing

A second counter interceptor technique is *spoofing*. This consists of transmitting an approximation of an adversary waveform interleaved with the LPIS waveform. The spoofing waveform power is chosen to be substantially above the power required of the LPIS for its mission, typically 10 to 100 times. The antenna dwell time, T_D, is also chosen to approximate the same adversary system. The high effective radiated power may result in a preliminary spatial location of the LPI system. Even though the LPIS platform may be in a location inconsistent with the emulated adversary system deployment, multipath is common enough for one or two dwells that the spoofing will not be detected. The stealthy system then goes into a quiet (off) mode long enough for the tentative track files to be dropped (see Chapter 2, Section 2.2.4).

Figure 3.51 shows one possible implementation of the spoofing strategy. In this example, a high-peak-power narrow pulse is transmitted at the adversary frequency, F, while a low-power waveform at a different set of frequencies, F_1 through F_n, is transmitted and received during the interpulse intervals of the simulated adversary waveform. Depending on the relative PRFs of the two waveforms, there may be multiple LPI pulses during the interval between spoof pulses. The LPIS may listen on several of the frequencies simultaneously, depending on the range of interest and the capabilities of the front-end diplexing. The dwell time, T_D, typically will be longer than desirable for an LPIS, but this is part of the spoof. Obviously, the adversary simulated signal must be plausible relative to the stealth platform. For example, a ground attack fighter LPIS should not try to appear to be a fire control system on a naval cruiser.

Another spoofing strategy is the use of an apparent data link waveform as a radar waveform, or vice versa. This can induce incorrect emitter recognition and, consequently, the wrong countermeasures by the adversary. Waveforms described in Chapter 5 can be made to appear to be either data link or radar.

3.6   TYPICAL DEPLOYED INTERCEPT RECEIVERS

Hundreds of different intercept receivers have been deployed worldwide. Often, the actual performance parameters are classified. Table 3.15 is a compilation of unclassified data from *Janes*,[18–23] *ECM Handbook*,[24] manufacturers data,[25,26] and guesses. Systems from communist China, Czech Republic, France, Israel, Russia, United Kingdom, and the U.S.A. were selected on the basis of available information, representative type, and wide usage. The RF coverage is the reference stated coverage or

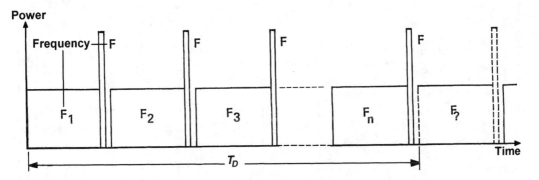

Figure 3.51   Emulation of adversary waveform interleaved with LPI waveform.

Table 3.15 Typical Deployed Intercept Receivers[18-26]

System	RF (GHz)	Azimuth (°)	Elevation (°)	Sensitivity (dBm)	Ant. gain (dB)	Az accuracy (°)	Update time (sec)	Type	Deployment
China									
DZ 9001	1–18	100	20	−70	10–30	3	2	Superhet	Ground
BM/KJ 8602	0.7–18	360	60	−40	0	15	1	CVR	Airborne
Czech									
MCS–93	0.8–18	100	45	−80	10–20	3	1	Superhet	Ground
France									
Strategie	0.8–18	180	45	−80	6	1	1	Interferometer	Ground
Phalanger	1–18	360	45	−50	3	1	1	Int./Transform	Airborne
Israel									
CR2700	0.5–18	360	20	−80	20–40	1	4–8	Superhet	Ground
Kingfisher	2–18	360	40	−60	0	2	4–8	Int./IFM	Airborne
Russia									
Sirena/SPO–10/15/23	6–21	360	45	−55	0	45	1	CVR	Airborne
NRS–1/pole dish	2–4, 8–17	360	45	−70 to −35	24–36	0.3	minutes	Superhet	Ground
RPS–1, 2, 3	0.5–37.5	360	45	−70 to −35	20–35	0.3	minutes	Superhet	Ground
RPS–5/twin box	0.5–10	360	45	−80 to −50	10–20	5	2–8	Superhet	Ground
U.K.									
Weasel	0.7–18	360	45	−80	10, 30	1	4	Hybrid	Ground
Zeus	0.5–18	360	90	−50	0	20/quadrant	2–8	IFM/Hybrid	Airborne
U.S.A.									
ALR–52	0.5–18	360	15–35	−70	13–26	2	1	IFM	Airborne
ALR–56	0.5–20	360	30	−50	0	20/quadrant	2–8	Superhet	Airborne
ALR–69	0.5–18	360	30	−50	0	20/quadrant	2–8	CVR	Airborne
WLR–11	0.5–18	360	45	−70	0	20/quadrant	1	IFM	Ship

numerical equivalent of the letter bands. Azimuth coverage may be instant or mechanically slewed/scanned. Elevation coverage is reference-stated coverage or a guess based on the size and appearance of the antenna in photos. As will be seen in the next chapter, sensitivities are dominated by the environment and signal processing rather than thermal or self-noise. Antenna gain is reference stated gain or a guess based on the size and appearance of the antenna in photos. Angle accuracies are very high signal-to-noise ratio cases, i.e., typically more than 40 dB (see Section 3.3). Update times are for high intercept traffic but are usually quite variable. The author has personally observed variations of 10:1 for what were considered well designed intercept receivers. Type and deployment are inferred from reference stated descriptions. None of this data should be taken too seriously, except that it is representative of what can be achieved. Often intercept and jamming systems do not do nearly as well as they could (including U.S. systems). The author has been in tests in which one could not tell if the threat system was even turned on, based on the observed performance. Many systems are very personnel dependent and, with fresh, experienced operators and maintainors, they work dramatically better than under "normal" conditions.

3.7 EXERCISES

1. Calculate the required transmitter powers for minimum power and minimum dwell and from this, the probability of intercept using the intercept receiver of Table 3.3, for the SMS mode of Table 2.5, 7 frequencies, $\sigma = 1$ m^2, range 100 km, 38×154 inch antenna of Table 2.11, and all other parameters as in the Chapter 2 examples.

2. Calculate the false alarm rate, probability of pulse overlap, and probability of detection by an interceptor with a 16-dB SNR threshold and an LPIS SNR of 0 dB, for a 300 kHz PRF, a 30 msec dwell, and the interceptor performance of Table 3.2.

3. Calculate the AOA accuracy for an SNR of 0 dB, the intercept parameters of Table 3.6, and the missile data link of Tables 2.5 and 2.8.

4. Calculate the cumulative probability of intercept for the example of Fig. 2.11 for T_{OT} of 40 msec using the parameters of Table 3.4.

5. Using the intercept parameters of Table 3.7, the array size of 58×154 from Table 2.11, and all other parameters for the emitter from Table 2.5 for the synthetic aperture spotlight mode, calculate the cumulative probability of intercept.

6. Using the parameters of Exercise 3 above, calculate the interceptor range accuracy for an emitter at 200 mi.

3.8 REFERENCES

1. Bao-Yen Tsui, J., *Microwave Receivers with Electronic Warfare Applications*, John Wiley & Sons, 1986.

2. Feller, W., I., *An Introduction to Probability Theory and Its Applications*, John Wiley & Sons, 1957.
3. Schleher, D., *Introduction to Electronic Warfare*, Artech House, 1986.
4. Skolnik, M., *Introduction to Radar Systems*, 2nd ed., McGraw-Hill, 1980, pp. 28, 410.
5. Rice, S., "Mathematical Analysis of Random Noise," *Bell System Technical Journal*, Vol. 23, 1944, pp. 282–332, and Vol. 24, 1945, pp. 46–156.
6. Gordon, G., "Low Probability of Intercept Radar Evaluation," *RDA-TR-173500-Ml*, December 1978.
7. Sokolnikoff and Redheffer, *Mathematics of Physics and Modern Engineering*, 1st ed., McGraw-Hill, 1958, p. 360.
8. Skolnik, M., Ed., *Radar Handbook*, McGraw-Hill, 1970, pp. 4-4 through 4-8.
9. Barton, D., *Radar System Analysis*, Artech House, 1976, pp. 37–63, 275–310, 32–34.
10. Peebles, P., *Radar Principles*, John Wiley & Sons, 1998, pp. 454–455, 396–406.
11. Papoulis, A., *Probability, Random Variables, and Stochastic Processes*, McGraw-Hill, 1965, pp. 116–199.
12. Brigham, E., *The Fast Fourier Transform*, Prentice-Hall, 1974, pp. 110–120.
13. Blackman, S., *Multiple Target Tracking with Radar Applications*, Artech House, 1986, pp. 65-73
14. Skolnik, M., Ed., *Radar Handbook*, 2nd ed., McGraw-Hill, 1990, pp. 2.28–2.29.
15. Lawson, J. and Uhlenbeck, G., *Threshold Signals*, Boston Technical Publishers, 1964, p. 107.
16. Berkowitz, R., *Modern Radar*, John Wiley & Sons, 1965, pp. 418–422, 187–190.
17. ITT Corporation, *Reference Data For Radio Engineers*, 6th ed., Howard W. Sams & Co., 1975, p. 33-2.
18. Streetly, M., Ed., *Janes Radar and Electronic Warfare Systems 1999–2000*, Janes Information Group, 1999, pp. 29–37, 41–64, 377–391, 435–530.
19. *Janes Weapon Systems 1984-85*, Janes Information Group, 1984, pp. 581–678.
20. Brinkman, D., Ed., *Janes Avionics 1990-91*, Janes Information Group, 1990, pp. 134–199.
21. Downs, E., Ed., *Janes Avionics 2001-2002*, Janes Information Group, 2001, pp. 347–524.
22. Cullen, T. and Foss, C., Eds., *Janes Land-Based Air Defence 2001–2002*, Janes Information Group, 2001, pp. 49–103, 109–194, 273–340.
23. Williamson, J., Ed., *Janes Military Communications 2001–2002*, Janes Information Group, 2001, pp. 163–377, 459–480, 715–770.
24. Horizon House, *International Electronic Countermeasures Handbook*, 2001 ed., pp. 307–337.
25. Argo Systems, Data Sheets, WLR-11.
26. Lockheed Martin, web page.
27. Schlessinger, M., *Infrared Technology Fundamentals*, Marcel Dekker, Inc., 1995, pp. 1–75.
28. Hughes Aircraft, "LPIR Phase 1 Review," unclassified report, 1977.
29. Tsui, J., Thompson, M., and McCormick, W., "Theoretical Limit on Instantaneous Dynamic Range of EW Receivers," *Microwave Journal*, January 1987, pp. 147–152.
30. A few photos, tables, and figures in this intellectual property were made by Hughes Aircraft Company and first appeared in public documents that were not copyrighted. These photos, tables, and figures were acquired by Raytheon Company in the merger of Hughes and Raytheon in December 1997 and are identified as Raytheon photos, tables, or figures. All are published with permission.
31. Schleher, D., *Electronic Warfare in the Information Age*, Artech House, 1999, pp. 133–199.
32. Sareen, S., "High Intercept Probability Improves ESM Performance," *Microwave Systems News*, July 1979.

4

Exploitation of the Environment

A significant part of stealth strategy is concerned with exploiting the environment in every way possible, consistent with the mission requirements. This consists of exploiting not only the natural environment (atmospheric attenuation, clutter, and so forth) but also the man-made environment [such things as the electronic order of battle (EOB) and electronic countermeasures (ECM)]. In the last part of this chapter, an example LPI-interceptor scenario analysis will be performed. This analysis shows the essential elements of most LPI mission assessments. In the third chapter, there was a survey of the performance and parameters of current intercept receivers. This chapter shows, among other things, that the environment limits passive and active detection of stealth platforms rather than raw sensitivity.

4.1 ATMOSPHERIC ATTENUATION

There is a strategy for an emitter in the atmosphere that will enhance low intercept probability. Figure 4.1 shows one-way atmospheric attenuation as a function of operating wavelength. There are several different contributors to atmospheric attenuation at micro and millimeter wavelengths. The first significant contributor is water, which has a resonance near 1.33 cm ($\approx$22.5 GHz). There is a second resonance near 0.16 cm ($\approx$183 GHz) in the millimeter-wave band. The oxygen in the atmosphere also has several resonances that have high levels of attenuation. There is an oxygen resonance that is very strong at approximately 0.5 cm ($\approx$60 GHz) and a second one at roughly 0.25 cm ($\approx$118 GHz). These resonances are widened by atmospheric pressure, so they are not like the ones astronomers often observe (even though they arise from the same physics).

A common LPI strategy is to operate near one of these atmospheric absorption peaks so that interceptors beyond the operating range will receive greatly attenuated signals from the LPIS. For example, this strategy suggests operation at Ku-band near the water line (❶ in Fig. 4.2), at Ka-band in the window between the water and the oxygen resonances (❹ in Fig. 4.2), or in the window centered about 95 GHz (❸ in Fig. 4.2). If one were operating a radar or data link in outer space and wished that the emitter not be intercepted on the Earth, then one of the oxygen resonances would be an excellent choice for operating frequency (❺ in Fig. 4.2). Alternatively, on Earth, the

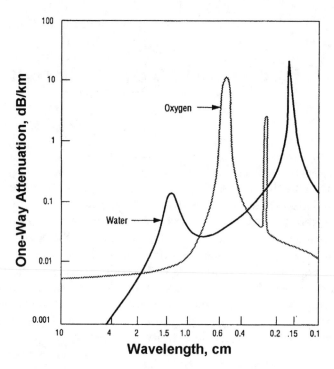

Figure 4.1 Atmospheric attenuation.

oxygen line might be an excellent radar operating frequency for very short-range applications (such as automotive active cruise control) while providing good LPI performance. Obviously, the density of constituents of the atmosphere, such as water and oxygen, is a strong function of altitude, weather conditions, and location on the Earth. Therefore, the right operating frequency may be different, depending on the mission of the stealth weapon system. For example, Fig. 4.2 shows the average atmospheric absorption of microwaves at two different altitudes for the composite of water and oxygen. Notice that the atmospheric attenuation in fractions of a decibel per kilometer is dramatically greater at sea level than it is at an altitude of 12,000 ft. This is because usually at 12,000 feet less oxygen and less water are present. The strategy of operating at or near one of these atmospheric resonances greatly simplifies the degree of difficulty for the design of the passive (RCS) part of a stealth signature, as well as improvement of the active signature.

A typical passband attenuation for a stealth antenna or radome is plotted on Fig. 4.2 (❶). Notice that the upper frequency rolloff can be shallower, because the water resonance provides additional protection. A related strategy is to operate at a frequency on the rising sides of one of the resonances and to adjust the operating frequency for more or less attenuation as a function of emitter mode (❷ in Figure 4.2). Table 4.1

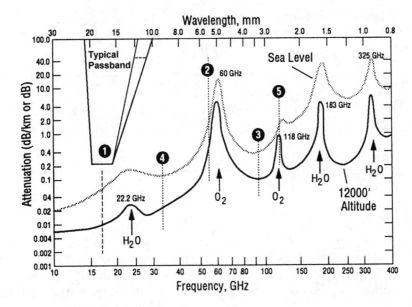

Figure 4.2 Average atmospheric absorption of radar waves.

Table 4.1 LPI Frequency Selection Matrix

Operating modes and conditions	Conventional frequency	Frequency masking	Low threat	Near O_2	Near H_2O	Higher frequency
High-altitude for LPIS and target		X		X	X	
Medium-altitude air-to-ground						X
Medium-altitude air-to-air		X				X
Interceptor closer than target		X				
Low-altitude long-range	X (No LPI)		X			
Low-altitude, dry medium-range					X	X
Low-altitude, over water, short-range		X	X			
High inband RCS, LPI			X	X (land)	X (water)	

summarizes some LPI frequency selection criteria for typical operating modes and locations. Another strategy is to operate in bands where there may be fewer threats. Table 4.2 tabulates relatively low threat regions of the spectrum.

These frequency bands are not currently approved for use for radars or data links but, under certain circumstances, might be used for LPI systems, because they are not likely to be detected, let alone cause interference.

Table 4.2   Low Threat Frequency Regions

Radar range performance rank	Frequency (GHz)	Typical current use
1 (longest range)	5.65–6.275	Amateur radio and telecommunications
2	10.48–11.96	Satellite downlink, amateur radio and telecommunications
3	13.37–14.4	Satellite uplink, and navigation
4 (shortest range)	16.2–20	Maritime and satellite communications

4.2   CLUTTER

Another attribute of the environment that can be exploited by an LPIS is clutter. Clutter can be used to mask both the RCS of a stealth vehicle as well as to confuse or mask the active emissions of an LPIS. As shown later, stealth vehicles are much better off in clutter. For example, when stealth aircraft operate at low altitudes, they will be embedded in clutter, and the probability of terrain masking is increased. This is true for both radar and electro-optical signatures. In general, the long-range threat will be at low grazing angles, even for spacecraft-based sensors. Table 4.3 gives land clutter reflectivity data for very shallow grazing angles between 0 and 1.5° for a variety of frequencies from UHF to Ka-band.

Table 4.3   Land Clutter Reflectivity, 0–1.58 Grazing Angle

Terrain type	UHF 0.5–1 σ_0	UHF σ_{84}	L 1–2 σ_0	L σ_{84}	S 2–4 σ_0	S σ_{84}	C 4–8 σ_0	C σ_{84}	X 8–12 σ_0	X σ_{84}	Ku 12–18 σ_0	Ku σ_{84}	Ka 31–36 σ_0	Ka σ_{84}
Desert	30		45	35	46	31	40		40	32				
Farmland	28		32V	26	34	24	33	25	33	20	23	10	18V	10V
Open woods			34H		33	26			30	22				
Wooded hills	24	18	28	20	28	15	27	14	26	13	20V	17	13V	8V
Residential					35	26	35	26	30	24				
Cities	22		25	15	23		21		20	>7	20V	10		

Reflectivity, dB below 1 m²/m² @ band, GHz*

*Average of both polarizations except where noted. σ_0 = median or mean backscatter coefficient in decibels below 1 m²/m² · σ_{84} = reflectivity that 84 percent of the cells are below. Composite of Refs. 1–3 and 8.

Similarly, Table 4.4 shows land clutter for 5 to 10° grazing angle. Tabulated data for various types of terrain are shown for the median and the upper one-standard-devia-

Table 4.4   Land Clutter Reflectivity, 5–10° Grazing Angle

Terrain type	UHF 0.5–1 σ_0	σ_{84}	L 1–2 σ_0	σ_{84}	S 2–4 σ_0	σ_{84}	C 4–8 σ_0	σ_{84}	X 8–12 σ_0	σ_{84}	Ku 12–18 σ_0	σ_{84}	Ka 31–36 σ_0	σ_{84}
Desert	40	34	39		36		33		30		18V	13V	25	
Farmland	36	30	30	26	28		26		26		22	11V	18V	10V
Open woods	23	20	23		33	26			30	22	22	15V		
Wooded hills	22	18	26	20	24		23		23		20	14V	13V	8V
Residential	23	17			23		35	26	30	24	18V	11		
Cities	9		15		18		18		16		9V	3V		

Reflectivity, dB below 1 m^2/m^2 @ band, GHz

*Average of both polarizations except where noted. σ_0 = median or mean backscatter coefficient in decibels below 1 m^2/m^2 · σ_{84} = reflectivity that 84 percent of the cells are below. Composite of Refs. 1–3 and 8.

tion point for both horizontal and vertical polarization, where available. These tables are the composites based on multiple sources, and the sources suggest that errors of 5 dB might be typical.

The sensor designer is forced to minimize false alarms. As a result, these clutter uncertainties lead to higher thresholds and more signature space in which to hide. Normalized reflectivity coefficients (in meters per square meter) are usually fairly low, especially for low grazing angles, but the number of square meters competing with a stealth platform usually is large; hence, operating in a high-clutter environment is an advantage. Operating at higher frequencies over areas that contain cities or heavily wooded terrain may provide advantageous performance by enhancing multipath and competing clutter.

Figure 4.3 shows a clutter summary as a function of depression angle at X-band for a wide range of terrain types. As the grazing or depression angle gets steeper, clutter backscatter typically rises. There are some notable exceptions for this, such as commercial and industrial areas that may contain high-rise buildings and many vertical features with large RCS. Furthermore, commercial and industrial areas typically have dramatically higher cross sections than areas that are largely devoid of cultural features. In addition, backscatter from clutter is conservative; most of the energy that a radar or data link places on the ground either intentionally or unintentionally is reflected, a fraction back in the direction of the radar observer but much scattered in other directions. This multidirectional scatter can be an advantage in commercial, industrial, and heavy vegetation areas, because it confuses angle-of-arrival intercept systems and direction finding. Angle of arrival differences of 10° with only 5 dB of attenuation over path lengths of 50 mi are common occurrences. In areas with many

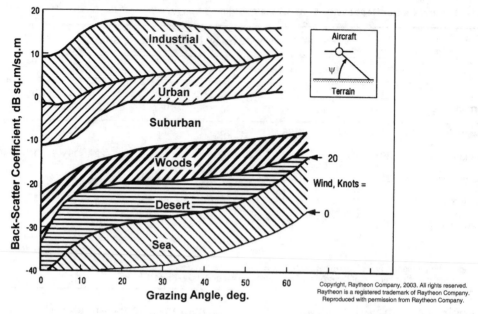

Figure 4.3 Summary of X-band clutter reflection data.[23]

cultural features, the scattering in other directions may be so strong as to punch through the sidelobes of the antenna on the intercept receiver and dramatically degrade dynamic range and signal discrimination characteristics. Over calm sea water or smooth desert sand, however, clutter is not as easily exploitable. One alternative is to enhance the clutter with ground-bounce ECM.

Another attribute of clutter that is an important property and beneficial for stealth vehicles is the probability density function of RCS with respect to size. For example, Ku-band data taken over rural terrain at medium resolution (40 × 40 ft cell) can be approximated by a normal probability density function, as shown in Fig. 4.4. The data shown is prior to pulse compression out of 6-bit A/D converters. A little over 20,000 sample points were used with a post-pulse compression dynamic range of 60 dB and a median reflectivity of −24.9 dBsm. The input is almost white noise, but the output is most decidedly not, as shown in the figures that follow.

Some representative statistics with and without smoothing for populated areas are given in Figs. 4.6 through Fig. 4.9. This data is from the USAF FLAMR program, which was declassified in 1987 and 1988.[8] One clutter feature important to stealth platforms is bright discretes. Figure 4.5 shows a summary collection of data for bright discretes in terms of density per square nautical mile versus equivalent radar cross section in square meters.

There are several important features to the clutter in populated areas. The mean and the median clutter reflectivity can differ by as much as 10:1 (very much non-

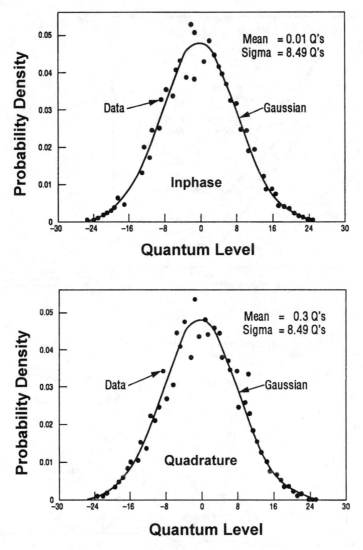

Figure 4.4 Rural terrain raw I-Q data histograms.[8]

Gaussian!). Commercial areas have large statistical tails and hence a significant probability of very large cross section. This means that there is a small but finite probability that there will be some very large RCS scatterers over any reasonable observation space. Similarly, an RCS probability density function over heavily populated terrain (e.g., Los Angeles) has a greatly distributed density function, which has an even larger probability of a few extremely large scatters in any observation space. The probability of a 10^5 m^2 scatterer can be as high as 0.001 in a city. There is a small but fi-

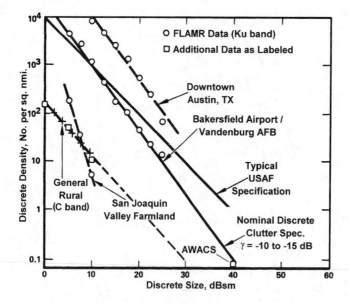

Figure 4.5 Density of bright discretes.[23]

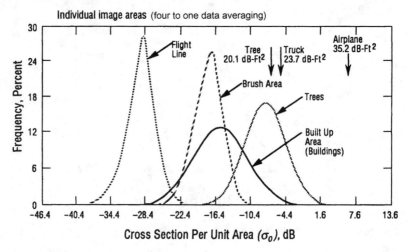

Figure 4.6 Clutter histograms for Vanenberg AFB vicinity.[8]

nite probability that a "bright discrete" scatterer will be in a range cell. A USAF specification used on many programs expects a bright discrete of 10^5 m^2 in every 10 nmi^2 and, similarly, a bright discrete of 10^6 m^2 in every 100 nmi^2, and so on. The sidelobes of these scatterers can be in hundreds of cells, which can mask a small target for quite an area over densely populated areas.

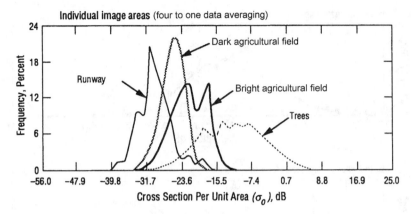

Figure 4.7 Clutter histograms for Point Mugu, CA, vicinity.[8]

The measurements of Ku-band reflectivity made with a synthetic aperture radar have high enough resolution that individual terrain features can be separated, compared to ground truth, and categorized as to type of clutter as shown in Figs. 4.6 through 4.9. These typically use thousands of data points to estimate the underlying statistics. Not withstanding the large sample sizes and averaging, the four figures still show significant variability for terrain that does not seem that different in the SAR imagery. The grazing angles in these cases were typically 5 to 10°. Clutter has been measured by many researchers since the 1930s. There are many approximate equations that characterize the main aspects of clutter as given in the following paragraphs, but they are by no means the final word on the subject, and the references throughout the book give many more examples.

Another aspect of clutter is from the point of view of its equivalent cross section per radar cell as a function of range for a fixed antenna size. For surface-based radars, the Earth's curvature dramatically reduces RCS with range. Such a typical graph is shown in Fig. 4.10 for various kinds of surface radar clutter.

This graph shows typical clutter returns and equivalent cross section for land clutter, several different sea states, and various amounts of rain and chaff as a function of range. Also plotted on the graph for reference is typical equivalent receiver noise and where it limits performance. As can be seen, surface clutter rapidly drops away with range because of radar horizon effects. However, volume clutter from such things as chaff and rain steadily increases as the range increases, because the cell size grows dramatically with range. Usually, the range at which a threat radar detects an LPIS, as limited by receiver noise, is far greater than its equivalent performance in the presence of surface and atmospheric clutter.

Land and sea clutter are proportional to illuminated area. They are strongly grazing-angle dependent. Often, grazing angle and depression angle are close enough in value to be used interchangeably but, for extended clutter purposes, grazing angle

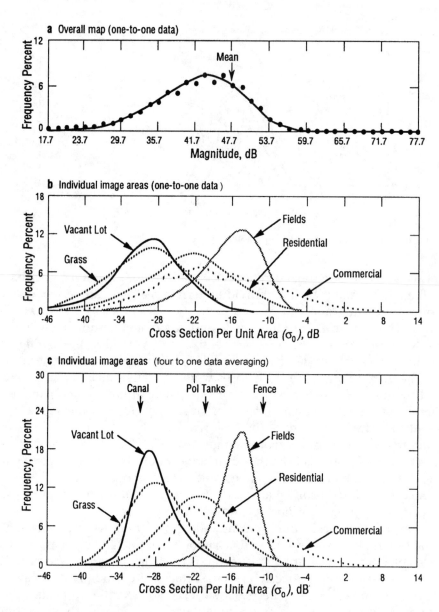

Figure 4.8 Clutter histograms for Bakersfield, CA, airport vicinity.[8]

must take into account the Earth's curvature. In this case, grazing angle and reflectivity are typically approximated when operating in the atmosphere as shown in Equations (4.1) and (4.2). The emphasis in the equations that follow is simple first-order approximations.

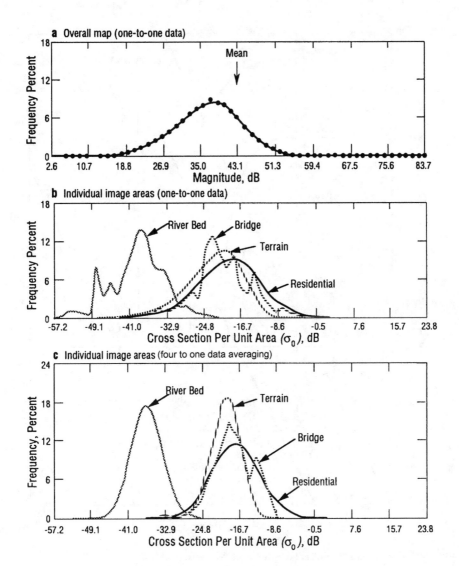

Figure 4.9 Clutter histograms for Austin, TX, vicinity.[8]

The grazing angle is

$$\psi \cong \sin^{-1}\left(\frac{h_r}{R_s} - \frac{R_s}{2 \cdot 4/3 \cdot R_e}\right) \tag{4.1}$$

The unit area radar cross section is

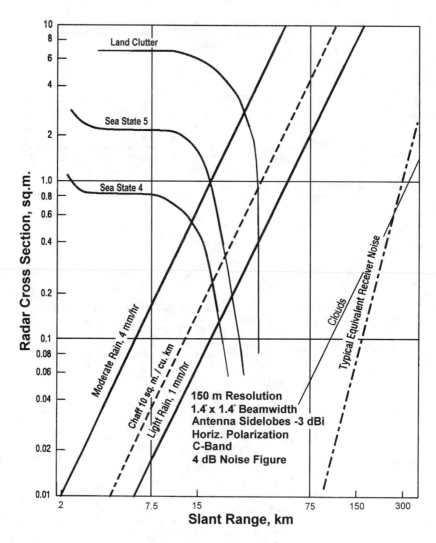

Figure 4.10   Typical surface radar clutter (adapted from Nathanson[2]).

$$\sigma_{0land} = \gamma \cdot \sin\psi \cong \gamma \cdot \left(\frac{h_r}{R_s} - \frac{R_s}{2 \cdot 4/3 \cdot R_e} \right) \qquad (4.2)$$

where
  $\gamma =$ a backscatter coefficient that is reasonably constant for a given frequency with a typical σ_{84} value of 3
  $h_r =$ height of the radar

R_s = slant range to the target
R_e = average Earth radius with a value of 6,370,880 m[1]

Sea clutter is strongly dependent on grazing angle, rms wave height, and operating wavelength. Wave height is related to surface winds and location on the Earth. Equation (4.3) is one of many approximations to unit area sea clutter.

$$\sigma_{0sea} = \exp\left[-8 \cdot \pi^2 \cdot \left(\frac{\sigma_h \cdot \psi}{\lambda}\right)^2\right] \qquad (4.3)$$

where σ_h is the rms wave height, typically 0.107 m for sea state 1, 0.69 m for sea state 4, 1.07 m for sea state 5, and, for the *perfect storm*, sea state 8, 6 m.

Similarly, rain clutter is proportional to illuminated volume. Equation (4.4) is one approximation to unit volume rain clutter.

$$\sigma_{1rain} \cong 7.03 \cdot 10^{-12} \cdot f^4 \cdot r^{1.6} \qquad (4.4)$$

where f is operating frequency in gigahertz, and r is rainfall rate in millimeters per hour.

Yet another aspect of clutter is its velocity distribution as a function of range. A clutter velocity model for various types of backscatter is shown in Fig. 4.11. This figure shows apparent radial velocity as a function of range. Notice that, as range increases,

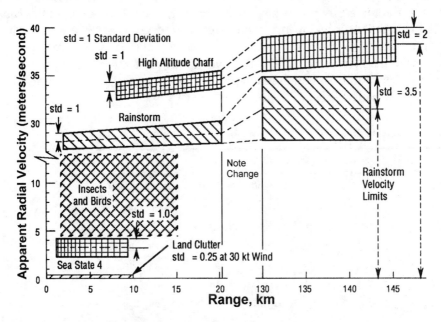

Figure 4.11 Clutter velocity model (adapted from Nathanson[3]).

the apparent radial velocity of chaff and rain gets larger and is determined by the velocity of either the storm cell or the chaff cloud. A similar condition exists at shorter range for sea and land clutter. Many land clutter items, such as rooftop ventilators and fans, have apparent velocities similar to aircraft and all land vehicles. The only way to sort these scatterers out is to track them long enough to determine that their location is unchanged. In addition, there is a region of velocities, occupied by insects and birds, that fills in additional parts of the range-velocity space at lower altitudes. Insects and birds can be especially problematic for radars attempting to detect and track stealth vehicles.

As suggested by Fig. 4.11 (adapted from Fig. 8.7, p. 371, of Ref. 3), at low altitudes, there is significant natural moving clutter arising from insects and birds. Table 4.5 tabulates the RCS, flock size, and aspect for three different radar bands for several common birds and insects. As can be seen from Table 4.5, even though the individual insect or bird cross sections are extremely low, their aggregate RCS can be significant, depending on the quantity in a radar cell. Grackles, for instance, flock in large numbers at night to feed on insects, and it is not uncommon for 1000 grackles to be in a single radar cell. A flock of grackles can have a 1 m^2 RCS in a single radar cell and have internal Doppler, which easily masks a stealth aircraft. These conditions have been observed regularly on radar programs such as the TPQ-36 and 37. The old saying "birds of a feather flock together" comes into play, in fact, forcing higher false alarms, more processing, or lower sensitivity. Furthermore, wing speed as well as velocity can be high enough, especially in certain kinds of hummingbirds and geese, that these birds can easily be confused with aircraft. These sources of clutter can be rejected, of course, but a major increase in signal processing is required and, in many cases, dramatically lower exposure times for a stealth system are the result.

The conclusions from Fig. 4.11 and Table 4.5 are that there is significant moving clutter at lower altitudes, which results in a major increase in threat processing and lower stealth vehicle exposure times.

Table 4.5   Reasons to Stay Low in a Stealth Vehicle

Type of bird or insect	Flock size per cell	Individual RCS (dBsm)			Typical net RCS per cell (dBsm)
		UHF	S–Band	X–Band	
Sparrow	10	−56	−28	−38	−18 to −46
Pigeon	100	−30	−21	−28	−1 to −10
Duck	10	−12	−30	−21	−2 to −20
Grackle	1000	−43	−26	−28	+4 to −13
Hawkmoth	10	−54	−30	−18	−8 to −44
Honeybee	1000	−52	−37	−28	+2 to −22
Dragonfly	10	−52	−44	−30	−20 to −42

4.3 TERRAIN MASKING

Another important issue is terrain and local feature masking. As grazing angles become very low, the amount of the terrain and the space immediately above the terrain actually visible decreases dramatically. Figure 4.12 shows terrain and local feature masking comparisons for two different depression angles at 8° and 4° for the Fulda Gap region in Germany. The black areas are masked from the observer's viewing direction. Note also that the city in the upper left corner of the images is also partially masked at lower grazing angles. It can be seen that significant parts of the terrain, and thus a stealth vehicle, will be masked if it is flying low or on the surface at shallow viewing angles.

A study of terrain masking versus elevation angle was performed at the University of Southern California under contract to Hughes Aircraft in the late 1970s for the Cooperative Weapon Delivery program.[10] One result of that study is summarized in Fig. 4.13. The figure shows the fraction of the surface terrain that is masked as a function of elevation or grazing angle for various types of terrain. What can be observed is that, for terrain that contains woods and urban areas, as much as 40 percent of the terrain is masked at a grazing angle of 5°. This study used hypsographic (topographical relief) map data for a wide range of areas in both Germany and the U.S.A. One can observe that a significant fraction of the terrain is masked because of the height of urban features or trees. There are a number of classified and unclassified terrain masking studies from which data can be obtained for more accurate modeling.[3,4] These terrain features can be used by an LO/LPI system not only to mask itself from conventional radar detection but also to scatter its emissions and prevent reception by an interceptor. One possible terrain-masking model that fits the experimental data reasonably well is a calculation of the cumulative probability of obscuration along the line of sight between the surveillor and the target or emitter. The instant probability of

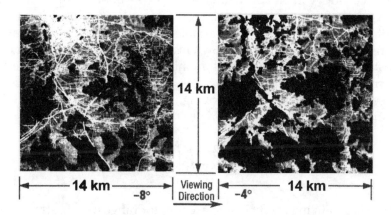

Figure 4.12 Terrain and local masking comparison.

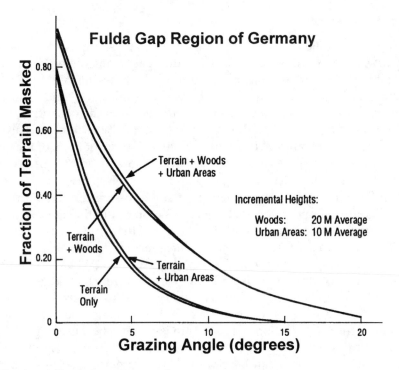

Figure 4.13 Terrain masking versus grazing angle.

obscuration can be modeled as a function of the apparent height of the line of sight (LOS) above mean terrain altitude. The cumulative terrain height in a range cell, ΔR, determines whether the observer can see beyond that cell to the next. The probable height in a cell is a function of the terrain roughness.

Cell height statistics can be approximated by a chi-squared amplitude distribution with two degrees of freedom. One minus the cumulative chi-squared distribution with LOS altitude as the independent variable determines the instant probability that the remaining terrain is obscured as shown in Equation (4.5). (This is not exactly true, but is almost true when most LOS cells are higher than the terrain.) The LOS height must be normalized by the terrain roughness or variance and the cell size in which the roughness was calculated. Summarizing, Equation (4.5) gives a simple approximation for the probability of obscuration per range cell.

$$\mathcal{P}_{ob} = 1 - \int_{0}^{h_a} \frac{x \cdot Cell^2}{C_r^2} \cdot \exp\left(-\frac{(x \cdot Cell)^2}{2 \cdot C_r^2}\right) dx \tag{4.5}$$

where *Cell* is a normalizing coefficient to account for range cell size (range cells are *not* range bins but, rather, correlation lengths), ΔR, and the terrain roughness statistic in that size range cell; h_a is the apparent height or altitude of the line of sight above mean terrain in a specific range cell; and C_r is the terrain roughness in the same units as the apparent height (typically feet). For example, for a range cell, ΔR, of one nautical mile and a cumulative experimental probable rms terrain height, *Cell* is 1.41, i.e., the square root of 2 sigma of a cumulative chi-squared function. Each incremental height for additional terrain features is square root sum squared with the basic terrain roughness to approximate overall roughness. Typical rms values for 1 nautical mile range cell size for C_r are 50 ft for steppe terrain (typical of Siberia, Canadian Territories, Kansas, and Florida—no glacial moraines, however, in Florida), 150 ft for grassy terrain (typical of Indiana, Oklahoma, and Northern Africa—no grass, however, in North Africa), 250 ft for wooded terrain (typical of Eastern New York and Canada, Germany, Austria, and Viet Nam) and 1000 ft for mountainous terrain (typical of the Rockies, Sierras, Switzerland, Urals, Afghanistan, Himalayas, and Columbia). Figure 4.14 shows plots of Equation (4.5) for the parameters mentioned above and a range cell size of 1500 ft.

The microwave horizons in nautical miles for the 4/3-Earth model for surveillor and target, respectively, are given in Equations (4.6).

$$R_{Hs} = (1.998 \cdot h_s)^{0.5} \cdot \frac{5280}{6076.1} \text{ and } R_{Ht} = (1.998 \cdot h_t)^{0.5} \cdot \frac{5280}{6076.1} \qquad (4.6)$$

where h_s is the height of the surveillor above the Earth's surface in feet, and h_t is the height of the emitter or target above the Earth's surface in feet. The maximum visible slant range typically is less than or equal to the sum of the horizon distances for the surveillor and the target/emitter as shown in Equation (4.7). If the target is inside the slant range to the horizon, then obviously the maximum range is the distance between the surveillor and target.

$$R_{s\,max} \le R_{Hs} + R_{Ht} \qquad (4.7)$$

The apparent altitude is the height of the line of sight between the surveillor and the target over a 4/3 Earth for every possible slant range. The apparent altitude in feet for a given slant range with all units in feet is shown in Equation (4.8).

$$h_a = \left\{ \frac{\left[\left(1 - \frac{R_s}{R_{smax}}\right) \cdot (R_{e43} + h_s)^2 + R_s^2 \right.}{\left. + [(R_{e43} + h_t)^2 - R_{smax}^2] \cdot \frac{R_s}{R_{smax}} \right]} - R_{e43} \right\} \qquad (4.8)$$

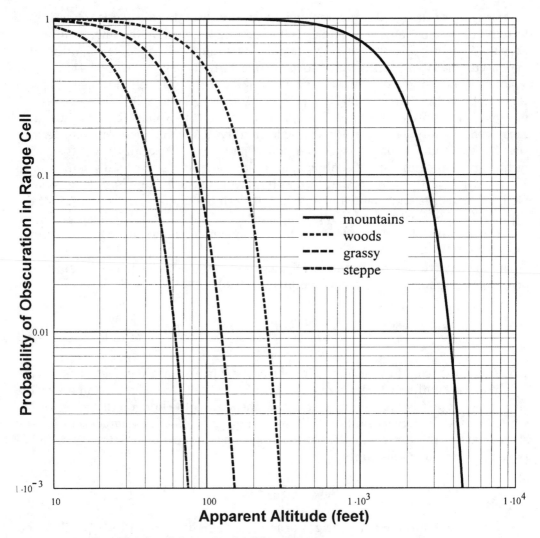

Figure 4.14 Instant probability of obscuration for different terrains.

where R_s is the slant range to the range cell at the height, h_a, and R_{e43} is the mean radius of the 4/3 Earth, generally taken to be 27,871,086 ft. [Yes, this value is slightly different from what one would calculate using the value in Equation (4.2).[2] The same controversy surrounds the velocity of light.]

Figure 4.15 shows the geometry for Equations (4.6) through (4.8). The surveillor, the target, and the center of the Earth form a triangle from which the apparent height can be estimated. The distance to the surveillor from the center of the Earth is $h_s + R_e$; so as to take atmospheric refraction into account, the Earth radius to be used is R_{e43}. The

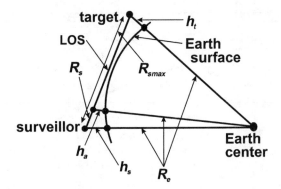

Figure 4.15 Apparent altitude geometry.

second leg of the triangle is the distance to the target from the center of the Earth, $h_t + R_e$; again, the Earth radius to be used is R_{e43}. The third leg is the line of sight between the surveillor and the target with slant range R_{smax}. Because the triangle is completely known, $h_a + R_e$ can be calculated from the slant range to a specific range cell R_s, and subtracting R_e yields h_a. Obviously, h_a is a small difference in large numbers, and one must be careful to prevent ridiculous results.

An example plot of apparent altitude versus range for a surveillor altitude of 6000 ft and a target altitude of 10 ft is shown in Fig. 4.16. Note that the apparent altitude drops to essentially zero, because the LOS grazes the limb of the Earth at the maximum range. Clearly, visibility is very improbable at maximum range. Even very smooth terrain, such as steppe, will not provide much visibility near maximum range.

Let $R_s = j \cdot \Delta R$, where ΔR is the length at which cells are 75 percent decorrelated (not radar or ladar range bins). Then, the cumulative probability of obscuration out to the jth range cell, R_s, is 1 minus the product of the instant visibilities (1 minus the probability of obscuration equals the probability of visibility) for all the range cells to the range R_s as shown in Equation (4.9).

$$\mathcal{P}_{cumob} = 1 - \prod_{i=1}^{i=j} \{1 - \mathcal{P}_{ob}[h_a(i \cdot \Delta R)]\} \tag{4.9}$$

Repeated calculation of the cumulative obscuration at successive values of R_s, $(j \cdot \Delta R)$, out to R_{smax} allows the generation of plots of obscuration versus range or grazing angle similar to the experimental data. For example, consider two cases

1. The surveillor altitude is 30,000 ft, and the target altitude is 100 ft (or vice versa) in wooded or grassy terrain typical of AEW aircraft detecting a cruise missile.

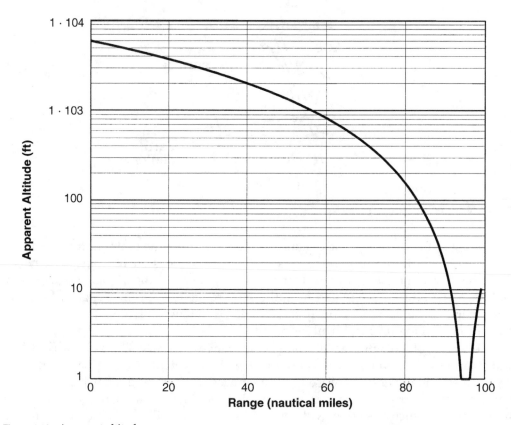

Figure 4.16   Apparent altitude versus range.

2. The surveillor altitude is 6000 ft, and the target altitude is 10 ft in wooded or mountainous terrain typical of mountaintop, UAV, or helicopter surveillance in covert military operations.

A graph of obscuration probability for case one using a ΔR size of 1 nmi and a range bin to cell normalizing factor of 0.248 is shown in Figs. 4.17 and 4.18 as a function of range and grazing angle, respectively. Note that the visibility is down to 50 percent for wooded terrain at 85 to 90 nmi, which is roughly 40 percent of maximum range. For grassy terrain, visibility is 50 percent at roughly 65 percent of maximum range.

Figures 4.19 and 4.20 show visibility for Case 2 wooded and mountainous terrain when the surveillor altitude is 6000 ft and the target altitude is 10 ft with a cell size, ΔR, of 1 nmi; Cell is 0.248; and C_r has rms values (i.e., 63 percent of all cells will not rise more than C_r in 1 nmi) of 50 ft for steppe, 150 ft for grassy, 250 ft for wooded, and 1000 ft for mountainous terrain. As can be seen, the visible range is almost nil for

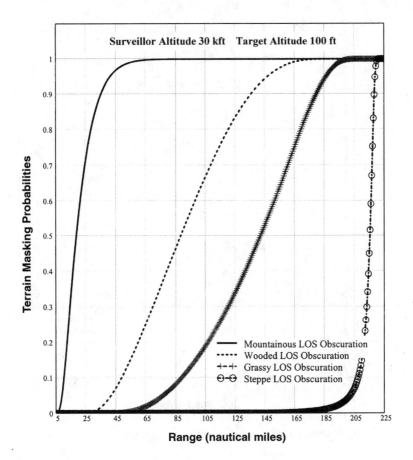

Figure 4.17   Terrain masking versus range—case 1.

mountainous terrain and low-altitude surveillance. For example, such conditions exist in the Balkans, Alps, and Dolomites in Europe. The maximum range for this case is approximately 98 nmi, so the 50 percent visibility range for wooded terrain is a poor 15 percent of maximum.

The grazing angle for low-altitude targets can be approximated by Equation (4.10).

$$\psi = \sin^{-1}\left[\left|\left(\frac{h_s}{R_{smax}}\right)\cdot\left(1 + \frac{h_s}{2\cdot R_e}\right) - \frac{R_{smax}}{2\cdot R_e}\right|\right] \tag{4.10}$$

One very rough rule of thumb from the foregoing is that visibility is typically down to 10 percent at approximately 80 percent of maximum range between the surveillor and the target for their respective operating altitudes.

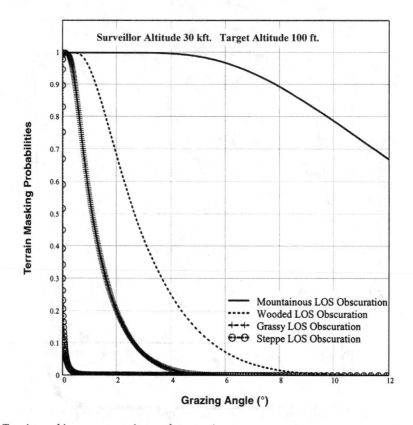

Figure 4.18   Terrain masking versus grazing angle—case 1.

Another aspect of the visibility issue for low-altitude targets is the persistence of visibility with travel in the cross-range dimension. Assuming that the target is detected at a specific range, R_s, what is the probability that it will stay visible for a cross-range distance of $R_c = (j - 1) \cdot \Delta R$? As the target moves a given number of cells laterally, it is repeatedly subject to the LOS visibility probability associated with the range, R_s. However, one cell lateral displacement means that most cells along the line of sight are still the same or almost the same. In fact, the new set of cells overlap the old set by more than 50 percent for 1/2 the range between the observer and the target. Clearly, for the highly correlated cells, if they permitted visibility before, they still permit visibility. An arbitrary offset is three cells in which less than 1/2 of the cells are correlated. The cumulative lateral obscuration per cell of offset on the basis of the foregoing assumption is shown in Equation (4.11).

$$\mathcal{P}_{cumobl} = 1 - (1 - \mathcal{P}_{cumob}(\Delta R))^{\frac{j-1}{3}} = 1 - \mathcal{P}_{cumvis}(\Delta R)^{\frac{j-1}{3}} \qquad (4.11)$$

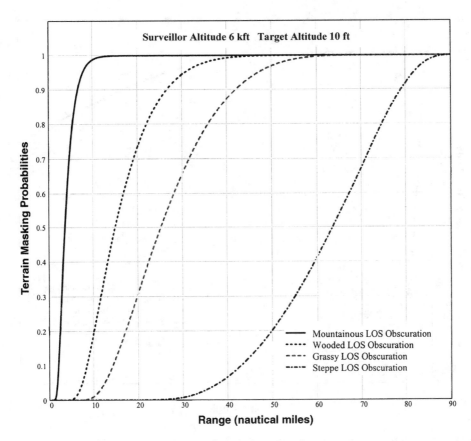

Figure 4.19 Terrain masking versus range—case 2.

where j is the cell index described above, ΔR is the cell length, and the cumulative visibility is $\mathcal{P}_{cumvis} = 1 - \mathcal{P}_{cumob}$. For wooded terrain with the surveillor at 30,000 ft and the target at 100 ft, lateral visible lengths are quite short until the target is less than 50 percent of maximum range as shown in Fig. 4.21. In other words, exposure times are less than 20 sec 50 percent of the time if the target is travelling at Mach 0.85 laterally to the surveillor at greater than 1/2 of maximum range. As surveillor altitude goes down, the situation gets even better for the low-altitude penetrator. Parenthetically, cross-range visibility for low-altitude targets makes one of the most compelling arguments for E-3 AWACS type surveillance aircraft, as visibility becomes dramatically better as the surveillor operates at higher altitudes.

Usually, a simple exponential approximation can be made to the obscuration curves of the form shown in Equation (4.12).

$$\hat{\mathcal{P}}_{cumob} = 1 - \exp(-a \cdot (R_s - b)^c) \qquad (4.12)$$

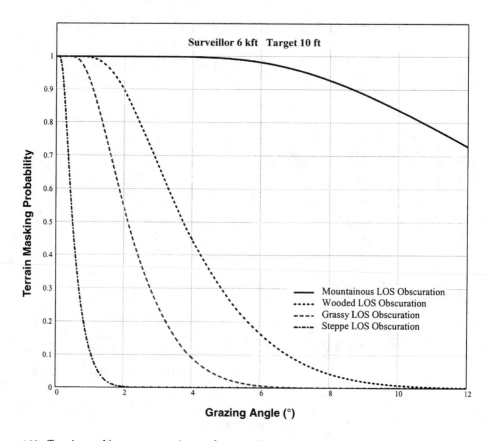

Figure 4.20   Terrain masking versus grazing angle—case 2.

where a, b, c are best fit values for an exponential regression to the curves generated from Equations (4.9) and (4.11). Terrain masking is beaten to death in the appendices.

The situation is significantly different if the terrain is gently rolling or grassy as shown in Fig. 4.22. For the same conditions mentioned above, the exposure time is 50 sec 50 percent of the time at 1/2 maximum range. Staying low in featureless terrain will not provide adequate masking, and other types of masking or signature reductions are a necessity.

Figure 4.23 may help to visualize the process that reduces cross-range visibility. The figure plots visibility on a log scale and, because visibility decreases are a power series, they appear as straight lines on a log scale.

Another question to ask is, "What is the effect of surveillance altitude on persistence of visibility for a given grazing angle?" Figures 4.24 and 4.25 compare 30-kft with 6-kft operating altitudes. Note that there is a 7.5:1 difference in range scales.

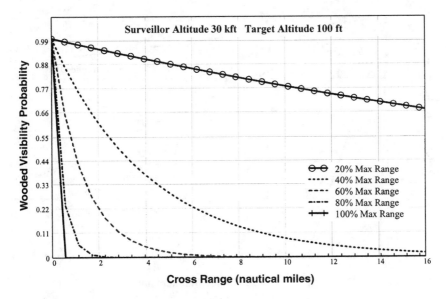

Figure 4.21 Typical wooded visibility versus cross range.

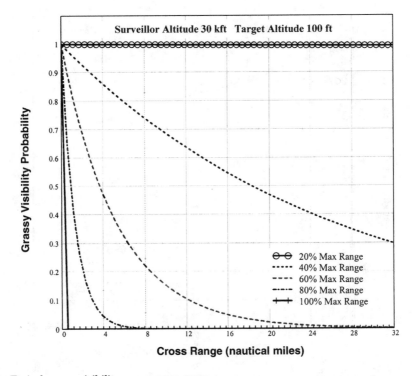

Figure 4.22 Typical grassy visibility versus cross range.

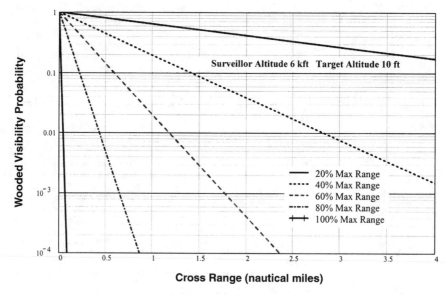

Figure 4.23 Low-altitude visibility versus cross range.

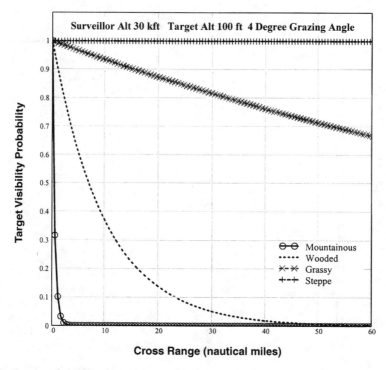

Figure 4.24 Persistence of visibility for 30-kft surveillor altitude.

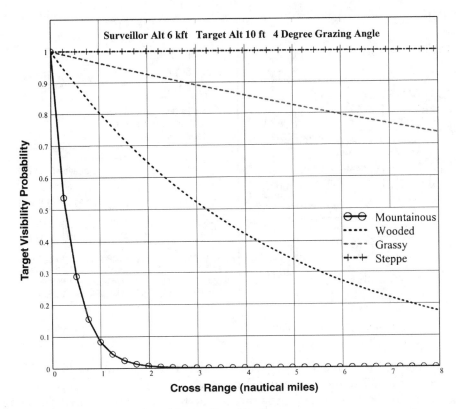

Figure 4.25 Persistence of visibility for 6-kft surveillor altitude.

4.4 ELECTRONIC ORDER OF BATTLE

The next important stealth consideration is exploitation of the electronic order of battle (EOB). The EOB is the deployment and utilization of various kinds of radar, ECM, and radar intercept equipment in adversary military doctrine and tactics. The examples given here are somewhat dated but show a method of assessing stealth designs and tactics, which is generally useful. The latest adversary training and doctrine are necessary to evaluate the efficacy of a given stealth design. The training and doctrine are typically obtained by long-term observation with national technical and other means. An overview of the EOB threat for ground forces is given in Fig. 4.26. As shown in panel 1 of Fig. 4.26, each front will have a hierarchy of military units that will have threat systems attached to them based on coverage and numbers deployed. These systems will range from air defense batteries to intercept and direction finding, to jamming, to early warning, to electronic intelligence, and ultimately to national technical means. A stealthy weapon system must concern itself first with the immediate threats and having dealt with those, later with the longer-term threat. For an air-

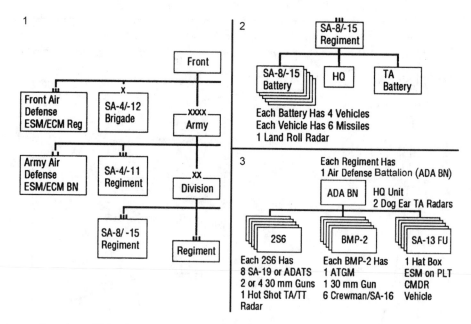

Figure 4.26 Overview of EOB threat—ground forces.

borne penetrator, the primary threat is air defense. Panels 2 and 3 of Fig. 4.26 show a typical regimental air defense complement.

In addition to air defense that may be organic to a regiment, there will be air defense regiments in a division. Usually, a division-level air defense regiment will have longer-range threats than the air defense battalion organic to a regiment. This may include surveillance, communications specific to targeting, and some jamming. At the combined arms army level, there will be air defense regiments and air defense electronic support measures (ESMs) and electronic countermeasures (ECMs). Finally, at the front or regional command level, there will be a SAM brigade and ESM/ECM regiment. Although this EOB example is quite specific, none of the details should be taken too seriously. What is important is the approach and manner of thinking about EOB threat scenarios.

As will be seen in the following EOB scenario description, the build-up of emitters results in surprisingly large numbers. These emitters, even if far away, are so numerous that the ambient power spectrum can be the limiting factor in interceptor sensitivity. Appendix A.4 contains a table that lists a large number of emitter types and their performance parameters. Section 4.7 provides a smaller table of typical emitters that could be expected on a battlefield. The parameters are gathered primarily from unclassified sources; however, in some cases, the author has made plausible guesses based on photos and observation of the equipments at international defense shows (be skeptical!).

The organization of a generic surface-to-air missile (SAM) regiment is given in Figure 4.27. SAM regiments will have a command post that includes battle management with data links and communications for target handoff (as well as other command and control communications). There will be long- and short-range acquisition radars. The missile batteries will have fire control radars and electro-optical systems (EOSs). Each complement of missiles will have a transporter-erector-launcher (TEL or TELAR) with data links for both the missile and target handoff. Usually, there will also be service vehicles for fuel replenishment and support personnel who have radios and data links.

The idea of EOB can be understood by schematically sketching out some hypothetical battle order based on tactics and doctrines. Figure 4.28 shows a typical EOB based on Warsaw Pact tactics and doctrines of 1990 for the utilization of radar, ECM, and radar intercept equipment on the central German front. The central German front is a much-studied area that has been subjected to many wars and war games over the last 150 years. Shown schematically is a 400-km battlefront, behind which are deployed first- and second-echelon armies. In the first 35 km behind the forward edge of the battle area (FEBA) across a 400 km front, 10 tank divisions and 10 motorized rifle divisions might be deployed and be moving toward or at the front. From 35 to 100 km behind the FEBA, reinforcements would be deploying and flowing toward the battle area. From 100 to 250 km behind the front, reinforcing second-echelon armies would be flowing toward the battle area as quickly as they could be transported into the battle.

The second-echelon armies might contain an additional five tank divisions and five motorized rifle divisions flowing as reinforcements toward the front. As part of these first- and second-echelon armies, large numbers of radar, ECM, radar intercept, and air defense systems are routinely deployed. The total area covered by the adversary side of the FEBA is 100,000 km^2 or roughly 38,500 mi^2. There is a comparable amount of territory on the friendly side of the FEBA with similar forces. As pointed out earlier

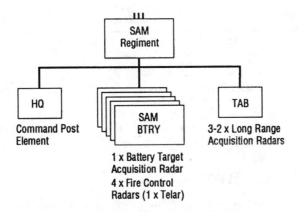

Figure 4.27 Generic SAM regiment.

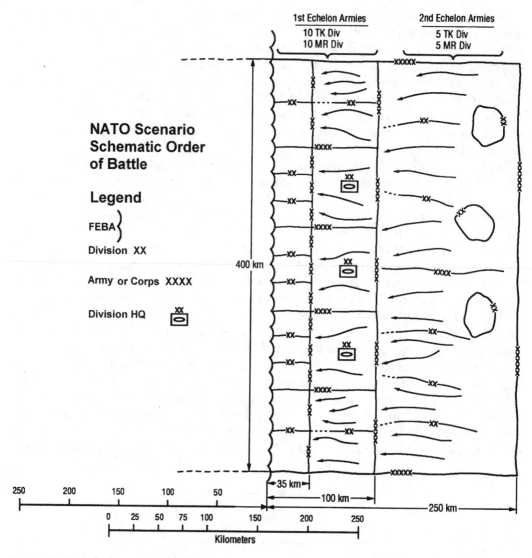

Figure 4.28 Typical schematic order of battle.

in the text, what is important here is a manner of thinking, not the specific accidentals of the electronic order of battle described in the following sections.

4.4.1   Radar and EW Intercept EOB

Each first-echelon division will have radar intercept and direction finding capability as shown schematically in Fig. 4.29. Each division has a reconnaissance battalion with

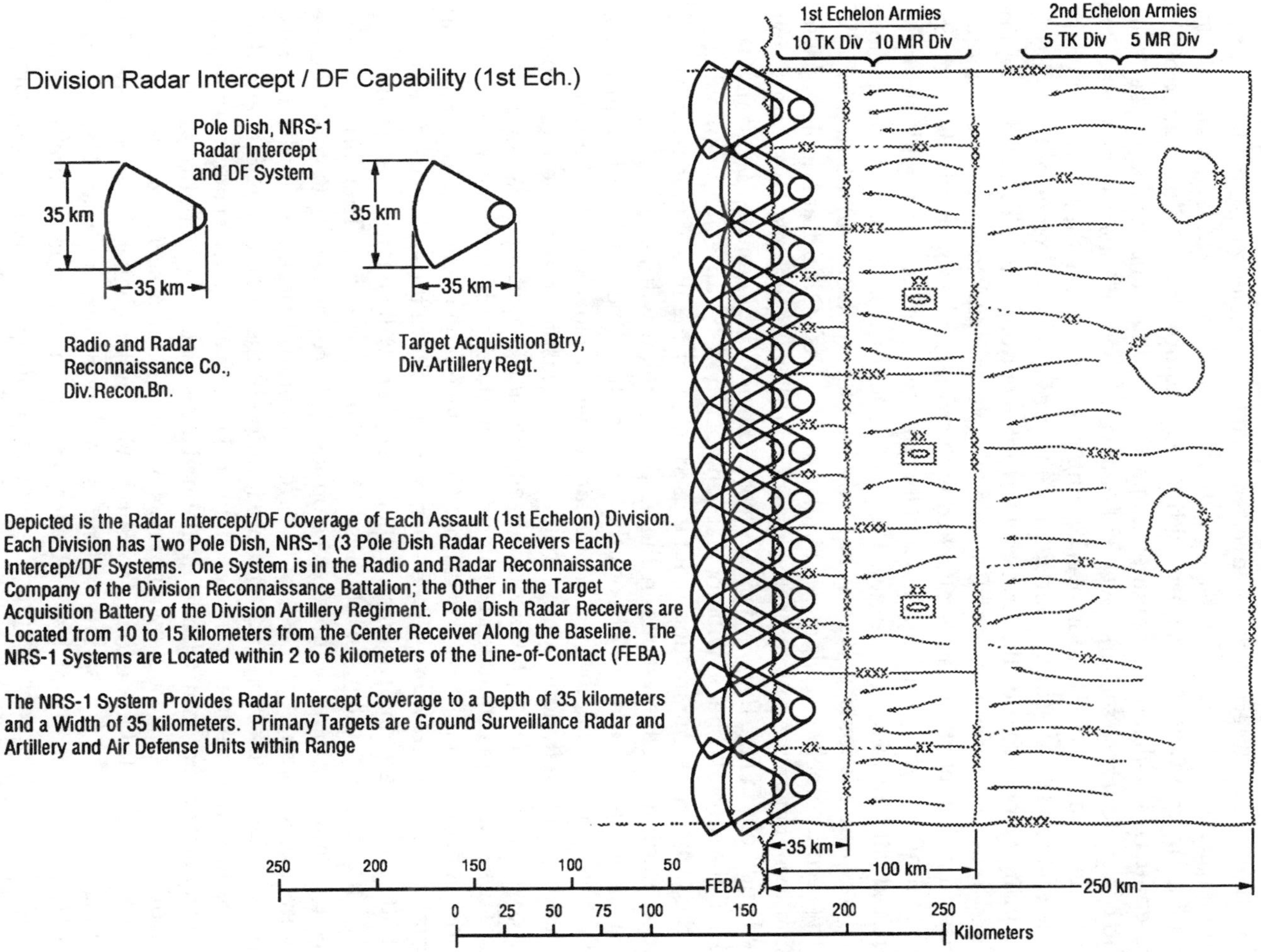
Division Radar Intercept / DF Capability (1st Ech.)
Pole Dish, NRS-1 Radar Intercept and DF System
35 km
35 km
35 km
35 km
Radio and Radar Reconnaissance Co., Div. Recon.Bn.
Target Acquisition Btry, Div. Artillery Regt.
Depicted is the Radar Intercept/DF Coverage of Each Assault (1st Echelon) Division. Each Division has Two Pole Dish, NRS-1 (3 Pole Dish Radar Receivers Each) Intercept/DF Systems. One System is in the Radio and Radar Reconnaissance Company of the Division Reconnaissance Battalion; the Other in the Target Acquisition Battery of the Division Artillery Regiment. Pole Dish Radar Receivers are Located from 10 to 15 kilometers from the Center Receiver Along the Baseline. The NRS-1 Systems are Located within 2 to 6 kilometers of the Line-of-Contact (FEBA)
The NRS-1 System Provides Radar Intercept Coverage to a Depth of 35 kilometers and a Width of 35 kilometers. Primary Targets are Ground Surveillance Radar and Artillery and Air Defense Units within Range
1st Echelon Armies
10 TK Div 10 MR Div
2nd Echelon Armies
5 TK Div 5 MR Div
35 km
100 km
250 km
FEBA
250 200 150 100 50
0 25 50 75 100 150 200 250
Kilometers

Figure 4.29 Division intercept and direction finding (DF).

a radio and radar reconnaissance company containing modern systems similar to the older Pole Dish/NRS-1 radar intercept and direction finding system. The system consists of three intercept receivers deployed on a 20 to 30 km long baseline for accurate two-dimensional localization by triangulation. Each division artillery regiment also contains a target acquisition battery with similar equipment to allow attack of detected emitters. Localization of emitters is partially manual and typically takes minutes to complete. These sensors cover 35 km beyond the FEBA and are deployed 2 to 6 km behind the FEBA. There is overlapping coverage across the entire line of the FEBA as shown. Twenty-six sites are required to provide full coverage of the FEBA.

The primary objective of the first-echelon intercept and direction finding equipments is to detect the location of battlefield surveillance and counter battery and air defense radars so that they can be attacked with artillery, rockets, or missiles (e.g., SS-21) or neutralized with jamming. As such, long range is not necessary, and sensitivity is required only for location accuracy. The more modern systems have some sidelobe interference cancellation capabilities, but only for a few (one or two) interferers.

In the second echelon, equipments similar to Pole Dish/NRS-1 are deployed with division reconnaissance battalions and division artillery regiments. These are shown dashed and bold in Fig. 4.30 for typical second-echelon deployments. Those units deployed in the second echelon will not all be in operation, because they are moving toward the front, but some fraction will be used to cover the line of march. Intercept/direction finding (I/DF) systems in the second echelon will have significantly degraded capabilities as a result of both the distance from the FEBA, terrain masking, and their semimobile deployment along the routes of movement to the front. Notwithstanding the limitations, some capabilities will exist for the intercept of airborne radar coverage of the deployment, movement, and targeting of second-echelon forces. These intercepts could lead to alerting of air defense, hiding, or jamming.

In addition, at the combined arms army (CAA) level, there will be additional radar intercept and direction finding capabilities over and above those that are attached to first- and second-echelon divisions. At this level, there will be longer-range intercept and direction finding systems in addition to the short-range units. Their primary mission is to counter air defense units of the opposing forces as well as to counter close air support attack aircraft and associated ground control. More specifically, the mission of the I/DF is detection and accurate location of the operating radars and data links of SAM batteries, counter battery field artillery, ground surveillance, ground attack/target designation control, and unmanned air vehicle (UAV) control stations.

The additional coverage is shown in Fig. 4.31 and, as can be seen, provides overlapping coverage to 150 km beyond the FEBA. One major challenge associated with longer-range detection is visibility of emitters at low or zero altitude above ground level (AGL). Additionally, at low altitudes, multipath and ducting are significant problems that limit location accuracy independent of signal-to-noise ratio. The primary methods of low-altitude visibility are siting at higher altitude AGL, if available, or from aircraft. The minimum altitude for 90 percent visibility is probably 5500 ft

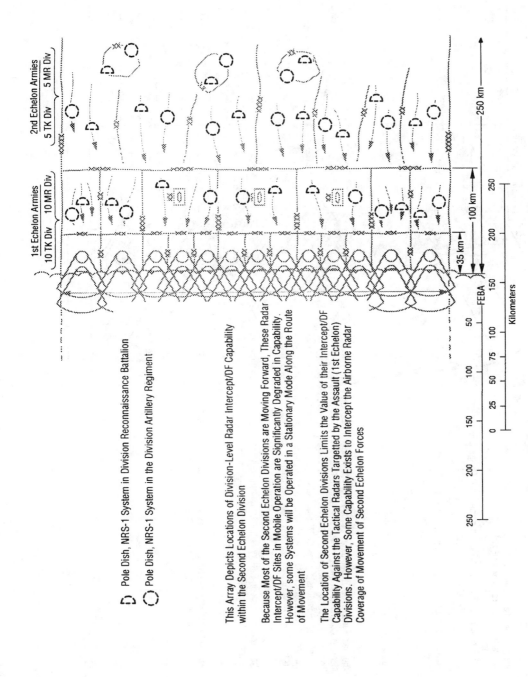

2nd Echelon Armies — 5 TK Div / 5 MR Div
1st Echelon Armies — 10 TK Div / 10 MR Div

Pole Dish, NRS-1 System in Division Reconnaissance Battalion

Pole Dish, NRS-1 System in the Division Artillery Regiment

This Array Depicts Locations of Division-Level Radar Intercept/DF Capability within the Second Echelon Division

Because Most of the Second Echelon Divisions are Moving Forward, These Radar Intercept/DF Sites in Mobile Operation are Significantly Degraded in Capability. However, some Systems will be Operated in a Stationary Mode Along the Route of Movement

The Location of Second Echelon Divisions Limits the Value of their Intercept/DF Capability Against the Tactical Radars Targetted by the Assault (1st Echelon) Divisions. However, Some Capability Exists to Intercept the Airborne Radar Coverage of Movement of Second Echelon Forces

250 km
100 km
35 km
FEBA
Kilometers

Figure 4.30   Second-echelon intercept and direction finding (DF).

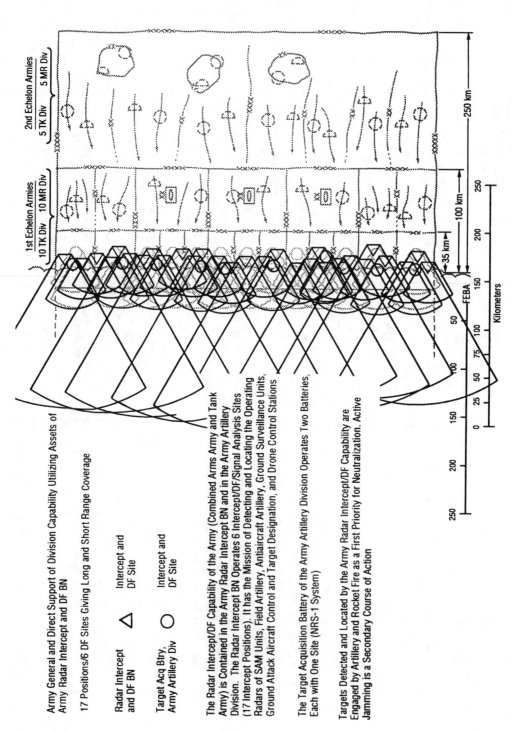

Army General and Direct Support of Division Capability Utilizing Assets of Army Radar Intercept and DF BN

17 Positions/6 DF Sites Giving Long and Short Range Coverage

Radar Intercept and DF BN △ Intercept and DF Site

Target Acq Btry, Army Artillery Div ○ Intercept and DF Site

The Radar Intercept/DF Capability of the Army (Combined Arms Army and Tank Army) is Contained in the Army Radar Intercept BN and in the Army Artillery Division. The Radar Intercept BN Operates 6 Intercept/DF/Signal Analysis Sites (17 Intercept Positions). It has the Mission of Detecting and Locating the Operating Radars of SAM Units, Field Artillery, Antiaircraft Artillery, Ground Surveillance Units, Ground Attack Aircraft Control and Target Designation, and Drone Control Stations

The Target Acquisition Battery of the Army Artillery Division Operates Two Batteries, Each with One Site (NRS-1 System)

Targets Detected and Located by the Army Radar Intercept/DF Capability are Engaged by Artillery and Rocket Fire as a First Priority for Neutralization. Active Jamming is a Secondary Course of Action

Figure 4.31 Army level intercept and DF.

AGL. Such emitter location can be accomplished from medium utility or attack helicopters, probably MI-24, or medium aircraft. Surface or airborne I/DF systems may not have a unique signature other than their large RCS created by the I/DF antennas or by their data link signatures. Whether ground based, airborne, or mixed, there will be a minimum of 17 sites to provide full coverage of the battle space.

At the next highest level, there are also radar intercept and direction finding sites to support army groups. There may be as many as 80 positions at 23 sites that contain radio and radar intercept regiments supporting multiple armies across a broad front. They are targeted against all military radars, data links, and communications within their coverage range. Their deployment is shown in Fig. 4.32 throughout the first-, second-, and third-echelon areas. In general, the frequency coverage of these multiple systems as tabulated in Fig. 4.32 ranges from approximately 500 MHz to a little less than 17 GHz. Some I/DF systems go all the way to 37.5 GHz but, at the higher frequencies, the intercept range is very limited by decreased sensitivity and increased atmospheric attenuation. In addition, some of these systems have excellent direction-finding accuracy—perhaps as low as a fraction of a degree in a few cases. The primary weaknesses of ground-based systems are the limitations imposed by low-altitude visibility and ducting.

Larger airborne I/DF and radar systems will be deployed in support of the full battlefront. These systems will include sensitive ELINT and COMINT equipment. They will be limited primarily by their signal-processing number-smashing capabilities, because thousands of emitters and tens of thousands of potential targets will be within their line of sight. One to three systems will be required continuously on orbits over the battle space. This, in turn, requires a total of 17 systems to allow refueling, replenishment, maintenance, crew replacement, travel to and from orbits, phase maintenance, and so on. Although the required resources are much smaller than for counterpart ground I/DF systems, the resources required in theater are still substantial (up to 1000 personnel). Examples of such a systems are the A-50, AN-70, AN-71, KA-31, TU-142, and MIG-25RBK/F surveillance/reconnaissance aircraft currently deployed with Russian military forces. Counterpart U.S. systems include E-2, E-3, E-8, RC-135, RC-12, and so forth. Additionally, at the theater level, space national assets may be dedicated to the battlefront.

4.4.2 Radar Emitter EOB

In addition to the intercept equipment described in the preceding sections, numerous electronic warfare, early warning radar, and air defense radar systems are deployed with the army across the front. Three separate classes of radar are normally deployed: early warning radars with reasonably long ranges, height-finding radars, and target acquisition/fire control radars. A CAA will have an early warning battalion, and within that battalion there will be early warning companies typically deployed as shown in Fig. 4.33. Each of these companies might have four radars for the

Front Radar Intercept/DF Sites in General Support of Armies

80 Positions 23 Sites – ☐ } Per Radio and Radar Intercept Regiment

The Front Organization Structure Contains One or Two Radio and Radar Intercept Regiments. Each Regiment Consists of Two Mobile Radio and Radar Intercept Companies and Four Heavy Radio and Radar Intercept Companies. The Regiment Uses About 23 Intercept/DF/Signal Analysis Sites Totalling 80 Separate Intercept Positions. This Scenario, Depicting the Soviet Central Front Offensive Operations (1990 Time Frame) Employs One Radio and Radar Intercept Regiment

The Regiment's Capabilities are Targetted Against all Military Radars within Range. The Regiment Provides General Support of the Deployment Armies. Exact Numbers and Types of Radar Receivers are Unknown. Most of the Regiment's Intercept and DF Capability is Deployed in the Forward Areas of Committed Divisions. Primary Radar Intercept/DF System Equipment is:

System	Accuracy (Ground Radar Targets)	Frequency (MHz)
Pole Dish / NRS-1 (Radar DF)	5 mr	2,500-3,750 8,100-16,700
Bar Brick (Radar Intercept)	±4°	800-4,000 4,000-10,000
Swing Box (Radar Intercept)	±3°	1,200-3,000
Big Ear (Radar Intercept)	±1°	2,600-2,900
Twin Box /RPS-5 (Intercept/DF)	±5°	500-10,000
Brick Round / RPS-1 (Intercept/DF)	±5°	6,500-11,500
Brick Square /RPS-1 (Intercept/DF)	±5°	2,700-6,800

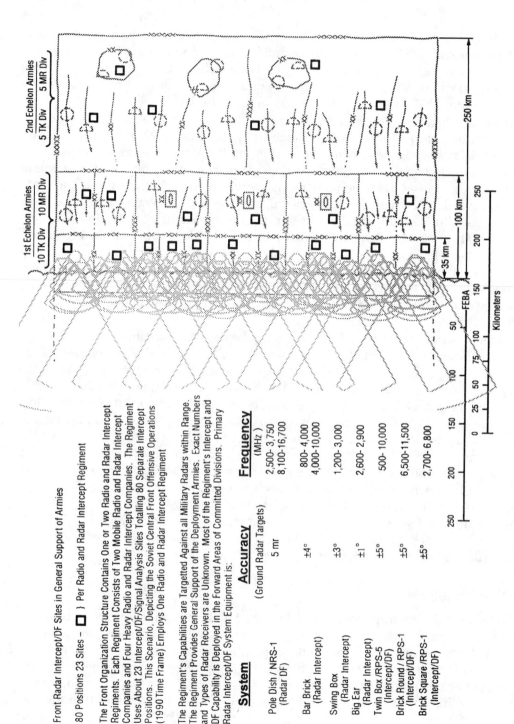

Figure 4.32 Front intercept and DF capability.

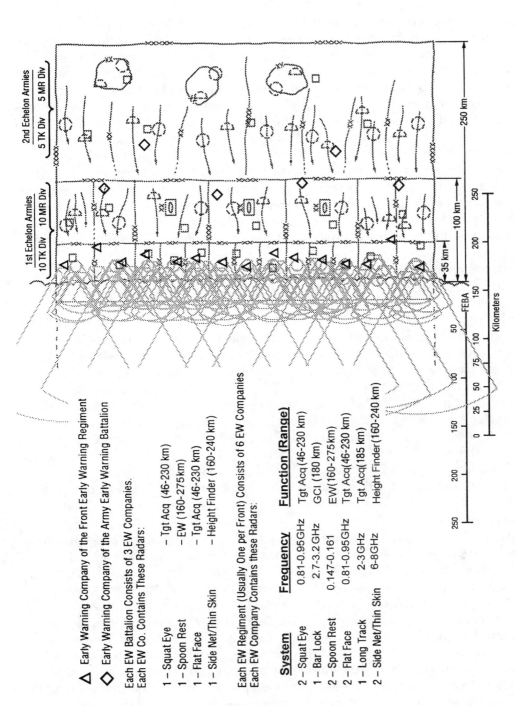

2nd Echelon Armies
5 TK Div   5 MR Div

1st Echelon Armies
10 TK Div   10 MR Div

250 km
100 km
35 km
FEBA

Kilometers
250   200   150   100   75   50   25   0   50   100   150   200   250

△ Early Warning Company of the Front Early Warning Regiment

◇ Early Warning Company of the Army Early Warning Battalion

Each EW Battalion Consists of 3 EW Companies.
Each EW Co. Contains These Radars:

1 – Squat Eye   – Tgt Acq (46-230 km)
1 – Spoon Rest   – EW (160-275km)
1 – Flat Face   – Tgt Acq (46-230 km)
1 – Side Net/Thin Skin – Height Finder (160-240 km)

Each EW Regiment (Usually One per Front) Consists of 6 EW Companies
Each EW Company Contains these Radars:

System	Frequency	Function (Range)
2 – Squat Eye	0.81-0.95GHz	Tgt Acq (46-230 km)
1 – Bar Lock	2.7-3.2 GHz	GCI (180 km)
2 – Spoon Rest	0.147-0.161	EW(160-275km)
2 – Flat Face	0.81-0.95GHz	Tgt Acq(46-230 km)
1 – Long Track	2-3GHz	Tgt Acq(185 km)
2 – Side Net/Thin Skin	6-8GHz	Height Finder (160-240 km)

Figure 4.33   Army and front early warning companies.

231

functions described above. These systems are capable of detecting aircraft and missiles out to 250 km, depending on radar site and aircraft operating altitudes. For an early warning radar to detect targets at 250 km, the sum of the aircraft/missile altitude and the radar site altitude must be greater than 12,000 ft. In addition, the Spoon Rest radar is capable of detecting stealth aircraft and missiles, because its operating wavelength of roughly 6 ft is near the vehicle resonance region. The good news is that range and angle resolution is so poor that target handoff is very difficult. Squat Eye and Flat Face also have some capability against stealth vehicles, and better resolution as well but, at a wavelength of roughly 1 ft, the effects of modern stealth techniques are fully in play. Long Track and Bar Lock are S-band target acquisition and GCI radars, and stealth limits their performance significantly. Site location is critical to the success of these radars and, in a major battle, the choices may be very limited. One purpose behind short-range air defense and other passive measures is to force attacking aircraft to higher altitudes where the early warning assets can actually give early warning.

In addition, there are early warning companies associated with a front early warning regiment, and each early warning regiment will have companies containing as many as 84 radars across the front. Siting for front early warning radars is easier, because they are farther from the FEBA and may be able to choose high ground. These limitations have given rise to airborne radars to surveil and control assets in the battle space. There may be up to 96 early warning class radars associated with a front. Not only are there early warning radars, but also there are radars associated with air defense guns and surface-to-air missile (SAM) systems. These systems usually operate at higher frequencies and have short radar ranges. However, because they typically are high-peak-power, low-duty-ratio systems, their intercept ranges can be extremely long. Air defense guns such as the ZSU-23 and mobile surface-to-air missiles such as the SA8 through SA20 will be deployed at the division level in both the first and second echelon. They will be deployed where there are troop concentrations as well as along lines of march. In addition, there will be the longer-range SAMs deployed at the Corp/Army level to defend command, control, and communications (C3) and leadership concentrations.

Within the front, there may be as many as 480 Gun Dish radars (ZSU-23) and 600 Land Role (SA-8) or Snow Drift (SA-11/17) radars. Furthermore, there are radars associated with SA4, SA6 (Straight Flush), SA-10 (Tin Shield, Flap Lid, Clam Shell, Tombstone), SA-13 (Dog Ear), and SA-12 (Bill Board, High Screen) fire units, which may contain as many as 212 radars across a front. One concept for the deployment within the example EOB is given in Fig. 4.34. The figure also tabulates the detection and engagement range performance of some (but not all) of the systems in the battlespace. Similar numbers of radar emitters will exist on the friendly side of the FEBA. All this "music" greatly complicates the interceptor's sorting problems and, as will be shown in a later section, limits usable sensitivity. The spectrum congestion can be so bad that it is not uncommon to see another's video marching through your radar window.

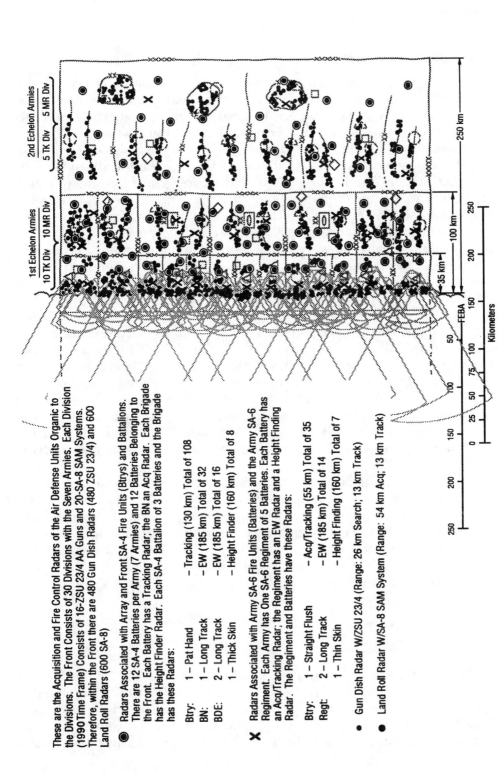

These are the Acquisition and Fire Control Radars of the Air Defense Units Organic to the Divisions. The Front Consists of 30 Divisions with the Seven Armies. Each Division (1990 Time Frame) Consists of 16-ZSU 23/4 AA Guns and 20-SA-8 SAM Systems. Therefore, within the Front there are 480 Gun Dish Radars (480 ZSU 23/4) and 600 Land Roll Radars (600 SA-8)

⊙ Radars Associated with Array and Front SA-4 Fire Units (Btrys) and Battalions. There are 12 SA-4 Batteries per Army (7 Armies) and 12 Batteries Belonging to the Front. Each Battery has a Tracking Radar; the BN an Acq Radar. Each Brigade has the Height Finder Radar. Each SA-4 Battalion of 3 Batteries and the Brigade has these Radars:

Btry:	1 – Pat Hand	– Tracking (130 km) Total of 108
BN:	1 – Long Track	– EW (185 km) Total of 32
BDE:	2 – Long Track	– EW (185 km) Total of 16
	1 – Thick Skin	– Height Finder (160 km) Total of 8

✕ Radars Associated with Army SA-6 Fire Units (Batteries) and the Army SA-6 Regiment. Each Army has One SA-6 Regiment of 5 Batteries. Each Battery has an Acq/Tracking Radar; the Regiment has an EW Radar and a Height Finding Radar. The Regiment and Batteries have these Radars:

Btry:	1 – Straight Flush	– Acq/Tracking (55 km) Total of 35
Regt:	2 – Long Track	– EW (185 km) Total of 14
	1 – Thin Skin	– Height Finding (160 km) Total of 7

• Gun Dish Radar W/ZSU 23/4 (Range: 26 km Search; 13 km Track)

● Land Roll Radar W/SA-8 SAM System (Range: 54 km Acq; 13 km Track)

Figure 4.34 Radars associated with air defense guns and SAMs.

4.4.3 Electronic Countermeasures EOB

On top of the radar emitters described above, significant numbers of jammers are used to deny opposing forces' use of their radar sensors, navigation, and communication equipment. Some of the jammers are designed to deny accurate navigation data to attacking aircraft or missiles and may be oriented toward such systems as TACAN and GPS. As many as 54 jammers may be deployed across the front. Some will have effective radiated powers high enough to reach 200 km to the friendly side of the FEBA at higher altitudes (greater than 8000 ft). These cover X- and Ku-band primarily but do reach down to C-band (E-, I-, and J-band in modern nomenclature). The typical deployment of these jammers is shown in Fig. 4.35. In addition, there will be some number of airborne standoff jammers to reach lower-altitude targets.

Jamming can be of many types, including broadband noise, spot or swept noise, set-on (requires detection first), repeater, deception (requires detection first), and so on. Usually, an LPIS has so much coherent gain that broadband, spot, and swept noise, as well as deception jamming, is invisible. The type that usually works best against both LPIS and GPS is repeater jamming, because detection is not required, and the repeater sends the complex LPI or GPS signal with a small delay that, by definition, will be match filtered at the receiver. Because the SNR at the repeater jammer will be less than unity, the repeated signal will return with a high level of noise, which can be detected and then nulled with an antenna canceller. Because sidelobe/mainlobe cancellers are typically limited by the available number of degrees of freedom in the antenna, they can be overwhelmed if there are enough repeater jammers (usually tens). For the threat interceptor, jamming greatly complicates detection of LPI emitters, as the jamming is usually blind as to LPI location. Adversary jammer sidelobes will usually mask an LPIS.

It can be seen that large numbers of both interceptors and radar and data link emitters are deployed along a front. In summary, possibly 100 to 200 intercept receivers will be deployed in the battle area. Fortunately for a stealth system, there will be 2000 to 5000 friendly and adversary force emitters potentially visible to a high-sensitivity intercept receiver. These make up the environment that can be exploited by a stealth system. These emitters can be used to mask and confuse by careful choice of stealth waveforms and emission strategy.

4.5 RF SPECTRUM MASKING

4.5.1 Example Ambient Spectra

Ambient RF spectra can be another form of masking for LPI systems. Figures 4.36 and 4.37 show typical ambient raw low-frequency spectra at two different altitudes: 18,000 feet above ground level (AGL) and 3000 feet AGL. These plots were made in an aircraft flying over the Antelope Valley in northern Los Angeles county. The only smoothing applied is that inherent in the chart recorder servo. Note that the total am-

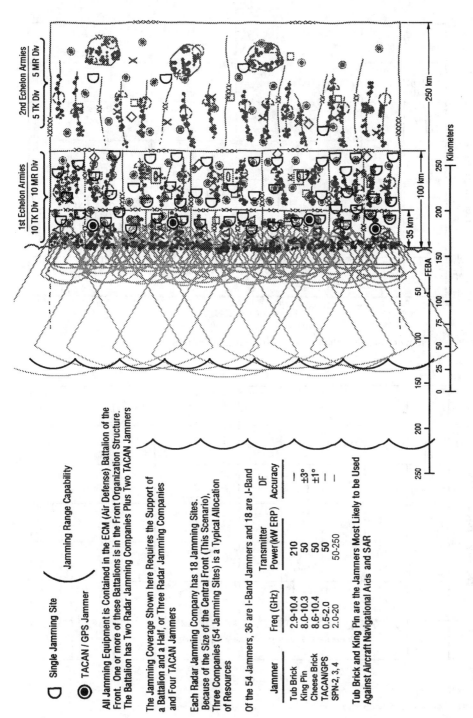

Single Jamming Site

Jamming Range Capability

TACAN / GPS Jammer

All Jamming Equipment is Contained in the ECM (Air Defense) Battalion of the Front. One or more of these Battalions is in the Front Organization Structure. The Battalion has Two Radar Jamming Companies Plus Two TACAN Jammers

The Jamming Coverage Shown here Requires the Support of a Battalion and a Half, or Three Radar Jamming Companies and Four TACAN Jammers

Each Radar Jamming Company has 18 Jamming Sites. Because of the Size of the Central Front (This Scenario), Three Companies (54 Jamming Sites) is a Typical Allocation of Resources

Of the 54 Jammers, 36 are I-Band Jammers and 18 are J-Band

Jammer	Freq (GHz)	Transmitter Power (kW ERP)	DF Accuracy
Tub Brick	2.9-10.4	210	—
King Pin	8.0-10.3	50	±3°
Cheese Brick	8.6-10.4	50	±1°
TACAN/GPS	0.5-2.0	50	—
SPN-2, 3, 4	2.0-20	50-250	—

Tub Brick and King Pin are the Jammers Most Likely to be Used Against Aircraft Navigational Aids and SAR

Figure 4.35 Front jamming capabilities.

235

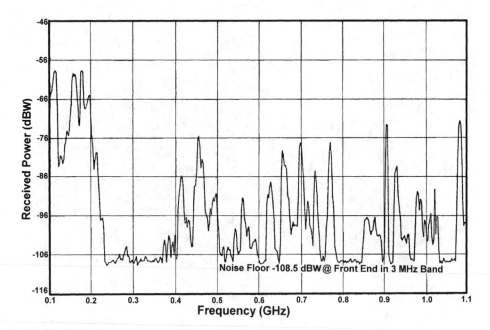

Figure 4.36   Ambient power spectrum: 18,000 ft AGL, southern California.

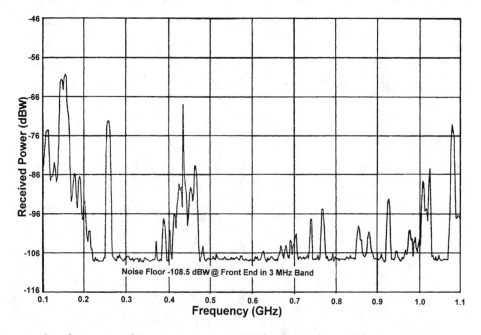

Figure 4.37   Low-frequency ambient power spectrum: 3000 ft AGL, southern California.

bient spectral power decreases with decreasing altitude, but local emitter power increases with lower altitude. The minimum discernible signal for these two figures is –108.5 dBW in a 3-MHz bandwidth. Some spectral masking is provided by the mountains, which surround the Antelope Valley and reach an altitude of almost 12,000 ft, so this spectrum may be optimistic.

Figure 4.38 shows highly smoothed data for the ambient power spectrum from ground emitters taken by the Institute for Telecommunications Sciences spectrum survey in Los Angeles, California.[6] The data used for Fig. 4.38 was collapsed into 100-MHz bins, and mean power was estimated and scaled to power in a 3-MHz band. The data were gathered with several measurement strategies, and the compromise estimate for the minimum discernible signal is –117 dBW. A small amount of smoothing was applied for the plot. One could expect this level of civilian spectral emissions over a large metropolitan area, even during wartime. In the next section, Fig. 4.41 shows a typical estimated battlefield spectrum for 18,000 feet AGL. All of these spectra are plotted as power in decibels below 1 W referenced to the front end in a 3-MHz bandwidth for an omnidirectional antenna versus frequency.

The figures show that there are many segments of the spectrum that have power far above the noise-limited sensitivity. The ambient will mask and break up LPIS emissions that are lower than the ambient spectrum, as described in Chapter 3. During a battle in which jamming may also be used, the ambient power will be much higher and will further limit interceptor sensitivity.

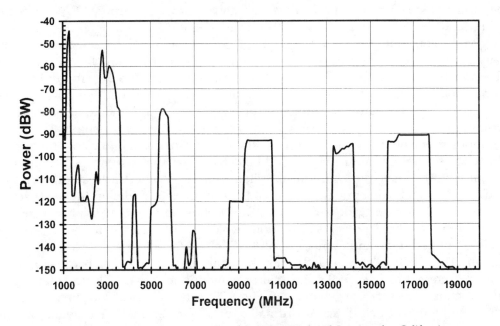

Figure 4.38 Microwave ambient power spectrum: 0 ft AGL, 3 MHz band, Los Angeles, California.

4.5.2 Estimating Ambient Spectra

The ambient spectrum for a battlefield can be estimated based on an assumed EOB as presented in the previous sections. Using the extended Table 4.7 and some assumed EOB (i.e., some set of ground and airborne emitters), the average ambient power spectrum over a battlefield at an interceptor with an omnidirectional antenna can be calculated. The spectrum will also contain the emissions of many civilian sources that will be present in any modern EOB as well (Hughes regularly received cell phone calls from the battle area and theater during Desert Storm, requesting both equipment and personnel help). This ambient power limits the usable interceptor sensitivity and can mask a stealth system. Many studies have been made of forecast pulse traffic in the European theater.[18] Once an EOB scenario is chosen, pulse traffic can be estimated using the models provided in this section. The basic assumption is that there are emitters distributed uniformly and randomly over the battle space. Furthermore, a small fraction of the emitter antenna mainbeams are pointed in the direction of the interceptor, and they are also random.

There are four elements of the total spectrum. The first element, and the largest in number, is ground emitter sidelobes. The second element, and the largest in instantaneous power, is ground emitter mainlobes. Only some mainlobes are pointed at the interceptor at any time. An emitter dwell time on the interceptor is proportional to mainlobe beamwidth and scan rate. When there are a large number of emitters, and because their scan rates are slow relative to interceptor detection time, the number of mainlobe emitters illuminating the interceptor is proportional to their beamwidth and the solid angle scanned. The third element is airborne and spaceborne emissions from sidelobes. The last element is airborne mainlobe emissions. Again, only a small fraction proportional to the beamwidth of the mainlobes is pointed at the interceptor at any time. The ground emitters exhibit an area density, and the airborne/spaceborne emitters exhibit a volume density. Each emitter power as received at the interceptor is a function of its ERPP, instant and total bandwidth, distance from the interceptor, atmospheric attenuation, and wavelength. The actual locations of the emitters with respect to the interceptor are random to a first order. There are several techniques that could be used to estimate average power per unit of bandwidth under these circumstances, including Monte Carlo simulation and probability-weighted summation. One of the examples provided in the next section is based on a Monte Carlo simulation. The technique selected for this section is probability weighted power summation, which is adequate if there are enough emitters of each type and enough resolution cells (which is usually true in a major battle). The Earth is assumed flat for the analysis that follows. Because emitter power decreases as $1/R^2$, a flat-Earth approximation makes very little difference in the estimate. Consider the geometry of Fig. 4.39 using spherical coordinates for airborne emitters and cylindrical coordinates for surface emitters. We define the following variables:

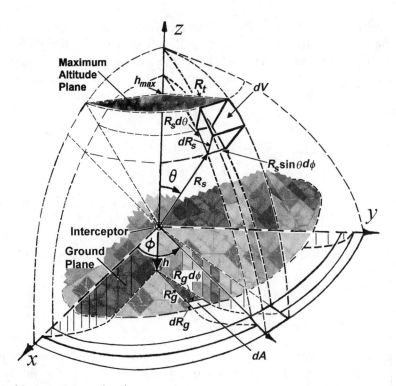

Figure 4.39 Ambient spectrum estimation geometry.

R_g = ground range
R_s = slant range
h = interceptor altitude
θ = angle from z or up axis to R_s
ϕ = angle from x or east axis to the projection of R_s in the x-y plane
N_m = the number of the mth emitter type
IBW_m = the instant bandwidth of the mth emitter type
f_{um} = the upper end of the operating band of the mth emitter type
f_{lm} = the lower end of the operating band of the mth emitter type
PS_{erpm} = the effective radiated peak power of the mth emitter type in its sidelobes
PM_{erpm} = the effective radiated peak power of the mth emitter type in its mainlobe
$PUL(f,m) = 1$ if $f_{lm} \leq f \leq f_{um}$
$PUL(f,m) = 0$ if $f \leq f_{lm}$ or $f_{um} \leq f$, i.e., the pulse function to allow summation of multiple emitters in multiple bands
$Q(m) = 1$ if emitter is airborne and $= 0$ otherwise
Δf = the analysis bandwidth
P_{erpm} = mainlobe or sidelobe effective radiated peak power

At each emitter site, an emitter's probable frequency is assumed to be uniformly random. The emission spectra will be spaced at intervals roughly the instant bandwidth apart. The apparent power density is the interceptor analysis bandwidth divided by the instant bandwidth. Its contribution to the average ambient spectral power density assuming all frequencies in the operating band are used or visited is approximately

$$\mathcal{P}_{Fm} = \frac{IBW_m \cdot PUL(f,m) \cdot \Delta f}{(f_{um} - f_{lm}) \cdot IBW_m} \tag{4.13}$$

Of course, the ambient power density at the interceptor from each emitter in its operating band is subject to the range Equation (1.3) given in Chapter 1 with G_{IP} and G_I equal to 1 and $P_T \cdot G_{TI}$ equal to P_{erpm}, as well as the probability, $\mathcal{P}_{Em}$, of an emitter of the mth type at range, R_s; therefore, the probable differential power is

$$dP_m(f,R_s) = \frac{P_{erpm} \cdot \lambda_m^2 \cdot L_m \cdot d\mathcal{P}_{Em} \cdot \mathcal{P}_{Fm}}{(4 \cdot \pi)^2 \cdot R_s^2} \tag{4.14}$$

The probability of an emitter in a differential cell in this case is approximately the volume or area of coverage times the density of emitters, D_{Em}, of the mth type,

$$d\mathcal{P}_{Em} = D_{Em} \cdot dA \text{ or } (D_{Em} \cdot dV) \tag{4.15}$$

where the density is the total number of emitters of that type divided by the total volume or area considered in the battlespace,

$$D_{Em} = N_m/V_{bat} \text{ or } N_m/A_{bat} \tag{4.16}$$

The area, A_{bat}, and volume, V_{bat}, of the battlespace can be related to some maximum range, R_{max}, which is related to the horizon or to the size of the front as

$$A_{bat} \cong 2 \cdot \pi \cdot R_{max}^2 \text{ and } V_{bat} \cong \pi \cdot h_{max} \cdot R_{max}^2 \tag{4.17}$$

Note that, in Fig. 4.39, slant range, R_s, and ground range, R_g, as well as the mean operating wavelength, λ_m, and operating frequency range between f_{um} and f_{lm} are related by

$$R_s = \sqrt{h^2 + R_g^2} \text{ and } \lambda_m = (2 \cdot c)/(f_{um} + f_{lm}) \tag{4.18}$$

where c is the velocity of light and is approximately 0.983 ft/nsec. Then, for the ground emitter elements, sidelobes and mainlobes of the total spectrum in cylindrical coordinates are

$$PG_m(f) = \int_{A_{max}} \int \frac{P_{erpm} \cdot \lambda^2 \cdot L_m \cdot \boldsymbol{P}_{Fm}}{(4 \cdot \pi)^2 \cdot R_s^2} d\boldsymbol{P}_{Em} \qquad (4.19)$$

We refer to Fig. 4.39 and note that dA and A_{max} in cylindrical coordinates are

$$dA = R_g \cdot d\phi \cdot dR_g \text{ and } A_{max} = \pi \cdot R_{gmax}^2 \qquad (4.20)$$

Substituting for λ_m, $\boldsymbol{P}_{Fm}$, and $d\boldsymbol{P}_{Em}$, using Equations (4.15) through (4.20), assuming that loss is a constant (for simplification), and integrating in angle, the ambient peak power for the mth type ground emitter as a function of frequency is

$$PGS_m(f) = PUL(f,m) \cdot \frac{D_{Em} \cdot PS_{erpm} \cdot \Delta f \cdot L_m \cdot c^2}{2 \cdot \pi \cdot (f_{um}^2 - f_{lm}^2) \cdot (f_{um} + f_{lm})} \int_0^{R_{gmax}} \frac{R_g}{h^2 + R_g^2} dR_g \qquad (4.21)$$

Therefore, integrating again and substituting for R_{gmax} using Equation (4.18), the ambient peak power attributable to ground emitter sidelobes is

$$PGS_m(f) = PUL(f,m) \cdot \frac{D_{Em} \cdot PS_{erpm} \cdot \Delta f \cdot L_m \cdot c^2 \cdot \ln(R_{smax}^2/h^2)}{4 \cdot \pi \cdot (f_{um}^2 - f_{lm}^2) \cdot (f_{um} + f_{lm})} \qquad (4.22)$$

For clarity, define the above equation in two parts. The first part is all the characteristics of the mth emitter type and is independent of geometry. The second part is the interceptor geometry, which is independent of emitter type. For a given altitude and presumed maximum range, part 2 is constant over all emitters.

Then, part 1 as a function of frequency is

$$KS_m(f) = \frac{PUL(f,m) \cdot D_{Em} \cdot PS_{erpm} \cdot \Delta f \cdot L_m \cdot c^2}{2 \cdot \pi \cdot (f_{um}^2 - f_{lm}^2) \cdot (f_{um} + f_{lm})} \qquad (4.23)$$

and combining with part 2, then

$$PGS_m(f) = KS_m(f) \cdot \ln(R_{smax}/h) \qquad (4.24)$$

Element 2 of the total ambient power frequency spectrum uses the same formula except that the number of emitters is reduced to those that are pointed at the interceptor, and PM_{erpm} is the peak mainlobe ERP. For surface emitters, one can assume that the mainlobes are uniformly distributed in azimuth and will rapidly visit or continuously cover elevations up to 30° (antenna patterns often are csc^2 or multibeamed in elevation). Total coverage is a little over π steradians or roughly 1/4 of isotropic. The probability of mainlobe illumination will be approximately proportional to beamwidth.

$$dP_{Em} = \frac{D_{Em} \cdot 4 \cdot dA}{G_{MLm}} \tag{4.25}$$

Substituting the mainlobe probability density Equation (4.25) into the total spectrum integral for the mth emitter type Equation (4.19) yields the ambient peak power density attributable to ground emitter mainlobes for the mth type in its operating band.

$$PGM_m(f) = 2 \cdot PUL(f,m) \cdot \frac{D_{Em} \cdot PM_{erpm} \cdot \Delta f \cdot L_m \cdot c^2 \cdot \ln(R_{smax}^2/j^2)}{\pi \cdot G_{MLm} \cdot (f_{um}^2 - f_{lm}^2) \cdot (f_{um} + f_{lm})} \tag{4.26}$$

Again, for clarity, define KM_m and substitute

$$KM_m(f) = \frac{2 \cdot PUL(f,m) \cdot D_{Em} \cdot PM_{erpm} \cdot \Delta f \cdot L_m \cdot c^2}{\pi \cdot G_{MLm} \cdot (f_{um}^2 - f_{lm}^2) \cdot (f_{um} + f_{lm})} \tag{4.27}$$

And thus the power for ground emitter mainlobes is

$$PGM_m(f) = KM_m(f) \cdot \ln(R_{smax}/h) \tag{4.28}$$

Similarly, for the airborne emitters that are distributed according to a volume density as summarized in Equation (4.29) below,

$$PA_m(f) = \iiint_{V_{max}} \frac{P_{erpm} \cdot \lambda^2 \cdot L_m \cdot P_{Fm}}{(4 \cdot \pi)^2 \cdot R_s^2} dP_{Em} \tag{4.29}$$

The above integration is best carried out in three sectors. The first sector is a cone centered on the zenith axis with apex at the interceptor/surveillor altitude, h, with base at the maximum altitude, h_{max}, and with base radius, R_{tmax}. This sector is integrated in cylindrical coordinates. The second sector is a sphere with conical boring centered at the surveillor altitude, h, with radius, R_{smax}. This sector is best integrated in spherical coordinates for angles in the range of

$$\theta = \pi/2 - \text{asin}\left(\frac{h_{max} - h}{R_{smax}}\right) \text{ to } \pi/2 + \text{asin}\left(\frac{h}{R_{smax}}\right) \tag{4.30}$$

The third sector is a cone centered on the zenith axis with apex at the interceptor/surveillor altitude, h, with base at the ground level and with base radius, R_{gmax}. This sector is integrated in cylindrical coordinates. Substituting and noting that, in the two coordinate systems, dV for spherical and cylindrical coordinates, respectively, is

$$dV_{sphere} = R_s \cdot d\phi \cdot R_s \cdot \sin\theta \cdot d\theta \cdot dR_s$$

or

$$dV_{cylinder} = R_{g \text{ or } t} \cdot d\phi \cdot dZ \cdot dR_{g \text{ or } t} \tag{4.31}$$

We also note that R_{gmax} and R_{tmax} are

$$R_{gmax} = \sqrt{R_{smax}^2 - h^2} \text{ and } R_{tmax} = \sqrt{R_{smax}^2 - (h_{max} - h)^2} \tag{4.32}$$

And, in anticipation of the integration to follow, define two intermediate variables, a and b,

$$a = \frac{h^2}{\sqrt{R_{smax}^2 - h^2}} \text{ and } b = \frac{(h_{max} - h)^2}{\sqrt{R_{smax}^2 - (h_{max} - h)^2}} \tag{4.33}$$

Then, the three segment integrals are

$$PAS_m = \int_0^{2\cdot\pi} d\phi \int_{\pi/2 - \text{asin}((h_{max}-h)/R_{smax})}^{\pi/2 + \text{asin}(h/R_{smax})} \sin\theta \cdot d\theta \int_0^{R_{smax}} \frac{PS_{erpm} \cdot \lambda^2 \cdot L_m \cdot \boldsymbol{P}_{Fm} \cdot R_s^2}{(4 \cdot \pi)^2 \cdot R_s^2} dR_s$$

$$+ \int_0^{2\cdot\pi} d\theta \int_0^{h} dZ \int_0^{(1-Z/h)\cdot R_{gmax}} \frac{PS_{erpm} \cdot \lambda_m^2 \cdot L_m \cdot \boldsymbol{P}_{Fm} \cdot R_g}{(4 \cdot \pi)^2 \cdot (R_g^2 + h^2)} dR_g$$

$$+ \int_0^{2\cdot\pi} d\phi \int_h^{h_{max}} dZ \int_0^{(Z/(h_{max}-h))\cdot R_{tmax}} \frac{PS_{erpm} \cdot \lambda_m^2 \cdot L_m \cdot \boldsymbol{P}_{Fm} \cdot R_t}{(4 \cdot \pi)^2 \cdot (R_t^2 + (h_{max} - h)^2)} dR_t \tag{4.34}$$

After integrating and much algebra,

$$PAS_m(f) = KS_m(f) \cdot \left[\begin{array}{c} \dfrac{h_{max}}{2} \cdot \ln\left(\dfrac{h_{max}^2 + b^2}{b^2}\right) + \dfrac{h}{2} \cdot \ln\left(\dfrac{b^2 \cdot (h^2 + a^2)}{a^2 \cdot (h^2 + b^2)}\right) \\[3mm] + b \cdot \tan^{-1}\left(\dfrac{h_{max}}{b}\right) - b \cdot \tan^{-1}\left(\dfrac{h}{a}\right) + a \cdot \tan^{-1}\left(\dfrac{h}{a}\right) \end{array} \right] \qquad (4.35)$$

For all cases of practical interest, $R_{smax} \gg h$, so the arctangent terms in the interceptor geometry portion of Equation (4.35) are very small relative to the other terms. Thus,

$$PAS_m(f) \cong KS_m(f) \cdot \left[\begin{array}{c} \dfrac{h_{max}}{2} \cdot \ln\left(\dfrac{h_{max}^2 + b^2}{b^2}\right) \\[3mm] + \dfrac{h}{2} \cdot \ln\left(\dfrac{b^2 \cdot (h^2 + a^2)}{a^2 \cdot (h^2 + b^2)}\right) \end{array} \right] \qquad (4.36)$$

Furthermore, the second term in Equation (4.36) can also be dropped with a maximum geometry error penalty of 18 percent. The author recommends dropping the second term in most cases, because the EOB will not be known within 20 percent accuracy.

Similarly, the mainlobe differential probability density of airborne emitter pointing directions is approximately distributed over 1/4 of isotropic, but for completely different reasons (aircraft scan volumes are usually small, but they are typically maneuvering even when on station, which expands the covered volume). Thus,

$$d\mathcal{P}_{Em} = \frac{D_{Em} \cdot 4 \cdot dV}{G_{MLm}} \qquad (4.37)$$

Integrating similarly to the above using Equation (4.29) and breaking the integration space into three regions as done in Equation (4.34) but for mainlobe emitter pointing probabilities, making the same substitutions and approximations, the airborne contribution to the ambient spectrum from the mth emitter is given in Equation (4.38) below.

$$PAM_m(f) \cong KM_m(f) \cdot \left[\begin{array}{c} \dfrac{h_{max}}{2} \cdot \ln\left(\dfrac{h_{max}^2 + b^2}{b^2}\right) \\[12pt] + \dfrac{h}{2} \cdot \ln\left(\dfrac{b^2 \cdot (h^2 + a^2)}{a^2 \cdot (h^2 + b^2)}\right) \end{array} \right] \qquad (4.38)$$

Recall from Equations (4.6) and (4.7) that R_{smax} is approximately

$$R_{smax} \leq (h_I^{0.5} + h_E^{0.5}) \cdot 7463 \text{ feet} \qquad (4.39)$$

where h_I is the interceptor altitude and h_E is the emitter altitude. Therefore, $h = h_I + h_E$ is always very small compared to R_{smax}, which is why the approximations of Equations (4.36) and (4.38) are so accurate. Because most ground emitters are at relatively low altitude, h_E is often assumed to be 20 ft. Most airborne emitters are above 1000 ft and, often, h_E is assumed to be 1000 ft. If these assumptions are made, then h and R_{smax} are different for the airborne and ground contributions to the overall spectrum. (Figures 4.40 and 4.41 make this assumption.)

Recall that each of the individual spectral contributions is a function of frequency. Accumulating all the emitters in both the sidelobes and mainlobes, the overall ambient spectrum can be estimated for a set of systems in a given scenario. Then,

$$P_{amb}(f) = \sum_{m=1}^{m_{max}} \left[\begin{array}{c} (1 - Q(m)) \cdot (PGS_m + PGM_m) \\[6pt] + Q(m) \cdot (PAS_m + PAM_m) \end{array} \right] \qquad (4.40)$$

Using parameters from Table 4.7 or a larger table in the CDROM appendix, a typical battlefield spectrum can be estimated as shown in Fig. 4.40. That spectrum, when combined with a typical civilian spectrum (Fig. 4.38), shows what might be expected over some modern battlefield. Whether this spectrum is representative of a real battlefield is not important but, rather, the method used should be applied to a predicted EOB to obtain the expected spectrum. Interestingly, as noted in the civilian spectrum discussion, the ambient power may go up as the altitude goes down, because most airborne emitters are still visible and surface emitters are closer.

4.5.3 Estimating Ambient Pulse Density

Several studies have been reported that forecast intercept receiver pulse traffic.[18,19] Pulse traffic can be estimated using the model presented in the previous section with a few additions. The mth emitter type has an assumed population on the battlefield as represented by N_m and a characteristic range of average, PRF_m. Thus, the total num-

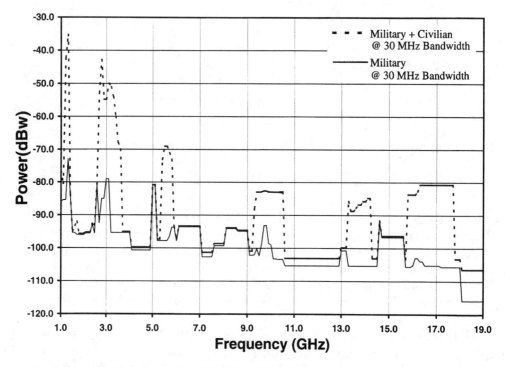

Figure 4.40   Estimated battlefield spectrum, 18,000 ft AGL in 30-MHz band.

ber of pulses emitted is the product $PRF_m \cdot N_m$, and, if the interceptor had infinite sensitivity, it would detect and be forced to process all of them. Because interceptors have limited sensitivity (often intentionally), a lower number of emitted pulses weighted by their power probability will be detected and must be processed. If one assumes a biased detector with exponential characteristic, then the number of pulses detected can be modeled as given in Equation (4.41).

$$PPD(f) = \sum_m N_m \cdot PRF_m \cdot \left[1 - \exp\left(\frac{-P_{amb}(f)}{TH}\right)\right] \tag{4.41}$$

where
   TH = threshold power bias
   N_m = total quantity of emitters of the mth type in the battlespace
PRF_m = average pulse repetition frequency of the mth emitter type

The use of a soft threshold as shown in Equation (4.41) is consistent with multipath and anomalous propagation typical of real environments. Similarly to Section 4.5.2, using the table in Section 4.7 or Appendix A.4, the pulse density can also be estimated

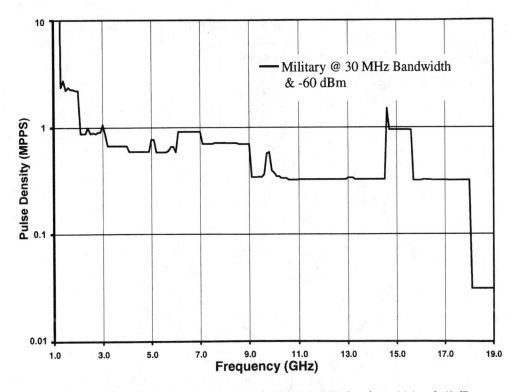

Figure 4.41   Estimated battlefield pulse density, 18,000 ft AGL in 30-MHz band, sensitivity of –60 dBm.

for a given sensitivity as shown in Fig. 4.41. The threshold, altitude, and analysis bandwidth in this case have been chosen to be –60 dBm, 18,000 ft AGL, and 30 MHz, respectively.

4.6   EXAMPLE SCENARIO ANALYSIS

4.6.1   Classification Usable Sensitivity

Two stealth scenarios will be analyzed in this section. The first provides a typical LPI and intercept receiver scenario. It will derive an average number of emitters visible for given sensitivity and make classification software loading estimates. Classification-limited present and ultimate usable sensitivity will be discussed. The second scenario uses conventional attrition analysis to estimate the benefits of stealth system survival rates as a function of observables.

Figure 4.42 shows a typical LPIS scenario similar to the early days of Desert Shield/ Desert Storm. A penetrating stealthy aircraft must fly from a base somewhere in friendly territory to the target somewhere in enemy territory. The total one-way travel might be 400 mi, of which some fraction is over enemy-controlled or threat territory.

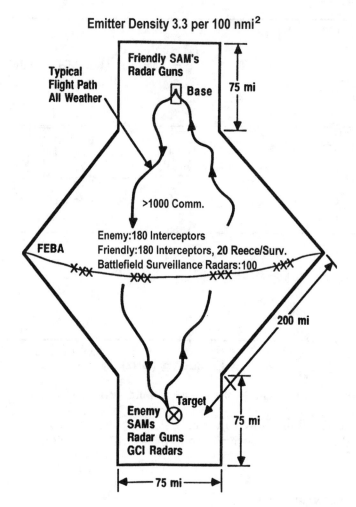

Figure 4.42 Example stealth penetrator scenario.

During such a flight, the penetrator will be in the engagement range or presence of perhaps 100 each friendly and enemy air defense systems with radar and data link emitters. In addition, there will be possibly 100 battlefield surveillance radars. In the air, there will be up to 400 aircraft emitters within the line of sight, as suggested in the figure. Some or all of these units will be emitting at any time. The total number of emitters could be a few less than 2000, with an average density of 3.3 per 100 nmi^2 or roughly 1 per 30 nmi^2. The emitters will be distributed on the average over the battle area 1 every 5.5 nmi in any direction, and their antenna mainlobes will be almost uniformly distributed in azimuth. The stealth penetrator might operate at altitudes from 30,000 to 200 ft AGL at various times during a mission.

Extending the scenario, suppose the enemy interceptors have a 1-GHz instant bandwidth receiver with sensitivities including antenna gain of –60, –40, –30 dBm and quadrant angle sorting (i.e., a typical RWR). The enemy interceptors are operating at an altitude of 5000 ft AGL (higher altitude hardly makes any difference). In this example, during the time in enemy territory, the stealthy penetrator is flying at night at an altitude of 1000 ft AGL (lower altitude improves penetrator stealth). The unfriendly aircraft and surface threats must operate and detect the stealth aircraft in this environment. The threat systems must detect, sort, and classify LPI emissions in the presence of all the other emitters. The natural question is, "How well can they do?"

To determine the answer, the ambient spectrum needs to be estimated, and the corresponding intercept receiver loading must be calculated. Using Equations (4.33) through (4.40) with the parameters assumed, the ambient spectrum can be calculated with and without the sidelobe contribution to power. The mainlobes of the emitters will be visible to the threat intercept receivers all the way to the RF horizon. The sidelobes of the emitters will be visible out to a ground range equivalent to the interceptor detection threshold signal level. For example, the typical enemy fighter in the table below will be detected in its sidelobes at 18 km with a –60 dBm sensitivity receiver. Table 4.6 summarizes the assumed parameters for the example scenario. Obviously, not all these systems would have identical parameters, but the values in the table are quite common and represent a good average. One could argue about any of the numbers in the example; the important idea is the method.

Table 4.6   Battlefield Emitter Parameters

Type*	Q	N	PS	PM	G_{ML}	G_{SL}	f_u	f_l	IBW	PRF
Friendly fighter	1	180	2500	10^8	5000	0.125	10	9	3	50
Reece/surveillance	1	20	5000	10^8	5000	0.25	9	8	500	10
Enemy fighter	1	180	$5 \cdot 10^4$	$5 \cdot 10^8$	5000	0.5	8	7	1	50
Air defense	0	200	$5 \cdot 10^4$	10^9	10,000	0.5	17	5	1	5
Bomber	1	20	5000	10^8	5000	0.25	16.5	16	5	5
Satcomm	0	50	10^5	$3 \cdot 10^8$	1500	0.5	5	4	5	50
Battle surveillance/GCI	0	100	10^5	$6 \cdot 10^8$	3000	0.5	4	3	1	3
Communications	0	200	500	10^6	1000	0.5	3	2	5	50
Short-range comm.	0	1000	83.3	$5 \cdot 10^3$	30	0.5	2	1	0.5	10
Long-range comm.	0	20	$1.7 \cdot 10^5$	10^9	3000	0.5	1	0.5	5	0.3

*Note: Q = 1 if airborne, N = number of systems in theater, PS = sidelobe *ERPP* in watts, PM = mainlobe *ERPP* in watts, G_{ML} = mainlobe gain, G_{SL} = sidelobe gain, f_u = upper edge of the operating band in gigahertz, f_l = lower edge of the operating band in gigahertz, IBW = instantaneous bandwidth in megahertz, PRF = average PRF in kilohertz.

With these parameters, the ambient pulse density can be calculated for several different intercept sensitivities as shown in Fig. 4.43. The sensitivities chosen are the probable hardware limited sensitivity and two probable software limited sensitivities. The actual calculation of the graph below is given in the CDROM Appendix A.4.

With estimates of the pulse density in each band as a function of sensitivity, the equivalent intercept processor loading can be calculated. Because a radar warning receiver (RWR) that gives a warning too late has no military utility, the maximum usable sensitivity can be estimated. Returning to the loading Equation of Section 3.1.7, which is restated as Equation (4.42), the intercept processor loading can be estimated for the various intercept receiver sensitivities given in Fig. 4.43.

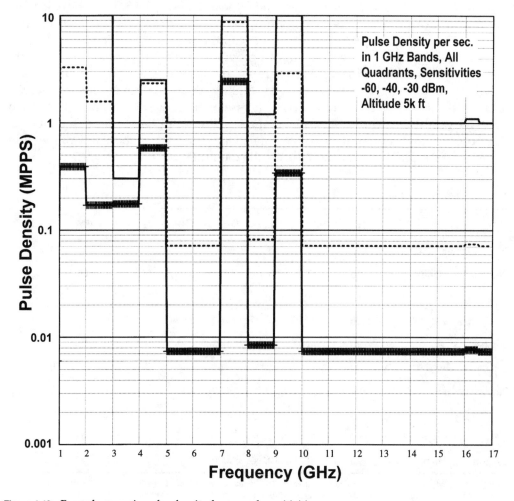

Figure 4.43 Example scenario pulse density for several sensitivities.

$$OPS/\text{sec} = 225 \cdot N_p \cdot \log_2(N_p) \cdot 10^3 + 2.9 \cdot 10^6 \qquad (4.42)$$

where N_p is the number of pulses intercepted per millisecond.

The highest sensitivity shown in Fig. 4.43, −60 dBm, has a total pulse traffic from 1 to 17 GHz of roughly $5 \cdot 10^7$ pulses per second. The traffic is made up of 4 bands of 10^7 pulses/sec and 12 bands of roughly 10^6 pulses/sec. Because the RWR is assumed to have quadrant angle separation and millisecond processing, the four bands each require

$$OPS/\text{sec} = 225 \cdot 2500 \cdot \log_2(2500) \cdot 10^3 + 2.9 \cdot 10^6 = 6.35 \cdot 10^9 \qquad (4.43)$$

Similarly, the 12 other bands each require

$$OPS/\text{sec} = 225 \cdot 2500 \cdot \log_2(250) \cdot 10^3 + 2.9 \cdot 10^6 = 0.32 \cdot 10^9 \qquad (4.44)$$

This requires a total number of processor operations for the entire band of $29 \cdot 10^9$. Assuming the best militarized processor currently deployed, the update time would be tens of seconds to several minutes.

When the processing time exceeds the emitter dwell time, emitters appear to sparkle or scintillate in angle and are much more difficult to track if they are moving. Mainlobe dwell times are typically 30 to 300 ms, and the interceptor must complete its processing in some fraction of that minimum time, which is why the 1 ms assumption was used in the analysis.

So what sensitivity is usable in the assumed battlefield scenario? The two other sensitivities, −40 and −30 dBm, shown in Fig. 4.43, require $11.9 \cdot 10^9$ and $3.83 \cdot 10^9$ OPS, respectively. Using the best available military processors, −30 dBm would require several seconds for updates. This update rate would be acceptable for surface targets that either don't move or are moving slowly as well as airborne emitters that are far away. For stealth vehicles that move, even slowly, low update rates prevent tracking and may prevent detection as a result of scan on scan discussed in Chapter 2. The intercept receiver designer has several options: increase complexity dramatically, limit sensitivity, or accept very low update rates. Most deployed RWRs have chosen low update rates and reduced sensitivity. Typical intercept sensitivities in a dense signal environment are limited to −35 to −50 dBm, and update rates exceed 5 sec. The objective is to limit intercepts to a few emitters per band. Normally, RWRs focus primarily on threats to the carrying platform limiting intercepts to mainlobe only. This is not the case, of course, for air defense suppression with antiradiation missiles (ARMs). Present classification is limited to about 5 to 10 intercepts in the same bin (RF, AOA).

Figures 4.44 and 4.45 show the number of emitters as opposed to pulses detectable in each band. Returning to Figs. 4.40 and 4.41, one can see the attraction for scanning

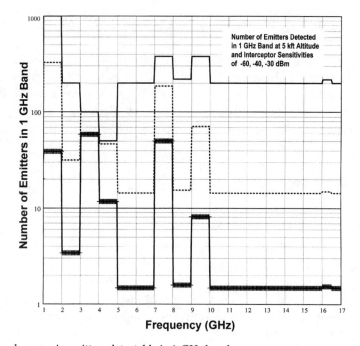

Figure 4.44 Example scenario emitters detectable in 1-GHz band.

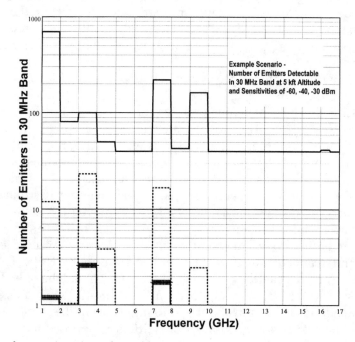

Figure 4.45 Example scenario emitters detectable in 30-MHz band.

superhet intercept receivers. Even though they don't have instant band coverage, the lower per-band pulse density and ambient power make classification software loading much easier, which produces a more balanced hardware and software design. In turn, the narrower instant bandwidth allows higher sensitivity if more processing is available.

In summary, most current intercept receivers have more intrinsic sensitivity than can be used with current processing. Dramatic reductions in bin size are required to match existing intrinsic sensitivity to existing processing in a dense signal environment. In turn, stealth systems must use waveforms that neutralize small bin sizes and increase processing load. The actual usable sensitivity is dependent on the friendly/ threat environment and the interceptor processing strategies. Therefore, an analysis similar to the foregoing must be performed for each new design and weapon system mission.

4.6.2 Monte Carlo Simulations

The second example will cover superficially a low-altitude penetrator simulation (LAPSIM) software program based on the engagement balloon concept introduced in Chapter 1. Simulations of this type attempt to model more accurately engagement geometries in terms of terrain, threat deployment, threat performance in the radar equation sense, threat response times, and flight trajectories. This simulation does not cover other emitter masking or penetrator end-game countermeasures but does include electronic order of battle, threat strategy, terrain masking, stealth platform cross section, flight trajectory, and multilayer engagement time lines. The fundamental notions involve a random laydown of air defense threats, a random penetrating aircraft flight trajectory, typical threat time line performance, and a terrain data base. The visibility limitations of Section 4.3 are taken into account in an encounter. The resulting "missile intercepts" are thus just the missile arriving in the vicinity of the target penetrator while still in track. Figure 4.46 shows the overall block diagram for a LAPSIM-type program. The stealth penetrator independent variables are altitude, radar cross section, and speed. Terrain data bases can be specific world locations, and threat performance/deployment can be as realistic as computing will allow. The environment functions of sensors, communications, command and control, and missile batteries can be everything from simple time lines to functional block simulations. Functional simulations allow not only time delay factors but also designation accuracies and battle degradation evaluation. The outputs are encounter statistics as well as typical visibility-time histories.

Figure 4.47 shows a typical threat laydown for the much-studied Fulda Gap region of Germany. Normally, there is a topographic map underlay for Fig. 4.47, but the composite does not reproduce well and confuses understanding of the basic geometry. Each box in the figure represents an air defense threat. Flight paths can be chosen at random and, for short periods, can be thought of as segments of constant-altitude

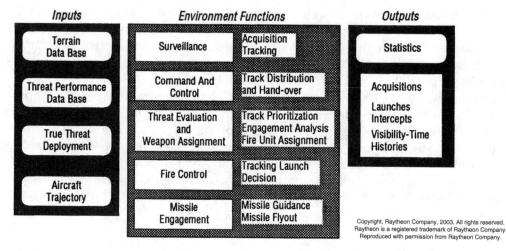

Figure 4.46   Low-altitude penetrator simulator function diagram (Raytheon[23]).

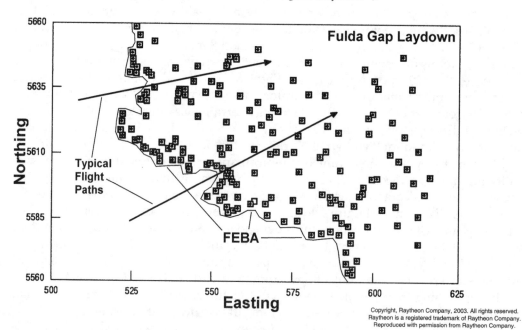

Figure 4.47   Typical random laydown of air defense threats (Raytheon[23]).

great-circle routes. Although encounters on some flight paths could be simultaneous, mathematically, they will be dealt with individually and then summed (a good first approximation). The specific results for real aircraft in various real-world locations are classified, but the basic models provide good first-order approximations.

The simulation operates by calculating the encounter probability to each threat as the penetrator travels along its flight path. That probability is made up of several parts: first, the probability that the penetrator is visible; second, the probability that the penetrator could be detected in the radar equation sense; and third, that the detection lasts long enough for acquisition, handoff, launch, and intercept. As mentioned earlier, each encounter is modeled as an independent one-on-one encounter, and then all the encounters are summed to arrive at an overall statistic. The number of detections, acquisitions, launches, and intercepts are accumulated over the entire flight path. This process is repeated for multiple trajectories, and statistics can be built up.

When or if the detection probability goes above some preset threshold, an acquisition timer is started. If the detection probability drops below a second threshold, acquisition is abandoned. If the detection probability stays above the acquisition threshold long enough, a launch timer is started, which runs until intercept, missile flight time-out, or the penetrator detection probability drops below the second threshold. The acquisition timer includes surveillance frame time, acquisition designation and track stabilization, target threat assessment, and fire unit assignment through missile launch. The launch timer includes launch, missile stabilization, and flyout.

The only math is the continuous calculation of the range between each air defense site and the penetrator as a function of time, the apparent altitude, and thus visibility along that line of sight [Equations (4.8), (4.9), (4.10)], and the range equation for a characteristic radar type as given in Table 1.5. In Chapter 1, Figs. 1.13 and 1.14 assumed detection thresholds set for $\mathcal{P}_D = 90$ percent and $\mathcal{P}_{FA} = 10^{-12}$; thus, detection was either certain if inside that range or zero if outside that range. For the current case, a Rayleigh fading single-frequency channel will be assumed with a radar square law detector with unknown phase in both inphase and quadrature channels followed by noncoherent integration. (It's not exactly correct, but it's easy to calculate.) The probability of detection is

$$\mathcal{P}_D(R_s) = \Gamma_{Eu}\left(1, \frac{-\ln(\mathcal{P}_{FA})}{1 + SNR(R_s)}\right) \tag{4.45}$$

where $\Gamma_{Eu}(1, y)$ is the incomplete Euler gamma function, with y the lower limit of integration and ∞ the upper integration limit. All other variables are as defined before.

The probability of acquisition is

$$\mathcal{P}_{acq} = C_{acq} \cdot \mathcal{P}_D(R_s) \cdot [1 - \mathcal{P}_{cumobl}(R_s, h_a, \Delta R_{c1})] \tag{4.46}$$

where

C_{acq} = a constant to account for acquisition probability and timing

$\mathcal{P}_D(R_s)$ = probability of detection in the radar equation sense and is a function of range, R_s

$\mathcal{P}_{cumob}, \mathcal{P}_{cumobl}$ = as defined in Equations (4.9) and (4.11), respectively

ΔR_{c1} = cross-range length necessary for acquisition

Similarly, the probability of launch depends on the probability of acquisition, the probability of continued detection, and an extended cross range length.

$$\mathcal{P}_{launch} = C_{launch} \cdot \mathcal{P}_{acq} \cdot \mathcal{P}_D(R_s) \cdot (1 - \mathcal{P}_{cumobl}(R_s, h_a, \Delta R_{c2})) \tag{4.47}$$

where C_{launch} is a constant to account for intercept probability and timing, and ΔR_{c2} is the additional cross-range length necessary for launch.

$$\mathcal{P}_{intercept} = C_{intercept} \cdot \mathcal{P}_{launch} \cdot \mathcal{P}_D(R_s) \cdot (1 - \mathcal{P}_{cumobl}(R_s, h_a, \Delta R_{c1} + \Delta R_{c2} + \Delta R_{c3})) \tag{4.48}$$

where $C_{intercept}$ is a constant to account for intercept probability and timing, and ΔR_{c3} is the additional cross-range length necessary for intercept. Because apparent altitude and range may change dramatically during an encounter, the visibility and detection probabilities must be continuously calculated. The independent variables in Equations (4.44) through (4.46) don't seem to match the need, as the above probabilities are usually time dependent not displacement dependent. For Equations (4.43) through (4.46) to be useful, a penetration geometry and speed must be invoked. Assume that a penetrator flying "constant speed, straight and level" is travelling a segment of a great-circle route. Assuming such a route with an arbitrary offset with respect to an observer near the Earth's surface allows the displacements to be converted to time lines.

Although the results shown in Figs. 4.50 through 4.54 are accumulation of many Monte Carlo runs, the basic idea can be understood by considering the geometry of Figs. 4.48 and 4.49. A whole-Earth cutaway view of the threat-penetrator geometry in the plane intersecting the threat, the penetrator, and the center of the Earth is shown in Fig. 4.48. Assume that the altitudes h_t and h_s don't change during the time of the encounter, and the penetrator flies a segment of a great-circle route. The distance between the threat and the penetrator is R_s.

Assuming no winds and a constant speed along a great circle route implies a constant angular velocity about the Earth's center. Then, arbitrarily setting $t = 0$ at the time that the penetrator crosses the x-z plane in Fig. 4.48 yields the following form for the penetrator motion:

$$y = (R_e + h_t) \cdot \sin(\omega \cdot t)$$

$$x = (R_e + h_t) \cdot \cos(\omega \cdot t) \cdot \sin(\theta)$$

$$z = (R_e + h_t) \cdot \cos(\omega \cdot t) \cdot \sin(\theta) \tag{4.49}$$

where

$$\omega = \frac{S_{AC}}{R_e + h_t} \quad \text{and} \quad \theta = \operatorname{atan}\left(\frac{R_{off}}{R_e}\right) \tag{4.50}$$

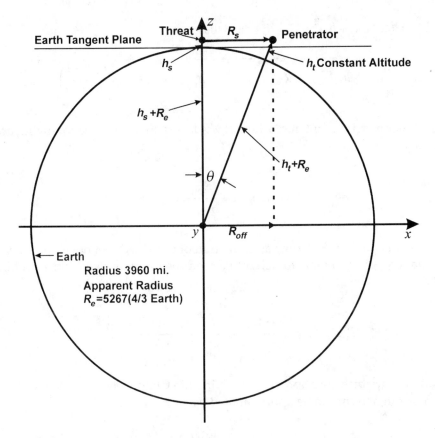

Figure 4.48   Whole-Earth view of penetrator geometry.

where S_{AC} is the penetrator speed, and R_{off} is the range at the point of closest approach ($t = 0$). The y, x, z speeds are

$$S_y = \frac{dy}{dt} = S_{AC} \cdot \cos\left(\frac{S_{AC} \cdot t}{R_e + h_t}\right)$$

$$S_x = \frac{dx}{dt} = S_{AC} \cdot \sin(\theta) \cdot \sin\left(\frac{S_{AC} \cdot t}{R_e + h_t}\right)$$

$$S_z = \frac{dz}{dt} = S_{AC} \cdot \cos(\theta) \cdot \sin\left(\frac{S_{AC} \cdot t}{R_e + h_t}\right)$$

(4.51)

The slant range between the penetrator and the surveillor is

$R_s^2 = (z - (R_e + h_t)^2 + x^2 + y^2),$ and thus

$$R_s(t) = \sqrt{(R_e + h_t)^2 - 2 \cdot (R_e + h_s) \cdot (R_e + h_t) \cdot \cos(\theta) \cdot \cos\left(\frac{S_{AC} \cdot t}{R_e + h_t}\right) + R_e^2} \tag{4.52}$$

For most geometries of interest where R_{off} is less than 500 mi, the following approximation is useful:

$$\cos(\theta) \approx 1 - \frac{R_{off}^2}{2 \cdot R_e^2} \tag{4.53}$$

Now that an equation for range as a function of time has been developed, a version of the radar equation suitable to determine the probability of detection with time can be stated.

$$SNR(R_s, \sigma) = \frac{P_T \cdot G_T^2 \cdot Duty \cdot T_D \cdot \lambda^2 \cdot L_R}{(4 \cdot \pi)^2 \cdot k_B \cdot T_0 \cdot NF_R} \cdot \frac{\sigma}{R_s^4} \tag{4.54}$$

where all the variables are as previously defined in Chapter 1.

The cross-range term can be approximated by

$$\Delta R_c \approx \frac{R_s(t_1) - R_s(t_2)}{\tan\left(a\cos\left(\frac{2 \cdot R_{off}}{R_s(t_1) + R_s(t_2)}\right)\right)} \tag{4.55}$$

where t_1 is the start of an acquisition, launch, or intercept interval, and t_2 is the end of the same interval. With these basic equations, it is now possible to calculate individual encounters.

Figure 4.49 shows a plan view of the penetrator geometry, and the trajectory projected on a tangent plane under the surveillor is a segment of an ellipse. Some segment of the trajectory ΔR must have good visibility and good SNR for a full engagement. The penetrator will appear to rise above the horizon in a segment of an ellipse. The point of closest approach is R_{off}. All of the curvatures are exaggerated to illustrate the geometrical effects.

For example, consider the case of an encounter between a penetrator flying at 200 ft over wooded terrain at a speed of 850 ft/sec, with radar cross section of –20 dBsm and the X-band fire control radar of Table 1.5 with an offset range of 5 mi. Assume acquisi-

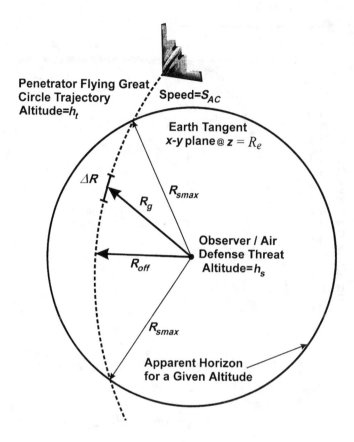

Figure 4.49 Plan view of penetrator geometry.

tion time of 30 sec, launch time of 30 sec, and maximum missile time of flight of 15 sec. Then, using the equations of Section 4.3, the probability of terrain masking is given in Fig. 4.50. The penetrator is not visible in wooded terrain for ranges greater than 2 nmi, as shown in Fig. 4.50, and thus $\mathcal{P}_D$ in the radar equation sense doesn't make any difference. At this offset range, all encounters result in *no* acquisitions, launches, or intercepts. Note, however, that grassy or steppe (not shown) terrain can result in an acquisition. If all offsets between 5 and 0 nmi are tested in a regular or Monte Carlo sense, then 7.5 percent of the total will be visible, and those visible will be well within the 95 percent probability of detection envelope. The transit time will range from 12 to 0 sec, and *no* launches will occur, as 30 sec has been assumed. If, however, acquisition time of 5 sec is assumed (some systems make this claim), then there could be 7.5 percent $\times (12 + 0)/(2 \times 5) = 9$ percent launches. No launch would result in an intercept, because the cross-range visibility is too short as shown in Fig. 4.51. But if the launched missile had an autonomous seeker that could detect a stealth

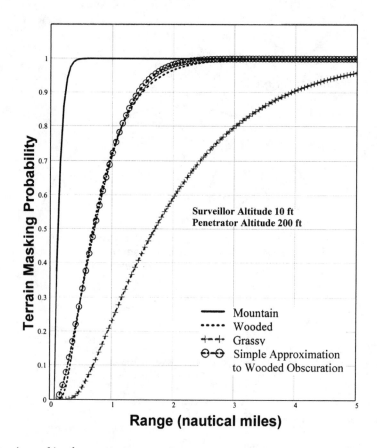

Figure 4.50   Terrain masking for penetrator geometry.

target in clutter, and if the missile had a lofted trajectory (5000 ft altitude) so that the grazing angle was greater than 6° as shown in Fig. 4.20, for example, then an intercept might occur. In most cases, the penetrator would be an opening target at the time of missile launch, and the penetrator must be more than 3.5 nmi from the missile launcher at the time of launch (e.g., 6-mi missile range in 15 sec versus 3.5-mi at missile launch + 2.5-mi penetrator travel during missile time of flight). Line-of-sight missiles would have no chance, because of cross-range loss of visibility. The coefficients for the simple approximation in Fig. 4.50 using Equation (4.12) are $a = 1.52$, $b = 0.11$, and $c = 1.56$.

Figure 4.52 through 4.55 show a summary of multiple runs of the type just described for a lofted trajectory missile and 5-sec acquisition time for various cross sections and penetrator altitudes. The absolute numbers in this example mean nothing—only the ratios between columns, because the numbers are a function of the number of Monte Carlo runs.

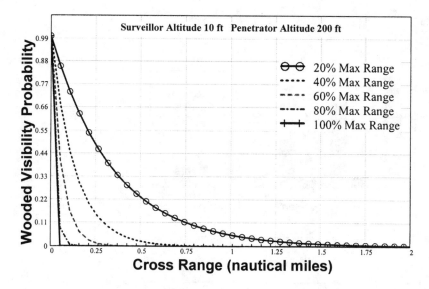

Figure 4.51 Cross-range visibility for penetrator geometry.

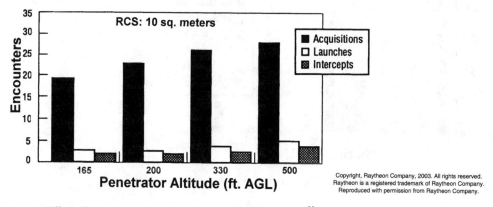

Figure 4.52 Effect of altitude on threat missile intercepts (Raytheon[23]).

4.7 TYPICAL DEPLOYED EMITTERS

There exist formidable, combat proven, and numerous airborne and surface emitters (e.g., the SA-8, MIG-29, and Roland shown in Fig. 4.56) that will fill the battlefield spectrum. These systems emit from L-band to Ku-band. There also are many civilian emitters that will be found in or near a battle area.

Table 4.7 contains typical parameters for deployed emitters. All are taken from open literature and thus may contain some disinformation. It was used to make some of the spectrum power estimates.

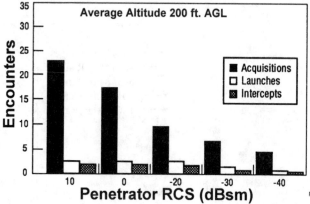

Figure 4.53 Radar cross section versus threat missile intercepts (Raytheon[23]).

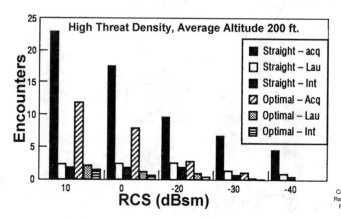

Figure 4.54 Straight and optimal penetration route versus threat (Raytheon[23]).

4.8 EXERCISES

1. Using the parameters of Table 1.5, LREW column, calculate the probability of detection, $\mathcal{P}_D$, using Equation (4.43) with R_s equal to 100 nmi and σ equal to –10 dBsm with $\mathcal{P}_{FA}$ equal to 10^{-10}.

2. Calculate the received power from sea clutter and rain clutter for a sea state 4 storm at a range of 100 nmi using the parameters of the JAS-39 from Table 4.7.

3. Find the pulse density in a 300-MHz band at 16 GHz for an intercept receiver at 30 kft and –85 dBm total sensitivity using the example parameters of Section 4.6.1.

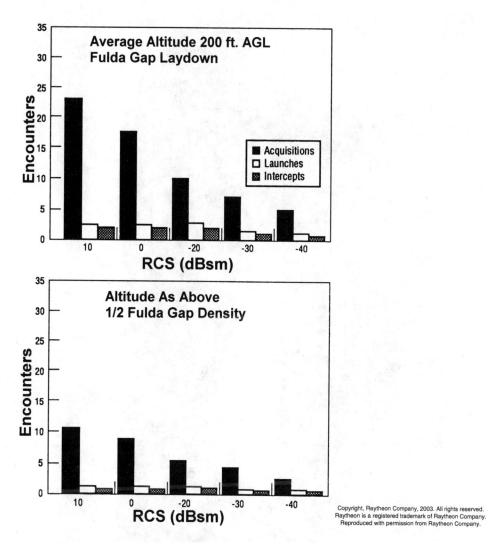

Figure 4.55   Effect of penetrating at lower threat density (Raytheon[23]).

4. Calculate the ambient spectrum in 1-GHz bands using the EOB from Fig. 4.35, including both the quantities and coverage bands given in the figure.

5. Assume an observer at 20 kft. What is the apparent minimum altitude for an emitter at 1000 ft, 240 nmi away?

6. What is the cross-range 60 percent visibility distance for steppe terrain for observer and target both at 1000 ft and separation of 80 nmi range?

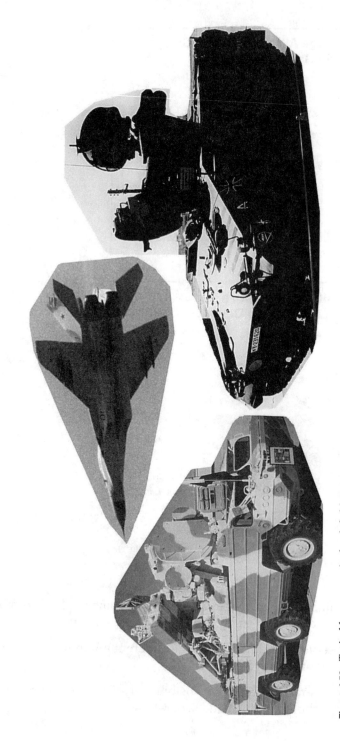

Figure 4.56 Typical large-quantity battlefield emitters (author photos).

Table 4.7 Typical Deployed Emitter Parameters[11-17]

System type	Operating band, low (GHz)	Operating band, high (GHz)	Average instant band-width (MHz)	Antenna mainlobe gain (dB)	Peak power (KW)	Mainlobe ERPP (dBW)	Antenna side-lobe gain (dBi)	Sidelobe ERPP (dBW)	Average power (W)	Sidelobe ERAP (dBW)	PRF, high (KHz)	PRF, low (KHz)	No. of systems in theatre
Air radar													
Communist China													
F–7	9.5	10.0	1	30	5	67	–3	34	40	13	3.0	1.0	200
F–8	9.6	9.9	1	32	8	71	–6	33	160	16	3.0	1.0	200
FBC–1	16.0	16.2	3	30	75	79	–3	46	300	22	5.0	0.5	30
France													
Mirage	9.5	10.0	1	35	200	88	–6	47	200	17	5.0	0.5	100
Entendard	9.5	10.0	1	33	50	80	–3	44	50	14	3.0	1.0	25
Jaguar	9.5	10.0	1	33	50	80	–3	44	50	14	3.0	1.0	25
Rafale	9.5	10.0	1	35	5	72	–10	27	500	17	20.0	1.0	25
Germany													
F4G	9.6	10.2	5	35	4	71	–10	26	400	16	20.0	1.0	50
Tornado	9.6	9.8	3	36	10	76	–3	37	400	23	5.0	0.5	50
EFA	9.6	10.2	5	35	4	71	–10	26	400	16	20.0	1.0	50
International													
F–5	9.6	9.8	1	32	50	79	–3	44	50	14	5.0	1.0	200
F–4	9.6	9.8	1	35	150	87	–3	49	150	19	5.0	1.0	100
F–104	9.6	9.8	1	34	100	84	–3	47	150	19	3.0	1.0	100
Mirage	8.0	10.0	1	35	200	88	–6	47	200	17	5.0	1.0	100
F–16	9.6	9.8	1	33	17	75	–6	36	200	17	30.0	1.0	200
Russia													
MIG–21	12.9	13.2	1	30	8	69	–3	36	150	19	3.0	1.0	200
MIG–23	12.9	13.2	1	35	100	85	–3	47	300	22	3.0	1.0	100
MIG–29/33	9.5	10.5	3	33	5	70	–10	27	1000	20	20.0	1.0	100

Table 4.7 Typical Deployed Emitter Parameters[11-17] (continued)

System type	Operating band, low (GHz)	Operating band, high (GHz)	Average instant band-width (MHz)	Antenna mainlobe gain (dB)	Peak power (KW)	Mainlobe ERPP (dBW)	Antenna side-lobe gain (dBi)	Sidelobe ERPP (dBW)	Average power (W)	Sidelobe ERAP (dBW)	PRF, high (KHz)	PRF, low (KHz)	No. of systems in theatre
MIG–31	9.5	10.2	2	37	10	77	–10	30	3000	25	300.0	5.0	100
SU–27	9.5	10.5	3	33	5	70	–10	27	1000	20	5.0	1.0	100
TU–16/22	8.0	15.0	1	33	100	83	–3	47	100	17	3.0	1.0	10
A–50	0.5	1.5	0.5	37	10	77	–3	37	1000	27	1.5	0.3	5
Sweden													
JA–37	9.5	10.0	1	34	5	71	–6	31	500	21	20.0	5.0	25
JA–39	9.5	10.0	1	34	10	74	–10	30	1000	20	20.0	5.0	25
U.K.													
Tornado	9.6	9.8	3	36	10	76	–3	37	400	23	5.0	0.5	50
AV–8	9.6	10.2	1	32	100	82	–3	47	300	22	5.0	1.0	25
EFA	9.6	10.2	5	35	4	71	–10	26	400	16	20.0	1.0	50
U.S.A.													
E–2	0.5	0.6	0.5	29	500	86	–3	54	20	10	1.0	0.2	3
E–3	2.0	3.0	1	38	750	97	–10	49	25	4	5.0	0.2	3
E–8	8.5	9.0	12.5	40	25	84	–3	41	5	4	5.0	0.5	3
P–3	9.0	9.5	150	37	100	87	–3	47	300	22	5.0	0.5	5
F–14	9.5	10.2	1	37	10	77	–6	34	5000	31	300.0	1.0	25
F–15	9.6	10.2	3	37	5	74	–10	27	1200	21	250.0	1.0	100
F–16	9.7	9.9	1	33	17	75	–6	36	200	17	30.0	1.0	100
F–18	9.6	10.2	3	35	4	71	–10	26	400	16	300.0	0.5	100
F–22	8.5	10.5	3	36	15	78	–10	32	3000	25	30.0	0.5	100
F–111	16.0	16.4	1	37	100	87	–3	47	100	17	5.0	0.5	25
AC–130	9.6	9.8	10	37	5	74	–10	27	1200	21	30.0	1.0	10
AV–8	9.5	10.2	3	35	4	71	–10	26	400	16	300.0	0.5	50
AH–64	90.0	100.0	5	35	5	72	–3	34	50	14	50.0	5.0	50

Table 4.7 Typical Deployed Emitter Parameters[11-17] (continued)

System type	Operating band, low (GHz)	Operating band, high (GHz)	Average instant band-width (MHz)	Antenna mainlobe gain (dB)	Peak power (kW)	Mainlobe ERPP (dBW)	Antenna side-lobe gain (dBi)	Sidelobe ERPP (dBW)	Average power (W)	Sidelobe ERAP (dBW)	PRF, high (KHz)	PRF, low (KHz)	No. of systems in theatre
U-2	8.4	8.9	50	40	3	75	-6	29	400	20	5.0	1.0	2
V-22	9.3	9.4	1	33	10	73	-6	34	5	1	2.0	0.2	5
C-5	9.3	9.4	1	33	10	73	-6	34	5	1	2.0	0.2	5
C-141	9.3	9.4	1	33	10	73	-6	34	5	1	2.0	0.2	4
C-17	9.3	9.4	1	33	10	73	-6	34	5	1	2.0	0.2	10
C-2	9.3	9.4	1	33	10	73	-6	34	5	1	2.0	0.2	5
C-130	9.3	9.4	4	30	10	70	-6	34	5	1	1.6	0.2	50
B-52	16.0	16.4	1	35	100	85	-3	47	150	19	3.0	0.5	20
B-1	9.6	9.8	3	37	15	79	-6	36	150	16	20.0	0.5	20
Aerostat	9.6	9.9	5	42	17	84	-6	36	200	17	5.0	0.5	5
Commercial	9.3	9.4	4	30	10	70	-6	34	5	1	1.6	0.2	100
Air EW													
Communist China													
BM/KG 8601	2.0	8.0	20	3	25	47	-3	41	25	11	100.0	1.0	300
BM/KG 8605/6	8.0	18.0	20	3	50	50	-3	44	50	14	100.0	1.0	300
France													
Remora	6.0	18.0	20	9	50	56	-3	44	50	14	100.0	1.0	50
ABD 2000	6.0	18.0	20	9	25	53	-3	41	25	11	100.0	1.0	300
Germany													
Sky Buzzer	6.0	18.0	20	9	100	59	-3	47	100	17	100.0	1.0	300
U.S.A.													
ALQ-99	1.0	18.0	20	20	1000	80	-3	57	1000	27	100.0	1.0	30
ALQ-136	8.0	18.0	25	10	500	67	-3	54	250	21	500.0	0.5	50
ALQ-165	0.80	35.0	25	15	2000	78	-3	60	500	24	500.0	0.5	50
ALQ-176	0.80	15.5	25	7	400	63	-3	53	400	23	500.0	1.0	10

Table 4.7 Typical Deployed Emitter Parameters[11-17] (continued)

System type	Operating band, low (GHz)	Operating band, high (GHz)	Average instant band-width (MHz)	Antenna mainlobe gain (dB)	Peak power (KW)	Mainlobe ERPP (dBW)	Antenna side-lobe gain (dBi)	Sidelobe ERPP (dBW)	Average power (W)	Sidelobe ERAP (dBW)	PRF, high (KHz)	PRF, low (KHz)	No. of systems in theatre
Air communications													
International													
SatComm	7.9	8.4	1.5	27	0.1	47	−3	17	100	17	1500	1500	30
ARC–840	0.2	0.4	0.025	6	0.025	20	−3	11	25	11	25	1.0	300
U.S.A.													
ATX–2740	11.0	14.5	275	30	0.025	44	−3	11	25	11	275000	275000	30
JTIDS	1.0	1.2	4	9	1	39	9	39	1000	39	3700	3700	30
JTIDS	1.0	1.2	4	9	0.2	32	9	32	100	29	3700	3700	500
SatComm	7.9	8.4	1.5	27	0.1	47	−3	17	100	17	1500	1500	30
GPS	1.2	1.6	0.5	50	100	100	−3	47	100	17	500	500	1
Tacan	1.0	1.2	0.1	0	0.75	29	0	29	100	20	5.4	5.4	20
AXQ–14	1.5	1.8	5	10	0.05	27	−3	14	40	13	500	0.0	100
Link 16	1.5	1.7	0.5	10	0	26	−3	13	50	14	500	1.0	100
Air IFF													
International													
Typical	1.0	2.0	1	15	2	48	−3	30	10	7	20	0.3	1000
Typical	8.0	10.0	1	35	1	65	−6	24	10	4	20	0.3	1000
Surface radar													
France													
Crotale	2.3	2.4	0.25	32	5	69	−3	34	200	20	6	5	50
France													
Crotale	16.0	16.4	1	40	20	83	−3	40	200	20	9	8	50
Shahine	2.3	2.4	0.25	32	5	69	−3	34	200	20	6	5	50
Shahine	5.0	16.4	1	40	20	83	−3	40	200	20	9	8	50
Roland	1.4	1.4	0.25	32	4	68	−3	33	200	20	9	8	50

Table 4.7 Typical Deployed Emitter Parameters[11-17] (continued)

System type	Operating band, low (GHz)	Operating band, high (GHz)	Average instant bandwidth (MHz)	Antenna mainlobe gain (dB)	Peak power (kW)	Mainlobe ERPP (dBW)	Antenna sidelobe gain (dBi)	Sidelobe ERPP (dBW)	Average power (W)	Sidelobe ERAP (dBW)	PRF, high (KHz)	PRF, low (KHz)	No. of systems in theatre
Roland	5.7	5.9	1	30	50	77	-3	44	100	17	6	5	50
Roland	16.0	16.4	3	40	10	80	-3	37	200	20	6	5	50
Germany													
Gepard AA	2.3	2.4	0.25	27	5	64	-3	34	200	20	6	5	100
Gepard AA	16.0	16.4	3	37	10	77	-3	37	200	20	9	8	100
Roland	1.4	1.4	0.25	32	4	68	-3	33	200	20	9	8	50
Roland	5.7	5.9	1	30	50	77	-3	44	100	17	6	5	50
Roland	16.0	16.4	3	40	10	80	-3	37	200	20	6	5	50
Wildcat	8.0	10.0	3	27	10	67	-3	37	100	17	10	9	100
International													
Cobra	5.7	5.9	5	36	20	79	-10	33	2000	23	10	5	20
Fire Can	2.7	2.9	1.5	33	300	88	-3	52	500	24	1.9	1.8	20
Super fledermaus	15.9	17.1	5	39	65	87	-3	45	50	14	8	7	50
Crotale	16.0	16.4	1	40	20	83	-3	40	200	20	9	8	50
Russia													
Dog Ear SA-9/13	3.0	6.0	0.5	37	200	90	-3	50	400	23	4	1	30
Gun Dish AA	14.6	15.6	3	37	135	88	-3	48	200	20	9	8	500
Big Fred	34.6	35.3	3	40	50	87	-3	44	50	14	4.4	2.5	500
Pat Hand SA-4	6.4	6.9	1	35	250	89	-3	51	200	20	2	1	200
Thick/Thin Skin	6.0	7.0	0.5	32	500	89	-3	54	500	24	0.5	0.3	100
Land Roll SA-8	6.0	7.0	0.5	32	100	82	-3	47	100	17	2	1	20
Land Roll SA-8	8.0	9.0	3	27	100	77	-3	47	100	17	4	2	500
Land Roll SA-8	14.5	14.6	3	38	100	88	-3	47	100	17	6	4	500
Bar Lock	2.7	3.1	0.5	36	650	94	-3	55	650	25	0.4	0.4	6
Spoon Rest	0.1	0.5	0.1	30	350	85	-3	52	700	25	0.4	0.3	15

Table 4.7 Typical Deployed Emitter Parameters[11-17] (continued)

System type	Operating band, low (GHz)	Operating band, high (GHz)	Average instant band-width (MHz)	Antenna mainlobe gain (dB)	Peak power (KW)	Mainlobe ERPP (dBW)	Antenna side-lobe gain (dBi)	Sidelobe ERPP (dBW)	Average power (W)	Sidelobe ERAP (dBW)	PRF, high (KHz)	PRF, low (KHz)	No. of systems in theatre
Flat Face	0.8	0.9	0.3	34	900	94	-3	57	2500	31	0.8	0.2	15
Side Net	2.6	2.7	0.2	36	650	94	-3	55	650	25	0.5	0.4	6
Big Back	1.0	2.0	0.2	40	900	100	-3	57	2500	31	0.3	0.2	5
Straight Flush SA-6	8.0	9.0	1	38	300	93	-3	52	300	22	100	50	35
Straight Flush SA-6	5.0	6.0	1	35	300	90	-3	52	300	22	1.5	1	35
Flap Lid SA-10	9.5	10.5	2	40	300	95	-10	45	5000	27	10	5	100
Hot Shot AA	2.0	3.0	0.5	33	25	77	-3	41	200	20	10	5	200
Hot Shot AA	14.6	15.6	3	37	200	90	-3	50	200	20	9	8	200
Snow Drift SA-11/17	8.00	9.0	1	38	300	93	-3	52	300	22	1.5	1	50
Fan Song, A and B SA-2	2.9	3.1	0.5	33	600	91	-3	55	600	25	2.8	0.8	100
Fan Song C and D SA-2	4.9	5.1	0.5	35	1500	97	-3	59	1500	29	2.8	0.8	100
Gauntlet SA-15	4.0	5.0	1	33	10	73	-3	37	500	24	8	7	100
Gauntlet SA-15	34.6	35.3	3	40	50	87	-10	37	50	7	10	5	100
Snap Shot SA-13?	14.6	15.6	3	35	100	85	-3	47	100	17	10	5	500
Bill Board SA-12	3.0	4.0	0.5	38	500	95	-10	47	7000	28	6	3	100
Grill Pan SA-12	7.5	8.5	3	42	300	97	-3	52	5000	34	6	3	50
Zoopark	6.0	7.0	5	40	10	80	-10	30	2000	23	10	5	100
Sweden													
Track Fire	15.9	17.1	5	39	65	87	-3	45	50	14	8	7	30
9KA500	15.9	17.1	5	39	65	87	-3	45	50	14	8	7	30
9KA500	8.6	9.5	3	36	200	89	-3	50	150	19	8.1	4.8	30
Sky Guard	8.6	9.5	3	36	26	80	-3	41	250	21	8.1	4.8	50
Leopard AA	8.6	9.5	3	36	26	80	-3	41	250	21	8.1	4.8	100
Leopard AA	34	35	5	40	10	80	-3	37	50	14	10	5	100

Table 4.7 Typical Deployed Emitter Parameters[11-17] (continued)

System type	Operating band, low (GHz)	Operating band, high (GHz)	Average instant band-width (MHz)	Antenna mainlobe gain (dB)	Peak power (kW)	Mainlobe ERPP (dBW)	Antenna side-lobe gain (dBi)	Sidelobe ERPP (dBW)	Average power (W)	Sidelobe ERAP (dBW)	PRF, high (KHz)	PRF, low (KHz)	No. of systems in theatre
U.K.													
Rapier	3.0	4.0	0.5	33	100	83	−3	47	100	17	5	4	50
Rapier	13.4	14.0	3	38	100	88	−3	47	100	17	10	5	50
Cymbeline	8.0	10.0	5	36	100	86	−3	47	100	17	10	5	50
TPS–32	2.9	3.1	0.3	41	2200	104	−10	53	22000	33	0.3	0.1	4
Watchman	2.8	3.1	0.5	34	58	82	−3	45	1300	28	1.0	0.4	10
Martello	1.0	2.0	1	41	132	92	−10	41	5000	27	0.3	0.1	4
U.S.A.													
TPQ–36	9.0	9.5	3	42	25	86	−10	34	500	17	10	5	50
TPQ–37	2.9	3.1	3	40	125	91	−10	41	5000	27	6	3	15
MPQ–64	9.0	9.5	3	40	5	77	−10	27	500	17	10	5	50
Hawk PAR	1.2	1.4	0.3	30	50	77	−3	44	1000	27	1.0	0.5	50
Hawk CWAR	9.0	9.5	0.1	36	5	73	−3	34	5000	34	100	100	50
TPQ–32, MPQ–49	1.2	1.4	0.3	27	15	69	−3	39	150	19	4.8	4.2	50
TPS–32	2.9	3.1	0.3	41	2200	104	−10	53	22000	33	1.0	0.3	4
TPS–43/70/75	2.9	3.1	0.3	36	4000	102	−10	56	6700	28	0.3	0.2	10–20
TPS–73	2.7	2.9	3	36	10	76	−3	37	1100	27	1.0	0.5	5
ASR–9	2.7	2.9	1	33	1000	93	−3	57	600	25	1.2	0.7	5
FPS–8/88	1.3	1.4	0.3	27	1000	87	−3	57	1100	27	0.4	0.3	5
FPN–66	2.7	2.9	1	37	10	77	−3	37	1000	27	1.2	0.8	2
PPS–5	16.0	16.5	4	32	1	62	−3	27	1	−3	4	3	100
MPN–25	9.0	9.2	1	38	5	75	−10	27	500	17	4.5	2.5	5
Condor 2	1.0	1.1	5	27	2	60	−10	23	100	10	5.0	0.5	5
TPN–18	9.0	9.6	5	39	200	92	−3	50	250	21	1.2	1.1	10
VPS–2	9.2	9.3	5	34	1.4	65	−3	28	10	7	20	15	100

Table 4.7 Typical Deployed Emitter Parameters[11-17] (continued)

System type	Operating band, low (GHz)	Operating band, high (GHz)	Average instant band-width (MHz)	Antenna mainlobe gain (dB)	Peak power (KW)	Mainlobe ERPP (dBW)	Antenna side-lobe gain (dBi)	Sidelobe ERPP (dBW)	Average power (W)	Sidelobe ERAP (dBW)	PRF, high (KHz)	PRF, low (KHz)	No. of systems in theatre
TPN-22	9.0	9.2	1	43	50	90	-6	41	300	19	5	4	5
TPN-24, GPN-22	2.7	2.9	1	37	500	94	-3	54	500	24	1.2	0.8	5
TPN-25	9.0	9.2	1	45	12.5	86	-10	31	40	6	4.3	2.7	5
FPS-117	1.2	1.4	0.5	36	25	80	-10	34	2000	23	0.5	0.2	10
Nexrad, WSR-88D	2.7	3	0.8	45.5	750	104	-3	56	1500	29	1.3	0.3	3
Wind shear	0.4	0.4	0.5	30	2000	93	-3	60	1000	27	0.1	0.1	20
ASR-3	1.3	1.4	0.5	34	5000	101	-3	64	3600	33	0.4	0.3	5
Surface EW													
Communist China													
BM/DJG8715	8.0	16.0	100	32	0.2	55	-3	20	200	20	100	1	50
970	8.0	12.0	120	32	0.2	55	-3	20	120	18	100	1	50
Russia													
Tub Brick	2.9	10.4	50	30	0.2	53	-3	20	200	20	50	1	10
King Pin	8.0	10.4	50	35	0.15	53	-3	19	150	19	50	1	20
Cheese Brick	8.6	10.4	50	35	0.15	57	-3	19	150	19	50	1	20
Tacan/GPS	0.5	2.0	10	20	0.5	57	-3	24	500	24	50	1	10
SPN-2.3.4	2.0	20.0	50	30	0.25	47	-3	21	250	21	50	1	20
U.S.A.													
MGARJS	0.4	18.0	100	30	0.25	54	-3	21	250	21	300	0	20
Surface communications													
France													
ST-701 lowband	0.6	0.9	2	3	0.005	10	-2	5	0.005	-25	2000	2000	100
ST-701 midband	1.4	1.9	2	9	0.003	14	-3	2	0.003	-28	2000	2000	100
ST-701 highband	4.4	5.0	2	13	0.002	16	-3	0	0.002	-30	2000	2000	100
Alcat. 179	4.4	5.0	2	30	1	60	-3	27	1000	27	2000	2000	10

Table 4.7 Typical Deployed Emitter Parameters[11-17] (continued)

System type	Operating band, low (GHz)	Operating band, high (GHz)	Average instant band-width (MHz)	Antenna mainlobe gain (dB)	Peak power (kW)	Mainlobe ERPP (dBW)	Antenna side-lobe gain (dBi)	Sidelobe ERPP (dBW)	Average power (W)	Sidelobe ERAP (dBW)	PRF, high (KHz)	PRF, low (KHz)	No. of systems in theatre
Russia													
R-423	4.4	5.0	2	33	1.5	65	−3	29	1500	29	2000	2000	10
U.S.A.													
GRC-201	4.4	5.0	2	30	1	60	−3	27	1000	27	2000	2000	100
TRC-170	4.4	5.0	2	33	2	66	−3	30	2000	30	2000	2000	10
Typical MLS	5.0	5.1	3	30	500	87	−3	54	50	14	5	5	10
Cellular	0.8	1.8	0.005	0	0.001	0	0	0	1	0	5	5	10000
Typ. MilSat–Uplink	7.9	8.4	40	42	0.5	69	−3	24	500	24	45000	45000	25
TCM-600	7.0	8.4	45	30	0.005	37	−3	4	5	4	45000	45000	100
USC–60	11.0	14.5	45	43	0.5	70	−3	24	500	24	45000	45000	100
Surface IFF													
International													
Typical	1.0	2.0	1	25	4	61	−3	33	200	20	10	0.3	1000
Typical	8.0	10.0	1	40	4	76	−6	30	200	17	10	0.3	1000

4.9 REFERENCES

1. Skolnik, M., *Introduction to Radar Systems,* 2nd ed., McGraw-Hill, p. 449.
2. Nathanson, F., *Radar Design Principles,* McGraw-Hill, pp. 33, 15, 169.
3. Nathanson, F., *Radar Design Principles* 2nd ed., McGraw-Hill, pp. 317–320, 371.
4. Biron, S. and Francois, R., "Terrain Masking in the European Soviet Union," *MIT Lincoln Laboratory Report,* CMT-31, Vol. 2, March 9, 1983.
5. Craig, D. and Hershberger, M., "Operator Performance Research," *Hughes Report* P74-504, December 1974.
6. Craig, D. and Hershberger, M., "FLAMR Operator Target/OAP Recognition Study," *Hughes Report* P75-300, 1975, declassified 12/31/1987.
7. Pearson, J. et al., "SAR Data Precision Study," *Hughes Report* P75-459, December, 1975, declassified 12/31/1987.
8. Stipulkosky, T., "Statistical Radar Cross Section Measurements for Tactical Target and Terrain Backgrounds—Phase I and Phase II," *Hughes Reports* P76-482, December 1976 and P77-251, pp. 3-1 to 3-30, August 1977, pp. 4-1 to 4-116, declassified 12/31/1987.
9. Sanders, F., Ramsey, B., Lawrence, V., "Broadband Spectrum Survey of Los Angeles, CA," *Institute For Telecommunication Sciences—NTIA Report* 97-336, May 1997, pp. 9–104.
10. Air Force Avionics Laboratory, "Cooperative Weapon Delivery Study," *Technical Report AFAL-TR-78-154,* October 1978, pp. 2-1 to 2-29, 3-1 to 3-71, declassified 12/31/1991.
11. Streetly, M., Ed., *Janes Radar and Electronic Warfare Systems 1999–2000,* Janes Information Group, 1999, pp. 29–37, 41–64, 377–391, 435–530.
12. *Janes Weapon Systems 1984–85,* Janes Information Group, 1984, pp. 84–125, 157–197, 229–285, 424–499, 543–580, 581–678.
13. Brinkman, D., Ed., *Janes Avionics 1990–91,* Janes Information Group, 1990, pp. 20–89, 134–199.
14. Downs, E., Ed., *Janes Avionics 2001–2002,* Janes Information Group, 2001, pp. 347–524.
15. Cullen, T. and Foss, C., Eds., *Janes Land-Based Air Defence 2001–2002,* Janes Information Group, 2001, pp. 109–196.
16. Williamson, J., Ed., *Janes Military Communications 2001–2002,* Janes Information Group, 2001, pp. 163–377, 459–480, 715–770.
17. Horizon House, *International Electronic Countermeasures Handbook,* 2001 ed., pp. 307–337.
18. Peot, M. A., "Electronic Warfare Signal Processing in the Year 2000," *Microwave Journal,* February 1987, pp. 169–176.
19. Schleher, D. C., *Introduction to Electronic Warfare,* Artech House, Inc., 1986, pp. 33–35.
20. Selby, S. M., Ed., *Standard Mathematical Tables,* CRC Press, 1968, p. 394, Eq. 70.
21. Gradshteyn, I. S. and Ryzhik, I. M., *Table of Integrals, Series, and Products,* Academic Press, 1965, p. 205, Eq. 2.733-1.
22. ITT staff, *Reference Data for Radio Engineers,* 6th ed., pp. 28-12 to 28-15.
23. A few photos, tables, and figures in this intellectual property were made by Hughes Aircraft Company and first appeared in public documents that were not copyrighted. These photos, tables, and figures were acquired by Raytheon Company in the merger of Hughes and Raytheon in December 1997 and are identified as Raytheon photos, tables, or figures. All are published with permission.

5

Stealth Waveforms

5.1 WAVEFORM CRITERIA

The primary unique criterion for stealth waveform design is reasonably flat, total operating frequency band coverage. This objective isn't always compatible with best data link or radar mode performance. Some obvious criteria are stated in this section. These criteria are then applied to various spread spectrum strategies such as frequency diversity, discrete phase codes, linear FM, and hybrid waveforms. LPI requirements, in addition to *SNR* considerations, dictate the use of high duty cycle waveforms. This result has two implications: (1) That the transmitted pulse period must be incrementally variable and (2) that large expansion/compression ratios usually are involved.

Another result based on stealth requirements is that the instantaneous (not just the average) bandwidth of the transmitted signal be as large and as uniform as possible. For each geometry, the power must be managed to the lowest level consistent with acceptable performance or *bit error rate (BER)* as mentioned in Section 2.1.4. It obviously is desirable to keep the preprocessing and the bandwidth to a minimum; therefore, the waveforms chosen should result in the lowest possible data rate prior to compression or decompression. Lastly, LPI time and frequency constraints described in Sections 2.2.4 and 2.2.5 require noncontinuous or burst transmission for both data links and radars. Some systems naturally operate in a burst mode such as the JTIDS, which uses a TDMA format. As mentioned in Section 2.2.4, the stealth platform, whether aircraft, ship, or vehicle, must move between transmissions to create an uncertainty volume.

These requirements outlined above are summarized below:

1. Incrementally variable transmit period

2. Power management

3. Large compression ratios

4. Wide instantaneous bandwidth

5. Uniform instantaneous bandwidth

6. Minimum preprocessing

7. Minimum required data rate

5.2   FREQUENCY DIVERSITY

There are a number of methods that have been used to achieve a wide bandwidth while minimizing hardware complexity. The most common stealth bandwidth expansion schemes are summarized in Fig. 5.1. They are listed in no particular order in the figure. The simplest and least effective is frequency hopping, because it requires the highest ERPP (notwithstanding some authors' arguments to the contrary). In this scheme, a single modulated center frequency is transmitted for each coherent array, and a single receiver channel is tuned to that frequency. The frequency is chosen pseudorandomly, it should be known only to the transmitter and the receiver, and the next frequency must not be easily derivable from the current frequency. The transmitted power should be programmed to be just enough for the chosen path length (one-way or two-way). The interceptor is required to cover the entire hop band and thus is limited by total noise bandwidth.

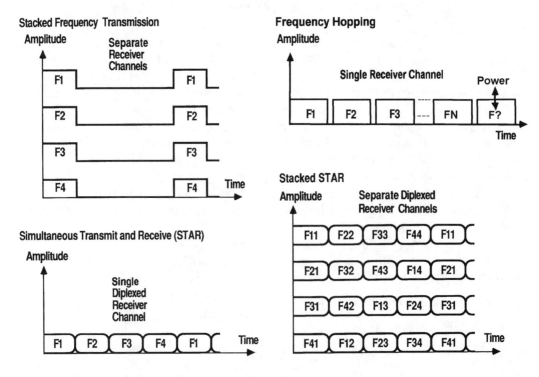

Figure 5.1   Alternative frequency diversity strategies.

The next scheme in complexity is stacked frequency transmission. Multiple modulated center frequencies are simultaneously transmitted and separately received in multiple independent channels. The channel spacing and modulation should be chosen to cover the entire band of operation in a reasonably uniform manner. The frequency set is chosen pseudorandomly, it should be known only to the transmitter and the receiver, and the next frequency set must not be easily derivable from the current frequency set. As in frequency hopping, the transmitted power should be programmed to be just enough for the chosen path length (one-way or two-way). The interceptor is required to cover the entire operating band and thus is limited by total noise bandwidth. The total ERPP is the same as frequency hopping but, for a channelized interceptor, the per channel power is less. This reduces the effectiveness of channelization. Because all channels transmit at once, channel-to-channel crosstalk is a minor problem, and the individual channels can be "shoulder to shoulder," almost completely filling the band. The natural question to ask is: "Why not create this same spectral spread with direct chip coding of a single channel?" The answer is that, often, multipath interference and antenna bandwidth make multiple channels easier to separate and process than direct encoding. The main weakness is that the duty cycle is limited typically to 33 percent or less.

The next most complex scheme is *simultaneous transmit and receive (STAR)*. It is like frequency hopping in which reception occurs on every frequency not being transmitted. Transmit frequencies are visited in a pseudorandom manner. All other frequencies are received in diplexed separate channels. This allows the transmitter to obtain almost 100 percent duty cycle while allowing all other frequencies to be received with maximum processing gain. The ERPP is lowest; however, the band of operation is incompletely covered at any one time, because channel to crosstalk is difficult to control. Pseudorandom frequency hopping must be used to fill the entire operating band.

The most complex scheme is a combination of stacked transmission and STAR. Almost 100 percent transmit duty cycle and wider frequency coverage is achieved. The price is very precise and difficult frequency diplexing so that the transmit frequency of one band does not leak into the receivers of the other bands, thus limiting the dynamic range. Again, the ERPP is lowest; however, the band of operation is incompletely covered at any one time. Pseudorandom frequency hopping must be used to fill the entire operating band. This scheme typically requires octave bandwidth, which is good from an intercept point of view but a challenge in terms of antenna and receiver complexity. Each of these schemes has been tested. Cost and complexity are the trade-offs.

5.2.1 Simultaneous Transmit and Receive Crosstalk

Frequency hopping has been described many times and needs minimal discussion.[1] The other three schemes are not as well known. Their essential or unique design elements are described in the next three sections. The first element to be described is crosstalk between elements. STAR is the stressing case for crosstalk. A simple STAR example is shown in Fig. 5.2. A low frequency, f_L, is transmitted with a 50 percent duty cycle, and a low-frequency receiver listens during the low-frequency transmitter off time. A high frequency, f_H, is transmitted with a 50 percent duty cycle, and a high-frequency receiver listens during the high-frequency transmitter off time. The high- and low-frequency channels alternate with a small switching guard time to allow diplexer settling. The transmit duty cycle approaches 100 percent. The high and low frequencies, including their modulation sidebands, must be sufficiently separated so that the leakage through the diplexer doesn't limit performance. This requires side-lobe weighting and a frequency guard band, so frequency band filling is not complete for any single transmission set, f_L and f_H. The band gaps can be filled on subsequent coherent arrays by selecting a different channel set.

A block diagram of a STAR transmitter and receiver for multiple frequency bands is shown in Fig. 5.3. To get sufficient isolation, there must be high-power filtering on both the transmitter output and the receiver inputs as well as high-power switching on both. Each receiver channel is subject to high-power transmitter leakage and must be protected from burnout as well as dynamic range limitations. Each receiver channel is processed independently until magnitude detection and then combined noncoherently. The exciter input which contains all the waveform modulation may be a

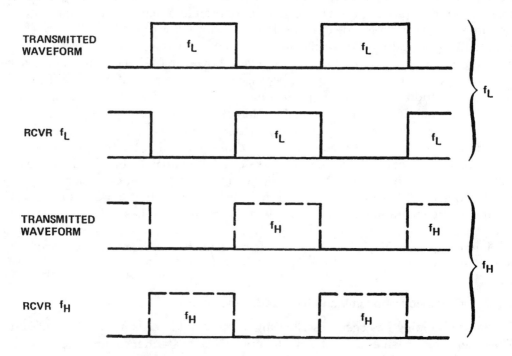

Figure 5.2 Simultaneous transmit and receive (STAR) concept.

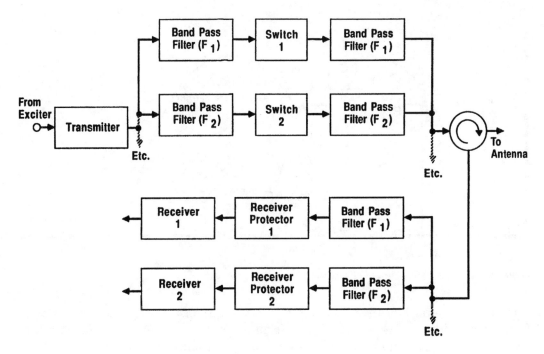

Figure 5.3   Multifrequency transmit-receive block diagram.

single frequency or multiple frequencies. The filters are four to six poles each and re-quire approximately linear phase for the pulse code.

Each transmitted frequency must not limit sensitivity in the other channels, and each received frequency must not suppress a signal in an adjacent channel. Consider the spectral situation in Fig. 5.4. The adjacent channel can interfere in two different ways. First, power from the adjacent channel can come through the front-end filter and influence the gain control settings limiting the channel dynamic range. Second, the smaller processed region filters and pulse compression can have their noise floors set by adjacent channel interference.

The total power in the receiver that may influence the gain control is sometimes called *near end crosstalk (NEXT)*. It is the integral of adjacent channel waveform power spectrum times the transmitter filter times the receiver filter.[7]

$$P_{NEXT} = \int S(f - f_n) \cdot H_T(f - f_n) \cdot H_R(f) \cdot df \tag{5.1}$$

where

$S(f)$ = the transmitted spectrum
$H_T(f)$ = the frequency response of the transmitter output filter
$H_R(f)$ = the frequency response of the front end receiver filter

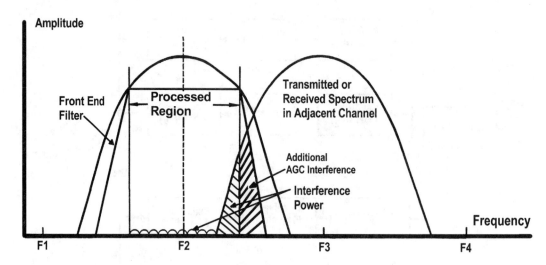

Figure 5.4 Adjacent-channel interference concept.

For example, consider the case of a power spectrum for a 13-chip Barker code with 50-ns chip width, t_c. This waveform is transmitted through an n-pole transmitter filter and received through an n-pole filter in an adjacent channel (more details in the appendices and Section 5.5). The normalized transmitted spectrum for a Barker code through a six-pole Butterworth filter is shown in Fig. 5.5. Figure 5.6 shows a multiline spectrum made up of adjacent channels, each with a binary phase code. Depending on the spacing between channels and filter skirt selectivity, a significant fraction of the spectrum will appear in the adjacent channel. The adjacent channel center frequency in Fig. 5.7 is $2/t_c$ or 40 MHz away from the transmit channel. The receiver crosstalk spectrum and the six-pole Butterworth receive filter frequency response are shown in Fig. 5.7. The value of the total P_{NEXT} calculated using Equation (5.1) normalized to 1 is –43 dB. If the transmitter power were 100 W peak, then the power in the receiver channel would be approximately –23 dBW, and it would very likely be setting the front-end gain and sensitivity. The noise floor in the receiver with the filter described above would probably be –121 dBW. A typical receiver of this type might have a spur free dynamic range of 75 dB and a 1-dB compression point of –20 dBW. Thus. the desensitization would be

$$NEXT \text{ Degradation } = -121 \text{ dBW} - (-75 - 23 \text{ dBW}) = -23 \text{ dB} \tag{5.2}$$

Such a reduction probably would be unacceptable, and thus more filter skirt selectivity must be used to prevent performance limitations. The maximum allowable leakage that will not limit performance is

$$\text{Maximum Allowable } NEXT = -121 \text{ dBW} - (-75) = -46 \text{ dBW} \tag{5.3}$$

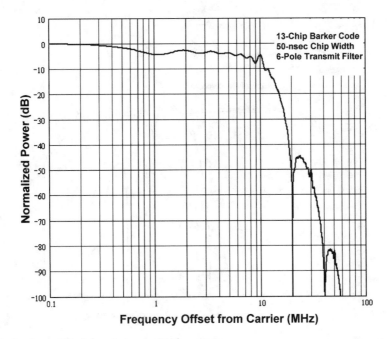

Figure 5.5   Single-channel Barker code transmitted spectrum.

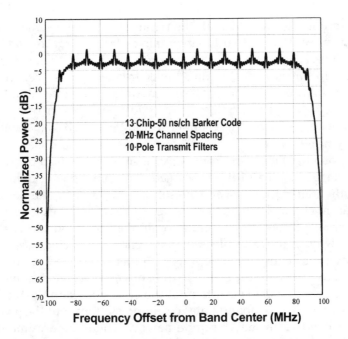

Figure 5.6   Barker coded stacked frequency spectrum.

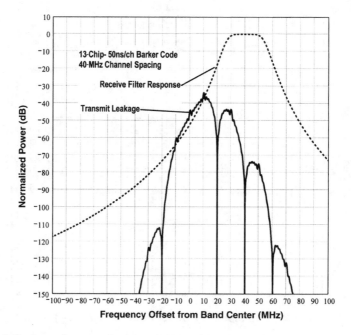

Figure 5.7 Example Barker code transmitted spectrum in adjacent receiver channel—six-pole filters.

Consider a second case in which all the parameters are the same, but the transmit and receive filters are 10 poles (hard to maintain over temperature and time but possible). Then, the NEXT spectrum would be as shown in Fig. 5.8. The integrated sidelobe ratio is –66 dB. In both cases, the filter break frequency has been chosen to be $1/(1.35\ t_c)$ to allow bit transitions to be 40 percent of the chip time (about the maximum allowable). In the case of 10-pole filters and 100 W peak power from the transmitter, the power in the adjacent receiver channel would be –47 dBW, just below that maximum allowable of –46 dBW.

The second potential limitation from NEXT is in the processed band noise floor. As can be seen in Fig. 5.8, parts of the processed band have power densities as high as –65 dB below the transmitter power. One solution when using binary phase codes is to choose adjacent channel center frequencies to be offset by exactly multiples of $1/t_c$. This offset results in perfect decorrelation in the pulse compressor and thus zero output. That is why, in the example above, the channel spacing was chosen to be $2/t_c$. Usually, however, if the dynamic range problem is solved, then the noise floor limit is also solved. In the example given, transmit and receive channels are the same bandwidth. For data links, this is often not the case. For example, SAR mapping radars may transmit very broadband data to ground stations, but the ground station to mapping radar bandwidth may be quite narrow. There are arguments for making uplinks more broadband to improve signal-to-interference ratios, because the anten-

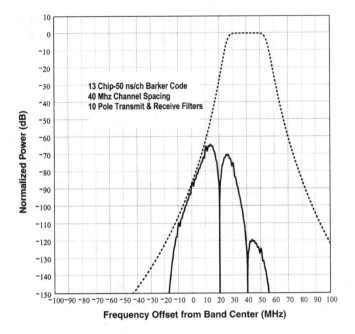

Figure 5.8 Barker code spectrum in adjacent receiver channel—10-pole filters.

nas may be smaller (as they often are on missiles or aircraft). The same general analysis applies.

In multifrequency STAR, stacked frequencies, and frequency hopping, there can also be crosstalk between receive channels in which a strong received signal in one channel interferes with a weak signal in another channel. This is sometimes called *far end crosstalk (FEXT)*. Dynamic range differences of 60 dB frequently occur. Note that the skirt selectivity in Fig. 5.8 at the upper 3-dB point of the next lower band is –80 dB, and so the interferer could be roughly even in power to the weak signal before compression. Again, by selecting channels in multiples of $1/t_c$, perfect or good decorrelation can be achieved in pulse compression and bit detection. The result is that, once the transmit problem is solved, the other crosstalk problems usually are solved. The spectrum filling shown in Fig. 5.6 for stacked frequencies will not work with any version of STAR, because the crosstalk and pulse compression requirements prevent a realizable transmit and receive filter set.

5.2.2 Low-Noise Adaptive Multifrequency Generation[10]

All of the frequency diversity approaches described in Section 5.1 still require very low-noise, stable frequency references. The brute force approach of adding together multiple stable oscillator outputs results in the noise sidebands growing proportionally to the number of simultaneous channels. This would be an unacceptable degradation in performance. Fortunately, there is a way to generate multiple simultaneous reference signals that will go through saturated power amplifiers and result in a noise degradation of at most 3 dB. The scheme was invented by Gregory and Katz more than 25 years ago.[10,22]

It is possible to generate an adaptive pure FM multitone waveform using the method diagrammed in Fig. 5.9. A low-noise signal generator (reference oscillator) output is split into two paths. One path is used to generate the third harmonic of the signal, which is then adjusted in amplitude and phase. It is summed with the reference signal in the other path, which is then used to frequency modulate a voltage controlled oscillator (VCO). Long-term stability for the VCO output is obtained by phase locking it to the low-noise signal generator (not shown in Fig. 5.9).

The simplest basic equation governing approximately equal amplitude multifrequency FM waveforms is given in Equation (5.4). The object is to generate a number of frequencies with very low FM noise of substantially equal amplitude with a pure FM waveform. The pure FM is necessary so that the waveform can pass through saturated amplifiers with low distortion to preserve pulse compression sidelobes. The modulation strategy used in Equation (5.4) generates 3^{N+1} central spectral lines.

$$v(t) = \sin\left(2 \cdot \pi \cdot f_0 + \beta \cdot \sum_{n=0}^{N} \sin(3^n \cdot 2 \cdot \pi \cdot f_m \cdot t + \alpha_n)\right) \qquad (5.4)$$

where f_0 is the carrier or center frequency of the multiline spectrum, f_m is the modulation frequency, β is the modulation index, which is chosen so that $J_0(\beta) = J_1(\beta)$. For example, if $N = 1$, $\alpha_0 = 0$, and $\alpha_1 = \pi/2$, then

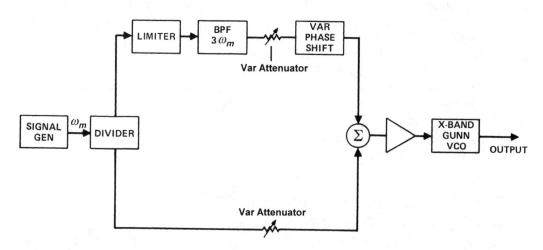

Figure 5.9 Example pure FM multifrequency generation.

$$v(t) = \sin(2 \cdot \pi \cdot f_0 + \beta \cdot [\cos(6 \cdot \pi \cdot f_m \cdot t) + \sin(2 \cdot \pi \cdot f_m \cdot t)]) \tag{5.5}$$

A typical hardware implementation of a multiline spectrum generator is shown in Fig. 5.10. This particular example produces an output at X-band and is based on a low-noise fifth-overtone crystal oscillator at roughly 106 MHz with modulation at 11.8/12 MHz and its harmonics. Those signals are generated in the dual divide-by-three circuit followed by bandpass filtering and variable attenuation. The modulation is then amplified and applied to the VCO after the LO phase comparator. The VCO is then multiplied up to the operating frequency (×12). Not shown are all the other outputs to the receiver channels to provide local oscillator signals, pulse compression, and A/D clocking. There is a sweep circuit inside the LO phase comparator for initial lockup of the phase locked loop, because such loops are narrowband. The diagram shows an FM modulation on-off switch for noise measurements with and without modulation.

A typical spectrum analyzer output from the circuit of Fig. 5.10 is shown for a nine-spectral-line FM modulation of 11.8 and 35.4 MHz and center frequency of 9.7 GHz in

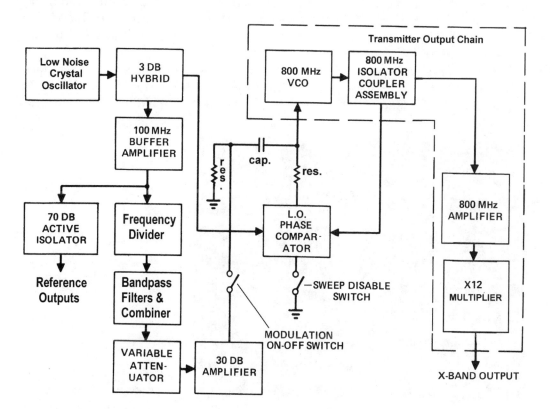

Figure 5.10 Example multifrequency hardware block diagram.

Fig. 5.11. The vertical scale was 2 dB/cm, and the horizontal scale was 20 MHz/cm. Spectral line spacing is, of course, 11.8 MHz. Final output frequency spacing is determined by the multiplier which follows the modulation scheme. The FM noise also is increased by the square of the multiplication ratio, and so starting FM noise must be very low.

The modulation Equation (5.4) generates the set of frequencies shown in Table 5.1. A more general theory indicates that any odd number of roughly equal tones can be generated, but it requires the selection of two modulation indices. An example spectrum for the two modulation index scheme is given in Fig. 5.14. Explanation of the mathematical criteria for two or higher modulations is beyond the scope of this text.

An important question is whether the multiline spectrum preserves the noise performance of the low-noise reference oscillator. Figs. 5.12 and 5.13 show the noise spectrum with no modulation and with multiline modulation. Figure 5.12 is a plot of the single-sideband FM noise with a calibration scale (because most noise-measuring test sets are nonlinear) for signal-to-noise ratio in a 1-kHz band relative to the center carrier. Note that the noise hovers around 92 dB below the carrier with no modulation. This noise level will allow detection of weak targets after filtering, compression, and *space-time-adaptive-processing* (*STAP*, see Section 7.11) down to about –135 dBW. This 3-dB degradation is the lowest multiline noise spectrum achievable for a given master oscillator noise performance.

A similar noise plot with modulation is shown in Fig. 5.13. At larger separations from the carriers, the signal-to-noise ratio hovers around 89 to 90 dB. This is no greater than a 3-dB degradation as expected. The explanation of the better *SNR* near

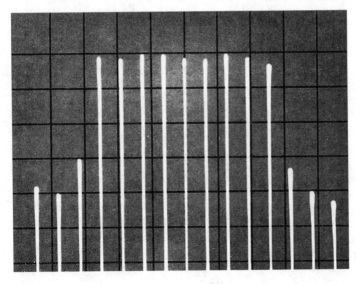

Figure 5.11   Nine-line spectrum (Raytheon[12]).

Table 5.1   Example Multifrequency Modulation

Frequency	Power relative to f_0	dB
f_0	1	0
$f_0 \pm f_m$	1.17	0.7
$f_0 \pm 2f_m$	1.14	0.6
$f_0 \pm 3f_m$	1.02	0.1
$f_0 \pm 4f_m$	1.02	0.1
$f_0 \pm 5f_m$	0.312	−5.1
$f_0 \pm 6f_m$	0.167	−7.8
$f_0 \pm 7f_m$	0.156	−8.1
$f_0 \pm 8f_m$	0.034	−14.7
$f_0 \pm 9f_m$	0.011	−19.6

the carriers is that the modulation noise is correlated to the master oscillator, and the correlated noise cancels.

As previously mentioned, an example of the more general modulation strategy is shown in Fig. 5.14. The spectrum analyzer display on the left in the figure has a horizontal scale of 50 MHz/cm and a vertical scale of 10 dB/cm. The display on the right in the figure has a horizontal scale of 20 MHz/cm and a vertical scale of 10 dB/cm. Three modulation frequencies are used: 11.8, 23.6, and 35.4 MHz. The noise performance is similar, because all the modulation is derived from the same master oscillator. The beauty of this scheme is that the spacing can always be matched to the chip width of a pulse code selected for operating mode reasons, not hardware limitations.

5.2.3   Detection of Multifrequency Waveforms

Multifrequency waveforms provide improved detection performance in the presence of fading and jamming, providing the desired detection performance probability is high or the required bit error rate, BER, is low. As more frequencies are used, the power per frequency goes down. At low signal-to-noise ratios, this results in worse detection or BER performance, as some frequencies are faded or jammed, and the

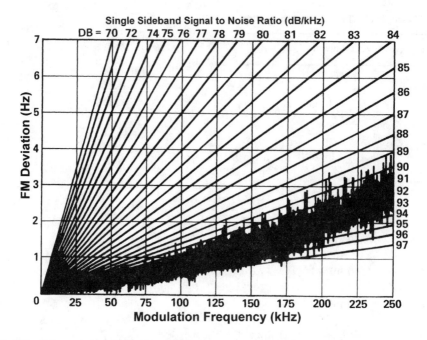

Figure 5.12   FM noise measurement—no modulation.

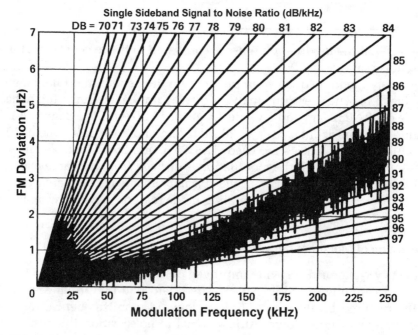

Figure 5.13   FM noise measurement—with modulation.

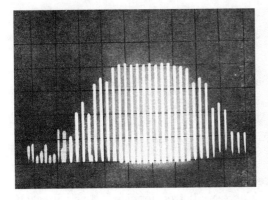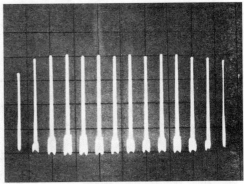

Figure 5.14   Thirteen-line spectrum (Raytheon[12]).

power is wasted. At high required SNRs, there is already enough power for detection, and the issue is the probability of a fade or jamming at any one frequency. In this case, more frequencies improve detection performance or reduce BER. Equation (5.6) gives one possible approximation for Rayleigh fading frequency channels and fixed total energy transmission across all frequencies with a square law detector in inphase and quadrature channels and unknown arrival phase followed by noncoherent integration of the individual frequency channels.[1,21,23,24]

$$\mathcal{P}_D(N_{Freq}) = \frac{\Gamma_{Eu}\left(N_{Freq}, \dfrac{P_{TH}(N_{Freq})}{1 + SNR/N_{Freq}}\right)}{(N_{Freq} - 1)\,!} \tag{5.6}$$

where $\Gamma_{Eu}(x, y)$ = the incomplete Euler gamma function, and y = the lower limit of integration with an upper limit of ∞. All the other variables are as defined before. N_{Freq} is the number of frequency channels, SNR is the rms signal to noise power ratio, P_{TH} is the power threshold. One example of Equation (5.6) is given in Fig. 5.15. For simplicity, an approximation to the optimum threshold is used as given in Equation (5.7).

$$P_{TH} = N_{Freq} - N_{Freq}^{0.11} \cdot \ln\left(\frac{\mathcal{P}_{FA}}{N_{Freq}^2}\right) - 1 \tag{5.7}$$

where $\mathcal{P}_{FA}$ is the desired probability of false alarm as previously defined.

Similarly, the multifrequency bit error rate, BER, for binary signalling and a Rayleigh fading channel, fixed total energy across all frequencies, matched filter bit waveform correlation, square law detection, unknown phase (i.e., center frequency phase is not recovered from signal), and noncoherent integration of each frequency channel is given in Equation (5.8).

$$BER(N_{Freq}) = F(N_{Freq})^{N_{Freq}} \cdot \sum_{i=0}^{N_{Freq}-1} \left(\frac{(N_{Freq}-1+i)!}{i! \cdot (N_{Freq}-1)!} \right) \cdot (1 - F(N_{Freq}))^{i} \quad (5.8)$$

where $F(N_{Freq})$ is a convenient intermediate variable as given in Equation (5.9).

$$F(N_{Freq}) = \frac{N_{Freq}^2}{2 \cdot N_{Freq}^2 + SNR} \quad (5.9)$$

For example, consider the radar case of a required $\mathcal{P}_{FA}$ of 10^{-6} and a Rayleigh fading target where the same total detection energy across all frequencies is used in each case. The probabilities of detection as a function of SNR for 1, 3, 5, 9, and 13 frequencies are given in Fig. 5.15. The frequency sets chosen are easily generated with multi-line FM as described in Section 5.2.2. Up to the point where $\mathcal{P}_D$ is 0.6, a single frequency has the best performance, even in the presence of fading.

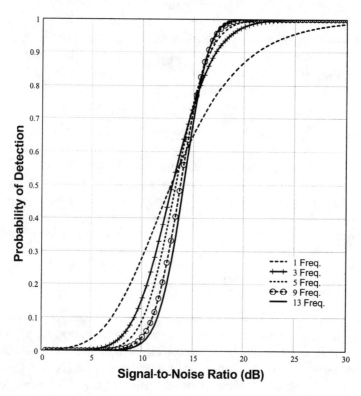

Figure 5.15 Multifrequency detection performance.

Similarly, consider the same situation and parameters for a data link as shown in Fig. 5.16. These curves may look different from many texts, because typically *SNR* is plotted on a per-frequency basis rather than on a total *SNR* basis across all frequencies. Again, the single-frequency *BER* is lower until the *SNR* is 17 to 25 dB. Of course, all these performances are significantly better with coherent integration and fine Doppler tracking. Because of the nature of the waveforms described, they are all mutually coherent and could allow coherent integration if Doppler can be sorted out. Resolving Doppler on a short-term basis as required for most stealth transmissions is problematical but possible.

These somewhat counterintuitive results emphasize the problem with all diversity schemes, because they generally don't have perfect integration. What they do have is more robust performance in the presence of jamming and better stealth performance. As in almost everything, there is a price to pay, which is significant unless *SNR* is quite high.

5.3 POWER MANAGEMENT

As mentioned in Section 5.1, power management (or transmitting exactly the power necessary to achieve the required *BER* or detection performance) is an essential ele-

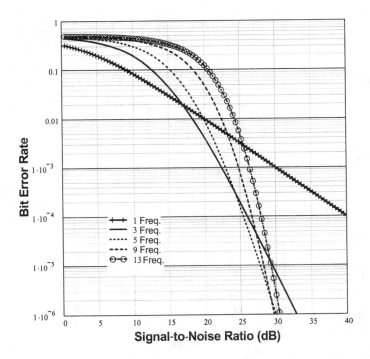

Figure 5.16 Bit error rate for fading multifrequency channels.

ment of stealthy waveforms. Hardware implementation of power management is really quite simple. The power management design follows ideas first used in the AN/TPS-32 and described in 1976. A typical power managed transmitter block diagram is shown in Fig. 5.17. It consists of both input and output attenuators that are controlled by a microprocessor in a closed-loop fashion. In the subsystem diagram shown, the TWT amplifier (TWTA) is operated primarily in its linear or unsaturated mode. Maximum power still is delivered from the TWTA saturated state. The input attenuation to the TWT is steadily applied as the required output power drops. This improves power efficiency, but at the expense of noise figure. Usually, the noise figure of a TWT is about 40 dB, but that value doesn't limit radar or data link performance if it's not allowed to deteriorate too much further. Once the input attenuator is decreased as far as is reasonable based on allowable noise performance, the power out of the TWT is low enough that power dividers can then be used at the TWT output to further reduce output power. Finally, when very low power levels are required, and the attenuation exceeds the TWTA gain, there is usually an RF switch at the TWT input and output to allow the amplifier to be bypassed altogether; thus, the 1 to 1000 mW that comes from the exciter can be attenuated by the cascaded power dividers and attenuators to provide the output. The directional coupler in the duplexer is used to verify that the com-

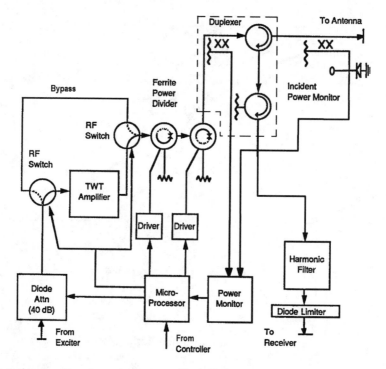

Figure 5.17   Typical transmitter power management block diagram.

manded power output is within the required tolerance. Attenuation is not set on a closed-loop basis, because experiment has shown such loops susceptible to counter-measures and loop instability.[20,25]

A TWT is reasonably nonlinear; therefore, an output power monitor must measure power from the TWT at all times. This measurement is utilized in a microprocessor that contains a premeasured and stored profile of attenuation versus output power as a function of frequency using the incident power monitor. The strategy is to apply the attenuation in such a way that the SNR at the output of the TWTA is maintained as close to optimum as possible and, at the same time, power efficiency degradation is minimized. This typically means that the first 10 dB of power management is applied from the diode attenuator at the input to the TWT. The power out of the TWT drops to 0.1 of its peak power, at which point ferrite power dividers can be utilized at the TWT output, and the next 30 dB is applied with the output power dividers. Even at 10 percent of maximum output, the power that must be dissipated in the power dividers is significant, and they are usually liquid cooled. Typical TWTAs have somewhere between 40 and 50 dB of gain, and, once the attenuation equals the gain of the TWTA, the amplifier is switched out, and the attenuation is applied to the local oscillator signal. In this way, very large power management ranges can be achieved (typically 70 dB or more) while maintaining the overall transmitter noise figure somewhere between 40 and 50 dB. The receiver protection and diplexing must be adequate to handle the leakage and broadband noise from the TWT, even when the transmit power is low.

A partially assembled power management unit from the Pave Mover program is shown in Fig. 5.18. (It is missing a running time meter and test connector. The loose fittings are for coolant from the system heat exchanger.) It is made up entirely of very conventional assemblies, attenuators, microprocessors, directional couplers, and so forth. The only important challenge is the calibration, which is done periodically during radar or data link operation.

5.4 PULSE COMPRESSION

5.4.1 Linear FM/Chirp

Linear FM or "chirp" is one of the oldest pulse compression methods. The basic process is shown in Fig. 5.19. The waveform output from each functional (lettered) block is shown in the bottom of the figure. A rectangular pulse of length, T, is generated by the pulse generator timing circuit. The pulse enables or selects a section of a steadily increasing or decreasing FM waveform. The linear FM or chirp bandwidth, B, from the mixer output is amplified by the transmitter and launched by the antenna, and it propagates to the target. The wavefront scattered from the target is received and amplified. This signal is then mixed in the demodulator with the opposite slope FM. This approximate matched filter results in an output pulse whose amplitude is propor-

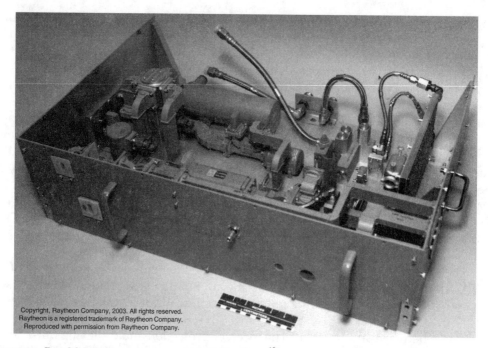

Figure 5.18   Pave Mover power management unit (Raytheon[12]).

tional to the square root of the time bandwidth product BT. For large bandwidths and unweighted matched filtering, the null-to-null width of the output pulse is $2/B$. For high-resolution systems, this leads to very high bandwidth front-end signal processing. Because the processed range swath in such systems is small relative to the PRF, this high bandwidth was very inefficiently used, and systems were invented to stretch the processing over the entire interpulse interval. Those approaches are called "stretch" processing.

To simultaneously achieve long-range detection and high range resolution requires a very long pulse and large frequency excursion. The long, linear FM waveform in which the pulse length exceeds the swath width has some very nice processing characteristics, which can be exploited by stretch processing.

The chirp transmitted signal is of the form shown in Equation (5.10).

$$v(t) = A \cdot \text{PUL}(t,T) \cdot \cos\left(2 \cdot \pi \cdot \left(f_0 \cdot t + \frac{\Delta f}{2 \cdot T} \cdot t^2\right) + \varphi_0\right) \tag{5.10}$$

where

$\qquad A$ = the transmitted amplitude
$\text{PUL}(t,T)$ = a rectangular pulse of length T
$\qquad f_0$ = the center of the operating band

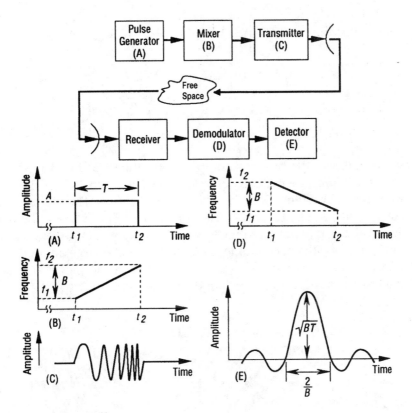

Figure 5.19 Basic chirp process.[14]

Δf = the total frequency excursion
φ_0 = the starting phase

The output power from a filter matched to this waveform is given in Equation (5.11).

$$P_{match}(t) = c_{match} \cdot SNR \cdot \exp(j \cdot 2 \cdot \pi(f_0 - f_D) \cdot (t - t_0)) \cdot X((t - t_0), (f_0 - f_D)) \quad (5.11)$$

where
c_{match} = a constant representing the transmitted amplitude, range equation and the receiver performance including the matched filter
f_D = the received signal Doppler offset from the carrier f_0
t_0 = the time of the matched filter sampling instant
X = a chirp waveform output from its matched filter as shown in Equation (5.12)

Define $\Delta f_D = f_0 - f_D$ and $\Delta T = t - t_0$, then

$$X(\Delta T, \Delta f_D) = T\left(1 - \left|\frac{\Delta T}{T}\right|\right) \cdot \exp(-j \cdot \pi \cdot \Delta f_D \cdot \Delta T) \cdot$$

$$\frac{\sin\left(\pi \cdot (\Delta f \cdot \Delta T + \Delta f_D \cdot T) \cdot \left(1 - \left|\frac{\Delta T}{T}\right|\right)\right)}{\pi \cdot (\Delta f \cdot \Delta T + \Delta f_D \cdot T) \cdot \left(1 - \left|\frac{\Delta T}{T}\right|\right)}$$

$$\text{(5.12)}$$

For example, consider the case of a compression ratio of 20:1 and 100-ft resolution with Δf of 5 MHz. Then, the matched filter response as a function of time delay and Doppler offset using Equation (5.12) is as shown in Fig. 5.20. The figure is a top view contour plot (i.e., like a topographic map) with 10-dB contour lines. The dark bands are very low sidelobes that are rapidly varying. The diagonal large central lobe shows one of the principal weaknesses of linear FM, which is the large ambiguity between range and Doppler. Another stealth weakness shown in the plot is that any cut in Doppler has significant power for much of the range/time axis.[15]

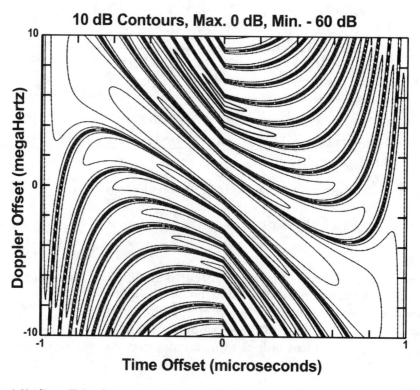

Figure 5.20   A 20:1 linear FM ambiguity contour plot.

5.4.2 LPI Performance Loss Incurred by Use of Chirp

Unfortunately, the instantaneous bandwidth is narrow for a long, linear FM waveform, concentrating all the output power in one interceptor detection filter for a significant time. For example, for a pulse compression ratio of 2000 and 10-ft resolution, the transmitted pulse width is 40 μsec, and the FM slope is 1.25 MHz/μsec, for a total excursion of 50 MHz. The transmit signal dwells in a single, 5-MHz wide detection filter for 4 μsec, or about 20 filter time constants. Not only is the detection filter fully rung up, but also this action takes place sequentially along the filter bank. Clearly, conventional linear FM is not a good LPI waveform. Figure 5.20 shows the LPI performance loss for a chirp waveform. The peak power associated with a pulse of fixed average power, without regard to the type of pulse compression used, is shown in the upper straight line in Fig. 5.21.

If a single pulse, or binary code, is used, the peak power will be instantaneously spread over 50 MHz, and a 5-MHz wide intercept receiver would pass only one-tenth of the peak power as shown by the lower line. However, if a chirp waveform is used, the filter in the receiver passes the full peak power unless the chirp rate is very fast (curved line). Very fast chirp rates occur only with low pulse compression ratios. Be-

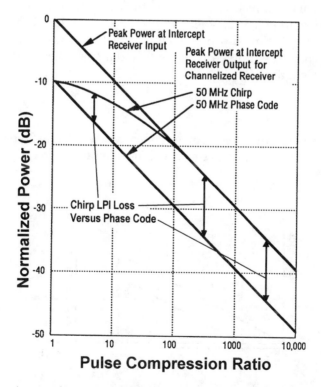

Figure 5.21 Peak power input and output in 5.0 MHz intercept receiver for 50-MHz chirp and phase code.

cause LPI requires high pulse compression ratios, chirp is not usually an acceptable waveform. The LPI loss inherent in using chirp instead of a spread-spectrum waveform is the difference between the two curves in Fig. 5.21.[17]

A typical hybrid pulse compression code required for good LPI might consist of a 50-MHz bit-rate cyclic phase code with a slow superimposed chirp to resolve the ambiguities of the cyclic code. Hybrids will be discussed in later sections. However, sophisticated receiver techniques other than the ones currently in use can counter binary phase codes.[1] Therefore, if more sophisticated threats are to be countered, a polyphase code should be implemented for best LPI.

5.4.3 Stretch Processing

If linear FM is such a poor LPI waveform, why ever use it? The answer is that there are processing means available for linear FM that are so simple and advantageous that the LPI performance loss against channelized intercept receivers is often balanced by this simplicity. Furthermore, chirp waveforms are effective against many wide-open receiver types. This simple processing method is known as stretch.[13]

Some discussion of the fundamentals of stretch processing is appropriate to lay the groundwork for what follows. The basis for the derivation of the stretch processing parameters is shown in Fig. 5.22, where T_{SW} is the swath width time, and T_E is the effective pulse length (the time during which returns from the entire swath come in simultaneously; hence, the time during which data is taken).[2,9,13,28,29] The actual transmitted pulse length is

$$T = T_{SW} + T_E$$

Similarly, the actual frequency excursion is Δf_T, while the effective frequency excursion over T_E is Δf. From simple geometrical arguments, the required IF bandwidth,

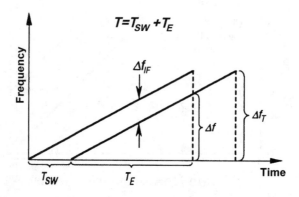

Figure 5.22 Stretch processing basic parameter derivation.

Δf_{IF} could be as small as shown in Equation (5.13). Because the total frequency excursion can be resolved into range bins equal to the reciprocal of Δf, then $\Delta f = 1/T_{rb}$. Therefore,

$$\Delta f_{IF} = \frac{T_{SW}}{T_E} \cdot \Delta f \text{ and } \Delta f_{IF} = \frac{T_{SW}}{T_E \cdot T_{rb}} = \frac{N_{rb}}{T_E} \qquad (5.13)$$

where

N_{rb} = number of range cells across the swath
T_{rb} = range resolution or cell time

A typical stretch mechanization is shown in Fig. 5.23(a). The received signal is deramped with a center frequency matched to the center of the swath minus the IF frequency. The output of a narrowband IF matched to the frequency excursion over the swath width is heterodyned to baseband. The baseband is I/Q sampled at a sample rate of $N_s = 1/(2\Delta f_{IF})$ to meet the Nyquist criterion. This is typically followed by an FFT in two dimensions (Doppler and range).

The total number of I *and* Q samples across T_E is then

$$N_s = T_E \cdot \Delta f_{IF} = N_{rb} \qquad (5.14)$$

This demonstrates one advantageous processing feature of stretch pulse compression processing. In particular, the number of samples required prior to range compression

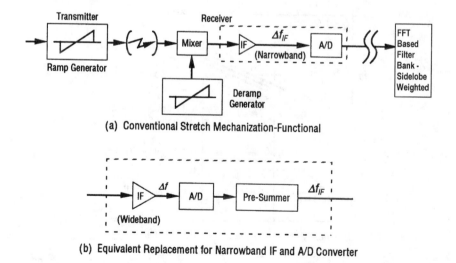

(a) Conventional Stretch Mechanization-Functional

(b) Equivalent Replacement for Narrowband IF and A/D Converter

Figure 5.23 Functional stretch mechanization.

(which can be accomplished efficiently with an FFT) equals the number of samples after range compression, thus minimizing the precompression data rate. In addition, the pulse length is easily varied, and preprocessing is minimal. The range compression can be accomplished by a simple FFT because, after deramping, targets at a given range within the swath are characterized by an almost constant frequency within the Δf_{IF} band.

Two functional approaches to mechanizing stretch are shown in Fig. 5.23. The first approach [5.23(a)] is the one conventionally described. The second approach [5.23(b)] shows that the narrowband IF and A/D converter, whose bandwidth and sample rate are Δf_{IF}, can be replaced by a wider band IF of bandwidth, Δf, followed by an A/D converter and a digital presummer. With the advent of higher-speed A/D converters, it often turns out that greater dynamic range can be achieved by digital filtering after conversion rather than analog filtering before conversion in the IF.

More details of the stretch technique are shown in the example of Fig. 5.24. In the example given, the reference ramp from the deramp generator is a frequency of f_0 + 1500 MHz and has a frequency excursion of 500 MHz. The received signal is at f_0 and three point targets (A, B, C) separated by the limiting resolution are represented by corresponding FM ramps. The time extent of the transmitted pulse in the example is 10 μsec. The limiting resolution is the reciprocal of the time extent or 100 kHz. Because the transmitted pulse covers 500 MHz in 10 μsec, the FM chirp slope is 50 kHz/

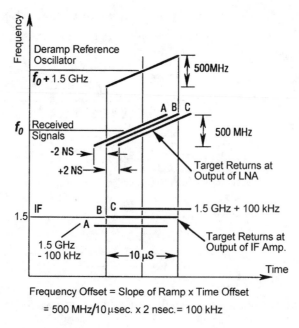

Figure 5.24   Stretch technique converts time delay into frequency offset.[13]

nsec. The dechirped signal at IF consists of continuous tones centered about the IF center frequency of 1.5 GHz. Note that the IF or presummer bandwidth determines the range swath length. The subsequent IF/presummer output is Fourier transformed into separate targets in range. This processing is very simple, because it allows a large transmitted bandwidth, a small receiver processing bandwidth, a small range window, and a wide range of usable PRFs (usually medium or low). How to get simple processing and good LPI? One method, complementary coding, is described in a later section.[13–15,28]

An important issue with any pulse compression waveform is the sidelobe performance, because it limits the dynamic range in adjacent bins just as shown in Section 5.2.1 for multifrequency operation. There are several important pulse compression figures of merit, including the ratio of the peak compressed signal to the peak sidelobe (PSLR), the ratio of the peak signal to the rms sidelobes (RMSLR), and the ratio of the peak signal to the integral of all the sidelobes (ISLR). If $X(t)$ is the output from the pulse compressor, then Equation (5.15) defines the ISLR. The RMSLR is the ISLR divided by the pulse compression ratio.

$$ISLR = \frac{\int_{-\frac{T}{2}}^{\frac{T}{2}} |X(t)|^2 dt - \int_{(-2.3 \cdot T_{rb})}^{2.3 \cdot T_{rb}} |X(t)|^2 dt}{\int_{-\frac{T_{rb}}{2}}^{\frac{T_{rb}}{2}} |X(t)|^2 dt} \tag{5.15}$$

Equation (5.15) can often be integrated from 0 to the upper limit only if the sidelobes are symmetrical or hermitian. With stretch processing, time is converted to frequency (if one ignores the Doppler effects), and the ISLR is often calculated in the frequency domain. As a practical matter, weighting must be applied to reduce sidelobes in the analysis filters, so individual bandwidths are usually significantly larger than $1/T_{rb}$.

For a chirp waveform, the swept frequency residue after deramp must lie in a single filter, and the spurious FM noise covering the entire analysis band must be low enough that its integrated power doesn't limit the dynamic range of adjacent channels, and vice versa. Therefore, linearity for the combination of the transmit waveform and its receiver demodulation must be on the order of T_{rb}/T usually split evenly between transmit and receive. For the case of Fig. 5.24, this requires 2 nsec/$(2 \cdot 10 \ \mu sec) = 10^{-4}$ linearity each for transmit and receive. Furthermore, the integral of the FM noise and clutter in the analysis filter mainlobe and sidelobes must be as given in Equation (5.16); otherwise, distributed clutter, broadband noise jamming or FM noise will limit the desired dynamic range, DR_d, in every other bin, as the FM spreads this signal over the entire band.

$$Dynamic\ Range \le \frac{P_{match}(t)|_{max}}{\left(\frac{\int_{-T_E}^{\frac{T_E}{2}} (s_{FM}^2 + s_{clutter}^2) \cdot |h(t)|^2 dt}{2}\right)} \tag{5.16}$$

where s_{FM}^2 is the FM noise power spectrum and $s_{clutter}^2$ is the clutter power spectrum. In addition, the pulse compression filter bank ratio of minimum signal to peak sidelobes for each filter must be greater than the desired dynamic range; otherwise, bright discretes (Fig. 4.5) or repeater jammers will limit dynamic range. Thus,

$$Dynamic\ Range \le \frac{P_{match}(t)|_{min}}{\left(\frac{\int_{-T_E}^{\frac{T_E}{2}} |BrtDis(\Delta T)*h(t)|^2 dt}{2}\right)} \tag{5.17}$$

where $BrtDis(\Delta T)$ is an impulse of strength proportional to ΔT and the cross range patch width. For example, if the desired dynamic range is 10^6, then the product of the filter *ISLR* and the noise variance must be 10^{-6}. The *ISLR* of an unweighted FFT using the parameters of Fig. 5.24 and 3-dB mainlobe width is –4.1 dB. So the FM noise must be approximately $2.5 \cdot 10^{-6}$ below the carrier if there is no interfering clutter. Assuming a weighted filter bank using a fourth-power Parzen window and using the 10-dB mainlobe width yields an *ISLR* of –15.3 dB, which would provide some margin for interfering clutter as well as allowing some relaxation of the FM noise requirement. Parzen window far sidelobes are low enough that a bright discrete 20 bins away will be attenuated by more than 60 dB. Over the bandwidth of the chirp in Fig. 5.24, these requirements are very difficult to achieve.[15]

5.5 DISCRETE PHASE CODES

Discrete phase codes include binary phase codes such as Barker and compound Barker, random, pseudo-random, and complementary codes; and polyphase codes such as Frank codes and discrete chirp. Typically, the pulse compression ratio (*PCR*) of a discrete phase code is the ratio of the chip time, t_c, to the pulse time, T, or, alternatively, the number of chips in the code.[4–8,11,14,15]

The major problem with long, discrete phase codes, such as those required by typical LPI criteria, is that the number of range samples required before range compression is large; in fact, it is equal to the number of range cells plus the pulse compression ratio (*PCR*). For example, for a swath width of 512 range cells and a *PCR* of 2000:1, 2512 range samples are needed, which is five times the number required with stretch. This problem can be resolved for some codes, such as the complementary codes by using the Hudson-Larson decoder.

If range pulse compression can be accomplished in a single ASIC, then this large number of samples is not a problem. Simple processing is possible for the more structured codes such as the Barker, compound Barker, and complementary codes, but they do not lend themselves to incremental variation in pulse length. On the other hand, the random-like codes, which can be chopped off or extended to any desired pulse length, require complicated, inverse filter processing to improve their relatively poor integrated sidelobe ratios (*ISLRs*) resulting from matched filtering. (It is easy to show that the *ISLR* of a random discrete phase code approaches 0 dB as the code length increases.) Such processing could be done in a surface acoustic wave device for a few code lengths but, for large numbers of code lengths, this approach is inappropriate. For a 10-to-1 pulse length variation, for example, 25 different code lengths would be required to cover this range in approximately 10 percent increments.

5.5.1 Barker Codes

The basic notions of discrete phase code can be illustrated with a 13-chip Barker code as shown in Figs. 5.25 and 5.26. Figure 5.26 shows the waveform generation block diagram of a phase code. Figure 5.26 shows the waveforms at three points, A, B, and C, in Fig. 5.25. The phase code waveform shown at A in Fig. 5.26 is applied to an RF phase shifter with a net phase difference between the two states of 180° at the source frequency.

The phase reversal CW signal at point C in Figs. 5.25 and 5.26 is shifted upward in frequency in an RF amplifier and transmitted to the target. Usually, the overall transmitted pulse envelope is determined separately, and the modulation at B in Figs. 5.25 and 5.26 is used to turn the RF amplifier on and off.

On reflection from the target and receipt by the radar, the signal is amplified, demodulated, and presented to a pulse compressor, as shown in Fig. 5.27. The pulse compressor is often a "matched filter" and has the property that the output pulsewidth, T_{rb}, is the chip width, t_c, and the amplitude is equal to the number of chips (or bits) in the phase code. For many reasons, such as sidelobe suppression and straddling loss, the compressed output only approaches the matched filter.

One example of a binary phase code pulse compressor is shown in more detail in Fig. 5.28. The example shows a five-chip Barker code decoder in which a shift register slides the received detected signal past a correlator, which multiplies the detected signal by phase weights of 1 or −1 corresponding to the transmitted phase waveform and

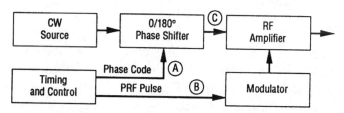

Figure 5.25 Phase-coded transmitter.

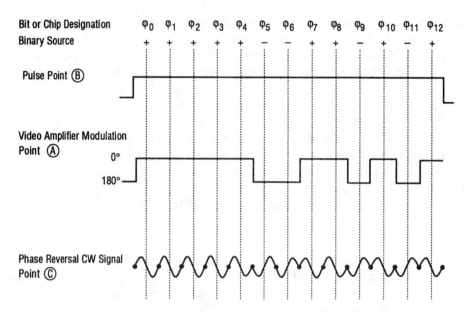

Figure 5.26 Waveforms for a Barker code of 13 chips.

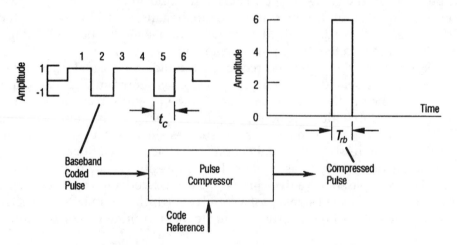

Figure 5.27 Phase code pulse compressor.

sums the weighted results to produce a discrete output as shown by the dots (connected for visualization). This process is the discrete autocorrelation function, of course, whose general form is shown in Fig. 5.29. Because the returned waveform is not the transmitted waveform, the pulse compressor is performing a cross correlation, which digs out scattering from point targets in the range bin.

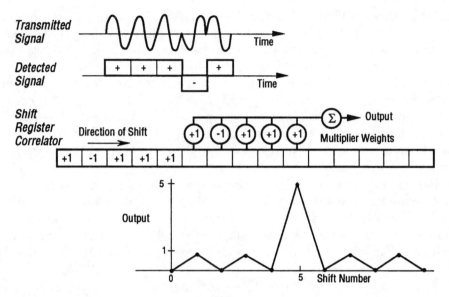

Figure 5.28 Example binary phase code pulse compressor.

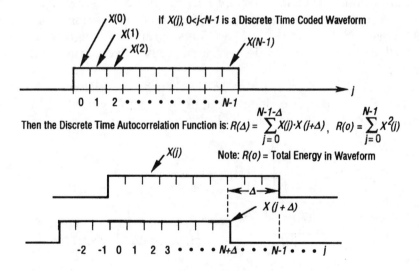

Figure 5.29 Autocorrelation function for discrete time-coded waveform.

The correlation is just the chip-by-chip product of the prototype waveform and the received waveform summed over all chips with successive offsets of an integral number of chips. The zero-offset case is just the sum of the square law detection of all the chips.

Returning, then, to the extended example of Figs. 5.25 and 5.26, the output of a pulse compressor for a 13:1 Barker code is shown in Fig. 5.30. Originally, Barker code compressors were analog devices made up of delay lines and summing networks and, as such, were completely inflexible. With the advent of high-speed A/D converters in the 1960s, Barker codes were some of the earliest examples of digital compression. The incoming waveform is shown idealized at point 1 in Fig. 5.30. Each phase weight φ_0 through φ_{12} is at a delay line tap one chip time apart. The outputs are summed and appear as shown at point 2; this output was quite noisy, and so it was lowpass filtered with a bandwidth matched to the chip. The resulting output was as shown at point 3 in Fig. 5.30.

The Barker code cross-correlation calculation follows the correlation model of Fig. 5.29. Note that the sidelobes for an ideal received waveform are identically 1 or 0 and extend ±12 elements from the central response. Each successive time delay T advances the waveform one chip, and the product of the tap weights times the ideal signal when summed yields the successive summation outputs to the right in Fig. 5.31.

A more representative example of the sort of waveform at the receiver input to a 13-chip Barker code compressor is shown in Fig. 5.32. Three curves are shown: the in-phase waveform, the quadrature waveform, and the corresponding magnitude multiplied by 2 to separate it from its two components. The only waveform that is remotely recognizable is the magnitude wave. The chip time is 50 nsec, and the overall pulse length is 650 nsec. Note that the front-end filters significantly degrade the rise and fall times as well as produce amplitude overshoot. These effects degrade total detection performance.

A summary of all the Barker codes and their peak to sidelobe ratios is given in Fig. 5.33. The *ISLR* for Barker codes are roughly $-10 \log(N)$, with the 13:1 code obtaining a best value of -11.5 dB. No other codes do as well at any length. Barker codes are

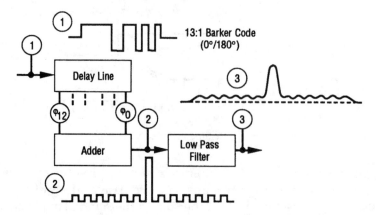

Figure 5.30   Discrete phase Barker code pulse compression.

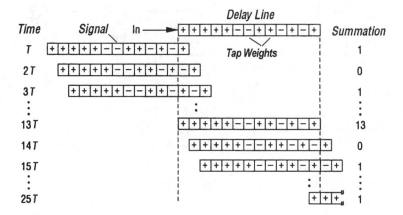

Figure 5.31   13:1 Barker compression calculation.

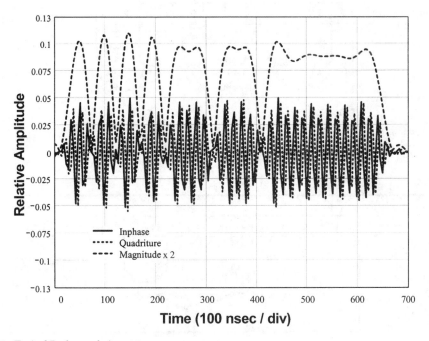

Figure 5.32   Typical Barker code input to compressor.

commonly used in many magnetic media synchronization, radar, and communication applications.

Numerous simplifications have been devised for Barker pulse compression, such as the one shown in Fig. 5.34. This particular design was used in a single integrated circuit pulse compressor. The result is a very inexpensive and very fast pulse compres-

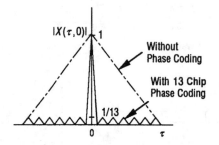

Length of Code N	Code Elements	Peak Sidelobe Ratio, dB (-20 Log N)
2	+ −, + +	-6.0
3	+ + −	-9.5
4	+ + − +, + + + −	-12.0
5	+ + + − +	-14.0
7	+ + + − − + −	-16.9
11	+ + + − − − + − − + −	-20.8
13	+ + + + + − − + + − + − +	-22.3

Figure 5.33 Barker code summary.

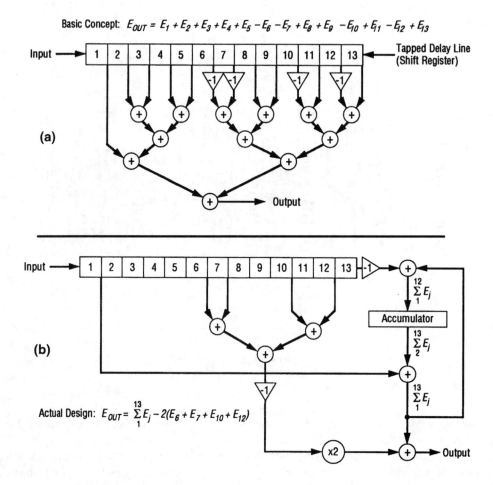

Figure 5.34 Typical pulse compression design: (a) canonical and (b) actual.

sor that can be used alone or in cascade to create longer code compressors with chip times to the order of 1 nsec.

There are no known Barker codes beyond 13, but some of the advantages of Barker codes can be obtained by creating longer codes that are products of Barker codes. The inner Barker code is phase modulated by an outer Barker code (Fig. 5.35) such as that shown for 169:1 in Fig. 5.36. Code compounding can produce quite an array of performance parameters as shown later in this chapter.

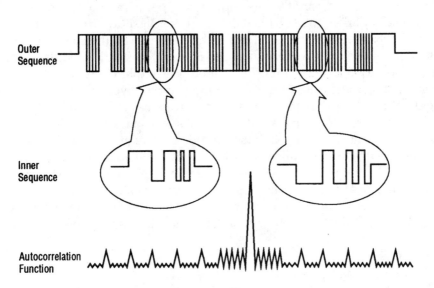

Figure 5.35   Schematic of compound or concatenated Barker sequences.

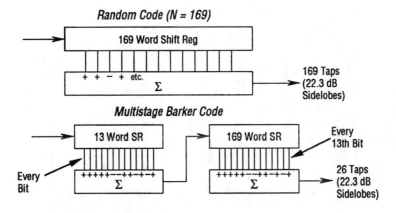

Figure 5.36   A 169:1 multistage Barker code comparison.

In addition to reasonably good sidelobes, a compound Barker code is often much simpler to implement than the same length random code. Sidelobe suppression filters for compound Barker codes are also very simple.

The central portion of a compound or multistage Barker code is shown in Fig. 5.37. Most of the sidelobes are of height 0 or 1 with 24 lobes of height 13. The overall code sequence and autocorrelation function are shown schematically in Fig. 5.35. The very regular sidelobe pattern of compound Barker codes lends itself to simple sidelobe suppression filters, which is explained later in this section.

It is interesting to compare the close-in power spectrum of a 13-chip Barker code and a compound Barker code of 169 chips. Figure 5.38 shows the normalized power spectrum of both codes. Obviously, for a given peak power, the compound code will have 11 dB more energy. The normalization is useful to see the relative similarity of the two in the frequency domain. Their differences are obvious in the time domain but, depending on the mechanization of an intercept receiver and its binning process, one might easily misidentify the waveform and response strategy. Clearly, the Barker codes discussed here and polyphase Barker codes to be presented in later sections can be concatenated to create almost any length code and frequency occupancy.

Often, the sidelobe performance of Barker codes is not adequate, and sidelobe suppression is used to improve the peak and integrated sidelobe performance at the expense of some loss in SNR. The basic idea can be most easily understood by describing the pulse compressor output in the Z transform domain and then observing that there is an ideal desired part and an undesired part (the sidelobes), which

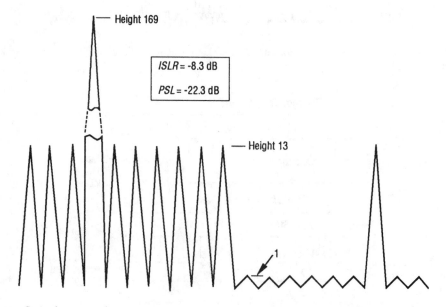

Figure 5.37   Central portion of compound Barker autocorrelation function.

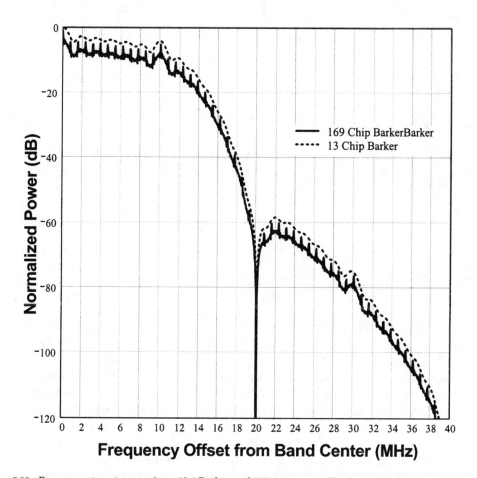

Figure 5.38 Power spectrum comparison, 13:1 Barker and 169:1 compound Barker with chip time, t_c, 50 nsec.

should be completely cancelled. This approach can be used with almost any binary code. For example, the autocorrelation function of a 13-bit Barker code in the Z domain is as given in Equation (5.18) below.

$$R(Z) = 12 \cdot \left(\underbrace{1}_{\text{Desired}} \right) + \left(\underbrace{\frac{C_B \cdot X(Z)}{}}_{\text{Undesired}} \right)$$

$$\text{where } X(Z) = \left(\sum_{n=-6}^{6} Z^{2 \cdot n} \right) \text{ and } C_B = 1/12$$

(5.18)

The ideal sidelobe suppression filter transformation is then $H(Z)$, which can be expressed as a power series in $X(Z)$ as shown in Equation (5.19).

$$H(Z) = \frac{1}{(1 + C_B \cdot X(Z))} = 1 - C_B \cdot X(Z) + C_B^2 \cdot X^2(Z) - \ldots \qquad (5.19)$$

An approximate sidelobe suppression filter transformation contains the first two terms of the series.

$$H_A(Z) = 1 - C_{app} \cdot X(Z) \qquad (5.20)$$

where a near optimum $C_{app} = 0.047 = 1/32 + 1/64$ was determined by computer simulation and a simple binary weight approximation. Then, for convenience, H_A can be written as shown in Equation (5.21).

$$H_A(Z) = Z^{-13} - C_{app} \cdot Z^{-1} \cdot \sum_{n=0}^{n=12} Z^{2 \cdot n}$$

$$H_A(Z) = Z^{-13} - C_{app} \cdot Z^{-1} \cdot (1 - Z^{-26})/(1 - Z^{-2}) \qquad (5.21)$$

The approximate sidelobe suppression filter, H_A, is used in cascade with the matched decoder pulse compressor as shown in Fig. 5.39.

The filter, H_A, can be implemented either as a canonical finite impulse response (FIR) filter or as a recursive filter as given in Equations (5.20). Such sidelobe suppression filters can obviously be extended to compound Barker codes by applying the same type of filter to the outer code as well as the inner code, as suggested in Fig. 5.40. Of course, the weighting terms are offset by 13 chips rather than 1 chip in the outer code.

The advantage of the FIR form of this sidelobe suppression filter is that it can be pipelined and thus handle smaller chip times. The IIR version uses fewer adders and

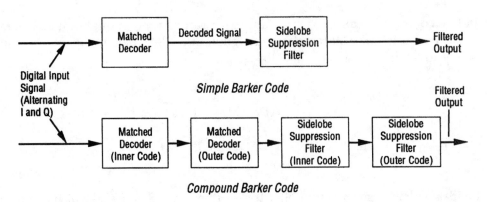

Figure 5.39 Sidelobe suppression filters.

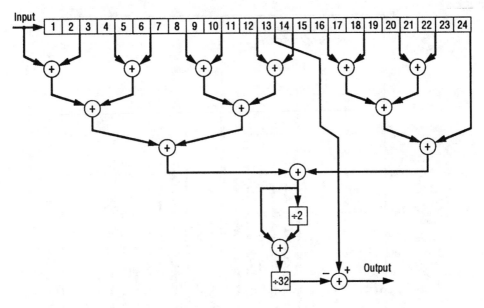

Figure 5.40 Sidelobe suppression filter FIR form.

may save cells in an ASIC or standard cell array. The 0 Doppler sidelobes for a 13-bit Barker code with and without sidelobe suppression are shown in Fig. 5.41. Note that the normal Barker sidelobes extend 12 bins on either side of the mainlobe whereas, with suppression, the sidelobes extend twice as far but are down more than −10 dB. Although not shown, suppression performance is also improved in the presence of uncompensated Doppler.

In summary, Table 5.2 shows the improvements to be gained in *ISLR* and peak-to-sidelobe ratio (*PSLR*) while sacrificing a small *SNR* reduction for both 13:1 and 169:1 codes. Obviously, similar improvements are obtainable with other Barker code lengths and code compounds. Appendix A.5 on the CDROM contains program files with more details on Barker code pulse compression sidelobe performance.

Table 5.2   Barker Code Sidelobe Suppression Summary

Code length	Sidelobe suppression	ISLR (dB)	PSLR (dB)	SNR loss (dB)
13	Without	−11.5	−22.3	0
	With	−24.9	−32.9	−0.55
169	Without	−8.3	−22.3	0
	With	−21.9	−32.9	−1.1

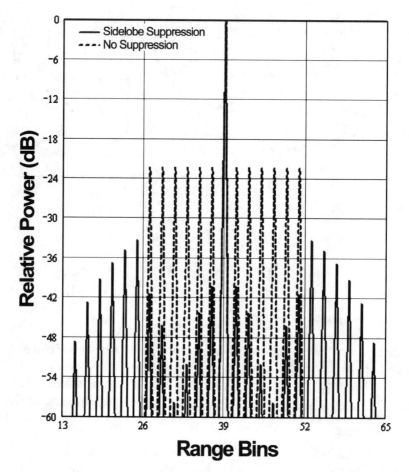

Figure 5.41 Barker code sidelobes with and without suppression.

Barker Code compression and sidelobe suppression can be performed in the chip shown in Fig. 5.42. The digital pulse compressor or correlator chip shown in Fig. 5.42 is a one-off design, not an interconnected general purpose array. This chip is partially analog and differential to improve *SNR*. The chip was designed to be cascaded to provide any size code length, but device technology has passed it by, and chips an order of magnitude faster and denser are now available.

Another important consideration in any binary phase code is its sensitivity to the Doppler shift. This can either be a result of relative motion between radar and target or between data link transmitter and data link receiver. Doppler sensitivity depends on the form of the code employed. Barker codes are often used as the synchronization code for simple data links. There are some Doppler invariant codes, but they have not been widely used in radar or military data links. Chirp waveforms are sensitive to

Figure 5.42 Digital pulse compressor (Raytheon[12]).

Doppler shift frequencies on the order of the bandwidth of the transmitted pulse. Binary phase codes are sensitive to Doppler shifts of the order of the reciprocal of the uncompressed pulse length or the transmitted bandwidth divided by the *PCR*. Frank codes, discussed in the following section, are less sensitive to Doppler shifts and are of the order of the reciprocal of the linear segment length or the transmitted bandwidth divided by the square root of the *PCR*. This is because Frank codes are approximations of chirp waveforms.

A way to visualize the result of Doppler shifting on a pulse compression waveform is shown in Fig. 5.43. The figure shows a plot of the crosscorrelation function between the transmitted waveform and its Doppler shifted return from a target at Doppler shift frequencies, $f_d \cdot T$, relative to the transmitted frequency of 1.2, 0, and 0.6. (These are three cuts through the ambiguity diagram at these offset Dopplers.) As is well known, the zero Doppler response is the waveform autocorrelation function, which is

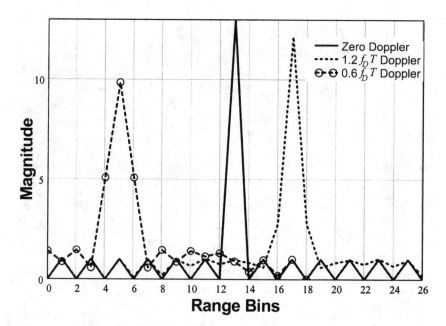

Figure 5.43 Binary code response with Doppler shift.

always greater than or equal to any other response. Doppler shifted responses typically have higher sidelobes and lower peak responses. If the objective is to reject Doppler shifted targets, this is ideal, of course; however, in most cases, the targets of interest are the Doppler shifted returns, and some compensation is required. These three cases represent *ISLRs* of –7, –11.5 (theoretical best), and –1.2 dB, respectively. For short phase codes like the Barker code shown here with 50 nsec chip widths, these Doppler offsets represent interplanetary velocities (20 kmph) and just don't arise in most cases. For very long uncompressed pulse widths and large compression ratio codes, however, more normal Dopplers will cause sidelobe degradation if there is no Doppler compensation.

A widely published comparison of ambiguity functions for a simple pulse, a chirped pulse, and a Barker code of equal time length and similar Doppler resolution is given in Fig. 5.44. Each has a 1-µsec pulse length in the example. One disadvantage of a chirp waveform is that Doppler offset can be misinterpreted as a different target range because of its large half-power ambiguity area.

The necessity for Doppler compensation is graphically demonstrated in Fig. 5.45. It shows the degradation in peak and integrated sidelobes for a 13:1 Barker code with sidelobe suppression as a function of uncompensated phase error across the transmitted pulse. This behavior of the sidelobes in the presence of uncompensated phase warp across the pulse is similar for all binary phase codes. Any successful binary code design must deal with Doppler compensation.

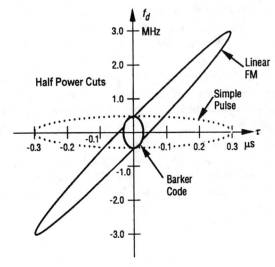

Half Power Cuts

- Doppler Axis ($\tau = 0$) Width Same Since Each Waveform has 1 μsec Time Envelope

- Barker Code has Smallest Width on Delay Axis ($f_d = 0$) Since it has Largest Bandwidth
  - Barker = 13 MHz
  - Linear FM = 10 MHz
  - Simple Pulse = 1 MHz

- Linear FM has Strong Doppler – Time Delay Coupling

  $$\text{Slope, } \left. df_d \middle/ d\tau \right. = \frac{B}{T} = \frac{10 \text{ MHz}}{1 \text{ μsec}}$$

Figure 5.44   Ambiguity function comparison.[6]

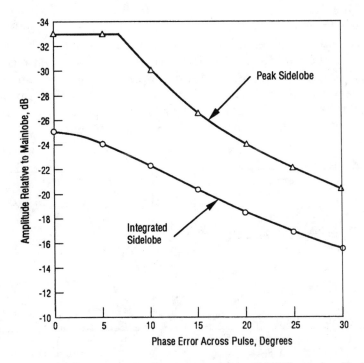

Figure 5.45   Sidelobe degradation versus Doppler phase error for a 13:1 Barker code with sidelobe suppression.

The simplest form of Doppler compensation is to apply binary phase pulse compression after Doppler filtering. Most modern radars apply Doppler filter bank processing to all target returns to improve SNR and false target rejection and so, by interchanging the order of pulse compression and Doppler filtering, Doppler compensation automatically occurs. If the code is long relative to the processed swath, compression is often done in two parts, before and after Doppler filtering. Significant sidelobe degradation occurs when uncompensated Doppler phase error is between 30° and 60° across the total pulse. Similarly, data links require Doppler compensation. The most stringent case occurs during acquisition when all possible Dopplers must be searched. If fast acquisition is required, then all Doppler and time offsets must be match filtered simultaneously. Fortunately, the number of bins to be processed is dramatically less than for a high-resolution radar. Usually, the required computing power is already available. (See Fig. 5.46.)

5.5.2 Frank and Digital Chirp Codes

Another useful class of codes is digital approximations of chirp waveforms. Such schemes have been implemented in hardware for at least 35 years with varying degrees of success. Their main attraction is the simplicity of processing coupled with the stability of digital techniques. When chirp waveforms are used for very high (greater than 10,000:1) PCRs, performance is usually limited by linearity, stability, and distortion in the transmitted and demodulating waveforms. This naturally leads to the use of polyphase digital techniques. Figure 5.47 shows a quadratic phase curve for chirp and compares this to a 13:1 Barker code and a 16:1 Frank code. Note that Barker code is not a bad approximation itself, but Frank codes are a better approximation for longer chirps. Frank codes of length N^2 have $2\pi/N$ phase states, which are visited first by 0 increments, then by 1, 2, 3, and so forth, as shown in Fig. 5.48. Obviously, conve-

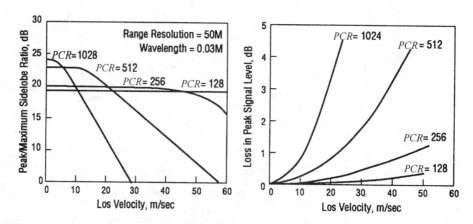

Figure 5.46 Most binary codes have some Doppler sensitivity.

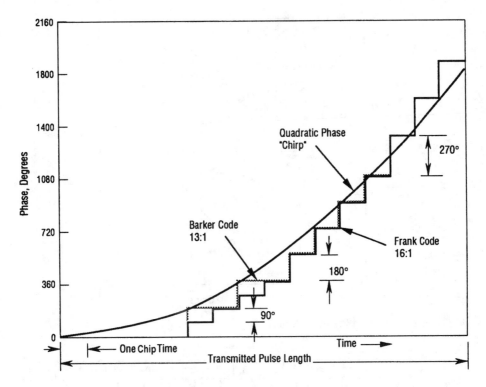

Figure 5.47 Digital code phase histories approximate a chirp.

nient length Frank codes have integer square roots. By phase modulating the transmitter carrier with digitally generated phase increments, large and very stable pulse compression waveforms can be generated and decoded.[5]

Digital chirp is similar to Frank codes except that, at each bit or chip interval, the closest approximation to the desired continuous chirp phase history is selected, limited only by the quantization accuracy of the digital phase modulator, the accuracy of the chip clock, and modulator phase accuracy (modulo 2π, of course). Digital chirp systems with compression ratios of 50,000:1 and greater have been implemented in modern radars with phase quantization of 10 to 12 bits. These large compression ratios are usually decoded by stretch techniques described earlier in this chapter. As was mentioned, chirp waveforms have poorer LPI performance for a given transmitted bandwidth, but processing simplicity often balances this performance loss.

Quite long Frank codes have been generated using a version of the algorithm in Fig. 5.48. Their sidelobes are somewhat worse than linear FM but, at large compression ratios, the differences are small.

The "brute force" decode of a length 16 Frank code is shown in Fig. 5.49. Figure 5.50 shows the tabular calculation of the correlation function of a Frank code of length 16.

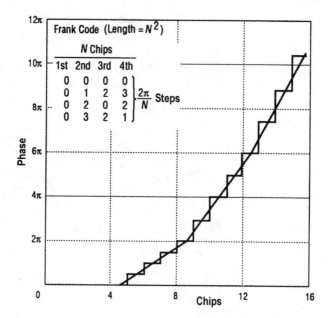

Figure 5.48 Frank code for length 16 phase history.

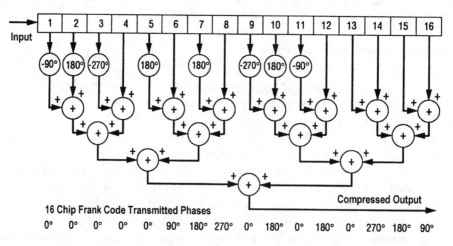

Figure 5.49 Frank code pulse compressor length 16.

Note that the compression coefficients are the complex conjugate of the transmitted code. One of the beauties of four-phase modulation and decode is that multiplication by increments of 90° consists of digital complements and swapping inphase for quadrature signal components and can be performed at very high speeds and with very little hardware.

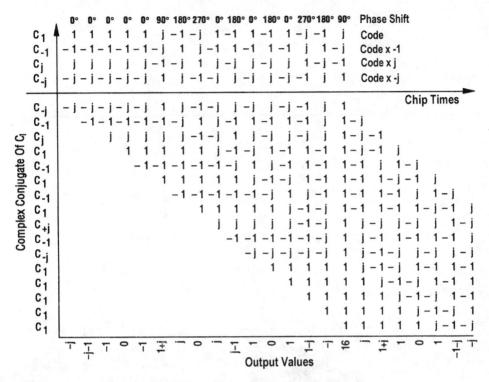

Figure 5.50   Tabular calculation of the correlation of a length 16 Frank code.

A plot of the correlation function of Fig. 5.50 with the points between the decoder outputs fitted with a curve for visualization is shown in Fig. 5.51. A photograph of such a Frank decoder that was used on the original USAF/DARPA LPI program in a stacked frequency waveform of the type described in Figs. 5.1 and 5.11 is shown in Fig. 5.52.

5.5.3 Complementary Codes

5.5.3.1 Complementary Code Overview.[11]   Complementary codes are a class of binary codes that have been widely used in LPI radars, because they are very easy to generate and decode. They have excellent properties, especially for air-to-ground waveforms. Complementary codes are defined as a pair of equal length codes having the property that the time sidelobes of the autocorrelation of one code are equal in amplitude and opposite in sign to that code's complement. Figure 5.53 shows an eight-chip code A, its complement $\overline{A}$, and their sum. Note that the sum has sidelobes that are identically zero. By alternately transmitting A and $\overline{A}$ and summing the range swaths

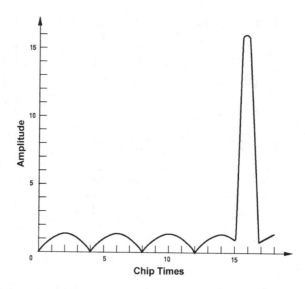

Figure 5.51 Frank decoder output.

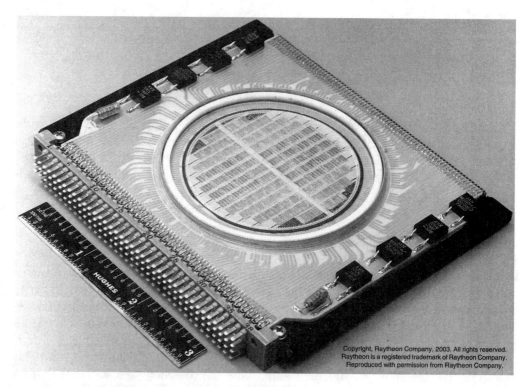

Figure 5.52 Frank code digital pulse compressor (Raytheon[12]).

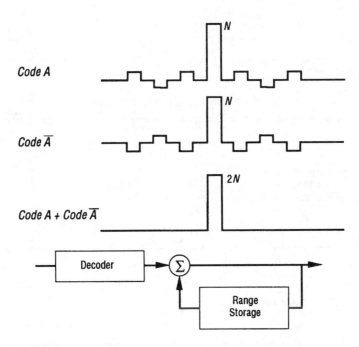

Figure 5.53   Complementary codes.

bin by bin, *PSLR*s that are infinite could theoretically be achieved. In actuality, implementation errors limit performance, but the sidelobes of error terms generally are also complementary and partially cancel. For example, a 0.2-radian cumulative phase error and a 1/8-chip total timing error over the total transmitted pulse envelope in the received waveform yields an uncancelled residue of approximately –45 dB on peak sidelobes for a 1024:1 complementary code pair. Actual field performance of LPI radars has demonstrated peak-to-far-sidelobe ratios between –50 and –60 dB and peak to near sidelobe ratios between –30 and –40 dB. This field performance is unequaled by any other pulse compression scheme. Furthermore, the performance holds up for very long codes. Achieving the same performance from chirp systems requires very expensive linearizing schemes or non-real-time "massaging" of the data. More details are in the software in the appendices.[11]

A number of complementary codes are known to exist. Golay[4] found codes of lengths $L = 2^n$ (e.g., 1, 2, 4, 16, 32...) as well as certain other lengths (10, 20, 26, 40...). Welti[3] found binary "D-codes" that are complementary pairs, have length 2^n, and are formed from shorter complementary sequences. The rule is that a complementary code can be formed from a complementary sequence by dividing the sequence in half, designating the first half A and the second half B, and then concatenating the sequence AB to the sequence $A\,\overline{B}$; i.e., $ABA\,\overline{B}$ and its complement is $AB\,\overline{A}B$. A D-code sequence is shown in Fig. 5.54.

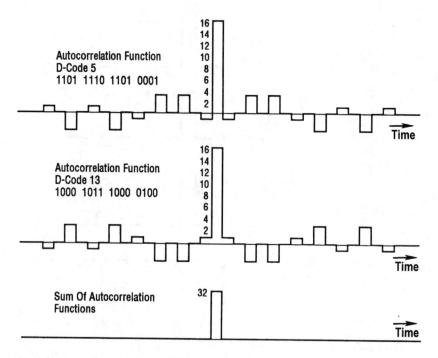

Figure 5.54 Length 16 D-code sequences and their corresponding idealized compressed outputs.

There are a large number of possible complementary codes. For example, there are 16 D-code sequences of length 8 chips. Equations (5.22) and (5.23) show the fifth and sixth D-codes of length 8 and their complements. These 8-chip codes, when concatenated, produce 16-chip codes. D-codes 5 and 6 are obvious complements. Less obvious, until one thinks carefully about it, are D-codes 13 and 14, which are complements of 5 and 6, respectively. D-code 13 is the mirror image complement of 5, because the autocorrelation function of a real finite waveform is real and even; therefore, $R(-t) = R(t)$. The best *PSLRs* are obtained from D-codes 5, 7, 14, and 16.

The fifth D-code sequence is

$$\underset{1101}{\underbrace{A}} \quad \underset{1110}{\underbrace{B}} \quad \Rightarrow \quad \underset{1101}{\underbrace{A}} \quad \underset{0001}{\underbrace{\bar{B}}}$$

(5.22)

The sixth D-code sequence is thus

$$\underset{1101}{\underbrace{A}} \quad \underset{1110}{\underbrace{B}} \quad \Rightarrow \quad \underset{0010}{\underbrace{\bar{A}}} \quad \underset{1110}{\underbrace{B}}$$

(5.23)

Another type of complementary code is called a C-code. For example, a length 16 C-code and its complement are shown in Equation (5.24).

$$
\begin{array}{cccc}
A & B & C & D \\
\underbrace{1011} & \underbrace{1110} & \underbrace{1000} & \underbrace{1101}
\end{array}
\Rightarrow
\begin{array}{cccc}
A & \bar{B} & C & \bar{D} \\
\underbrace{1011} & \underbrace{0001} & \underbrace{1000} & \underbrace{0010}
\end{array}
\qquad (5.24)
$$

Note that D is the mirror image of A, and C is the mirror image complement of B. C-codes have the property that every fourth bin after compression is 0, rather than every other bin, and they produce lower *PSLR*s. Again, there are a large number of possible codes. In addition, seen later in this section, there are a number of polyphase complementary codes.

Table 5.3 compares C and D codes of various lengths for each individual code; i.e., not cancelled. Note that C-codes have slightly better *PSLR*s. Without code cancellation, none of these codes come close to Barker codes in *ISLR*. Although *PSLR* is important, *ISLR* is the real limiter of performance in almost all pulse compression applications.

Table 5.3 Complementary Code Sidelobe Levels

Code type	Code length	Peak SLR (dB)[*]	ISLR (dB)[*]
D-codes	8	−8.5	−4.3
	16	−14.5	−5.1
	32	−11.0	−4.6
	64	−15.0	−4.8
	128	−15.7	−4.7
C-codes	8	−12.0	−4.3
	16	−14.5	−5.1
	32	−14.5	−4.6
	64	−16.1	−4.8
	128	−17.0	−4.7

[*]Individual codes, not pairs

What justifies the interest in complementary codes is the particularly simple decoding that can be used. Figure 5.55 shows one configuration of a pulse compressor for length 16.

Note that each unit delay can be replaced by a "seed" code, making the code an almost arbitrary length. By varying the seed code and the chip width, even though the basic decoding algorithm is simple, the number of parallel decoders required by the interceptor can be made arbitrarily large. The rule is extendable for any binary power or products of binary powers and smaller inner codes such as 10, 26, and so forth. Doubling the length of the code requires only the addition of one-half the code length

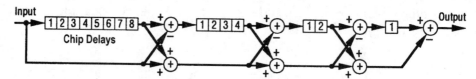

Figure 5.55   Length 16 Type II complementary code pulse compressor.

of storage or delay and two additional adders. It is possible to employ very long codes with minor amounts of hardware. This forces the interceptor to use square law detection across the entire operating band, which gives the emitter the $N^{0.5}$ advantage over the interceptor where N is the code length. To give a practical advantage, the codes must be very long but, because compression is simple, this is not difficult. Lengths over 10,000:1 are required.

In addition, for codes that are much longer than the processed range swath (a common occurrence in LPI radars), there is a further simplification, called the Hudson-Larson pulse compressor, that saves even more hardware. Even though these waveforms seems completely predictable, they can be made very difficult to intercept just by choosing long inner seed codes and selecting them at random at the time of the coherent array initiation. This forces the interceptor to have large numbers of decoders all going at once—a very uneconomical prospect. Further, by psuedorandomizing the code length and chip width, intercept receivers that using doubling and pulsewidth matching become so inefficient that it isn't worth the added complexity.

Figure 5.56 shows the actual return from a corner reflector with a length 64 C-code without cancellation. Even with the degradation associated with real hardware, *PSLR* and *ISLR* are close to theoretical.

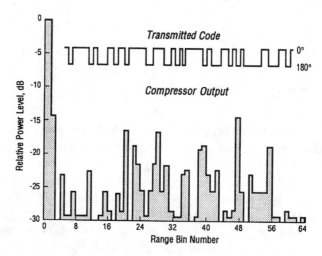

Figure 5.56   Length 64 C-code corner reflector measurement.

Another consideration for complementary codes is, "Do they cover the band reasonably uniformly, and does the spectrum somehow 'give away' the code?" The answer is, yes, they cover the band in spite of their obvious redundancies. And, no, the spectrum doesn't give the code away. The spectrum in Fig. 5.57 shows an alternating complementary code of length 64 chips other than the double nulls near zero frequency offset, even a short code looks like the transmitted chip width. As shown in subsequent figures, longer codes give little clue to the transmitted code.

Figures 5.58 and 5.59 show the central spectral lobe of 2048 chip length complementary codes. It can be proved mathematically, as shown in Fig. 5.59, that the local maxima and minima do not vary more than 3 dB from the local mean in a family of Type II codes. Thus, if the selection of complementary codes is agile, the spectrum gives away nothing.

Well, if complementary codes are so fabulous, why aren't they used more often? There are two main weaknesses. First, the uncancelled sidelobes are shifted to 1/2 the PRF. Second, like all binary and polyphase codes they do have Doppler sensitivity, which must be compensated in most applications. Figure 5.60 shows the sidelobe shift problem in the frequency domain.

Figures 5.61 and 5.62 show the time domain sidelobes of a 128 chip length complementary code and the resulting sidelobes of a complementary pair when summed. In essence, the sidelobes are shifted to the PRF/2. The simplest way around this weakness is to ensure that PRF/2 Doppler lies outside the mainlobe of the antenna as

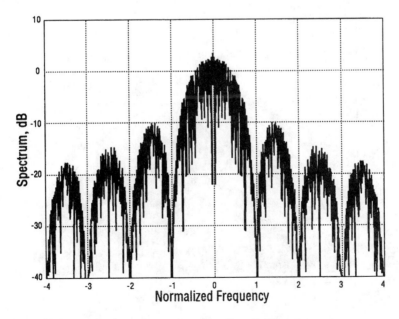

Figure 5.57 Spectrum of alternating complementary code of length 64.

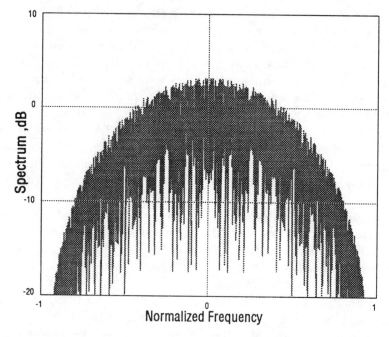

Figure 5.58   Spectrum of central lobe of complementary code of length 2048.

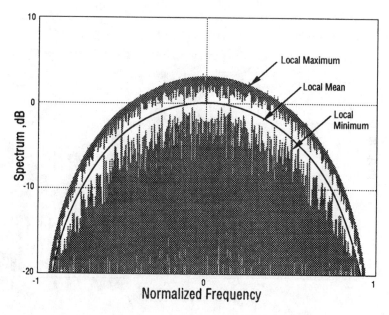

Figure 5.59   Central spectrum of alternating complementary codes of length 2048.

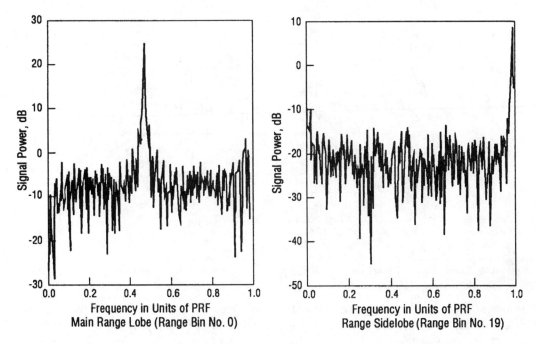

Figure 5.60 Frequency spectrum for alternating complementary codes.

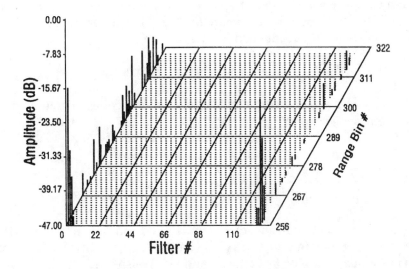

Figure 5.61 Either member of a complementary pair has poor range sidelobes.

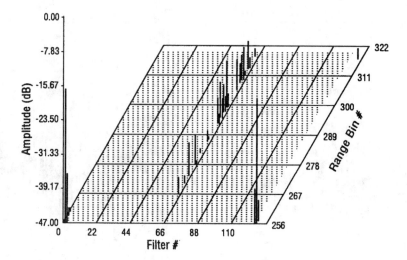

Figure 5.62 Complementary pulse compression shifts range sidelobes PRF/2.

shown in Fig. 5.69. The second way around this weakness is to transmit the code and its complement as simultaneous quadrature pairs. There is a small set of complementary codes in which the sidelobes of the A-code in the B-code compressor and vice-versa are also complementary. Thus, they can be transmitted simultaneously in quadrature and achieve the cancellation properties.

5.5.3.2 Type II Complementary Codes. There are many implementations of complementary codes, but the most popular today are the type II encoders and decoders. A type II code for a code of length ℓ is defined from a type I as

$$A = (a_1, a_2, \dots a_{\ell/2}) \text{ and } B = (b_1, b_2, \dots b_{\ell/2}) \tag{5.25}$$

i.e., if $A\,B$ is the code, and if A and B are complementary codes, then the bit-by-bit interleave are complements.

$$A_{II} = (a_1, b_1, a_2, b_2 \dots a_{\ell/2}, b_{\ell/2}) \text{ and}$$

$$B_{II} = (a_1, \bar{b}_1, a_2, \bar{b}_2 \dots a_{\ell/2}, \bar{b}_{\ell/2}) \tag{5.26}$$

A set of recursive equation forms for generation and compression of type II complementary codes are shown in Equations (5.27) and (5.28), respectively. One cannot help but notice the similarity to the fast Fourier and fast Walsh-Hadamard transforms and their recursive equations.

Type II complementary code generation

$$A_i(p + m) = B_{i-1}(p) \text{ and } A_i(p) = A_{i-1}(p)$$

$$B_i(p + m) = -B_{i-1}(p) \text{ and } B_i(p) = A_{i-1}(p)$$

where for code length $2N$ and recursion index i,

$$i = 0,1,2...\log_2(N); \ m = 2^i; \ p = 0,1,2...(m-1)$$

$$(5.27)$$

Type II complementary code compression

$$A_i(p) = A_{i-1}(p) + B_{i-1}(p)$$

$$B_i(p) = A_{i-1}(p + m) - B_{i-1}(p + m)$$

where for code length $2N$ and recursion index i,

$$i = \log_2(N), \log_2(N) - 1, \log_2(N) - 2,...1,0;$$

$$m = 2^i; \ p = 0,1,2,...2 \cdot N + m - 2$$

$$(5.28)$$

An example transmitted phase code and its counterpart compressed output for a 32-chip code is shown in Fig. 5.63. Codes as long as 2^{16} chips are easily generated and compressed with type II configurations. Usually, the innermost 16 chips are compressed brute force to allow maximum flexibility of the inner most segment of the code. For example, the inner code might be a Frank code, a quadrature binary complementary code, or a binary complementary code selected according to some rule.

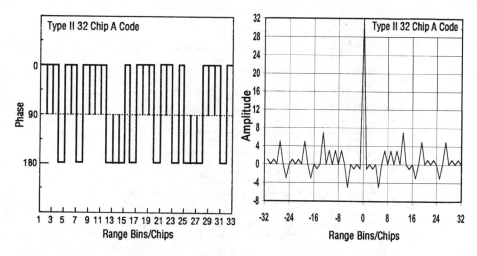

Figure 5.63 Type II complementary code example.

If one considers the simple 16-chip code flow diagram shown in Fig. 5.64, it is easy to see how a very large compression ratio and a small range swath can be accomplished with minimal processing using type II complementary codes.

The observation that many intermediate results in the pulse compression did not have to be calculated when the swath was a fraction of the transmitted pulse led to the invention of the Hudson-Larson pulse compressor shown below in Fig. 5.65. Most of the complexity in the block diagram of Fig. 5.65 is associated with the 4:1 multiplexing, which is only there to allow slower arithmetic to keep up with the small chip times. If the processing could keep up with the chips, only two of the eight 256 bin accumulators would be required. Similarly, the inner code accumulators would not have to be "ping-ponged" if the hardware was faster. Although the drawing may seem complex, this is a very simple compression scheme allowing extremely long codes.

5.5.3.3 Polyphase Complementary Codes. Polyphase complementary codes have several advantages. First, intercept exploitation is harder. Second, uncancelled sidelobes often are better; thus, cancelled sidelobe residue in the presence of hardware or Doppler degradation is better. Third, cancelled sidelobes in the presence of jamming and

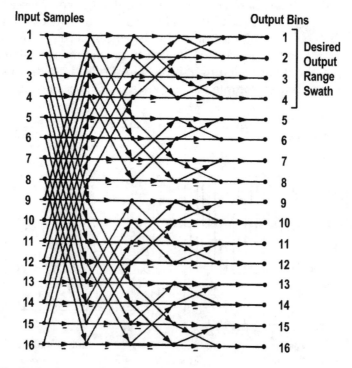

Figure 5.64 Type II complementary code compression flow diagram.

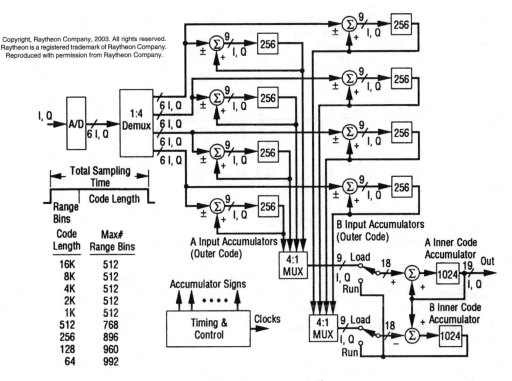

Figure 5.65 Hudson-Larson complementary code pulse compressor.[12]

front end nonlinearities are often better. The easiest polyphase codes to implement for both transmission and compression are four-phase (quadraphase) codes. Longer codes can be formed by concatenating shorter complementary pairs in which the second pair has a shift of 90° added to it to form a longer sequence pair as shown in Equations (5.29). The complement of the longer pair is formed by interchanging A for B, reversing the chip order of A and B, adding 180° to the new A complement and finding the complex conjugate of the new B complement as shown in Equations (5.29). Obviously, for even codes, this procedure can be continued indefinitely to create very long polyphase sequences.

$$\text{If} \qquad PC_2 = \begin{bmatrix} A_2 \\ B_2 \end{bmatrix} = \begin{bmatrix} 0 & 1 \\ 0 & 3 \end{bmatrix} \text{ and } \overline{PC}_2 = \begin{bmatrix} \overline{A}_2 \\ \overline{B}_2 \end{bmatrix} = \begin{bmatrix} 0 & 3 \\ 0 & 1 \end{bmatrix}$$

$$\text{then} \quad PC_4 = PC_2 \text{ Join } \overline{PC}_2 \cdot 90° = \begin{bmatrix} A_2 & \overline{A}_2 + 1 \\ B_2 & \overline{B}_2 + 1 \end{bmatrix} = \begin{bmatrix} A_4 \\ B_4 \end{bmatrix}$$

and $\quad \overline{PC}_4 = \begin{bmatrix} \overline{A}_4 \\ \overline{B}_4 \end{bmatrix} = \begin{bmatrix} B_4^* Reverse + 2 \\ A_4^* Reverse \end{bmatrix}$

then $\quad PC_4 = \begin{bmatrix} A_4 \\ B_4 \end{bmatrix} = \begin{bmatrix} 0 & 1 & 1 & 0 \\ 0 & 3 & 1 & 2 \end{bmatrix}$ and $\overline{PC}_4 = \begin{bmatrix} \overline{A}_4 \\ \overline{B}_4 \end{bmatrix} = \begin{bmatrix} 0 & 1 & 3 & 2 \\ 0 & 3 & 3 & 0 \end{bmatrix}$ (5.29)

where each matrix entry of $A_N = a_0\ a_1\ a_2\ \cdots\ a_{N-1}$, and the matrix value equals $\exp(j \cdot 2 \cdot \pi \cdot a_i)$ with $A_N(i) = 0°, 90°, 180°, 270°$, modulo 2π, or a_i is modulo 4. The rule above can be restated as

$$\overline{A}_N = -B_N^*(N - i); \quad \overline{B}_N(i) = A_N^*(N - i) \text{ for } i = 0, 1, \ldots N - 1 \qquad (5.30)$$

Equations (5.31) show a few short polyphase code kernals from which codes of almost arbitrary length can be formed. Using the rule suggested above, these codes can be expanded in integral multiples of their length. They can also be used as the inner code in which the outer code is a binary complementary code.

$$PC_3 = \begin{bmatrix} A_3 \\ B_3 \end{bmatrix} = \begin{bmatrix} 0 & 1 & 0 \\ 0 & 0 & 2 \end{bmatrix} \qquad \overline{PC}_3 = \begin{bmatrix} \overline{A}_3 \\ \overline{B}_3 \end{bmatrix} = \begin{bmatrix} 0 & 2 & 2 \\ 0 & 3 & 0 \end{bmatrix}$$

$$PC_5 = \begin{bmatrix} A_5 \\ B_5 \end{bmatrix} = \begin{bmatrix} 0 & 1 & 3 & 2 & 1 \\ 0 & 0 & 0 & 1 & 3 \end{bmatrix} \qquad \overline{PC}_5 = \begin{bmatrix} \overline{A}_5 \\ \overline{B}_5 \end{bmatrix} = \begin{bmatrix} 3 & 1 & 2 & 2 & 2 \\ 3 & 2 & 1 & 3 & 0 \end{bmatrix}$$

$$\text{Also} = \overline{PC}_5 + [1] = \begin{bmatrix} 0 & 2 & 3 & 3 & 3 \\ 0 & 3 & 2 & 0 & 1 \end{bmatrix}$$

$$PC_{13} = \begin{bmatrix} A_{13} \\ B_{13} \end{bmatrix} = \begin{bmatrix} 0 & 0 & 0 & 1 & 2 & 0 & 0 & 3 & 0 & 2 & 0 & 3 & 1 \\ 0 & 1 & 2 & 2 & 2 & 1 & 2 & 0 & 0 & 3 & 2 & 0 & 3 \end{bmatrix}$$

$$\overline{PC}_{13} = \begin{bmatrix} \overline{A}_{13} \\ \overline{B}_{13} \end{bmatrix} = \begin{bmatrix} 3 & 2 & 0 & 3 & 2 & 2 & 0 & 1 & 0 & 0 & 0 & 1 & 2 \\ 3 & 1 & 0 & 2 & 0 & 1 & 0 & 0 & 2 & 3 & 0 & 0 & 0 \end{bmatrix}$$

(5.31)

In addition, other lengths can be formed using combinations of the kernals as shown in Equations (5.32). The equation shows the procedure for codes of $2 \cdot N \cdot M$ in length. In all of these, the attraction is simple code generation and even simpler decoding than any other compression scheme. The simplicity is, of course, a weakness from a stealth point of view, and that is why many codes must be used and many chip and pulse lengths, so guessing will not be an advantage to the interceptor. In Equation (5.33), the example used is $M = N = 3$, and so the resulting code is 18 long. The complement for length 18 is the one previously stated, i.e.,

$$\bar{A}_N = -B_N^*(N - i); \bar{B}_N(i) = A_N^*(N - i) \text{ for } i = 0, 1, \dots (N - 1) \tag{5.32}$$

$$PC_{2 \cdot M \cdot N} = \begin{bmatrix} A_{2 \cdot M \cdot N} \\ B_{2 \cdot M \cdot N} \end{bmatrix} = \begin{bmatrix} (A_N \times A_M) & (B_N \times \bar{A}_M) \\ (A_N \times B_M) & (B_N \times \bar{B}_M) \end{bmatrix}$$

For example, for $M = N = 3$

$$PC_{18} = \begin{bmatrix} 0 & 1 & 0 & 1 & 2 & 1 & 0 & 1 & 0 & 0 & 2 & 2 & 0 & 2 & 2 & 2 & 0 & 0 \\ 0 & 0 & 2 & 1 & 1 & 3 & 0 & 0 & 2 & 0 & 3 & 0 & 0 & 3 & 0 & 2 & 1 & 2 \end{bmatrix}$$

$$\tag{5.33}$$

where $A_N = a_0 \, a_1 \, a_2 \dots a_{N-1}$ and $B_M = b_0 \, b_1 \, b_2 \dots b_{M-1}$ and $A_N \times B_M = a_0 \cdot (b_0 \, b_1 \dots b_{M-1})$ $a_1 \cdot (b_0 \, b_1 \dots b_{M-1}) \dots a_{N-1} \cdot (b_0 \, b_1 \dots b_{M-1})$.

Lastly, it is also possible to form polyphase codes in which four members are required to obtain perfect cancellation. One example for length 16 is given below in Equation (5.34). These are obviously Frank codes and have all the good features of those codes plus the complementary sidelobe cancelling feature. Such codes are often used in conjunction with a binary complementary outer code to allow Hudson-Larson decode. The inner code is then compressed brute force. The disadvantage of a four-phase code is that the ambiguities are at the PRF/4.

$$PC_{16} = \begin{bmatrix} 0 & 0 & 0 & 0 & 0 & 1 & 2 & 3 & 0 & 2 & 0 & 2 & 0 & 3 & 2 & 1 \\ 0 & 1 & 2 & 3 & 0 & 2 & 0 & 2 & 0 & 3 & 2 & 1 & 0 & 0 & 0 & 0 \\ 0 & 2 & 0 & 2 & 0 & 3 & 2 & 1 & 0 & 0 & 0 & 0 & 0 & 1 & 2 & 3 \\ 0 & 3 & 2 & 1 & 0 & 0 & 0 & 0 & 0 & 1 & 2 & 3 & 0 & 2 & 0 & 2 \end{bmatrix} \tag{5.34}$$

5.5.3.4 Self-Noise Performance of Complementary Codes. All pulse compression waveforms are subject to real-world degradations from hardware and software limitations. There are at least three significant internal sources of sidelobe cancellation degradation for complementary codes. First, transmit phase modulation may not be orthogonal so that one phase state is less than 90° from another. A phase modulation error of 10° total will not be completely cancelled on receive, thus raising sidelobes as summarized in Fig. 5.66. Similarly, transmitter power supply droop during high-power oper-

Complementary Code Pulse Compression Range Sidelobe Power Due To:

Code Length	10° Phase Modulation Error:	60° Phase Shift Across Pulse:	5% AGC Amplitude Modulation:
16	-26.2 dB	-19.5 dB	-44.0 dB
20	-28.0 dB	-12.5 dB	-43.7 dB
32	-23.9 dB	-13.0 dB	-37.0 dB
40	-22.3 dB	-14.8 dB	-43.6 dB

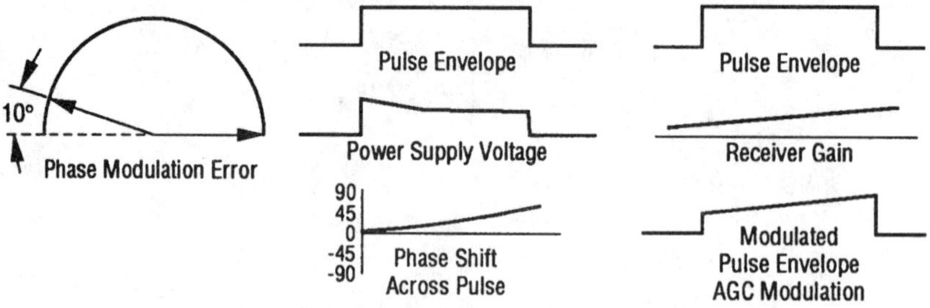

Figure 5.66 Many effects reduce the cancellation property.

ation can cause phase pulling of the transmitter during a pulse. This pulling causes phase warp across the code, which is similar to Doppler, which if not compensated, increases sidelobes and reduces mainlobe after compression. The transmitter has thousands of electrical degrees of delay, and so a small droop can easily cause 60° phase warp across the uncompressed pulse. Lastly, receiver AGC and STC can cause amplitude modulation across the received uncompressed pulses, which also gives rise to increases in uncancelled sidelobes. Such modulation, if not compensated, can easily be 5 percent across the pulse. The effect of all three of these sources for various code lengths is summarized in Fig. 5.66.

5.6 HYBRID WAVEFORMS

5.6.1 Hybrid Spread Spectrum Stretch (S-Cubed)

S-cubed combines the advantageous processing properties of stretch with the wide instantaneous bandwidth of discrete phase codes by superimposing a short, cyclically repeated, discrete phase code on a linear FM waveform. This process is analogous to concatenation of codes in that the total PCR is the product of the PCRs of the two waveforms. For example, a conventional stretch waveform with an effective frequency excursion of 6.25 MHz will compress to 160 nsec $[1/(6.25 \cdot 10^6)]$ or 80 ft. An S-cubed waveform of the same excursion, combined with an eight-bit binary phase code of 20 nsec/bit, compresses to 20 nsec, or $80/8 = 10$ ft.[16]

In the spectral domain, shown in Fig. 5.67, similar conclusions can be drawn. The figure shows the spectrum of an N-chip code, characterized by a bandwidth, ΔF, of about $1/T_{rb}$, where T_{rb} is approximately the chip duration, t_c, and the range resolution, d_r equals $(T_{rb} \cdot c)/2$. When an N-chip code is repeated cyclically, a line spectrum results, with the spacing between the lines equal to the repetition frequency, $1/(N \cdot T_{rb})$. A waveform characterized by this line spectrum would be ambiguous in range. However, by sweeping the signal slowly in frequency, the line spectrum can be filled in to give an approximation to the original N-chip code spectrum. Clearly, the minimum frequency sweep that accomplishes this result is $1/(N \cdot T_{rb})$. Although larger sweep excursions will do, a sweep of $1/(N \cdot T_{rb}) = \Delta F/N$ results in minimum processing complexity.

Another way to view the underlying S-cubed concept is stimulated by the preceding result. Conventional stretch requires an effective frequency excursion of $\Delta F = 1/T_{rb}$ so as to eliminate the ambiguities resulting from the fact that returns from every range bin occur in every receiver sample. With an S-cubed waveform as shown in Fig. 5.67, after decoding the discrete phase code, a given target shows up only in every Nth range sample. Zero cyclic sidelobes are required for the S-cubed phase code, which is somewhat different from conventional phase codes. Zero cyclic sidelobes are achievable as shown in the next section.

Therefore, the stretch portion of the waveform is required only to resolve targets separated in time by $N \cdot T_{rb}$, resulting in a required effective frequency excursion of $\Delta F/N$. Conventional ambiguity function analysis is required to verify that the ambiguities lie outside the bandwidths of the presum filters and the antenna beamwidth as shown in Fig. 5.69. This condition is easily achieved for most geometries. Of course, ambiguities and their interaction with geometry and antenna shape are not unique to S-cubed waveforms. These assessments must be performed no matter what waveforms are selected.

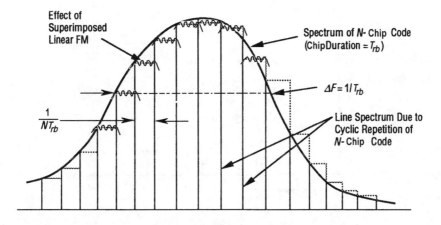

Figure 5.67   Many effects reduce the cancellation property.

5.6.2 Hybrid Spread Spectrum Stretch Processing

The preceding introduction gives some insight into the basic S-cubed concept. Typical signal processing/pulse compression and parameters are described in this section. Some discrete-phase cyclic codes that are required for successful operation also are discussed. Generation of an S-cubed waveform is shown in Fig. 5.70.

For illustration purposes, a four-bit binary phase code with good cyclic properties $(1, 1, -1, 1)$ is used. In the receiver shown in Fig. 5.71, the deramped signal is passed through a wideband IF, sampled, and presummed, as in the second stretch mechanization shown in Fig. 5.23. However, the decoder for the discrete phase code follows the A/D converter, and, because signals from a given set of ambiguous ranges appear in only every Nth sample as shown in Fig. 5.68, the decoder outputs are multiplexed cyclically between N presum channels. The word rate at each of these N presum outputs is $\Delta f/N$, which is $1/N$th of the equivalent conventional stretch IF bandwidth. Therefore, the data rate after presumming is equal to the stretch data rate, which is the minimum possible.

The N FFTs are each of order N_{rb}/N. This structure compares favorably with the conventional stretch requirement for a single, order N_{rb}, FFT. In fact, the required number of operations is reduced by a factor of $1 - \log(N)/\log(N_{rb})$. The N FFTs are sequentially offset across the frequency band to accommodate the offsets of the corresponding range cells in the swath. After FFT pulse compression, the FFT outputs must be demultiplexed to order them across the range swath. Specifically, the output of the first presum FFT channel contains range bins $1, N + 1, 2N + 1, 3N + 1, \ldots$. The second channel contains range bins $2, N + 2, 2N + 2, 3N + 2$, and so on. The multiplexing and demultiplexing is a simple data distribution function, equivalent functionally to a cyclic stepping switch.

Using the previously given four-chip binary phase code as an example, Fig. 5.72 shows how the discrete phase decoder might be mechanized. That figure also shows

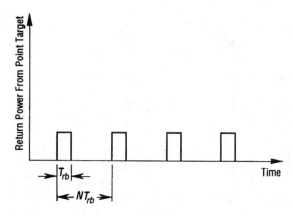

Figure 5.68 Ambiguous returns from point target resulting from a cyclically repeated N-chip code.

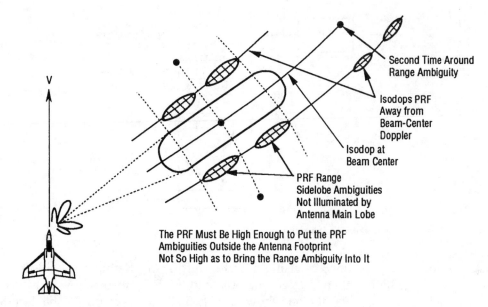

Figure 5.69 Ambiguity locations must be given careful consideration.

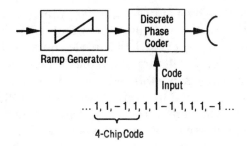

Figure 5.70 Generation of S-cubed waveform.

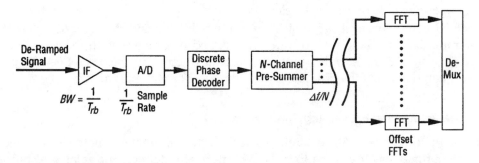

Figure 5.71 S-cubed processing.

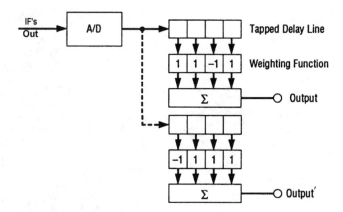

Figure 5.72   Decoder implementation (four-chip code)

via the dotted line path how, by using a cyclic shift of the N-chip code, additional samples at a rate higher than l/N_{rb} can be obtained. One S-cubed option is to use a two-channel decoder such as the one shown in Fig. 5.72. In this example, the decoder weights are identical to the N-chip code. This equivalence is not always so, as discussed later in this section.

Figure 5.73 shows how the data appear after deramping and decoding, in both frequency and time sample space. The figure shows that, with the frequency excursion chosen as previously developed, every Nth sample contains data from $1/N$th of the range cells, and that these N_{rb}/N cells are separable in frequency, although there is frequency overlap before the time separation achieved by decoding. If the frequency excursions were left at ΔF, the value for conventional stretch, then there would be no frequency overlap. For a reduction in frequency excursion to a value greater than $\Delta F/N$, the returns could be more separated in frequency than is shown in Fig. 5.73. Such a choice would require some oversampling and would allow for more flexibility in filter shaping. Such shaping for sidelobe control is important in both conventional stretch and S-cubed, because the near sidelobes are displaced by N cells. Some penalty in data rate is required for sidelobe control.

S-cubed waveform and processing parameters introduced throughout this section are summarized in Table 5.4. To compare S-cubed parameters with those of conventional stretch, one can simply set N, the number of chips in the discrete cyclic phase code, equal to one in Table 5.4, to get the equivalent stretch parameters.

5.6.3   Waveform and Processing Parameters

The criteria for choosing the discrete phase code are minimal mismatch loss in decoding and optimal spectral spreading. These criteria are defined and related to particular code choices in what follows. It can be shown that a length N, weighted, tapped

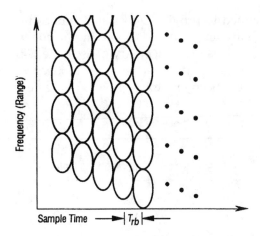

Figure 5.73 Time-frequency plot of S-cubed signal after dechirp and decode.

Table 5.4   S-Cubed Waveform and Processing Parameters

Symbol	Parameter	Relationship
T_{SW}	Swath width (seconds)	$T_{SW} = 2c/SW$
T	Transmitted pulse length (seconds)	
T_E	Effective transmitted pulse length (seconds)	$T_E = T - T_{SW}$
d_r	Range resolution (feet)	
SW	Swath length (feet)	
T_{rb}	Range resolution (seconds)	$T_{rb} = 2c/d_r$
N_{rb}	Number of range cells across swath	$N_{rb} = T_{SW}/T_{rb}$
PCR	Effective pulse compression ratio	$T_E/T_{rb} = PCR$
c	Velocity of light (feet/nanosecond)	0.983
N	Number of chips in discrete phase code	
ΔF	Effective frequency excursion (Hz)	$\Delta F = 1/(NT_{rb})$
ΔF_T	Transmitted frequency excursion (Hz)	$\Delta F_T = T\Delta F/T_E$
$f_{A/D}$	A/D sample rate (Hz)	$f_{A/D} \approx 1/T_{rb}$
Δf	Presum bandwidth (Hz)	$\Delta f = (T_{SW} \Delta F)/(T_E N_{rb})$ $\Delta f = N_{rb}/(NT_E)$
PSR	Presum ratio	$T_E/T_{SW} = PSR$
N_s	Number of range samples required after presumming	$N_s = N_{rb}$

delay line can be designed to provide zero sidelobe decoding of any cyclic discrete phase code. Some codes and the corresponding decoder weights are listed in Table 5.5. The length 4 code is the only binary phase code with no mismatch loss; i.e., the zero sidelobe decoder is also the matched filter. Notice that the decoder weighting vector for the length 16 Frank code is the complex conjugate on the code vector.

Table 5.5   Examples of Cyclic Discrete Phase Codes

Code	Length	Weighting factor	Mismatch loss (dB)
1, 1, −1, 1	(4)	1, 1, −1, 1	0
1, 1, 1, −1, −1, 1, −1, 1	(8)	1, 3, 1, −1, −1, 3, −1, 1	1.25
1, 1, 1, 1, −1, −1, 1, 1, −1, 1, −1, 1	(12)	1, 1, 2, 1, −1, −2, 1, 1, −2, 1, −1, 2	0.5
1, 1, 1, 1, j, −1, −j, 1, −1, 1, −1, 1, −j, −1, j, 1	(16) (Frank)	1, 1, 1, 1, −j, −1, j, 1, −1, 1, −1, 1, j, −1, −j, 1	0

The mismatch loss can be calculated from Equation (5.35).

$$L_{mm} = \frac{\left(\sum_{i=0}^{N-1} a_i \cdot w_i\right)^2}{N \cdot \sum_{i=0}^{N-1} |w_i|^2} \tag{5.35}$$

where the a_i are the code entries and the w_i are the decoder weightings. This equation gives the loss in SNR resulting from the use of a mismatched decoder, and Table 5.5 summarizes this loss for the example codes.

The second criterion, optimal spectral spreading, should be interpreted in terms of the intercept threat detection filter model. For example, a code of length 10, with a 20 nsec chip time, generates a line spectrum with lines $1/(10 \times 20 \text{ nsec}) = 5$ MHz apart, as suggested in Fig. 5.67. With 5-MHz interceptor detection filters, increasing the code length much beyond ten 20-nsec chips simply puts more lines in each detection filter. However, as shown in Section 5.4.2, a detection filter designed for conventional stretch could be much narrower than 5 MHz, and then a longer code would be effective. A shorter code, on the other hand, does not result in enough spectral lines within the example 50-MHz envelope to produce near-optimal spectral spreading, defined here as a near-uniform spreading of the instantaneous signal power over $1/T_{rb} = 50$ MHz and producing close to a 10-dB LPI improvement for a 5-MHz interceptor detection filter, relative to conventional stretch. For example, the power in the dc term for the length 4 code is down only 6 dB. This value can be calculated from Equation (5.36).

$$P_{LR} = \frac{\left| \sum\limits_{i=0}^{N-1} a_i \right|^2}{N^2} \tag{5.36}$$

Values of the power line ratio, P_{LR}, for other than the dc terms, can be obtained by applying a phase rotation across the code, corresponding to the frequency offset, and using the above formula. For comparison, the dc and higher frequency lines for length 12 code are down by 9.5 dB or more.

For an actual S-cubed system, both a 12-chip binary phase code and a 16-chip polyphase code are potential candidates. Because the binary code comes within 0.5 dB of the theoretically possible performance and is less costly to implement, it is often used. For many applications, the length 16, four-phase code looks attractive, based not only on the simple 5-MHz interceptor detection filter model but on the advantages of a polyphase code operated against more sophisticated detection filters, designed for binary phase codes.

One option to increase the spectral spreading beyond the band associated with a given resolution is to use bit times smaller than a range bin. For example, going to 5-nsec bit times would spread the transmitted power over 200 MHz, with the resulting higher range resolution returns added together in the receiver/processor to obtain the desired resolution. This approach, which is the analog of the spread spectrum communications technique, has been utilized in some systems. If the desired targets have correlation length greater than a range cell, this processing will result in improved *SNR*.

Further LPI improvement can be achieved by transmitting in multiple adjacent frequency bands as described in Section 5.2, possibly separated by small guard bands, to provide multiple looks in frequency. This data would be separated in the receiver by parallel channels and processed by functionally parallel processing.

In summary, the spread spectrum stretch approach has the highly desirable sampling and data rate properties of stretch, along with the instantaneous bandwidth properties of discrete phase codes. The basic S-cubed concept, which imbeds a cyclically repeated discrete phase code within a long stretch waveform, was explained in terms of stretch, ambiguity resolution, and processing sequence. Example S-cubed waveforms and processing parameters were shown for discrete phase codes and their zero sidelobe decoders. A near-optimal spectral spreading for LPI improvement is possible for many geometries.

5.7   Noise Propagation in Pulse Compressors

The ultimate performance and interceptability of all LPIS are determined by *SNR*. Although *SNR* is often determined by external noise, the effort to improve *SNR* by intensive pulse compression and other processing naturally gives rise to increased internal noise which if not managed can limit performance.

Figure 5.74 shows a block diagram of a pulse compressor with length M and input word length of $N + 1$. Each of the samples is weighted by the corresponding code chip value, a_i, and summed to form the pulse compressor output. The figure tabulates the input noise power, σ_n^2, input signal power, σ_s^2, the maximum or peak signal, and the corresponding output values for each of these parameters. B_c represents the chip bandwidth, and f_s represents the sampling rate. If the pulse compression code is an unweighted binary phase code, then the pulse compression ratio is equal to the code length, M, and the output signal power grows as M^2. Assuming that the noise is uncorrelated chip to chip, the noise grows as M (this is not true, but close enough for most first-order analysis). The signal-to-noise ratio, SNR, grows as

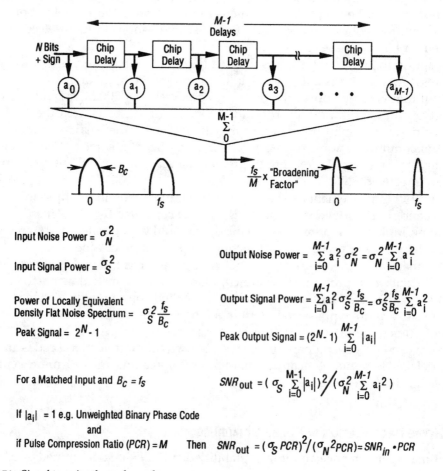

Input Noise Power = σ_N^2

Input Signal Power = σ_S^2

Power of Locally Equivalent
Density Flat Noise Spectrum = $\sigma_S^2 \dfrac{f_s}{B_c}$

Peak Signal = $2^N - 1$

For a Matched Input and $B_c = f_s$

If $|a_i| = 1$ e.g. Unweighted Binary Phase Code
and
if Pulse Compression Ratio $(PCR) = M$

Output Noise Power = $\displaystyle\sum_{i=0}^{M-1} a_i^2 \, \sigma_N^2 = \sigma_N^2 \sum_{i=0}^{M-1} a_i^2$

Output Signal Power = $\displaystyle\sum_{i=0}^{M-1} a_i^2 \sigma_S^2 \dfrac{f_s}{B_c} = \sigma_S^2 \dfrac{f_s}{B_c} \sum_{i=0}^{M-1} a_i^2$

Peak Output Signal = $(2^N - 1) \displaystyle\sum_{i=0}^{M-1} |a_i|$

$SNR_{out} = \left(\sigma_S \displaystyle\sum_{i=0}^{M-1} |a_i| \right)^2 \Big/ \left(\sigma_N^2 \sum_{i=0}^{M-1} a_i^2 \right)$

Then $SNR_{out} = (\sigma_S \, PCR)^2 / (\sigma_N^2 PCR) = SNR_{in} \cdot PCR$

Figure 5.74 Signal to noise through a pulse compressor.

$M^2/M = M$. Noise propagation in pulse compression and filters is covered in more detail in Chapter 7.

Summarizing Fig. 5.74, the output *SNR* grows proportionally to the *PCR* if unweighted. If weighted, the *SNR* grows approximately as the ratio of the uncompressed pulse width to the net compressed pulse width assuming matched filtering at the input.

Although, historically, pulse compression was often done by analog means, almost all modern pulse compression is done digitally after A/D conversion. The first and often largest contributor to pulse compression self-noise is the A/D converter. Figure 5.75 shows the noise performance of an A/D as a function of the ratio between the saturation level and the RMS level of the input signal for an A/D with the number of bits quantization as a parameter. The set of curves in the figure have been widely published at least since the early 1960s. The input statistics are assumed to be Gaussian with zero mean. It can be seen that there is a broad but obvious minimum noise optimum for a given quantization. For example, the optimum RMS level for the signal with a six-bit quantization is 0.305 times the saturation level ($K_{AD} = 3.28$ in the figure). This optimum will be derived in a later chapter.

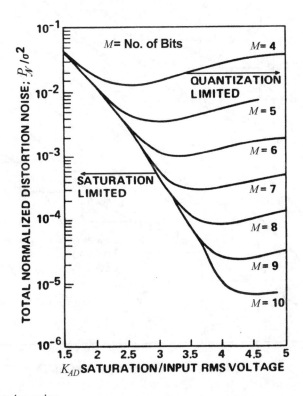

Figure 5.75 A/D conversion noise.

After quantization, the signal goes through a pulse compressor as depicted in Fig. 5.76, in which the signal and noise grow in amplitude in each stage of the compression.[18] Because large compression ratios are usually performed in multiple steps, not in a single step as shown in Fig. 5.74, one must track the growth in each stage to insure that the *SNR* is preserved. Figure 5.76 shows the value of the some hypothetical signal in bits as a function of a range of equivalent radar cross sections (1 to 10,000 m²)

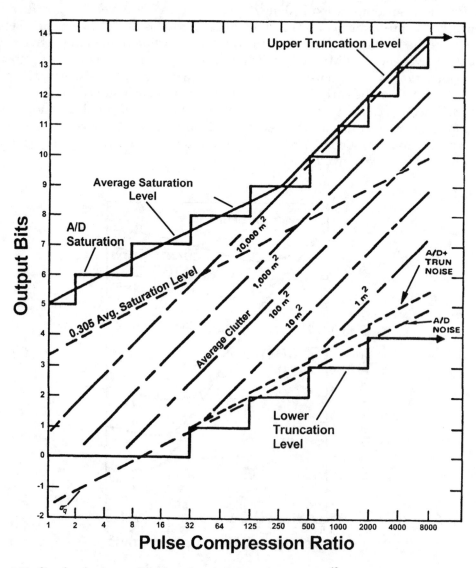

Figure 5.76 Signal and noise amplitude as a function of compression ratio.[18]

for various compression ratios. Not only must the signal be tracked through the compression network but also average clutter, A/D conversion noise and truncation noise. In this particular figure the A/D converter is assumed to have 6 bits. Each of these components is calculated and plotted in Fig. 5.76 along with some possible upper and lower truncation levels. The number of output bits plotted assumes no truncation since truncation is a designer's choice based on the filtering which follows.

5.8 WAVEFORM SUMMARY

Summarizing stealth pulse compression whether for data links or radars, the dominant characteristics are as follows:

1. Low interceptability for a given bandwidth coverage

2. Complete bandwidth coverage as possible

3. Lowest transmitted power for a given mode

4. Low self-noise on transmit and receive

5. Low sidelobes especially ISLR

6. Simplest generation and compression because codes are long

7. Code agility in chip width, total length and seed

Table 5.6 shows some example trade-offs for several different short codes including PSLR, ISLR, hardware complexity, modulation, and sidelobe weighting in the presence of typical hardware degradation. When the codes are very long (i.e., 4000:1 or greater), the only compression waveforms worth considering are chirp/stretch or

Table 5.6 Pulse Compression Trade–Offs

Code	PSLR (dB)*	ISLR (dB)*	Hardware complexity	Modulator	Weighting†
96:1 Discrete chirp	−33	−16	1.45	32 Phase shifts	1 lead/lag pair $C_{app} = 0.5$
128:1 Complementary	−39	−23	1	Binary phase	None
1024:1 Complementary	−62	−29	1.2	Binary phase	None
144:1 Frank	−31	−17	1.25	12 Phase shifts	1 lead/lag pair $C_{app} = 0.5$
169:1 Barker product	−26	−13	1.85	Binary phase	26 lead/lag pairs $C_{app} = 0.041$

*10° phase error and timing error of 1/8 chip across pulse.
†Transversal equalizer approach.

complementary codes coupled with some multichannel/multifrequency wave-form.[19,20]

A comparison of *ISLR* for some short codes in the presence of 5° of phase modulation error, 5 percent AGC modulation, and 5° of uncompensated phase warp is given in Fig. 5.77. In most systems, ISLR is the essential performance parameter.

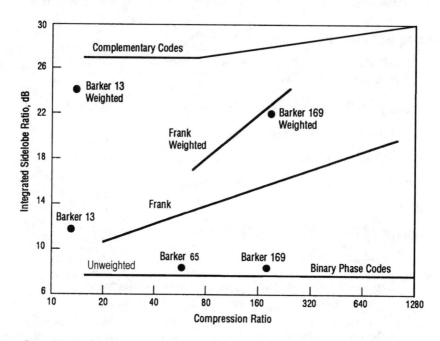

Figure 5.77   Comparison of pulse compression codes.

5.9   EXERCISES

1. Estimate the NEXT power for a simultaneous multifrequency waveform for 8-pole Butterworth filters, 10-nsec chip width, 150-MHz channel spacing, and 50-W peak transmit power.

2. Find the required modulation index, β, and phasings, a_n, for a multiline spectrum generation for $N = 3$ and $f_m = 50$ MHz.

3. Calculate the bit error rate, *BER*, for a 17-frequency multiline waveform in a fading channel with $SNR = 10$ dB.

4. Calculate the value of the ambiguity function for a linear FM waveform at a Doppler offset of 10 kHz, time offset of 0.5 nsec, with $\Delta f = 600$ MHz and $T = 25$ μsec.

5. Estimate the dynamic range in the presence of 1000-m^2 bright discretes for a stretch processed waveform using an unweighted FFT filter bank with T_E = 23 μsec and ΔT = 13 μsec.

6. Calculate the ISLR for a triple Barker code of length $13 \times 13 \times 7 = 1183$ chips with 20° phase shift across the total pulse.

7. Calculate the ISLR for 16-bit complementary Frank codes with 45° phase shift across the total pulse after summing four members of the set.

8. Calculate the output SNR for an input SNR of –10 dB, 4-bit input quantization, and pulse compression ratio of 16384 for a type II complementary code after summing two members of the set, assuming the rms signal level at the A/D input was set optimum.

5.10 REFERENCES

1. Simon, M. K., et al., *Spread Spectrum Communications Handbook*, rev. ed., McGraw Hill, 1994, pp. 1–38, 135–186, 605, 1031–1093.
2. Skolnik, M. I., Ed., *Radar Handbook*, McGraw Hill, 1970, pp. 15-6 to 15-27.
3. Fahey, Michael D., et al., *Interception of Bi-Phase Coded Radar Signals*, U.S. Army Missile Command Technical Report RE-77-e, originally classified confidential, declassified 12/31/1990.
4. Golay, M. J. E., "Complementary Series," *IRE Transactions on Information Theory*, Vol. IT-7, April 1961, pp. 82–87.
5. Frank, R., "Polyphase Complementary Codes," *IEEE Transactions on Information Theory*, Vol. 26, No. 6, November 1980, pp. 641–647.
6. Sinsky, A., Brookner, E., Ed., *Radar Technology*, Artech House, 1977, Chapter 7, pp. 132–137.
7. Weil, T., Brookner, E., Ed., *Radar Technology*, Artech House, 1977, Chapter 27, pp. 379–383.
8. Lewis, B., F. Kretschmer, and W. Shelton, *Aspects of Radar Signal Processing*, Artech House, 1986, pp. 7–63.
9. Cohen, M., and Nathanson, F., *Radar Design Principles*, 2nd ed., McGraw Hill, 1990, pp. 533–582, 587–634.
10. Hughes Aircraft, "LPIR Phase 1 Review," unclassified report, 1977.
11. Grettenberg, T., and Jeffers, R., "Complementary Codes for Hughes," *Applied Systems Laboratories Report*, unclassified report, April 1979.
12. A few photos, tables and figures in this intellectual property were made by Hughes Aircraft Company and first appeared in public documents that were not copyrighted. These photos, tables, and figures were acquired by Raytheon Company in the merger of Hughes and Raytheon in December 1997 and are identified as Raytheon photos, tables, or figures. All are published with permission.
13. Perdy, R. J., "Signal Processing Linear Frequency Modulated Signals," in *Radar Technology*, E. Brookner, Ed., Artech House, 1977, pp. 155–162.
14. Skolnik, M., *Introduction to Radar Systems*, 2nd ed., McGraw Hill, 1980, pp. 422–434.
15. Peebles, P., *Radar Principles*, John Wiley & Sons, 1998, pp. 288–318.
16. Sivaswamy, R., "Digital and Analog Subcomplementary Sequences for Pulse Compression," *IEEE Transactions on Aerospace and Electronic Systems*, Vol. AES-14, No. 2, March 1978, pp. 343–350.

17. Garmatyuk, D., and R. Narayanan, "ECCM Capabilities of an Ultrawideband Bandlimited Random Noise Imaging Radar," *IEEE Transactions on Aerospace and Electronic Systems*, Vol. AES-38, No. 4, October 2002, pp. 1243–1255.
18. Rivers, D., and A. Berman, "Pulse Compression Dynamic Range Analysis," *Hughes Aircraft IDC*, 15 May 1979.
19. Hughes Aircraft, "Atlas Radar Pulse Compression Tradeoffs," 1976.
20. Hughes Aircraft, "Pave Mover TAWDS Design Requirements," November 1979, unclassified specification.
21. Mooney, D., "Post-Detection STC in a Medium PRF Pulse Doppler Radar," U.S. Patent 4095222.
22. Katz, J., and E. Gregory, "Modulation System," U.S. Patent 4130811.
23. Dillard, R., and G. Dillard, *Detectability of Spread Spectrum Signals*, Artech House, 1989.
24. Skolnik, M., G. Andrews, and J. Hanson, "Ultrawideband Microwave-Radar Conceptual Design," *IEEE AES Systems Magazine*, October 1995, pp. 25–30.
25. White, H., and F. Bartlett, "A New Tactical Surveillance Radar for the U.S. Marine Corps," *International Defense Review—Air Defense Systems*, Interavia, Geneva, Switzerland, 1976, pp. 23–25.
26. Skolnik, M. I., Ed., *Radar Handbook*, 2nd ed., McGraw Hill, 1990, pp. 10.4 to 10.10.
27. Berkowitz, R., *Modern Radar*, John Wiley & Sons, 1965, pp. 216–243.
28. Haggarty, R., Meehan, J., and G. O'Leary, "A 1 GHz 2,000,000:1 Pulse Compression Radar-Conceptual System Design" in *Radar Technology*, E. Brookner, ed., Artech House, 1977, pp. 175–180.
29. Stimson, G., *Introduction to Airborne Radar*, 2nd ed., Scitech, 1998, pp. 163–176.

6

Stealth Antennas and Radomes

6.1 INTRODUCTION

The purpose of Chapter 6 is to summarize the results of antenna and radome technology that are of particular interest for antenna stealth. Shape is everything when it comes to both the active and passive signatures of antennas and radomes. Once the shape is right, then such things as element pattern, amplitude weighting, thickness, and edge treatment become important. First, the relationship between the radiation pattern (beamwidth, sidelobes, and so forth) and the current distribution across the antenna aperture is discussed. This is followed by descriptions of the various types of antennas that have been applied to radar and data links, including reflectors, lenses, and arrays. Several methods of pattern synthesis are discussed. The effect of broadband signals and errors in the aperture distribution on the radiation and radar cross section (RCS) patterns is considered. Both the active and passive signatures of antennas and radomes will be covered. The chapter closes with brief discussions of radome and antenna near field interaction.

6.2 ANTENNA PARAMETERS

6.2.1 Fundamental Definitions (after Skolnik[1] and Silver[2])

The material in this section is closely modeled on books by M. Skolnik and S. Silver and is included because it is so useful in what follows. The gain function, $G(\theta, \phi)$, is defined as the ratio of the power, $P(\theta, \phi)$, radiated in a direction, (θ, ϕ), per unit solid angle to the average power radiated per unit solid angle. The average power radiated per unit solid angle is equal to the total power, P_t, divided by 4π. Thus,

$$G(\theta, \phi) = \frac{P(\theta, \phi)}{P_t/(4 \cdot \pi)} = \frac{4 \cdot \pi \cdot P(\theta, \phi)}{\int\limits_{0}^{2 \cdot \pi} d\phi \int\limits_{0}^{\pi} P(\theta, \phi) \cdot \sin(\theta) \cdot d\theta} \tag{6.1}$$

The radiated directive gain can be written as

$$G_D = \frac{4 \cdot \pi \cdot (\text{maximum power per unit solid angle})}{\text{total power}} \tag{6.2}$$

The maximum power per unit solid angle is the peak of the radiation pattern, $P(\theta, \phi)$, and the total power radiated is the integral of the volume under the radiation pattern. Equation (6.2) can be written as

$$G_D = \frac{4 \cdot \pi \cdot P(\theta,\phi)_{max}}{\iint P(\theta,\phi) \cdot \sin(\theta) \cdot d\theta \cdot d\phi} = \frac{4 \cdot \pi}{\Omega} \tag{6.3}$$

where Ω is defined as the "cookie cutter" beam area described in Chapter 2.

$$\Omega = \frac{\iint P(\theta,\phi) \cdot \sin(\theta) \cdot d\theta \cdot d\phi}{P(\theta,\phi)_{max}} \tag{6.4}$$

The beam area is the solid angle through which all the radiated power would pass if the power per unit solid angle were equal to $P(\theta, \phi)_{max}$ over the disk (as shown in Figs. 2.13 and 2.14) that defines an equivalent zero sidelobe antenna pattern. If θ_{az} and θ_{el} are defined as the half-power beamwidths in two orthogonal surfaces, the beam area, Ω, is approximately equal to θ_{az} times θ_{el}. Substituting into Equation (6.3) gives

$$G_D \cong \frac{4 \cdot \pi}{\theta_{el} \cdot \theta_{az}} \text{ or } G_D \cong \frac{41,253}{\theta_{el} \cdot \theta_{az}} \tag{6.5}$$

The first or second directive gain equation above is used, depending on whether the half-power beamwidths are measured in radians or in degrees, respectively. For example, a $3° \times 3°$ beamwidth has approximately 36.6 dB directive gain. This is a very handy formula, as most system engineers know.

The definition of directive gain, G_D, is based primarily on the shape of the radiation pattern. It does not take account of losses due to heating, mismatch, and so on. The power gain, G, includes the effect of any losses that lower the antenna efficiency. The power gain is defined as

$$G = \frac{\text{maximum radiation intensity from antenna}}{\text{intensity from lossless isotropic radiator–same power input}} \tag{6.6}$$

The directive gain is always greater than the power gain. The power gain and the directive gain are related by a radiation efficiency factor, ρ_e, as

$$G = \rho_e \cdot G_D \tag{6.7}$$

The definitions of power gain and directive gain are described in terms of a transmitting antenna. There is a tacit assumption of reciprocity, which is true for simple antennas. Under certain conditions, the transmitting and receiving patterns of an antenna are almost the same. Total reciprocity is almost never true for monopulse and electronic scanned antennas, but they are often close. Thus, the gain definitions can apply equally well, whether the antenna is used for transmission or for reception. Because these notions will later be used to calculate antenna RCS, it is important to be aware that antenna systems usually aren't completely reciprocal. One must be careful to identify the transmit and receive gain differences.

Another antenna parameter related to the gain is the effective receiving aperture, or effective area. It is a measure of the effective area presented by the antenna to an incident wave. The gain, G, and the effective area, A_e, of an antenna are related by

$$ G = \frac{4 \cdot \pi \cdot A_e}{\lambda^2} = \frac{4 \cdot \pi \cdot A \cdot \rho_e \cdot \rho_a}{\lambda^2} \qquad (6.8) $$

where
   A = physical area of antenna
   ρ_a = aperture efficiency, which takes into account scattering, blockage, and so on

Figure 6.1 summarizes the directive gain and effective antenna area for some typical antenna elements and antennas.

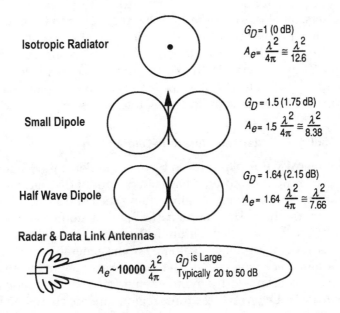

Figure 6.1 Some comparisons of gain and effective area.

The direction of polarization of an antenna is defined as the direction of the electric field vector. Most radar and data link antennas are linearly polarized; that is, the direction of the electric field vector is either quasivertical or horizontal. Linear polarization is most often used in conventional radar and data link antennas, as it is the easiest to achieve and generally results in lowest overall antenna losses and RCS.

Data link antennas sometimes transmit and receive on both polarizations. The polarization may also be elliptical or circular. Elliptical polarization may be considered as the combination of two linearly polarized waves of the same frequency, travelling in the same direction, which are perpendicular to each other in space. The relative amplitudes of the two waves and the phase relationship between them can assume any values. If the amplitudes of the two waves are equal, and if they are 90° out of (time) phase, the polarization is circular. Circular polarization and linear polarization are special cases of elliptical polarization. There is a limited range of weather and target recognition in which circular polarization is useful.

Low sidelobes are generally desired for radar and data link applications. If too large a portion of the radiated energy were contained in the sidelobes, there would be a reduction in the main-beam energy, with a consequent lowering of the maximum gain. More importantly, low sidelobes reduce clutter return, ECM susceptibility, and probability of intercept. No general rule can be given for specifying the optimum sidelobe level. Lower sidelobes are almost always better from a system standpoint. If the sidelobes are not low enough, strong discretes can enter the receiver and appear as false signals. A high sidelobe level makes jamming of the radar or data link easier. Also, high sidelobes cause radar or data links to be more subject to interference from nearby friendly transmitters. Sidelobes cover most of the space around the antenna and therefore often are the major contributor to exposure index and probability of intercept.

Sidelobes of the order of −35 to −40 dB below the main beam can be readily achieved with practical antennas. With extreme care it is possible to obtain sidelobes as low as −50 or −55 dB. However, considerably lower sidelobes are difficult to achieve, primarily as a result of manufacturing tolerances.

6.2.2   Antenna Radiation Pattern and Aperture Distribution

The electric field intensity $E(\theta, \phi)$ produced by the radiation emitted from the antenna is a function of the amplitude and the phase of the current distribution across the aperture. $E(\theta, \phi)$ may be found by vectorially adding the contribution from the current elements across the aperture. The mathematical summation of all the contributions from the current elements contained within the aperture gives the field intensity in an integral form. This integral cannot be evaluated in the general case. However, approximations to the solution are available by dividing the area about the antenna aperture into three regions as determined by the mathematical approximations that must be made. The demarcations between these three regions are not sharp and blend one into the other as shown in Fig. 6.2.

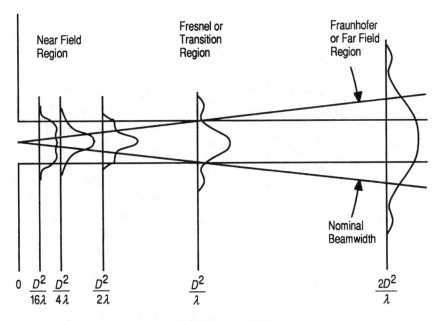

Figure 6.2 Transition from near-field to far-field radiation pattern.

The region in the immediate neighborhood of the aperture is the near field. It extends several antenna diameters from the aperture and, for this reason, is important for antenna cavity/radome interactions. Ray tracing is often useful here, but it is not definitive. The near field is followed by the Fresnel region. In the Fresnel region, rays from the radiating aperture to the observation point, receiving antenna, or target are not parallel, and the antenna radiation pattern is not constant with distance. The farthest region from the aperture is the Fraunhofer, or far-field, region. In the Fraunhofer region, the radiating source and the observation point are at a sufficiently large distance from each other so that the rays originating from the aperture may be considered parallel to one another at the target or observation point.

The "boundary" R_F between Fresnel and Fraunhofer regions is usually taken to be either the distance $R_F = D^2/\lambda$ or $R_F = 2D^2/\lambda$, where D is the width of the aperture in the observation plane, and λ is the wavelength. D and λ obviously are measured in the same units. At a distance given by D^2/λ, the gain of a uniformly illuminated antenna is 0.94 that of the Fraunhofer gain at infinity. At a distance of $2D^2/\lambda$, the gain is 0.99 that at infinity (phase curvature is 22.5°).

It is often true that the radiation pattern can be calculated from the aperture field, assuming scalar diffraction as shown in Figure 6.3. The plot of the electric field intensity, $|E(\theta, \phi)|$, is called the field-intensity pattern of the antenna. The plot of the square of the field intensity $|E(\theta, \phi)|^2$ is the power radiation pattern, $P(\theta, \phi)$, defined in the previous section. The scalar diffraction field is given in Equation (6.9) below.

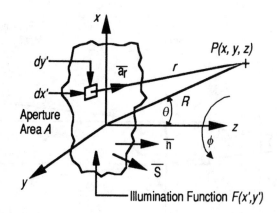

Figure 6.3   Calculation of radiation pattern in terms of aperture field.

$$E(\theta,\phi) = \frac{1}{4 \cdot \pi} \iint_{area} \left[\begin{array}{c} F(x',y') \cdot \dfrac{\exp(-j \cdot k \cdot r)}{r} \\ \cdot [(j \cdot k + 1/r) \cdot \vec{n} \bullet \vec{a}_r + j \cdot k \cdot \vec{n} \bullet \vec{s}] \end{array} \right] \cdot dx' dy'$$

(6.9)

where

$k = 2 \cdot \pi/\lambda$

$\vec{n}$ = the aperture normal

$\vec{a}_r$ = the unit vector along the line of sight, r, between the differential area $dxdy$ and the observation point $P(x,y,z)$

r = the distance between the differential area and the observation point $P(x,y,z)$

θ = the angle from the z axis to the observation point

ϕ = the rotation angle about the z axis to the observation point from the x-z plane

$F(x',y')$ = the phase and amplitude aperture distribution

$\exp(-j \cdot k \cdot r)/r$ = the phase and amplitude change to the observer

$(j \cdot k + 1/r) \cdot \vec{n} \bullet \vec{a}_r$ = the projected area of $dx \cdot dy$ and the rate of phase change in the r direction

$j \cdot k \cdot \vec{n} \bullet \vec{s}$ = the projected phase taper across the aperture

Equation (6.7) and most of the equations that follow in this section contain an implicit $\exp(j \cdot \omega \cdot t)$ factor, which represents the actual carrier frequency. Because it is everywhere, it is left out for clarity.

In the far field, the radiation pattern, $E(\theta, \phi)$, can be described approximately by the inverse Fourier transform of the aperture distribution. In the Fraunhofer or far-field region, the integral for electric field intensity in terms of current distribution across the aperture is given by a Fourier transform relation shown in Equation (6.10). The approximations that allow the simplified calculations of Equation (6.10) for the scalar diffraction field of Fig. 6.3 are given in Table 6.1. Although this set of approximations doesn't provide a completely accurate antenna pattern, it is usually quite good. It also allows the use of the whole body of Fourier transform methods to provide understanding of antenna synthesis and performance degradation.

Table 6.1 Far-Field Pattern Approximations

Extent of the aperture is small compared to range.

The aperture is large compared with λ.

Aperture field approximates a plane wave, i.e., only small phase variations across the aperture.

The difference between R and r is accounted for only in the phase term $\exp(-j \cdot k \cdot r)$ and a first-order expansion is used for r.

The radiated field of interest is restricted to the region near the antenna mainbeam axis.

Far-field $R > 2D^2/\lambda$, D = maximum aperture dimension.

The far-field pattern approximation is

$$E(\theta,\phi) \cong \frac{j \cdot \exp(-j \cdot k \cdot R)}{\lambda \cdot R} \cdot \iint_{area} \left[\begin{matrix} F(x',y') \cdot \\ \exp[j \cdot k \cdot (\sin\theta \cdot (x' \cdot \cos\phi + y' \cdot \sin\phi))] \end{matrix} \right] \cdot dx'dy'$$

(6.10)

Normally, three intermediate variables (u, v, w) are defined as shown in Equation (6.11).

$$u = \sin\theta \cdot \cos\phi, \quad v = \sin\theta \cdot \sin\phi, \quad w = \cos\theta$$

(6.11)

Substituting u and v, it becomes obvious that the integral of Equation (6.10) is just the two-dimensional Fourier transform of the aperture illumination.

$$E(\theta,\phi) \cong \frac{j \cdot \exp(-j \cdot k \cdot R)}{\lambda \cdot R} \cdot \iint F(x',y') \cdot \exp[j \cdot k \cdot (x' \cdot u + y' \cdot v)] \cdot dx'dy'$$

(6.12)

For example, consider the rectangular aperture and coordinate system shown in Fig. 6.3. The two cardinal surfaces of the far-field pattern are the xz and yz planes. The height of the aperture in the x dimension is a, and the angle measured from the z axis is θ and the angle around the z axis measured from the xz plane is ϕ. The far-field intensity in the xz plane ($\phi = 0$), assuming $a \gg \lambda$, is

$$E(\theta,\phi) = \int_{-a/2}^{+a/2} F(x') \cdot \exp\left(j \cdot \sin(\theta) \cdot \frac{2 \cdot \pi \cdot x'}{\lambda}\right) \cdot dx' \tag{6.13}$$

where $F(x')$ = current on the aperture at distance x', assumed to be flowing in the y' direction. $F(x')$, the aperture distribution, may be written as a complex quantity, including both the amplitude and phase distributions, as follows:

$$F(x') = |F(x')| \cdot \exp(j \cdot \xi(x')) \tag{6.14}$$

where
  $|F(x')|$ = amplitude distribution
  $\xi(x')$ = phase distribution

Equation (6.13) represents the summation, or integration, of the individual contributions from the current distribution across the aperture according to Huygens' principle.[1] At an angle θ, the contribution from a particular point on the aperture will be advanced or retarded in phase by $2 \cdot \pi \cdot \sin(\theta) \cdot (x'/\lambda)$ radians. Each of these contributions is weighed by the factor $F(x')$. The field intensity is the integral of these individual contributions across the face of the aperture. For $F(x')$ equal to a constant, the integral of Equation (6.13) is obviously

$$E(\theta,\phi = 0) = \frac{4 \cdot \pi \cdot a \cdot b}{\lambda^2} \cdot \left[\frac{\sin(\pi \cdot \sin(\theta) \cdot a/\lambda)}{\pi \cdot \sin(\theta) \cdot a/\lambda}\right] = \frac{4 \cdot \pi \cdot a \cdot b}{\lambda^2} \cdot \operatorname{sinc}(U) \tag{6.15}$$

One way of thinking about Equation (6.15) and most aperture and RCS patterns is that the area determines the maximum amplitude, and the edges determine the shape of the pattern. The function in the brackets in Equation (6.15) occurs so often that it has its own symbol $\operatorname{sinc}(U)$ where, in this case, $U = (\pi a / \lambda)\sin(\theta)\cos(0)$. Another important observation about the aperture in Fig. 6.4 is that there must be a discontinuity at the edges, sometimes called the *Gibbs phenomenon*. That edge phenomenon has been modeled in various ways; the simplest is to assume an independent wire loop traversing the perimeter of the aperture (more about this later in the chapter).

The aperture distribution can be defined in terms of the current i_y. It may also be defined in terms of the magnetic field component H_x for polarization in the y direction, or in terms of the electric field component E_x for polarization in the x direction, provided these field components are confined to the aperture.[3]

---

[1] Every incremental element on a wavefront gives rise to a secondary spherical wavefront, which when superimposed gives the field at the observation point

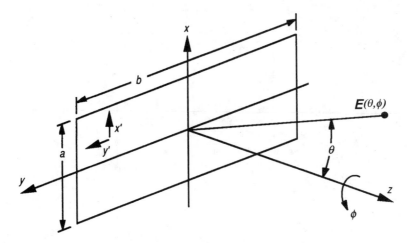

Figure 6.4 Aperture geometry for far-field calculation in cardinal plane.

6.3   SINGLE RADIATORS

6.3.1   The Electric Dipole (after Radiation Laboratories[2])

In the preceding section, a radiation field arising from an aperture distribution of time-varying currents was described. Now, some small-scale idealized current distributions and their associated electromagnetic fields will be discussed. These elements are useful models in real low-sidelobe and low-RCS antennas.

The simplest form of idealized radiator is the electric dipole shown in Figure 6.5. A dipole consists mathematically of a pair of equal and opposite charges, each of magnitude q, separated by an infinitesimal distance δ. If a vector $\vec{\delta}$ is directed from $-q$ to $+q$, then the dipole moment is a vector defined as

$$\vec{\mathbf{p}} = q_0 \cdot \exp(j \cdot \omega \cdot t) \cdot \vec{\delta} = I_0 \cdot \ell \cdot \frac{\exp(j \cdot \omega \cdot t)}{j \cdot \omega} \cdot \vec{\delta} \qquad (6.16)$$

where
I_0 = maximum rate of change of charge, q
ℓ = length of the dipole

The closest real-life example of such an antenna is the AM radio antenna embedded in the windshield in some cars. An antenna equivalent to a dipole also is shown in Figure 6.5. It consists of thin wires terminated in small spheres, and the assumed dimensions are very small as compared with a wavelength. The spheres form the capacitive element of the structure, and the charge at any instant can be considered localized to them. If the antenna is energized by RF applied across the

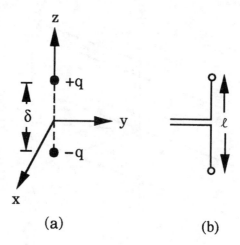

(a)   (b)

Figure 6.5 The electric dipole: (a) mathematical dipole and (b) dipole antenna.[2]

gap at the center, the charges on the spheres are given by the magnitude of the dipole moment q_0, which corresponds to the current flowing in the infinitesimal antenna of $I_0 \ell / j\omega$.

The electromagnetic field set up by a dipole can be described in spherical coordinates as shown in Fig. 6.6. The components of the **E** and **H** fields are as shown in the figure. Because the dipole is infinitesimal, the equations that follow are good every-

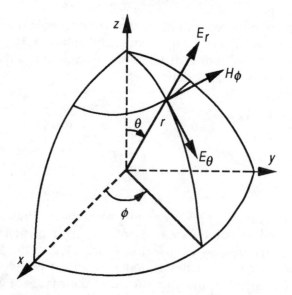

Figure 6.6 Geometry for electric dipole oriented along the z axis.

where not on the dipole. Although Silver references Stratton for the derivation of the dipole fields, a more accessible text is *Antenna Theory*, by Balanis.[4] The Equations (6.17) stated below are from that reference. The one inconvenient fact in this description is that arrays of dipoles usually want the z coordinate to be normal to the dipole not parallel with the dipole axis.

Assuming that the dipole is at the origin and oriented along the z axis, and dropping the exp($j\omega t$) because it is in all terms, then the far-field pattern in spherical coordinates is given in Equation (6.17).

$$E_r = \frac{\eta \cdot I_0 \cdot \ell \cdot \cos(\theta) \cdot \exp(j \cdot k \cdot r)}{2 \cdot \pi \cdot r^2} \cdot \left[1 + \frac{1}{j \cdot k \cdot r}\right]$$

$$E_\theta = \frac{j \cdot k \cdot \eta \cdot I_0 \cdot \ell \cdot \sin(\theta) \cdot \exp(-j \cdot k \cdot r)}{4 \cdot \pi \cdot r} \cdot \left[1 + \frac{1}{j \cdot k \cdot r} - \frac{1}{(k \cdot r)^2}\right]$$

$$H_\phi = \frac{j \cdot k \cdot I_0 \cdot \ell \cdot \sin(\theta) \cdot \exp(-j \cdot k \cdot r)}{4 \cdot \pi \cdot r} \cdot \left[1 + \frac{1}{j \cdot k \cdot r}\right]$$

$$H_r = H_\theta = E_\phi = 0$$

$$(6.17)$$

where the propagation constant, k, and the intrinsic impedance, η, are defined in Equations (6.18).

$$k^2 = \omega^2 \cdot \mu \cdot \varepsilon = (2 \cdot \pi/\lambda)^2 \text{ and } \eta = \sqrt{\mu/\varepsilon}$$

and the free-space impedance is $\eta_0 = \sqrt{\mu_0/\varepsilon_0} = 377 \ \Omega$ $\qquad$ (6.18)

As a consequence of the axial symmetry of the radiator, the field of the dipole is independent of ϕ. The dipole is a true point source, because the equiphase surfaces are spheres with centers at the origin; it is directive, because the intensity of the field varies with the direction of observation. In design specifications, it is customary to characterize such cuts in the three-dimensional polar diagram by two widths if they exist; the "half-power width," θ_{3dB}, which is the full angle in that cut between the two directions in which the power radiated is one-half the maximum value; and the "tenth-power width," θ_{10dB}, the angle between the directions in which the power radiated is one-tenth of the maximum. These two values and their counterparts in u, v space from Equation (6.9) are useful for estimating mainlobe and sidelobe antenna performance.

In the far field, $kr \gg \ell$, the Equations of (6.17) reduce to

$$E_\theta = \frac{j \cdot k \cdot \eta \cdot I_0 \cdot \ell \cdot \sin(\theta) \cdot \exp(-j \cdot k \cdot r)}{4 \cdot \pi \cdot r}$$

$$H_\phi = \frac{j \cdot k \cdot I_0 \cdot \ell \cdot \sin(\theta) \cdot \exp(-j \cdot k \cdot r)}{4 \cdot \pi \cdot r}$$

$$E_r \approx H_r = H_\theta = E_\phi = 0$$

(6.19)

The gain function of the dipole when oriented along the z axis is a shown in Equation (6.20). The gain is maximum at $\theta = 90°$. (The formula is slightly more complex for other orientations, but nothing has changed but the coordinates.)

$$G_e(\theta) = \frac{3}{2} \cdot \sin^2(\theta)$$

(6.20)

The power pattern of the dipole is independent of ϕ and is toroidal shaped. It can be represented by a normalized cut in any one plane containing the z axis, as shown in Fig. 6.7. Up to $\lambda/4$ dipole length, the θ_{3dB} beamwidth changes only by 3° from 90° to 87° and from $\lambda/4$ to $\lambda/2$ dipole length; the θ_{3dB} beamwidth changes only by 9° from 87° to 78°, so the small dipole is an excellent model for many real cases.

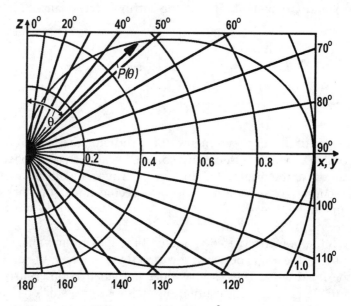

Figure 6.7 Power pattern of an electric dipole in cardinal plane.[2]

6.3.2 The Magnetic Dipole or Small Loop

The magnetic counterpart of the electric dipole antenna is a current loop with its radius small as compared with the wavelength as shown in Fig. 6.8. Such a current loop is equivalent to a magnetic dipole along the axis normal to the plane of the loop. As before, this vector will be assumed along the z axis in a spherical coordinate system (Fig. 6.6). As will be seen, the electric dipole E field in θ corresponds to the magnetic dipole H field in θ, and the electric dipole H field in ϕ corresponds to the magnetic dipole E field in ϕ. The complementary nature of the fields is useful, as it can simplify analysis and help visualization, e.g., replacing slots with dipoles and vice versa.

$$H_r = \frac{j \cdot k \cdot a^2 \cdot I_0 \cdot \cos(\theta) \cdot \exp(j \cdot k \cdot r)}{2 \cdot r^2} \cdot \left[1 + \frac{1}{j \cdot k \cdot r}\right]$$

$$H_\theta = \frac{-(k \cdot a)^2 \cdot I_0 \cdot \sin(\theta) \cdot \exp-(j \cdot k \cdot r)}{4 \cdot r} \cdot \left[1 + \frac{1}{j \cdot k \cdot r} - \frac{1}{(k \cdot r)^2}\right]$$

$$E_\phi = \frac{\eta \cdot (k \cdot a)^2 \cdot I_0 \cdot \sin(\theta) \cdot \exp(-j \cdot k \cdot r)}{4 \cdot r} \cdot \left[1 + \frac{1}{j \cdot k \cdot r}\right]$$

$$E_r = E_\theta = H_\phi = 0$$

$$(6.21)$$

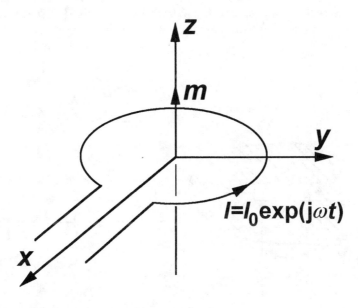

Figure 6.8 Magnetic dipole and equivalent current loop.[2]

The current in the loop is assumed to be a complex sinusoid with peak value I_0, just as for the electric dipole. The counterpart to the electric dipole length is the loop radius, a. The field components are given in Equations (6.21). The far-field approximations for $kr \gg 1$ should be obvious from the electric dipole Equations (6.19). Again, out to a dipole radius of $a = \lambda/5$, the beamwidth is almost unchanged. The far-field patterns of noncircular loops have almost the same field equations as the circular loop as long as the loop enclosed area is the same. This is also true of the loop as a scatterer.

6.3.3   Slot Radiators (after Blass[4])

The next important type of radiating element (or scatterer) is the small slot radiator. The metal surfaces in which the slots are cut will be large as compared with a wavelength, but the slots themselves are less that one wavelength in extent. Such slots may be excited by a cavity placed behind it, through a waveguide, or by a transmission line connected across the slot.

The simplest example of such a radiator consists of a rectangular slot cut in an extended thin flat sheet of metal with the slot free to radiate on both sides of this sheet, as shown in Fig. 6.9. The slot is excited by a voltage source from a transmission line connected to the opposite edges of the slot. The electric-field distribution in the slot can be obtained from the relationship between slot radiators and complementary dipole radiators. It has been shown that the electric-field distribution (magnetic current) in the slot is identical to the electric-current distribution on a complementary dipole. In the case of the rectangular slot of Fig. 6.9, the electric field is perpendicular to the long dimension, and its amplitude vanishes at the ends of the slot.

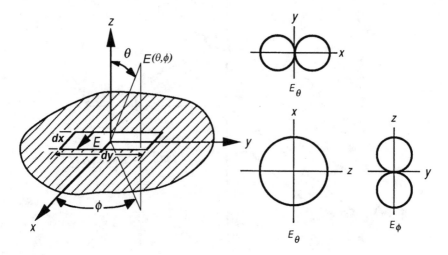

Figure 6.9   Principal plane field diagrams for thin rectangular slot (adapted from Blass[4]).

The electric field is everywhere normal to the surface of the slot antenna except in the region of the slot itself. The radiation of the currents in the sheet can be deduced directly from the distribution of the electric field in the slot. Consequently, the radiated field of an elementary magnetic dipole within the slot boundaries should include the contributions of the electric currents flowing on a metal surface. The field can be thought of as toroidal (donut) shaped with axis through the long dimension of the slot as suggested by the three cuts in Fig. 6.9.

Of course, most slot radiators are not free to radiate on both sides of the surface on which the slot is cut, because one side is either completely enclosed, e.g., the slotted waveguide antenna. In these cases, the influence of the enclosed cavity region on the excitation and impedance of the slot radiator is significant to the design.

6.3.4 Small Rectangular Slot in Infinite Ground Plane

The theoretical properties of a radiating slot in a flat sheet can be obtained from Booker's extension of Babinet's principle, which shows that the field on the slot can be deduced from those surrounding a dipole of the same dimensions by interchanging the electric and magnetic vectors. Alternatively, the field can be found from the equivalence principle. The two-sided radiation field of a small rectangular slot such as shown in Fig. 6.8 is given by

$$E_\theta = -j \cdot E_x \cdot \frac{\cos(\phi) \cdot dx \cdot dy \cdot \exp(-j \cdot k \cdot r)}{2 \cdot r \cdot \lambda}$$

$$E_\phi = j \cdot E_x \cdot \frac{\cos(\theta) \cdot \sin(\phi) \cdot dx \cdot dy \cdot \exp(-j \cdot k \cdot r)}{2 \cdot r \cdot \lambda}$$

$$(6.22)$$

where E_x is the x component of the electric field in the slot, and it is assumed that the electric field is parallel to the x axis, dx is the slot dimension in the x direction, and dy is the slot dimension in the y direction. The principal plane radiation patterns of this magnetic-current dipole also are shown in Fig. 6.9. It is seen that the radiation pattern in the xz plane is omnidirectional, and the pattern in the yz plane varies as $\cos(\theta)$. Note that the phase of the radiated field reverses on the two sides of the ground plane even though the amplitude patterns are identical.

6.3.4.1 Near Half-Wave Radiating Slot in Infinite Ground Plane. A rectangular slot cut in a flat sheet of metal will be resonant when it is approaching half wavelength. As in the case of the complementary dipole, the magnetic-current distribution for the thin slot is approximately cosinusoidal. The far-field radiation pattern of the near half-wave slot is given in Equations (6.23).

$$E_\theta = C_r \cdot \cos(\phi) \cdot \frac{\cos(V)}{V^2 - \pi^2/4} \cdot \frac{\sin(U)}{U}$$

$$E_\phi = -C_r \cdot \cos(\theta) \cdot \sin(\phi) \cdot \frac{\cos(V)}{V^2 - \pi^2/4} \cdot \frac{\sin(U)}{U}$$

$$E_r = H_r = 0 \text{ and } H_\phi = E_\theta/\eta, \quad H_\theta = -E_\phi/\eta$$

$$U = \frac{\pi \cdot a}{\lambda} \sin(\theta) \cdot \cos(\phi) = \frac{a \cdot k \cdot u}{2} \text{ and}$$

$$V = \frac{\pi \cdot b}{\lambda} \sin(\theta) \cdot \sin(\phi) = \frac{b \cdot k \cdot v}{2}$$

$$C_r = \frac{-j \cdot a \cdot b \cdot E_0 \cdot \exp{-(j \cdot k \cdot r)}}{2 \cdot r \cdot \lambda}$$

$$(6.23)$$

where $E_x = E_0 \cos(\pi\, y'/b)$ is the field strength of the cosine distribution in the slot used in Equation (6.22) to obtain Equation (6.23), and a is the x dimension and b is the y dimension of the slot. Of course, in all these equations, $\exp(j\omega t)$ has been dropped, because it is in all terms. For the case of a half-wave slot (i.e., $b = \lambda/2$ and $a \ll \lambda$), the cardinal plane cuts are

$$E_\phi(\phi = 0°) = E_\phi(\phi = 90°) = 0 \text{ and } E_\phi(\phi = 0°) = C_r$$

$$E_\phi(\phi = 90°) = \frac{-4 \cdot C_r}{\pi^2} \cdot \sec(\theta) \cdot \cos\left(\frac{\pi}{2} \cdot \sin(\theta)\right)$$

$$(6.24)$$

This corresponds to a θ_{3dB} beamwidth of 78°, the same as the electric dipole. There is one difference, however, which is that the slot radiator is usually single sided and fed by a waveguide network on the other side of the ground plane. This means that the peak gain relative to isotropic can be as high as 3. Thus, the slot or counterpart dipole gain is shown in Equation (6.25).

$$G_{e,\lambda/2}(\theta) = 1.64(\text{or } 3.28) \cdot \sec^2(\theta) \cdot \cos^2\left(\frac{\pi}{2} \cdot \sin(\theta)\right)$$

$$(6.25)$$

6.3.4.2 Slot Near Field. The near field of a slot radiator is useful in determining the coupling of radiators to each other. This is a common problem with stealth antennas, because element spacing is usually close. Although the radiation pattern of a slot appears to depend only on the slot shape in the far field, other slots in the near field can change the response dramatically. The near-field terms attenuate rapidly but are significant within one wavelength of the source. Experimental work carried out by Putnam, Russell, and Walkinshaw and reported in Blass[4] shows typical field distributions near a slot. Figure 6.10 illustrates the symmetrical measured electric field along the plane of the sheet. At a radiator spacing of 0.4λ, the field from an adjacent radiator is roughly 1/2 of the peak value, which has a major effect on the impedance and fields from each radiator. There are certain high off-angle conditions in which a combination of the mutual coupling term and a surface wave term can cause burnout in components in a radiator. Figure 6.10 is a plot of the phase front on the right side and amplitude distribution on the left side of a half-wave slot at resonance.

One can easily see in Fig. 6.10 that, at the radiator spacings of less than $\lambda/2$ common for high-gain antennas, there will be significant power at high phase shift at the adjacent slots. This coupling changes the apparent impedance seen at the next slot. If the antenna beam is canted from the normal (typical of stealth antennas), the impedance and power change is more dramatic. The good news is that the coupling can be used to peak the slot gain off normal and compensate for the loss in projected area caused by the beam cant in the first place. The bad news is that coupled and internally reflected power can damage front-end components if not carefully protected.

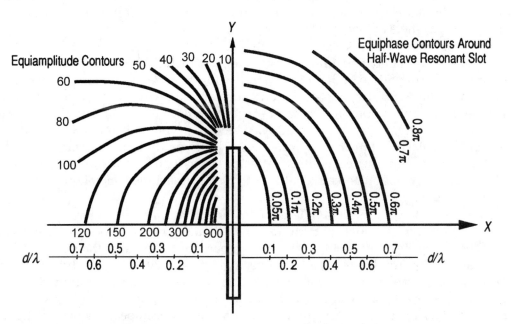

Figure 6.10 Phase and amplitude contours at half-wave slot (adapted from Blass[4]).

6.3.5 Broadband Radiators

The three radiator types just described are used extensively in radar and data link antennas. Unfortunately, the radiation impedance changes dramatically with wavelength for these simple types. For simple radiators, practical bandwidths that have good stealth features are limited to roughly 10 percent. A stealth antenna needs a good broadband match to free space both in and out of band. (If the match were perfect, an antenna would have no RCS!) A narrowband radiator on a stealth antenna places a greater burden on the radome. A number of broadband radiators have been explored for stealth antennas and are summarized here.[5] Figure 6.11 shows several broadband radiator types, including their typical bandwidth, beamwidth, size, and polarization capabilities. The size is stated in terms of the wavelength of the lowest design frequency, λ_L. The lowest design frequency for a stealth antenna usually is below the lower cutoff of the radome. Furthermore, there is often a rapid and not easily controllable phase change at the cutoff of both the radiator and the radome that can enhance the RCS unless carefully treated.

Figure 6.11 is adapted from C. G. Buxton.[5] One will notice that most broadband radiators typically are $\lambda_L/2$ in size at the lowest design frequency, which makes them vulnerable at the highest cutoff of the radome. So the best compromise is usually a radiator that allows less than 0.4 λ_L spacing, and the rest of the above band RCS control must be done with the radome or atmospheric resonances. Good RCS performance also demands very high reproducibility element to element. So, although high dielectric constants allow closer spacing, control of electrical dimensions dominates and often forces designers to machined elements and air dielectric.

Many of these radiators are specifically adapted to printed wiring board manufacture. Unfortunately, the lithography, water permeability, and dielectric constant of printed wiring board substrates generally do not have tight enough tolerances to support high-performance stealth antennas at microwave frequencies. Individual electrical dimensions for radiators usually must be reproducible to 0.0001λ over temperature, time, and humidity, which is a very tall order for anything made from plastic. With heroic measures, plastic or dielectric radiators can be manufactured, but the economies expected from batch fabrication evaporate. These facts have often forced stealth antennas to machined metal/air dielectric configurations.

Figure 6.12 shows a machined flared notch or Vivaldi radiator pattern in comparison to several typical dipole patterns (and you thought the "red priest" only worked on "The Seasons"). Note that the flared notch patterns are broader than all single dipoles and are approaching the patterns one might obtain with a double driven dipole. Furthermore, a radiating element embedded in an array has a significantly different pattern from that of an element in isolation. Typically, mutual coupling causes a broader pattern at band center, but it can also cause higher impedance mismatches at some element phasings, which can cause narrower bandwidth.

Radiating Element Type	Element Bandwidth	Beamwidth (degrees)	Electrical Size		Polarization
			Diameter	Height	
Archimedean spiral 	~10:1	75	$0.5\,\lambda_L$	$0.5\,\lambda_L$	Circular
Equiangular spiral 	~8:1	70	$0.5\,\lambda_L$	$0.5\,\lambda_L$	Circular
Sinuous 	~9:1	60-110	$0.4\,\lambda_L$	$0.4\,\lambda_L$	Dual linear
Foursquare	1.8:1	60-70	$0.35\,\lambda_L$	$0.11\lambda_L$	Dual linear
Microstrip patch 	3%	110	$\dfrac{0.49}{\sqrt{\varepsilon}}\,\lambda_L$	$\ll \lambda_L$	Dual linear
Stacked patch 	40%	65	$\dfrac{0.4}{\sqrt{\varepsilon}}\,\lambda_L$	$0.13\,\lambda_L$	Dual linear
Open Ended Ridged Waveguide 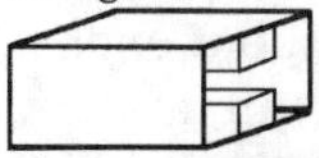	20%	150	$\dfrac{0.4}{\sqrt{\varepsilon}}\,\lambda_L$	0	Linear
Vivaldi or Flared Notch 	3:1	45-85	$\dfrac{1.2}{\sqrt{\varepsilon}}\,\lambda_L$	$\dfrac{2}{\sqrt{\varepsilon}}\,\lambda_L$	Linear
Flared Patch 	4:1	45-85	$\dfrac{0.35}{\sqrt{\varepsilon}}\,\lambda_L$	$0.1\,\lambda_L$	Dual linear

Figure 6.11   Summary of broadband radiators (adapted from Buxton[5]).

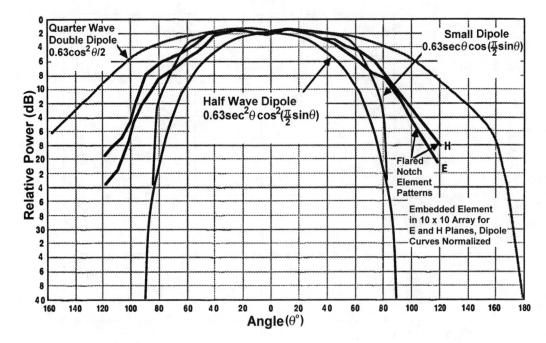

Figure 6.12 Comparison of broadband element with dipole/slot.

6.4 ANTENNA ARRAYS

6.4.1 Simple Apertures

Most stealth data link and radar antennas require high gain and low sidelobes and thus are arrays of radiators, not single elements. For example, consider the rectangular aperture and coordinate system given in Fig. 6.4 and repeated in Fig. 6.13. Recall that this aperture was first introduced in Chapter 2, Fig. 2.21, without explanation to calculate "cookie cutter" footprints. Define the height of the aperture in the x dimension as a, and the width of the aperture in the y dimension as b; the angle θ is measured from the z axis, and the angle ϕ is measured from the xz plane about the z axis. Then, integrating Equation (6.12) with uniform weighting in the x' and y' directions and calculating the equivalent power normalized over 4π steradians will yield the far-field gain. The normalized far-field gain pattern for uniform aperture E-field in the x direction is given in Equations (6.26).

$$G(\theta,\phi) = \frac{4 \cdot \pi \cdot a \cdot b}{\lambda^2} \cdot \frac{\sin^2(U)}{U^2} \cdot \frac{\sin^2(V)}{V^2} \cdot \frac{[1 + \cos(\theta)]^2}{4} \tag{6.26a}$$

or more commonly,

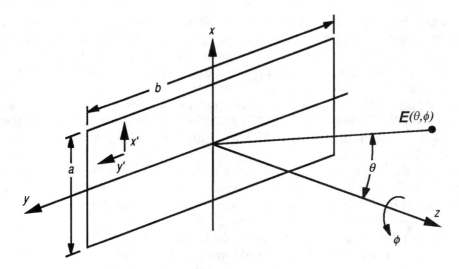

Figure 6.13 The illumination pattern of a rectangular aperture.

$$G(\theta,\phi) \cong \frac{4 \cdot \pi \cdot a \cdot b}{\lambda^2} \cdot \frac{\sin^2(U)}{U^2} \cdot \frac{\sin^2(V)}{V^2} \qquad (6.26b)$$

where

$$U = \frac{\pi \cdot a}{\lambda} \sin(\theta) \cdot \cos(\phi) \text{ and } V = \frac{\pi \cdot b}{\lambda} \sin(\theta) \cdot \sin(\phi)$$

Often, the $\cos(\theta)$ factor in Equation (6.26a) is dropped for pattern plots within 30° of the normal, because other edge effects make the pattern prediction poor at large off angles. In addition, if a purely scalar diffraction assumption (TEM) is made, then the raised cosine factor does not appear in the gain equation or corresponding electric fields for the rectangular aperture. Parallelogram apertures easily allow separable illumination functions that are multiplicative in the intercardinal space. One of the main features of a rectangular aperture is that, although the cardinal plane sidelobes decrease as $1/n^2$, in the intercardinal space, the sidelobes are decreasing as $1/n^4$, where n is the sidelobe number. The property of separable functions in two dimensions also is an advantage, because it limits implementation errors in one dimension from propagating into the other dimension.

The intercardinal spaces have orders of magnitude lower sidelobes because of this multiplicative property. Furthermore, the regular structure limits the effect of manufacturing errors to the dimension in which they occur. This is a very big advantage for a stealthy antenna, because manufacturing tolerances ultimately limit antenna RCS and sidelobe interceptability. A three-dimensional representation of the rectangular uniformly illuminated aperture was given in Fig. 2.13 and is repeated in Fig. 6.14.

The cardinal plane characteristics are summarized in Fig. 6.15. In the cardinal planes, the first sidelobe of such an aperture illumination is approximately only 13 dB down but, off plane, the first sidelobe is down 26 dB, and the second is down 32 dB and so on. The system designer can adjust the location of the high sidelobes so that they are in low-threat directions (away from the horizon, which is usually the highest threat). As seen later in this chapter, this pattern and other gain patterns are closely related to the RCS of these same antenna shapes.

To contrast the differences, next consider a circular aperture. This aperture was also introduced in Chapter 2, Fig. 2.20. The geometry of the illumination pattern is shown in Fig. 6.16, and a radial integration of illumination in ρ' and ϕ' is assumed. As one might expect, the resulting integral is a Bessel function of the form $J_1(x)/x$.

A uniformly illuminated circular aperture has a radiation pattern as shown in Fig. 6.17. Because it is circularly symmetric, any constant ϕ cut is a cardinal cut. A cardinal characteristic also is shown in Fig. 6.17. Although the first sidelobe is more than 17 dB down, that sidelobe is at all angles, ϕ, at constant θ. This characteristic is significantly worse from an interceptability and RCS point of view.

Similar to the uniformly illuminated rectangular aperture, the normalized far-field gain pattern for uniform illumination of a circular disk is given in Equations (6.27). Because z is normal to the disk, there is no ϕ dependence. D_u is the aperture diameter in u space.

$$G(\theta,\phi) \cong \frac{4 \cdot \pi \cdot a^2}{\lambda^2} \cdot \frac{J_1^2(D_u)}{D_u^2} \qquad (6.27)$$

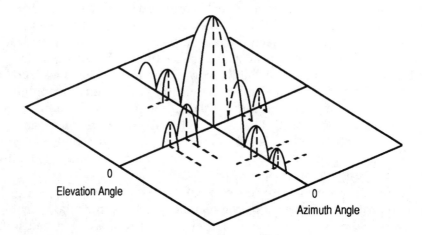

Figure 6.14 Gain pattern of uniformly illuminated rectangular aperture.

$$\textbf{Peak Gain} = \frac{4\pi\,(a \times b)}{\lambda^2} \qquad A = a \times b$$

<u>IN *x,z* Plane</u>

$$E(\theta,\phi = 0) \cong \frac{A \sin\left(\frac{\pi a}{\lambda}\sin\theta\right)}{\frac{\pi a}{\lambda}\sin\theta}$$

<u>IN *y,z* Plane</u>

$$E(\theta,\phi = \pi/2) \cong \frac{A \sin\left(\frac{\pi b}{\lambda}\sin\theta\right)}{\frac{\pi b}{\lambda}\sin\theta}$$

$$\text{3-dB Beamwidth} = 2\sin^{-1}\left(\frac{1.39\,\lambda}{\pi\,(a,b)}\right)$$

$$\cong 0.88\,\frac{\lambda}{(a,b)} \text{ Radians}$$

$$\cong 51\,\frac{\lambda}{(a,b)} \text{ Degrees}$$

1.0
0.1
0.01
0.001
Relative Gain
Half Power Point
U,V = 1.39
13 dB
-17.8 dB
0
4
8
12
U,V = π (a,b) / λ sinθ

Figure 6.15 Cardinal plane—uniformly illuminated rectangular aperture.

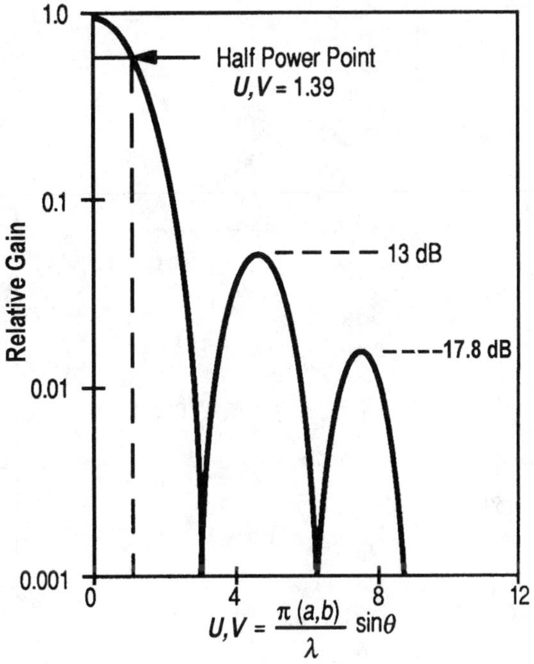

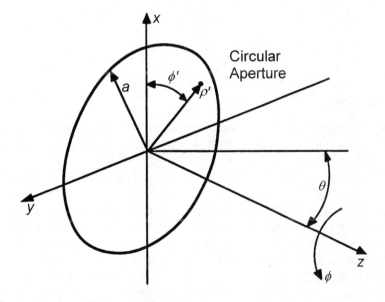

X
Circular
Aperture
a
φ'
ρ'
y
θ
z
φ

Figure 6.16 Illumination geometry for a circular aperture.

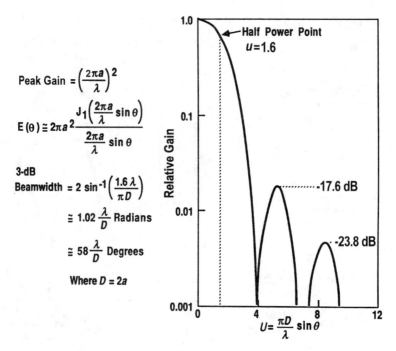

Figure 6.17   Cardinal cut of a uniformly illuminated circular aperture.

where

$$D_u = \frac{2 \cdot \pi \cdot a}{\lambda} \sin(\theta)$$

6.4.2   Sidelobe Reduction Functions

Obviously, the sidelobes of the simple arrays above are not adequate for stealth antennas. Although the use of parallelogram shapes and separable illumination functions aid in reducing sidelobes, more is required to achieve a stealthy antenna. Amplitude weighting across the aperture is used to reduce sidelobes at the expense of mainlobe gain. The amplitude weighting function also needs to be robust in the sense that small errors will not destroy the desired performance, and the weighting values are achievable with real hardware. Table 6.2 summarizes a number of well known weighting functions and compares them with an unweighted aperture (square). All of the apertures used for the table are square with a side equal to 25 wavelengths, with element spacing of 1/2 wavelength, and element pattern approximately equal to a 1/2-wave dipole imbedded in a large array. The degradation loss is the relative cookie cutter gain loss, i.e., $20 \log(BW_{3dB} \times Coherent\ Gain)$. The *ISLR* is as previously defined, except

2.3U is an arbitrary definition of the sidelobes, roughly the equivalent to the 18-dB down point. From a stealth point of view, *ISLR* and mainlobe gain/beamwidth (BW_{3dB}) are the most important parameters, as mentioned before.

Table 6.2   Antenna Weighting Functions

Name	Norm. BW_{3dB}	PSLR, dB	Sidelobe rolloff, dB/Oct.	Coherent gain	Degradation loss, dB	ISLR– 2.3U_{3DB}, dB
Rectangular	1.0	−13.3	−6	1.0	0.0	−10
Bartlet	1.44	−26.5	−12	0.5	−1.25	−23.2
Hanning	1.63	−31.5	−18	0.5	−1.76	−21.9
Hamming	1.48	−42.6	−6	0.54	−1.34	−22.4
Cosine^3	1.81	−39.3	−24	0.42	−2.39	−21.9
Blackman	1.86	−58.1	−18	0.42	−2.37	−23.7
Parzen	2.06	−53.1	−24	0.38	−2.83	−22.6
Zero Sonine	1.24	−24.7	−9	0.59	−2.71	−24.6
Mod. Taylor	1.22	−28	−6	0.7	−1.37	−22.2

Recall from Fig. 1.12 that the maximum safe range to prevent mainlobe intercept increases with antenna gain, but low sidelobes and ISLR are required to keep the intercept footprint/skyprint below the interceptor detection threshold.

If the aperture weighting function is symmetrical and amplitude only, then it can be represented by terms in a Fourier cosine series. These terms when transformed to the far-field (spacial frequency domain) are paired delta functions. For example, consider the amplitude weighting and its counterpart Fourier transform of Equation (6.28).

$$w(x) = \sum_i a_i \cdot \cos(x \cdot i \cdot \gamma) \Rightarrow W(\theta) = \sum_i \frac{a_i}{2} \cdot \delta(\theta \pm i \cdot \gamma) \tag{6.28}$$

For obvious reasons, these are called *paired echoes* because, for every weight, there are two impulses. Amplitude weighting of this form on some other illumination function at the aperture results in pairs of beams shifted in angle and summed with the mainbeam in the far field. The far-field pattern is of the form shown in Equation (6.29).

$$E(x) \cdot w(x) \Rightarrow a_0 \cdot E(\theta) + \frac{a_1}{2} E(\theta - \gamma) + \frac{a_1}{2} E(\theta + \gamma) + \dots \tag{6.29}$$

One of the weighting functions mentioned in the table is the Hamming weighting function. Figure 6.18 provides an explanation of that weighting using paired echo theory descriptions (this also applies to modified Taylor and Hanning weighting). The amplitude weighting can be thought of as the sum of three terms: the unweighted response, w_R, and two replica responses on either side of the mainlobe weighted by a suitable coefficient (for convenience, some of the weighting may be applied to the main response).

Raised cosine weighting of the type in Fig. 6.18 actually represents a family of simple weightings, which includes Hanning ($a = 0.5$, which is $\cos^2 x$), Hamming ($a = 0.54$), and modified Taylor. Figure 6.19 shows the idealized 25λ aperture far-field patterns for continuous weighting in the cardinal plane for Hamming and another cosine based amplitude weighting, cosine cubed ($\cos^3 x = 0.25 \cos(3\pi\, n/N) + 0.75 \cos(\pi\, n/N)$). Cosine cubed weighting has two sets of paired echoes, and ultimate sidelobes are much lower. Unfortunately, cosine cubed weighting, which works well for signal processing, is probably not realizable in a producible antenna, not to mention the mainlobe gain loss.

A summary of the effects of varying a is shown in Fig. 6.20. One caution is in order: as the weighting at the edge of the aperture approaches zero, the difficulty of implementation goes to infinity. Usually, the minimum edge weighting is greater than 0.1 for reproducible hardware. The figure also shows the definition of the integrated sidelobe region used to compute *ISLR*.

Some example idealized 25λ aperture far-field patterns for Taylor weightings based on Fig. 6.20 are shown in Fig. 6.21. The 53-dB sidelobe case is probably not realizable in most hardware, because the weighting at the edge of the array is quite small and corresponding manufacturing tolerances too expensive.

Another important consideration touched on in Chapter 2 was the percentage of the space covered by the sidelobes that are above some critical threshold. Table 6.3

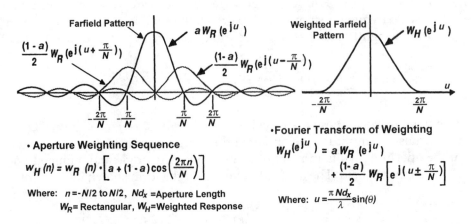

Figure 6.18   Aperture amplitude weighting example.

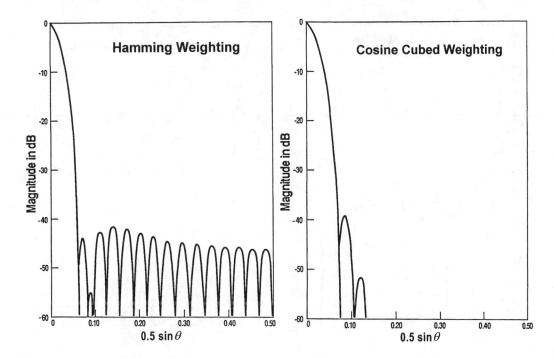

Figure 6.19 Aperture weighting examples.

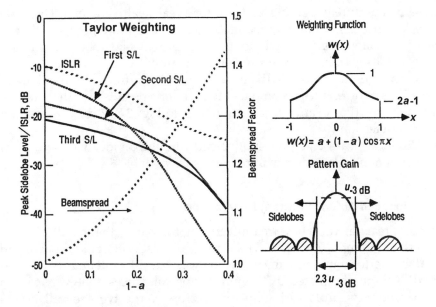

Figure 6.20 Modified Taylor weighting trade-offs.

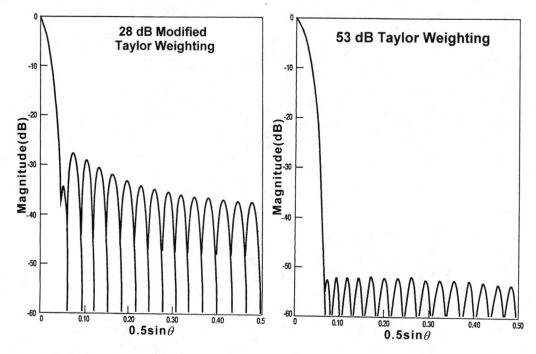

Figure 6.21   Taylor weighting examples.

summarizes an aperture weighting trade-off for a real stealth antenna. The trade-off is not simple, because degradation in mainlobe gain causes more power to be transmitted, increasing mainlobe interceptability while improving sidelobe interceptability. In this case, several separable two-dimensional weighting functions are compared for sidelobe percentages down to –55 dB below the mainlobe, from –55 to –60 dB, and below –60 dB. This performance is compared to mainlobe penalty, which increases interceptability directly. (For example, referring back to Fig. 2.8, 2 dB can easily double probability of detection.)

Figure 6.22 shows various representative cuts through an idealized stealth antenna pattern that might be selected from a trade-off such as Table 6.4.

6.4.3   Error-Induced Antenna Pattern Degradation

Unfortunately, real antennas have manufacturing tolerances that result in some overall rms phase and amplitude error. The errors give rise to degradations in sidelobe and mainlobe performance. In the final analysis, stealth antenna sidelobe performance and RCS are dominated by manufacturing tolerances. One way to represent the effect of random manufacturing tolerance errors on the average radiation pattern is given in Equation (6.30).[1,4,22,31]

Table 6.4 Example Stealth Antenna Trade-Offs[14,15]

Power relative to mainlobe peak (dB)	Percent of half space									
	Uniform weighting		31-dB Taylor weighting	Cosine squared weighting				Triangular weighting		
					Pedestal			Pedestal		
					Discrete			Continuous		
	Cont.	Discrete	Discrete	Cont.	0	0.1	0.316	0	0.1	0.316
$-55 < P \le 0$	8.15	12.82	6.44	0.84	0.98	1.36	4.89	1.64	4.06	5.87
$-60 < P \le -55$	4.09	7.18	2.46	0.24	0.26	1.47	0.94	0.75	1.2	3.05
$P \le -60$	87.76	80	91.1	98.92	98.76	97.17	94.17	97.61	97.74	91.08
One-way gain penalty (dB)	0	0	2.05	3.5	3.5	2.5	1.1	2.5	1.7	0.7

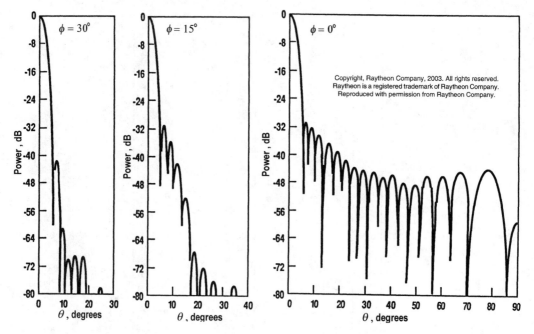

Figure 6.22 Example stealth antenna ideal patterns.[14,15]

$$\bar{G}(\theta,\phi) \cong G_0(\theta,\phi) + G_e(\theta,\phi) \cdot \varepsilon^2 \cdot \frac{\sum_M \sum_N I_{m,n}^2}{\left(\sum_M \sum_N I_{m,n}\right)^2} \qquad (6.30)$$

where

G_0 = "ideal" gain pattern

G_e = element gain pattern

ε^2 = total mean square error per element as defined in Table 6.4

$I_{m,n}$ = current in element m, n in an M by N array of elements

One observation that is always made is that, if manufacturing tolerances are constant, then a large array will have lower relative degradation, because the error gain grows only as $M \times N$. Also, a fixed dimensional tolerance represents a smaller electrical error as wavelength increases. There isn't much joy in this, as larger arrays at lower frequencies require larger platforms, are less mobile, and are subject to gravity and weather-induced distortion. The overall effect of random errors, regardless of source, are summarized in Table 6.4.

Table 6.4 Effects of Uncorrelated Errors

<table>
<tr><td>

Effect of Random Errors

$$\varepsilon^2 = \sigma_A^2 + \sigma_\varphi^2$$

where σ_A^2 = mean square amplitude error (volts/volt)^2 and σ_φ^2 = mean square phase error (radians)^2

</td></tr>
<tr><td>

On Sidelobe Level

$$\varepsilon^2 = G_0 \cdot G_{SL}$$

where G_0 = gain in pointing direction and G_{SL} = relative sidelobe level

</td></tr>
<tr><td>

On Directivity

$$G/G_0 \cong 1/(1 + \varepsilon^2)$$

</td></tr>
<tr><td>

On Pointing Accuracy

$$\Delta\theta/\theta_0 \approx 0.6\sigma_\varphi/(N \times M)^{0.5}$$

where θ_0 = pointing direction and $N \times M$ = number of elements in a 2-D array

</td></tr>
</table>

Variability from manufacturing processes, which is not random but correlated, is focused in the far-field and generally makes a tiny difference in sidelobe interceptability or RCS. On the other hand, uncorrelated errors are unfocused and cover the entire threat space and make a stealth system vulnerable in all directions. This is why features like separable amplitude weighting functions, rectangular or parallelogram antenna shapes, and repetitive assemblies restricted to one dimension are attractive. They typically lead to correlated errors. Although manufacturing dominates performance, there are other kinds of errors that are the result of design decisions, such as phasor quantization error and phase pulling due to mutual coupling.

The sidelobe degradation for the same percentage bands as Table 6.4 and for the design of Fig. 6.22 are shown in Fig. 6.23 as a function of rms phase error. It is obvious from the figure that only one or two degrees of error is allowable before ideal design is overwhelmed by the phase errors. What does this mean for manufacturing tolerances? Consider an antenna at X-band with a wavelength of 1 in and air dielectric; then, 1° electrical is roughly 2.8 mils in length. Because this overall tolerance may be made up of 100 electrical/mechanical features root sum squared together, the individual mechanical tolerance must be $2.8/(100)^{0.5} = 0.28$ mils. This is a manufacturing challenge.

For example, an antenna built to 0.5-mil tolerances based on the design of Fig. 6.22 is shown in Fig. 6.24. The 30° cut shown in the first panel in Fig. 6.22 should be com-

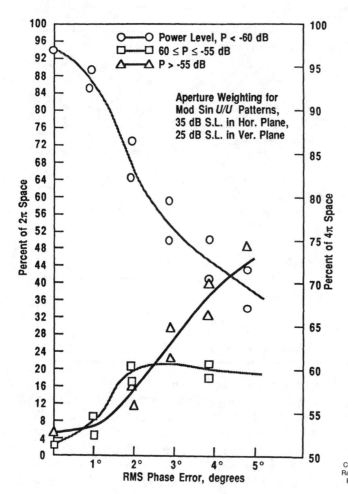

Figure 6.23   Error-induced sidelobe degradation.[14,15]

pared to Fig. 6.24. Note that the agreement is reasonably good down to approximately –50 dB. Beyond that level, sidelobes are dominated by manufacturing tolerances. The sidelobes give every appearance of being totally random (not quite, of course). In addition, such an antenna, when looking through a radome, will have its sidelobes degraded some more. It takes quite a good radome to keep the sidelobes within 3 dB of the values shown in Fig. 6.24.

Regular repetitive/periodic errors are focused and result in sidelobes in specific directions. For example, a four-element periodic error of 10 percent with element spacing of $1/2\lambda$ and total array width of 50 elements with Hamming weighting results in a sidelobe –30 dB below mainlobe peak at approximately 57°.

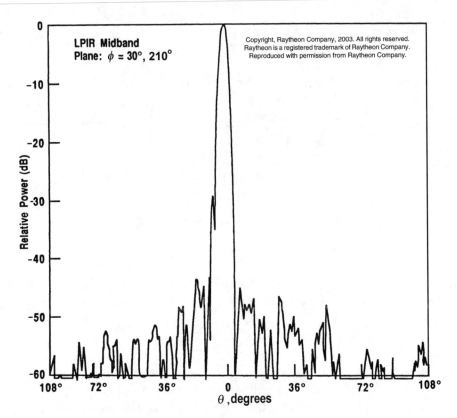

Figure 6.24 Actual realized sidelobe performance (Raytheon[14,15]).

6.4.4 Arrays of Elements

Of course, real arrays are made up of individual elements, not sheets of current. The next topic is the analysis of the radiation patterns of antenna arrays made up of individual elements. The basic assumption is that multiple elements can be linearly superimposed/summed to obtain the far-field pattern. This is almost true if the power isn't too high and scan angle isn't too great—and not even slightly true if it is. Consider an array of the elements in one dimension of the form shown in Fig. 6.25.

The radiation pattern, $E(\theta)$, is the summation of the amplitudes and their associated phases of the N elements in some "look" direction, θ, often called the array factor, multiplied by the individual element factor, E_e, as shown in Equation (6.31).

$$E(\theta) = E_e(\theta) \cdot \left[\sum_{n=0}^{N-1} a_n \cdot \exp\left(\frac{-j \cdot 2 \cdot \pi \cdot n \cdot d \cdot \sin(\theta)}{\lambda}\right) \right] \tag{6.31}$$

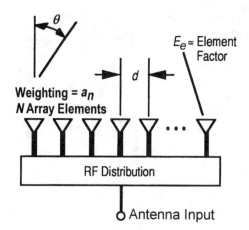

Figure 6.25 Uniformly spaced linear array.

where the variables are as previously defined or as defined in Fig. 6.25. For uniform illumination, the above equation has a closed-form solution as given in Equation (6.32).

$$E(\theta) = N \cdot a_0 \cdot E_e(\theta) \cdot \frac{\sin(\pi \cdot N \cdot d \cdot \sin(\theta)/\lambda)}{N \cdot \sin(\pi \cdot d \cdot \sin(\theta)/\lambda)} \tag{6.32}$$

where a_0 is the weighting of each element. Note that the element factor usually contains a $\cos(\theta)$ factor to account for obliquity. For small θ and large N, Equation (6.32) is approximately sinc(x). Some examples of this approximation are given in Fig. 6.26 for an array of length, $L = Nd$, made up of N elements spaced one-half wavelength, i.e., $d = \lambda/2$, apart. The power plot in Fig. 6.26 in a cardinal cut shows that, for a continuous array, the sidelobes drop off as sinc(x), whereas a finite array with only 10 elements, for example, has sidelobes that drop to a minimum and then begin to rise toward the grating lobe at $\sin(\theta) = 2$, i.e., not in real space. If the element spacing is λ, then the grating lobe would be $\sin(\theta) = 1$, i.e., endfire. Figure 6.26 also shows the relative error for the sinc(x) approximation for a 10-element array with $\lambda/2$ spacing.

Extending these notions for a two-dimensional array with a rectangular spaced grid of elements is shown in Fig. 6.27. A planar array is a system of N by M radiating elements in a linear distribution spaced dx and dy, respectively, as shown. The aperture width is thus M times dy, and the aperture height is N times dx. Note that, if dx and dy got arbitrarily small as N and M got arbitrarily large, such that their product was a constant, the radiation pattern in the limit should be the same as a uniformly illuminated rectangular aperture.

The radiation pattern is approximately equal to the product of the antenna excitation amplitude with the element radiation pattern with the two-dimensional spatial transform of the array. If the array has a uniform distribution, then it has the form shown in Equation (6.33).

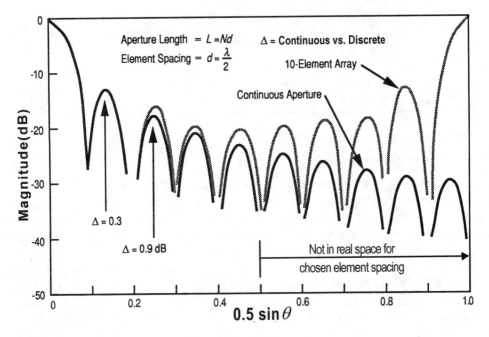

Figure 6.26 Comparison of radiation patterns (after Knittel[8]).

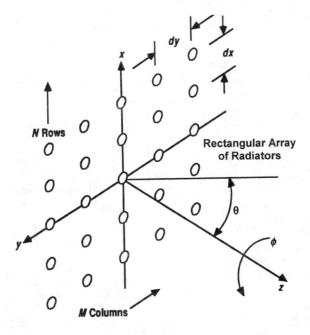

Figure 6.27 Geometry of a rectangular planar array of radiators.

$$E(\theta,\phi) \cong N \cdot M \cdot a_0 \cdot E_e(\theta,\phi) \cdot \frac{\sin(N \cdot U_x)}{N \cdot \sin(U_x)}\frac{\sin(M \cdot V_y)}{M \cdot \sin(V_y)} \tag{6.33}$$

where

$$U_x = \pi \cdot dx \cdot \sin(\theta) \cdot \cos(\phi)/\lambda$$
$$V_y = \pi \cdot dy \cdot \sin(\theta) \cdot \sin(\phi)/\lambda$$

6.5 ELECTRONICALLY SCANNED ARRAYS

6.5.1 Single-Beam Antennas

Electronically scanned arrays are made up of a number of components, most of which are not RF or microwave. Figure 6.28 shows a block diagram of a typical electronically scanned array. It consists of radiating elements, phase shifters, an RF distribution feed network, phase shifter drivers, array logic circuits, a beam-steering computer, power supply, cooling means, built in test and calibration, and miscellaneous sensors for temperature, arcing, failure detection, and so on. The beam-steering computer receives a look direction, the frequency of operation, and the operating mode of the data link or radar, which it uses in conjunction with the temperature of the array and calibration data to generate a set of phase commands to each phase shifter driver. An elaborate logic network distributes these commands to the phase shifter drivers as rapidly as possible. The phase shifter driver, in turn, generates a set of voltages (currents) that cause the phase shifter to produce the nearest phase shift to the commanded value. When the RF reaches the phase shifter through the feed distribution

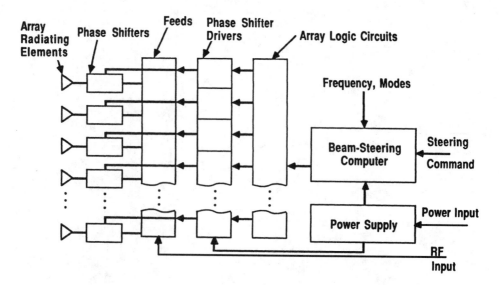

Figure 6.28 Electronically scanned array block diagram.

network, it is phase shifted by the appropriate amount and radiated. On receive, the process is similar. With very broadband arrays, the phase shifters must be spun both on transmit and receive to keep the beam pointed in the commanded direction.[25]

The simplest electronically scanned array has two elements. A two-element scanned array is shown in Fig. 6.29. The elements can be phased such that, for a plane wave propagating from or to the "look" direction, θ_0, the maximum signal response occurs in the direction, θ_0. The relative phase difference between elements, β_0, is adjusted so that the wavefront is focused in the direction θ_0. This requires the phase difference between elements to be $2 \pi d \sin(\theta_0)/\lambda$ as shown for the two-element array in the figure. The radiation pattern equation for such a two-element array in spherical coordinates in the x-z plane ($\phi = 0$) with the z axis normal to the radiators is given in the Fig. 6.29. The 3-D pattern of such an array is conical.

Generalizing this two-element array to a one-dimensional array of N elements, of the type shown in Fig. 6.29, leads to a summation similar to Equation (6.31) modified only by an element phasing term, β_0. A two-dimensional array is obviously similar. Referring to the system diagram in the top portion of Fig. 6.30, one can see the similarity to Fig. 6.25, differing only by a progressive phase shift as shown in Equation (6.34).

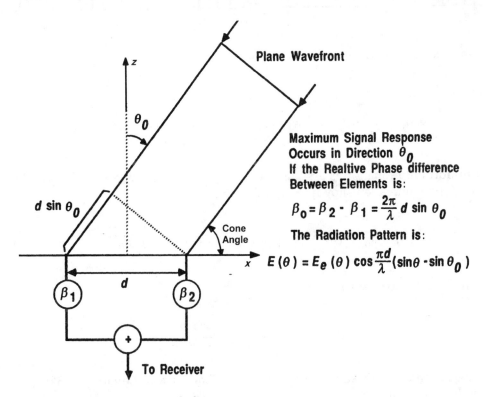

Plane Wavefront

Maximum Signal Response Occurs in Direction θ_0 If the Realtive Phase difference Between Elements is:

$$\beta_0 = \beta_2 - \beta_1 = \frac{2\pi}{\lambda} d \sin \theta_0$$

The Radiation Pattern is:

$$E(\theta) = E_e(\theta) \cos \frac{\pi d}{\lambda}(\sin\theta - \sin\theta_0)$$

To Receiver

Figure 6.29 Phase shifters used to scan the beam.

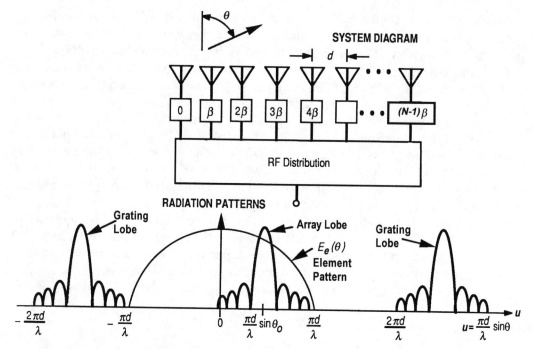

Figure 6.30 Radiation patterns of a linear phase scanned array.

$$E(\theta) = E_e(\theta) \sum_{n=0}^{N-1} a_n \cdot \exp(\,\mathrm{j} \cdot 2 \cdot \pi \cdot n \cdot d \cdot \sin(\theta)/\lambda - \mathrm{j} \cdot n \cdot \beta_0) \tag{6.34}$$

If uniform illumination is assumed, then Equation (6.34) becomes the well known closed form summation shown in Equation (6.35).[6,7]

$$E(\theta) = a_0 \cdot E_e(\theta) \sum_{n=0}^{N-1} [\exp(\,\mathrm{j} \cdot 2 \cdot \pi \cdot d \cdot \sin(\theta)/\lambda - \mathrm{j} \cdot \beta_0)]^n \tag{6.35a}$$

which is a geometric series whose closed form is

$$E(\theta) - a_0 \cdot E_e(\theta) \cdot \frac{1 - [\exp(\,\mathrm{j} \cdot 2 \cdot \pi \cdot d \cdot \sin(\theta)/\lambda - \mathrm{j} \cdot \dot{\beta}_0)]^N}{1 - [\exp(\,\mathrm{j} \cdot 2 \cdot \pi \cdot d \cdot \sin(\theta)/\lambda - \mathrm{j} \cdot \dot{\beta}_0)]} \tag{6.35b}$$

This equation can be restated in the more familiar form of Equation (6.36), which is given in many texts.

$$E(\theta) = E_e(\theta) \cdot a_0 \cdot N \cdot [\exp(j \cdot (N-1) \cdot (\pi \cdot n \cdot d \cdot \sin(\theta)/\lambda - \beta_0/2))]$$

$$\cdot \left[\frac{\sin\left(\dfrac{N \cdot \pi \cdot d}{\lambda}(\sin(\theta) - \sin(\theta_0))\right)}{N \cdot \sin\left(\dfrac{\pi \cdot d}{\lambda}(\sin(\theta) - \sin(\theta_0))\right)} \right]$$

(6.36)

where $\beta_0 = 2\pi \cdot d \sin(\theta_0)/\lambda$, and θ_0 is the "look" direction. If the exact center of the array (whether there is an element there or not) rather than one end is chosen to be zero phase, then $[\exp(j \cdot (N-1) \cdot (\pi \cdot d \cdot \sin(\theta)/\lambda - \beta_0/2)]$ vanishes, and the familiar form of Equation (6.37) emerges.

$$E(\theta) = E_e(\theta) \cdot a_0 \cdot N \cdot \left[\frac{\sin\left(\dfrac{N \cdot \pi \cdot d}{\lambda}(\sin(\theta) - \sin(\theta_0))\right)}{N \cdot \sin\left(\dfrac{\pi \cdot d}{\lambda}(\sin(\theta) - \sin(\theta_0))\right)} \right]$$

(6.37)

The function in the brackets in Equation (6.37) occurs so often that it is convenient to give it a symbol: $\text{sins}(N\theta) = \sin(N\theta)/N\sin\theta$. As was previously pointed out, a $\text{sins}(N\theta)$ distribution gives rise to grating lobes in the radiation pattern as shown in radiation patterns at the bottom of Fig. 6.30. Those grating lobe responses are weighted by the radiation pattern of the individual elements and thus are usually somewhat or very much smaller than the mainlobe pattern.

A generally similar pattern arises from two-dimensional arrays. Recalling Fig. 6.27 and Equation (6.33) and inserting an element to element phase shift yields the rectangular phased array equation of Equation (6.38). The z axis is normal to the plane of the array.

$$E(\theta,\phi) \cong N \cdot M \cdot a_0 \cdot E_e(\theta,\phi) \cdot \frac{\sin(N \cdot (U_x - \beta_0/2))}{N \cdot \sin(U_x - \beta_0/2)} \cdot \frac{\sin(M \cdot (V_y - \beta_1/2))}{M \cdot \sin(V_y - \beta_1/2)}$$

(6.38)

where

$$U_x = \pi \cdot dx \cdot \sin(\theta) \cdot \cos(\phi)/\lambda \text{ and } V_y = \pi \cdot dy \cdot \sin(\theta) \cdot \sin(\phi)/\lambda \text{ and}$$

$$\beta_0 = 2 \cdot \pi \cdot dx \cdot \sin(\theta_0) \cdot \cos(\phi_0)/\lambda \text{ and } \beta_1 = 2 \cdot \pi \cdot dy \cdot \sin(\theta_0) \cdot \sin(\phi_0)/\lambda$$

where θ_0 and ϕ_0 are the look directions in spherical coordinates with the z axis normal to the array. All other variables are as previously defined. Sometimes, the two sins terms are defined as shown in Equation (6.39).

$$\text{sins}(N \cdot U_x - \beta_0/2, M \cdot V_y - \beta_1/2) = \frac{\sin(N \cdot (U_x - \beta_0/2))}{N \cdot \sin(U_x - \beta_0/2)} \cdot \frac{\sin(M \cdot (V_y - \beta_1/2))}{M \cdot \sin(V_y - \beta_1/2)} \quad (6.39)$$

Clearly, the look directions, θ_0 and ϕ_0, lie on two orthogonal cones, and their joint pointing direction is the intersection of those cones.

Another property of an electronically scanned array is the fact that the beam broadens as the array is scanned to an angle, θ_0, away from broadside. This is the obvious result of the fact that the aperture, as projected in the direction θ_0, is smaller than at the normal.

As the electronic scan angle goes to endfire (90°), the beam does not go to zero, because of surface waves and diffraction (however, the element factor easily goes to zero, depending on radiator orientation). A good approximation of the scan performance is $\cos(\theta_0) + c_{min}$, where c_{min} is typically 0.1. The beamwidth is thus broadened by $1/(c_{min} + \cos\theta_0)$. This property is depicted in Fig. 6.31. Equation (6.40) summarizes the beamwidth and gain dependence on $\cos\theta_0$.

$$G(\theta,\phi) \cong \frac{4 \cdot \pi \cdot A \cdot (c_{min} + \cos(\theta_0))}{\lambda^2} \quad \text{and}$$

$$\Omega(\theta,\phi) \cong \frac{BW_0}{(c_{min} + \cos(\theta_0))}$$

$$(6.40)$$

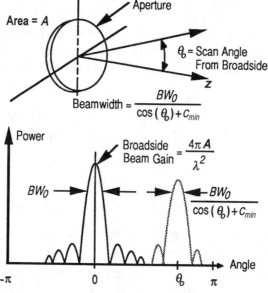

Figure 6.31 Beam broadening with scan angle.[23]

where A is the area or $E_e(\theta)a_0 NdxMdy$ for a discrete element array, and BW_0 is θ_{az} or θ_{el} for example. By the way, with the right type of radiating element, end-fire arrays can have high gain even though the projected area is small (see Balanis, pp. 229–234). Unfortunately, the elements are very hard to achieve in a stealth platform.

As previously mentioned, examination of Fig. 6.26 and Equation (6.32) shows an important relationship between the element spacing, d, the array operating wavelength, λ, and the location of grating lobes. Grating lobes will appear at endfire when the interelement phase shift $\beta_0 + 2\pi d/\lambda$ equals multiples of 2π, as shown in Fig. 6.32.[8] When $d/\lambda = 1$, the grating lobes just enter real space when the mainlobe is normal to the array. When $d/\lambda = 0.5$, the grating lobe just enters real space when the mainlobe is at endfire. Typically, array element spacings are chosen depending on maximum scan angle and shortest operating wavelength so that d/λ is between 0.5 and 1. As will be seen later, the active grating lobe condition is often more benign than the requirement for the RCS grating lobe. Stealth element spacings less than half a wavelength are usually required.

Because spacings tighter than 0.5λ are often required, triangular lattices are used to obtain better spacing at the expense of slightly more complicated electronic scan commands. Some typical triangular element layouts and their actual detail shapes are shown in Fig. 6.33. These depict radiating elements whose spacing is limited by the

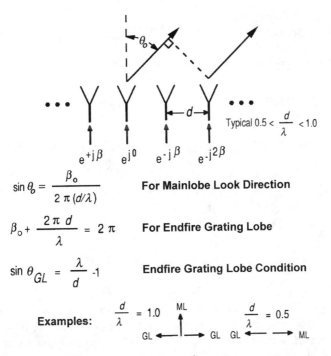

Figure 6.32 Element spacing and grating lobes (after Knittel[8]).

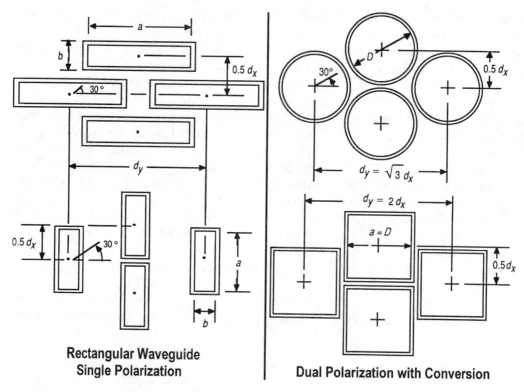

Figure 6.33 Waveguide radiators in triangular lattice (after Knittel[8]).

size of the feeding waveguide. The near broadside radiation pattern of such arrays is given in Equation (6.41).

$$E(\theta,\phi) \cong N \cdot M \cdot a_0 \cdot E_e(\theta,\phi) \cdot (1 + \exp(-j \cdot (U_x + V_y)))$$

$$\cdot \frac{\sin(0.5 \cdot N \cdot U_x)}{0.5 \cdot N \cdot \sin(U_x)} \cdot \frac{\sin(0.5 \cdot M \cdot V_y)}{0.5 \cdot M \cdot \sin(V_y)}$$

(6.41)

where

$$U_x = \pi \cdot (dx/\lambda) \cdot \sin(\theta) \cdot \cos(\phi) - \beta_x/2 \text{ and}$$

$$V_y = \pi \cdot (dy/\lambda) \cdot \sin(\theta) \cdot \cos(\phi) - \beta_y/2$$

(6.42)

Clearly, for electronic scanning the only changes in Equations (6.32) are the addition of a phase increment β_x and β_y to U_x and V_y to find the appropriate patterns for a rectangular electronically scanned array. The other factor in Equation (6.41) is to account for the fact that the triangular lattice is actually two offset interleaved arrays. Unfortunately, a triangular lattice has poorer RCS performance, and thus radiators require offsets from the feed locations, which limits usefulness.

Arrays, whether fixed or electronically scanned, are primarily fed in one of four ways. The first two ways are quasioptical in the sense that the array is space fed. Figure 6.34 shows two optical feed schemes. The reflect array has a feed horn in front of the array that feeds all radiating elements in the near field. The RF enters the radiator, goes through the phase shifter, is reflected by an RF short, passes through the phase shifter again, and is radiated into free space. The advantages are a simple feed, smaller feed losses, and half as much RF phase shift circuitry. The disadvantages are spillover and reflection lobes that raise sidelobes and a feed in front of the array, which results in higher RCS and aperture blockage.

The lens array has the feed behind the aperture. The RF enters the rear radiator, goes through the phase shifter, and is radiated into free space by the front radiator. The advantages are a simple feed, no blockage, and smaller feed losses. The disadvantages are reflection lobes that raise sidelobes and RCS and more phase shift and radiator circuitry. Raytheon, Hughes, and Emerson Electric (among others) built electronically scanned arrays with optical feeds. Both of these antenna types can have very large amounts of power reflected from the lens/reflector, which must be carefully treated or it can damage components receiving the reflected energy (including

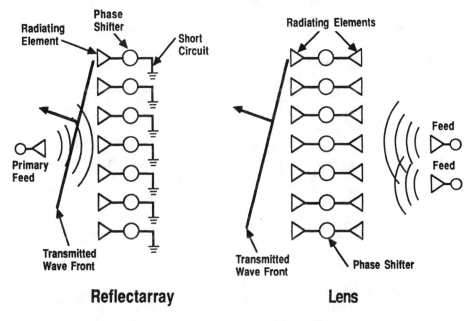

Figure 6.34 Electronically scanned array configurations (optical feeds).[23]

complete incineration). This same reflection problem can also occur between antenna and radome.

Two other principal feed schemes use constrained feed networks, as shown in Fig. 6.35. The RF delivered to each phase shifter and radiator is entirely in transmission line form (i.e., waveguide, coax, stripline, microstrip, squareax, and so forth) or some combination. These feeds are usually quite complex but have the advantage of small feed volume as well as complete control of leakage power at each phase shifter and RCS. The corporate feed attempts to keep the path length, phase shift, and transmission losses to each phasor and radiator the same (amplitude weighting will result in different powers at each phasor). The advantages are large instantaneous bandwidth, very good sidelobe control, simpler beam scan calculations, very low RCS, and very low inadvertent RF leakage. The disadvantages are design and manufacturing complexity, higher losses, and slightly larger feed physical volume.

The branchline or travelling wave feed has unequal path lengths to each phasor and radiator. For a fixed set of phase shifts, the beam will shift or "squint" with changing frequency, which can be either an advantage or a disadvantage, depending on use. Travelling wave feeds are often used to provide simple low-loss electronic scanning in one dimension by frequency shift. The advantages of the branchline feed are simplicity in manufacture, small volume, low losses, low inadvertent leakage, low sidelobes, low RCS, and separable amplitude weighting functions. The disadvantages

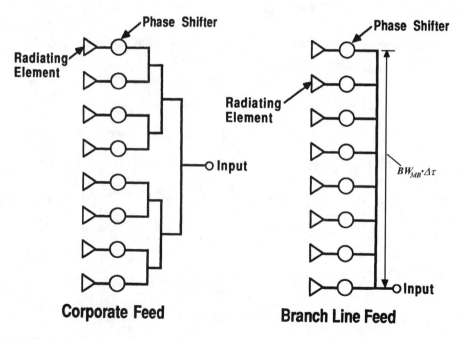

Figure 6.35 Electronically scanned array configurations (constrained feeds).[23]

are lower instantaneous bandwidth, more complex beam scan calculations, and more losses than optical feeds.

6.5.2 Multibeam Antennas

There are several ways to obtain multiple simultaneous beams from the same aperture. Obviously, with optically fed arrays, multiple primary feeds can achieve several adjacent beams with reasonable sidelobe performance as shown in the lens array of Fig. 6.34. In addition, if the primary feed can support two polarizations, two beams with reasonable isolation can be obtained if the secondary lens can handle or scan two polarizations. Constrained feed designs can also support multiple beams either with separate feeds, subarrays, or a single dispersive feed. Multibeam antennas have been in use for at least 50 years, with monopulse being the most common type. Dispersive or frequency scanning feeds provide a simple way to obtain simultaneous beams at offset frequencies, which allow high levels of beam-to-beam isolation.[24]

As a simple example, consider the branch line feed of Fig. 6.35 with total time delay, $BW_{3dB} \, \Delta\tau$. If a uniform and constant time delay is introduced in the main feed between branches, the natural result is a beam scanned off the normal that is related to the product of the operating frequency and the time delay. Returning to Equation (6.35) and expanding the definition of β to include a fixed element-to-element time delay yields a form shown in Equation (6.43).

$$E(\theta) = E_e(\theta) \sum_{n=0}^{N-1} a_n \cdot [\exp(\, j \cdot 2 \cdot \pi \cdot d \cdot \sin(\theta)/\lambda - j \cdot \beta)]^n \qquad (6.43)$$

where

$$\beta = 2 \cdot \pi \cdot d \cdot \sin(\theta_0)/\lambda - \Delta, \quad \Delta\tau = 1/\Delta f \ \text{ and } \ \Delta = BW_{3dB} \cdot \left(\frac{2 \cdot \pi}{N}\right) \cdot \left(\frac{f - f_0}{\Delta f}\right)$$

There are two terms in β: the beam phase scan term containing the "look direction," θ_0, and the time delay term, which contains the normalized beamwidth coefficient similar to Table 6.2, the uniform weighting phase shift per element required to scan one beamwidth and the normalized frequency offset from band center to achieve frequency scan. Also note that total beam scan is the combination of the conventional phase scan and the frequency scan created by the time delay. For example, a 50-element linear 10-GHz array with 1/2 wavelength spacing, 14.8-nsec total time delay, and Hamming weighting will produce adjacent beams every 100 MHz. Figure 6.36 shows three beams of the squint compensated example above offset by 100 MHz each, phase scanned to 22.5°. In the absence of compensation, the beams will be squinted away from the normal to the array, which is desirable for stealth antennas.

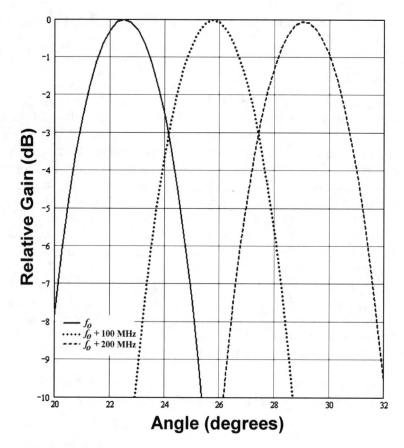

Figure 6.36 Example frequency and phase scanned multibeam antenna.

Also note that the total time delay in the array required to achieve the example scan-ning is 148 wavelengths at the center frequency, f_0. Although antennas of this type are made routinely, small regular errors are multiplied by a large number of wavelengths. These antennas usually require compensation after assembly and temperature/time compensation during normal operation. Additional details are in the appendix CDROM.

6.6 ANTENNA SCATTERING[4,7,16–19]

6.6.1 Basic Notions

Antenna RCS is really made up of several contributors. The three major contributors are antenna-mode scattering, antenna-structural scattering, and RCS grating (Bragg) lobes. Of course, antenna RCS is also strongly dependent on polarization for both types of major scattering components. Some antenna types such as dipoles provide very little backscatter for polarizations, which are orthogonal to the dipole axis. Aperture antennas such as horns, slotted arrays, and so forth can look totally reflective for an orthogonal incident polarization unless some provisions have been made to provide an impedance match for the orthogonal illumination. The backscatter is also influenced by the antenna geometry and the type of material from which is it constructed.[4,7,16–19] The antenna is always embedded in something, and the combined RCS is what is most important. A perfectly absorbing antenna in a bad installation may be worse than one "matched" to its surroundings. Summarizing, the essential design considerations are the following:

- Structural RCS

  - Antenna face: shape, orientation, edges
  - Interface to vehicle
  - Cavity and internal surfaces

- Antenna-mode RCS, i.e., reflections from inside antenna

  - Radiating elements
  - Isolation of major internal reflectors
  - Reduction and cancellation of minor reflectors
  - Uniformity across array

- Grating lobes, i.e., above RF band spikes

  - Higher operating frequencies
  - Filtering surfaces (radome or in antenna)

Figure 6.37 shows the scattering components for a conceptual antenna array. Observe from the figure that an incident RF signal impinges on the aperture and then gives rise to a scattered RF signal, which is made up of several components. One class of components is called the *structural mode scatter*. One component is reflected at the complement of the incident angle relative to normal. It is called the *optical* or *faceplate scatter component*. This optical scattering often results in multiple bounces both inside the antenna cavity and from other surfaces on the platform and can give rise to a retroreflected component of considerable magnitude. The structural mode antenna radar cross section is all the other cross section contributors from the antenna when the actual antenna circuits are invisible. Structural RCS consists of scattering from the aperture plate, supports, edges, multiple cavity bounces, and so on. Although experience and conventional RCS modeling provides insight into possible structural mode scattering, usually it must be measured and corrected with techniques described later in this chapter. The true structural mode may only be visible in cross polarization. All polarizations should be measured/estimated to find the structural RCS. The structural mode response of an antenna is not as easily estimated as the antenna mode.

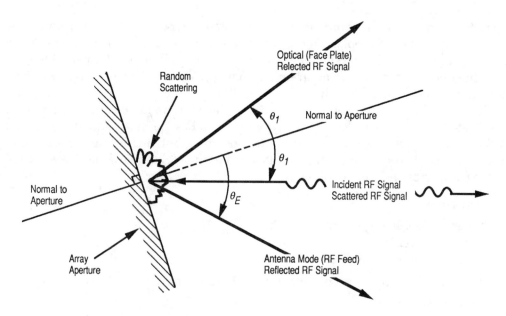

Figure 6.37   Antenna scattering components.

A second class of components of the scattered RF signal is the *antenna-mode scatter-ing*. Antenna-mode scattering is the radiation mode response that is associated with those currents that exist in the antenna when it operates in its normal antenna mode. In its most simple form, one could visualize the incident energy being absorbed by the antenna, coming out at the antenna input terminal, meeting a mismatch there, being reflected and, finally, being reradiated through the normal antenna pattern. The am-plitude of the antenna mode response may be estimated if the reflection factor of the antenna is known from the antenna gain characteristics. There are two ways that the antenna mode RCS can be visualized: either the RCS goes as the square of the fields times cosine squared, or it goes as the square of the fields but at double the spatial fre-quency. This duality arises because most patterns are made up of sinc functions, and the trigonometric identity $\text{sinc}(2\theta) = \cos(\theta)\text{sinc}(\theta)$ is sometimes substituted. Another major set of antenna mode contributors is from random scattering from individual ra-diating elements of the antenna caused primarily by manufacturing tolerance errors. As previously mentioned, in well designed low stealth antennas, manufacturing toler-ance errors ultimately dominate the active and RCS performance of the antenna at most angles.

The fundamental definition of radar cross section (RCS) is given in Equation (6.44).

$$\sigma = 4 \cdot \pi \lim_{R \to \infty} R^2 \frac{|\mathbf{E_s}|^2}{|\mathbf{E_i}|^2} = 4 \cdot \pi \lim_{R \to \infty} R^2 \frac{|\mathbf{H_s}|^2}{|\mathbf{H_i}|^2}$$

(6.44)

where

R = range to the observer

$\mathbf{E_s}, \mathbf{H_s},$ = scattered fields, which can be calculated from Equation (6.7)

$\mathbf{E_i}, \mathbf{H_i},$ = incident fields

The challenge is to find the relationship between $\mathbf{E_i}(x,y)$ and $\mathbf{F}(x,y)$, the radiating aperture illumination. One possible common assumption is to assume that $\mathbf{E_i}(x,y)$ at the aperture is a plane wave and the projected area is uniformly illuminated with each element on the aperture surface differing only in phase.[26–30]

If an object is complex and made up of multiple surfaces for which the RCS is known with some accuracy (spheres, cylinders, polygonal plates, dihedrals, and so forth), then a first-order RCS estimate can be obtained from Equation (6.45).

$$\sigma = \left| \sum_{m=1}^{M} \sqrt{\sigma_m} \cdot \exp(j \cdot 4 \cdot \pi \cdot r_m / \lambda) \right|^2 \tag{6.45}$$

A very important concept for first-order understanding of antenna RCS is the physical optics approximation for RCS. If a perfectly conducting surface is assumed, and the physical optics far-field approximations of Table 6.1 are made, then the square root of the RCS including some polarization effects as well as bistatic observation can be defined as shown in Equation (6.46).

$$\sqrt{\sigma} = \frac{-j \cdot k}{\sqrt{\pi}} \int_S \vec{n} \bullet \vec{e}_\mathbf{r} \times \vec{h}_\mathbf{i} \cdot \exp(j \cdot k \cdot r \cdot (\vec{i} - \vec{s})) \cdot dS \tag{6.46}$$

where

$\vec{e}_\mathbf{r}$ = receiver polarization unit vector

$\vec{h}_\mathbf{i}$ = incident magnetic field polarization unit vector

$\vec{n}$ = local surface normal

$\vec{i}$ = incident field propagation unit vector

$\vec{s}$ = scattered field propagation direction unit vector

For monostatic RCS, $\vec{i} = -\vec{s}$, thus $(\vec{i} - \vec{s}) = 2\vec{i}$. The above surface integral can often be replaced with a line integral. More generally, the RCS is a combination of the aperture surface area and the edges of the aperture. At angles far off the normal, an edge diffraction term and a travelling wave term not explicit in the equation above dominate the observed RCS. The easiest way to visualize and model the edge effects is to assume that an insulated wire with currents independent of the currents in the antenna aperture follows the perimeter and scatters coherently. A simple form for RCS based on Equation (6.46) when the object is not polarization sensitive and the observation is monostatic is given in Equation (6.47).

$$\sigma = \frac{4 \cdot \pi}{\lambda^2} \left| \int_A \exp(-j \cdot 4 \cdot \pi \cdot r/\lambda) dA \right|^2 \tag{6.47}$$

where dA is the area of dS projected in the direction of the observer. This model is reasonably accurate if the observation angle is 60° or less from the surface normal, and the surface area and the radius of curvature are large with respect to a wavelength.

Another approach to RCS estimation related to Equations (6.45) through (6.47) is to approximate a complex object with multiple simple facets. Most surfaces can be approximated by triangular facets, which can then be summed to estimate the RCS of the complete object. Any triangle can redefined as two connected right triangles whose base is the longest side. Equation (6.48) gives the physical optics root RCS approximation for a single right triangular facet in the x-y plane with the right angle at the origin. This facet can be translated and rotated to any place on the object.

$$\sqrt{\sigma} = (a \cdot b \cdot \sqrt{\pi}/\lambda) \cdot \cos(\theta) \cdot \frac{\exp(2 \cdot V) \cdot \text{sinc}(2 \cdot V) - \exp(2 \cdot U) \cdot \text{sinc}(2 \cdot U)}{2 \cdot U - 2 \cdot V} \tag{6.48}$$

where
  U and V = as defined in Equation (6.26) or (6.56)
    a = x-axis side
    b = y-axis side of the right triangle

For obvious reasons, $U - V \neq 0$ in this approximation. The monostatic antenna radar cross section then is made up of several components as shown in Equation (6.49).

$$\sigma_{antenna} = \left| \sqrt{\sigma_{am}} \cdot (1 - \Gamma_{ap}) \exp(j \cdot 4 \cdot \pi \cdot r_{am}/\lambda) + \Gamma_{ap} \cdot \sqrt{\sigma_s} \right|^2 \tag{6.49}$$

where
  σ_{am} = square root of the RCS of the antenna mode
  r_{am} = equivalent distance inside the antenna to the reflection point
  σ_s = square root of the RCS of the structural mode of the antenna (i.e., faceplate, edges and radiator supports)
  Γ_{ap} = complex reflection factor of the aperture (i.e., its match to free space)

Because the antenna mode is absorptive for like polarization, it masks/attenuates the structural mode RCS. Unless provisions are made to attenuate/match cross polarization illumination, there may be blooming of the RCS for cross pol. The apparent structural cross section is reduced by the masking effect of the absorption of illuminating power by the antenna, which is not reflected. Equation (6.50) shows a very approximate antenna mode radar cross section (good within 20° of the mainbeam center). This relationship can be used to estimate the antenna mode RCS. Note that the antenna mode RCS disappears for cross polarization and endfire in most cases.

$$\sigma_{am} \cong G(\theta,\phi) \cdot A_e \cdot |\Gamma_{an}|^2 \cdot (\vec{\mathbf{n}} \bullet \vec{\mathbf{i}})^2 \cdot (\vec{\mathbf{i}}_\mathbf{p} \bullet \vec{\gamma}_\mathbf{p})^2 \qquad (6.50)$$

where

- • = vector scalar (dot) product
- $\vec{\mathbf{n}}$ = aperture normal unit vector
- $\vec{\mathbf{i}}$ = incident wave unit vector
- $\vec{\mathbf{i}}_\mathbf{p}$ = incident wave polarization
- $\vec{\gamma}_\mathbf{p}$ = antenna aperture polarization unit vector
- Γ_{an} = complex reflection factor for power that actually enters the antenna

Recall that the effective area is related to the gain by Equation (6.51).

$$G = 4 \cdot \pi \cdot A_e / \lambda^2 \quad \text{or} \quad A_e = \lambda^2 \cdot G / (4 \cdot \pi) \qquad (6.51)$$

Because antenna RCS is ultimately dominated by manufacturing tolerances and the resulting variation in the complex reflection factor at each element, a more complex formula, but still approximate (good to 20° off mainbeam) for antenna mode RCS, is required as shown in Equation (6.52).

$$\bar{\sigma}_{am} \approx \frac{\lambda^2}{4 \cdot \pi} (\vec{\mathbf{n}} \bullet \vec{\mathbf{i}})^2 \cdot (\vec{\mathbf{i}}_\mathbf{p} \bullet \vec{\gamma}_\mathbf{p}) \cdot \left[\begin{array}{l} G_0(U,V) \cdot \bar{\Gamma}^2 \cdot \cos^2(U) \cdot \cos^2(V) + \\[2mm] G_e(\theta,\phi) \cdot \cos^2(\theta) \cdot \varepsilon_\Gamma^2 \cdot \dfrac{\displaystyle\sum_M \sum_N \Gamma_{m,n}^2}{\left(\displaystyle\sum_M \sum_N \Gamma_{m,n}\right)^2} \end{array} \right] \qquad (6.52)$$

where

- G_0 = "ideal" gain
- $\bar{\Gamma}$ = mean complex reflection factor
- G_e = individual element gain
- ε_Γ^2 = complex reflection factor variance
- $\Gamma_{m,n}$ = complex reflection factor of the m,n element
- N = total number of elements in the x dimension
- M = total number of elements in the y dimension

There are two items to note in Equation (6.52): the pattern of complex reflection factors has gain, and the random error pattern rolls off as $\cos^2(\theta)$.

The RCS for the single radiators described in Section 6.3 using Equation (6.45) and the gains shown in Fig. 6.1 are summarized in Equations (6.53).

$$\sigma_e = \frac{\lambda^2 \cdot \Gamma^2}{4 \cdot \pi} = \frac{\lambda^2 \cdot \Gamma^2}{4 \cdot \pi} \qquad \text{isotropic radiator}$$

$$\sigma_e = \frac{2.25 \cdot \lambda^2 \cdot \Gamma^2 \cdot \cos^2(\theta)}{4 \cdot \pi} \cdot (\vec{i}_\mathbf{p} \bullet \vec{\gamma}_\mathbf{p})^2$$

$$\cong 0.179 \cdot \lambda^2 \cdot \Gamma^2 \cdot \cos^2(\theta) \cdot \cos^2(\Delta_p) \qquad \text{small dipole/slot}$$

$$\sigma_e = \frac{10.76 \cdot \lambda^2 \cdot \Gamma^2 \cdot \cos^2(\theta)}{4 \cdot \pi} \cdot (\vec{i}_\mathbf{p} \bullet \vec{\gamma}_\mathbf{p})^2$$

$$\cong 0.856 \cdot \lambda^2 \cdot \Gamma^2 \cdot \cos^2(\theta) \cdot \cos^2(\Delta_p) \qquad \text{1/2 wave dipole/slot}$$

$$(6.53)$$

For example, using Equations (6.53), a 1/2-wave single slot or dipole with $|\Gamma| = 0.5$ and $\lambda = 0.03$ m and like polarization, has a radar cross section of −37 dBsm, or about the same as a sparrow.

In the cases above, Δ_p is the angle between the incident wave polarization and the polarization of the radiating element. The E-field orientation of these elements is along the y axis, and the orientation of the arrays discussed is in the x-y plane. These element RCS patterns of Equations (6.53) are approximate and representative; more accurate gain patterns were provided in Section 6.3, from which RCS patterns can be obtained. Equations (6.53) seem to show that RCS decreases with decreasing wavelength. What must be remembered is that the dimensions of these radiators decrease proportionately to fulfill the initial assumptions and data link or radar performance requires aperture area, so more radiators are required to fill it.

Returning to Equations (6.50) and (6.52), note that the amplitude of the reflection can be quite large if the gain, G, and/or the complex reflection coefficient, Γ, of the antenna is large. The peak RCS response of an antenna appears to be proportional to the square of the wavelength but, because gain is inversely proportional to the square of the wavelength (Equation 1.24), it is area that counts. It is apparent from this equation, then, that high-gain antennas that do not exhibit good free-space impedance match will exhibit high RCS responses. However, because the RCS response is the product of the gain and the effective area, in the sidelobe region, the antenna mode RCS falls away very rapidly for modern well made antennas, and thus it is not usually a consideration very far away from the mainlobe of the antenna pattern.

Some example aperture shapes filled with dipole radiators and their RCS sidelobes plotted in *u-v* space are shown in Fig. 6.38. The apertures are shown inscribed in some hypothetical disk shaped installation volume as well as how the sidelobes might fall in some required field of regard. The field of regard may either be the active operating window or the high threat window. Which shape is best is, of course, a function of the mission and threat.

For example, estimate the first-order RCS pattern of an antenna by considering the one dimensional linear array of Fig. 6.25. The field of the active pattern for a linear array is given in Equation (6.31). By analogy, substitute an element by element summation into Equation (6.45) as shown in Equation (6.54).

$$\sigma_{am} = \left| \sqrt{\sigma_e} \cdot \left[\sum_{n=0}^{N-1} \Gamma_n \cdot \exp\left(\frac{-j \cdot 4 \cdot \pi \cdot d \cdot \sin(\theta)}{\lambda} \right) \right] \right|^2 \tag{6.54}$$

where Γ_n is the reflection pattern for each element. If the reflection factor is the same for each element, substitute the 1/2-wavelength element pattern of equation (6.53) into Equation (6.54). Then, using the closed-form solution of the above series analogous to Equation (6.32) yields Equation (6.55).

$$\sigma_{am} = 0.856 \cdot \lambda^2 \cdot \Gamma^2 \cdot N^2 \cdot \cos^2(\theta) \cdot \cos^2(\Delta_p) \cdot \left[\frac{\sin(2 \cdot \pi \cdot N \cdot d \cdot \sin(\theta)/\lambda)}{N \cdot \sin(2 \cdot \pi \cdot d \cdot \sin(\theta)/\lambda)} \right]^2 \tag{6.55}$$

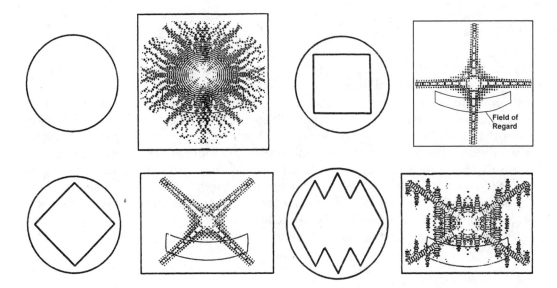

Figure 6.38 Effect of shape on RCS sidelobes.

This is just the element cross section times the linear array response at twice the spatial frequency in U space. The only term in this equation that may strike one as odd is the $\cos^2\theta$ term, which is the result of the dot product between the normal to the aperture face and the incident illumination direction, and the dot product, of course, is the $\cos\theta$. If somehow these elements were floating in space, this would be all the RCS. Obviously, there is a structural term that becomes completely visible in cross polarization.

Suppose for the moment that the structural-mode scattering from an antenna was only from the faceplate; then, one could compute the monostatic RCS for the faceplate by using the antenna of Fig. 6.13 with the uniform illumination rectangular aperture formula from Section 6.4.1, Equation (6.26), the gain for a rectangular antenna. This gain at double the spatial frequency in U, V space, when substituted in Equation (6.47), yields Equation (6.56). This same equation has been derived by other means many times. Furthermore, it has been verified that, for carefully made rectangular plates, the experimental radar cross section is, in fact, closely approximated by this equation out to about 60° off the normal. At large off angles, RCS is dominated by edge diffraction and travelling or surface waves for carefully made plates. As previously mentioned, the edge behavior can be modeled as an independent wire loop traversing the perimeter of the plate.

$$\sigma(\theta,\phi) \cong \frac{4 \cdot \pi \cdot a^2 \cdot b^2 \cdot \Gamma^2 \cdot \cos^2(\theta)}{\lambda^2} \cdot \frac{\sin^2(2 \cdot U)}{4 \cdot U^2} \cdot \frac{\sin^2(2 \cdot V^2)}{4 \cdot V^2}$$

where $\quad U = \dfrac{\pi \cdot a}{\lambda}\sin(\theta) \cdot \cos(\phi)$ and $V = \dfrac{\pi \cdot b}{\lambda}\sin(\theta) \cdot \sin(\phi)$ $\qquad$ (6.56)

Combining this with an antenna array term in Equation (6.49) would provide a good first-order estimate of total RCS.

Examination of the upper bounds of Equations (6.56) shows some important facts of stealth design. First, at the normal to a flat plate, $\theta = 0$, there is a large RCS "spike" proportional to $4\pi\, A_e^2/\lambda^2$, which is often 10^4 m^2. Note that, in the principal plane, when the sine functions go to 1, sidelobe peaks are independent of wavelength, because all the λs cancel out. Those peaks are dependent only on the dimension perpendicular to the line of sight (LOS). Furthermore, the intercardinal plane sidelobe peaks are *independent* of the size of the antenna (or facet, for that matter). No matter how large the plate is, its cross section is a function only of angle and wavelength. Equations (6.57) summarize these observations, which are an essential key to low-RCS systems.

$$\sigma(0,\phi) = 4 \cdot \pi \cdot A_e^2/\lambda^2; \quad \sigma_{peak}(\theta,0) = 4 \cdot b^2/(\pi \cdot \tan^2(\theta));$$

$$\sigma_{peak}(\theta, 90°) = 4 \cdot a^2 / (\pi \cdot \tan^2(\theta)) \text{ and}$$

$$\sigma_{peak}(\theta, \phi \neq 0° \text{ or } 90°) = \frac{16 \cdot \lambda^2}{(\pi^3 \cdot \sin^2(\theta) \cdot \tan^2(\theta) \cdot \sin^2(2 \cdot \phi))} \tag{6.57}$$

Assuming for the moment that all the elements of an array are identical with a complex reflection coefficient, Γ, then a rectangular planar array of radiators of the type shown in Fig. 6.27 will have an approximate RCS pattern as shown in Equations (6.58), which are related to the radiation pattern of Equations (6.33). This RCS pattern has cardinal axes in which the element RCS can be approximated by an average element reflection factor, Γ, which thus produces a very accurate approximation of the array RCS. Off the cardinal axes, RCS is dominated by surface waves, edge diffraction, and element random errors, and the average approximation doesn't work well. The following section deals with the RCS contribution from manufacturing errors.

Comparison of Equations (6.56) with Equations (6.58) suggests an important element of stealth antenna design: isomorphism. That is, the individual scatterers of an array must line up with the edges of the antenna aperture plate/structure or there will be two sets of cardinal surface sidelobes. The Chapter 1 introductory Figs. 1.5 and 1.6 describe the requirement for alignment of RCS "spikes." The practical effect of that requirement is that the axes of individual radiators must align with aperture edges. This is also true of stealth platform detail features and major platform edges. Structural-mode and antenna-mode scattering must align for best performance.

$$\sigma(\theta, \phi) \cong (4 \cdot \pi / \lambda^2) \cdot (M \cdot dy)^2 \cdot (N \cdot dx)^2 \cdot \Gamma^2 \cdot \cos^2(\theta)$$
$$\cdot \left[\frac{\sin(N \cdot 2 \cdot U_x)}{N \cdot \sin(2 \cdot U_x)}\right]^2 \cdot \left[\frac{\sin(M \cdot 2 \cdot V_y)}{M \cdot \sin(2 \cdot V_y)}\right]^2 \tag{6.58}$$

where

$$U_x = \pi \cdot dx \cdot \sin(\theta) \cdot \cos(\phi) / \lambda \text{ and } V_y = \pi \cdot dy \cdot \sin(\theta) \cdot \sin(\phi) / \lambda \tag{6.59}$$

If the array is made up of near 1/2-wave dipoles or slots with the E-field oriented along the x axis as defined in Equations (6.53), then the array RCS is approximately as given in Equation (6.60). Equation (6.60) suggests that the RCS is zero for cross polarization illumination. This is true for the antenna-mode but, unfortunately, the structural-mode RCS will still be present and, unless it is absorptive, it may be as large or larger than the antenna-mode. Sometimes, cross polarization loads are used on an array face to mitigate the structural RCS. Because a well made stealth antenna will have a small reflection factor, as mentioned before, the structural RCS will be masked by the inband, like polarization antenna mode loading of the scattering from edges and structure. The full structural-mode RCS can be found by calculating or measuring the RCS at all polarizations.

$$\sigma(\theta,\phi) \cong 0.856 \cdot \lambda^2 \cdot M^2 \cdot N^2 \cdot \Gamma^2 \cdot \cos^2(\theta) \cdot \cos^2(\Delta_p)$$
$$\cdot \left[\frac{\sin(N \cdot 2 \cdot U_x)}{N \cdot \sin(2 \cdot U_x)}\right]^2 \cdot \left[\frac{\sin(M \cdot 2 \cdot V_y)}{M \cdot \sin(2 \cdot V_y)}\right]^2$$
$$(6.60)$$

Another important issue for antenna array and radome RCS is grating lobes and their location. Because the grating lobe may approach the mainlobe in amplitude, it will also approach the mainlobe in RCS. It can be very large and at threat angles both directly and by reflection. The monostatic grating lobe RCS response of antenna arrays is closer to the array broadside when compared to the normal transmit/receive grating lobe response. This was alluded to in Section 6.5, Fig. 6.32. The normal grating lobe radiation angle is given by Equation (6.61) where θ_0 is the desired look direction, m is an integer grating lobe index, and d_x or d_y are the element spacings in the x or y dimension.

$$\theta_{GL} = \sin^{-1}\left[\sin(\theta_0) \pm \frac{\lambda \cdot m}{d_x \text{ or } d_y}\right] \tag{6.61}$$

The RCS grating lobe response angles, on the other hand, are determined primarily from element spacing, d_x or d_y, and usually not from element-to-element phasing, because the phase shifters are usually buried too deeply in the antenna to create a large amplitude response. Equation (6.62) gives the location of the RCS grating lobes. It can be seen that RCS grating lobes are twice as close in U, V space as regular grating lobes (Fig. 6.39).

$$\theta_{RCS\text{-}GL} = \sin^{-1}\left[\pm\left(\frac{\lambda \cdot m \cdot 0.5}{d_x \text{ or } d_y}\right)\right] \tag{6.62}$$

One solution for this difficulty is to fit the face of the antenna with a frequency selective surface (FSS) or radome that has tighter spacing and hence grating lobes at larger angles. Because the FSS is space fed in the near field of the antenna, radiating elements can be placed much closer. Such a system is shown in Fig. 6.40. Usually, these FSSs have other desirable properties incorporated (e.g., reflection cancelling) to enhance RCS performance.

6.6.2 Estimating Antenna RCS

The radome or FSS scattering components will be dealt with in the next section, but a conceptual view of how an antenna array is decomposed into progressively smaller

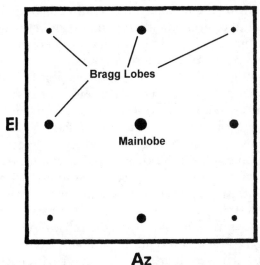

Bragg or RCS Grating Lobes Resulting from Illumination of Rectangular Lattice by a Wavelength, $\lambda_i < 2 \cdot d_x$ or d_y

Figure 6.39 Bragg or RCS grating lobes.

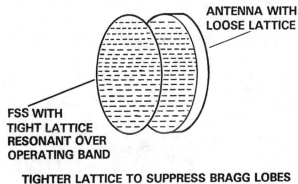

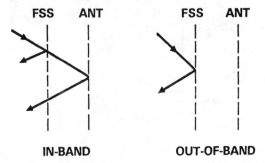

Figure 6.40 Frequency selective surface to suppress Bragg lobes.

elements for RCS analysis is shown in Fig. 6.41. Ultimately, each individual radiating element must be analyzed for its contribution to both antenna mode and structural scattering. The individual element is ultimately analyzed and decomposed as given in Fig. 6.48 for a typical electronically scanned array path from radiating element all the way to antenna input or to an isolator or circulator. The transmission lines and components between the antenna input and the radiating aperture may total several tens of wavelengths; consequently, small changes in the physical dimensions of these components can result in relatively large changes in the phase of the RF passing through them. In addition, small changes in the matching transformers between the various transmission line sections result in small amplitude errors. The analysis must determine the magnitude of the signal variations in both phase and amplitude. Much of the variation is the result of dimensions in the phase shifters, matching transformers, and waveguide between the input and the radiating aperture. The phase shifters may also have electrical variability caused by semiconductors, dielectrics, and magnetics. There are two levels of analysis. The first predicts the general shape of the RCS response for a given antenna architecture. The second assesses the complex reflection factor at each segment of the antenna, and its primary focus is component tolerance.[7,26-30]

To understand the first level of analysis, consider a small electronically scanned array (ESA) as shown in the functional diagram of Fig. 6.42. This particular antenna system has row and column phase steering with a circulator behind each vertical feed and a separate horizontal feed for transmit and receive. The antenna has an array of

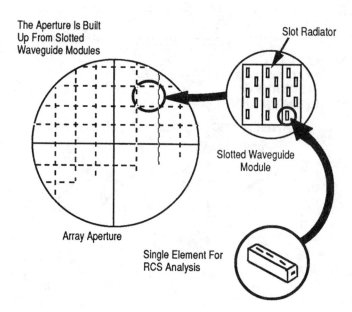

Figure 6.41 Front view of planar array.

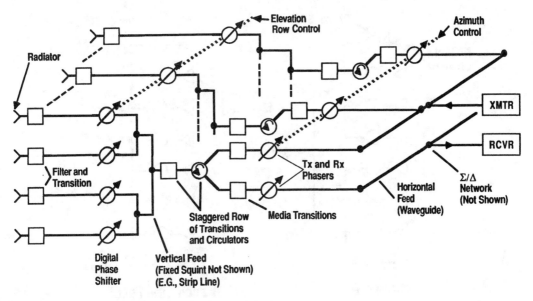

Figure 6.42   Example ESA block diagram for RCS calculation.

slot radiators with each row slightly offset so that the E-field cardinal surface is at an angle of 7.5° relative to vertical. The phasors are set to point to some angle off the normal, but there will still be some angle where there is a mainlobe RCS response. The particular example array is made up of 12 × 12 elements on near 1/2-wavelength spacing. The largest response will be inband like polarization.

As an incident wave gets deeper into the antenna, reflections that give rise to RCS get progressively more focused, and their effects on the signature are significantly different. Consider the corporate feed structure on the left of Fig. 6.43. A reflection at point 0 will be almost omnidirectional and, if correlated with all other point 0s, the RCS peak will be normal to the array face. The uncorrelated portion of the scatter will be broadly cosinusoidal. The reflection at point 1 will be returned through 2 phasors and elements, and even the random component will be focused by two elements. Similarly, a point 2 reflection will be focused by four phasors and radiating elements, and so on. Although this property will increase the RCS in the net beam direction, it has the effect of reducing the cross section everywhere else.

Obviously, large reflection lobes should be placed where structural lobes already exist. Figure 6.44 generally indicates the nature of the RCS from the major contributors in the block diagram. Out-of-band and cross polarization will have an RCS response normal to antenna aperture faceplate. Scattering from the radiator, filter, and transition are almost omnidirectional and cover most of space. Scattering from the structural aperture plate is sinc(x) in nature [Equations (6.56)] as shown in the figure. Scattering from the input to the vertical feed is focused in azimuth, because each row

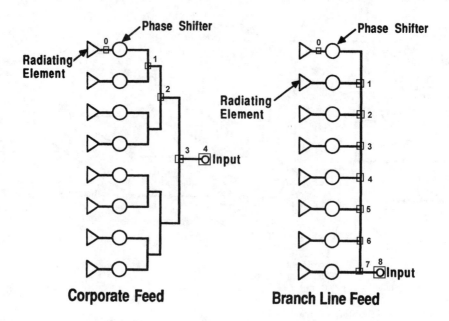

Figure 6.43 Scattering from internal antenna junctions.[23]

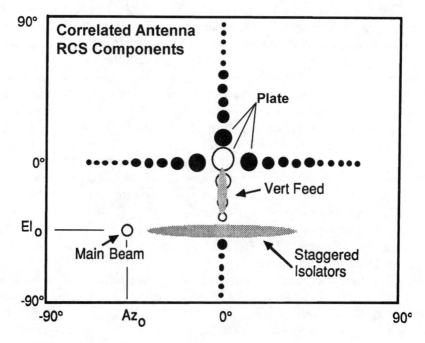

Figure 6.44 Location and nature of the antenna RCS response.

of phasors has the same phase shift by design. Scattering from the circulators at the front of the horizontal feeds is spread in azimuth but fully focused in elevation by the vertical feeds.

$$\sigma_{am} \cong \left| \begin{array}{l} \sqrt{\sigma_r}\exp(j \cdot 4 \cdot \pi \cdot r_r/\lambda) + \sqrt{\sigma_{ft}}\exp(j \cdot 4 \cdot \pi \cdot r_{ft}/\lambda) + \\ \sqrt{\sigma_{ep}}\exp(j \cdot 4 \cdot \pi \cdot r_{ep}/\lambda) + \sqrt{\sigma_{vf}}\exp(j \cdot 4 \cdot \pi \cdot r_{vf}/\lambda) \\ \sqrt{\sigma_{ctp}}\exp(j \cdot 4 \cdot \pi \cdot r_{ctp}/\lambda) + \text{everything else} \end{array} \right|^2 \tag{6.63}$$

The total RCS can be represented as the sum of the terms from each of the major scatters in the antenna as given in the general Equation (6.45). For the specific example here, the antenna mode RCS is given in Equation (6.63).

But the first term in equation is the scattering from the radiators in the array before being coupled into the antenna and hence is Equation (6.59). If the distance reference is taken as the front of the antenna, then $r_r = 0$ and the RCS of the first array scatter is given in Equation (6.64). The scatter is omnidirectional to a first order.

$$\sqrt{\sigma_r} \cong (2 \cdot \sqrt{\pi}/\lambda) \cdot (M \cdot dy) \cdot (N \cdot dx) \cdot \Gamma_r \cdot \cos(\theta)$$
$$\cdot \text{sins}(N \cdot 2 \cdot U_x) \cdot \text{sins}(M \cdot 2 \cdot V_y) \tag{6.64}$$

where Γ_r is the reflection factor at the input to the radiator, and all the other variables are as previously defined in Equation (6.59). Similarly, the scattering from the filter and transition that is reradiated with the element pattern is given in Equation (6.65).

$$\sqrt{\sigma_{ft}} \cong 0.925 \cdot \lambda \cdot (N \cdot M) \cdot (1 - \Gamma_r) \cdot \Gamma_{ft} \cdot \cos(\theta)$$
$$\cdot \text{sins}(N \cdot 2 \cdot U_x) \cdot \text{sins}(M \cdot 2 \cdot V_y) \cdot \cos(\Delta_p) \tag{6.65}$$

where Γ_{ft} is the reflection factor from the filter and transition to the phasor. Similarly, the scattering from behind the elevation phasor is given in Equation (6.66).

$$\sqrt{\sigma_{ep}} \cong 0.925 \cdot \lambda \cdot (N \cdot M) \cdot (1 - \Gamma_r) \cdot (1 - \Gamma_{ft}) \cdot \Gamma_{ep} \cdot \cos(\theta)$$
$$\cdot \text{sins}(N \cdot 2 \cdot U_x) \cdot \text{sins}(M \cdot 2 \cdot (V_y - \beta_0)) \cdot \cos(\Delta_p) \tag{6.66}$$

where Γ_{ep} is the reflection factor from the back of the elevation phasor, and β_0 is the phase shift in the elevation phasor. The reflected signal makes two passes through the phasor, so the total phase shift is $2\beta_0$. There is obviously a reflection back into the antenna as each main reflection returns to free space, but that second-order effect is ignored for this analysis.

Finally, lumping the remaining elements whose contributions to scattering are minimal (namely, the elevation feed, circulators, transition, and everything else) yields Equation (6.67).

$$\sqrt{\sigma_{vf \text{ and } ctp}} \cong 0.925 \cdot \lambda \cdot (N \cdot M) \cdot (1 - \Gamma_r) \cdot (1 - \Gamma_{ft}) \cdot (1 - \Gamma_{ep})$$

$$\Gamma_{vf \text{ and } ctp} \cdot \cos(\theta) \cdot \sins(N \cdot 2 \cdot (U_x - \beta_1))$$

$$\cdot \sins(M \cdot 2 \cdot (V_x - \beta_0)) \cdot \cos(\Delta_p)$$

(6.67)

where β_1 is the equivalent phase shift resulting from the position stagger of the circulators, $\Gamma_{vf \text{ and } ctp}$ is the composite reflection factor from the vertical feed and circulators, and so on. In this case and normally, the circulator stagger is chosen to place the response on top of some other unavoidable structural scatterer.

Table 6.5 tabulates the major RCS contributors. Each element has its own contribution to RCS and, although their phase interaction is important, a first-order estimate can be obtained by root sum squaring (RSS) each as though independent. Based on the specific antenna architecture, the angular pattern of the RCS can be characterized as shown in the second column. Scattering from the radiator is approximately omnidirectional. Scattering from the input to the filter and transition returns through the radiator and has the pattern of a dipole. Scattering from the elevation phasors will be focused in azimuth, creating a fan beam in elevation whose peak will be in the commanded elevation scan direction, ϕ_0 in the example here. Reflections from the vertical feed will be focused in azimuth and elevation but at the normal in azimuth and at ϕ_0 in elevation. The circulator, transitions, and azimuth phasor are the last of the uncorrelated error terms of significance. The remaining scatterers are full antenna gain mode terms and hence have very low sidelobes in the intercardinal space where the threat is designed to be.

Each major element has a characteristic source of error, and the table also elaborates on the manufacturing or design technique used to mitigate the absolute return loss and its random component. The typical return loss achieved by these control techniques is tabulated in the last column of the table.

Table 6.6 tabulates the calculation of each of the contributors to the intercardinal RCS and to the spikes for an example array of the type shown in Fig. 6.42. Both the random and the correlated components of the RCS are tabulated, and their general properties are as described in Fig. 6.44.

For the example here, assume that the observer is at a negative elevation angle of 35° with like polarization. The beam phasing is at an angle of $\theta_0 = 15°$, $\phi_0 = 39°$, which results in an actual pointing direction of 43° off the normal and 67.6° below the horizontal, i.e., lower right quadrant viewed from the antenna. The wavelength is 1.5 cm, the array has 12 × 12 half-wave elements, the return losses are as given in Table 6.5, and they have a random component of 2 percent. The RCS at the observer is the RSS

Table 6.5   Summary of RCs Contributors in Example

Element	Nature of scatter	Nature of errors	Error control	Return loss typ. value (dB)
Radiator	Omnidirectional	Random errors omnidirectional	EDM slots to 0.0001λ	-20
Filter and transition	$\cos^2\theta$	Correlated errors focused at normal $\theta = 0, \phi = 0$	Machine to 0.0001λ	-25
Elevation phaser	$\sim \dfrac{\sin^2 M\theta}{M^2 \sin^2\theta} \cdot$ above	Fan beam in el. focused at $\theta = 0$, $\phi = \phi_0$ el scan	Each row at same phase	-20
Vertical feed	$\sim \dfrac{\sin^2 N\phi}{N^2 \sin^2\phi} \cdot$ above	Focused at $\theta = 0$, $\phi = \phi_0$	Every junction terminated	-26
Circulator, transition, and az phaser	As above	Focused at $\theta = 0$ $\phi = \phi_0$	Tight VSWR match	-20
Power divider	As above	Focused at $\theta = \theta_0$ $\phi = \phi_0$ mainbeam	Every junction terminated	-20
Horizontal feed and switches	As above	As above	As above	-20
Σ/Δ network	As above	As above	Tight VSWR match	-20

of a random RCS component and the correlated RCS associated with the architecture. Summing the terms in decibel form for the first row in the table yields −54 dBsm for the random component scattered from the radiator. The correlated RCS for this row in the table is the scattering from the radiators that is focused at the normal to the array. Only the changes or deltas (Δs) to the terms in the uncorrelated columns are stated; i.e., element gain, $\lambda^2/4\pi$, and return loss don't change and are not restated for the summation. The observer is in the sidelobes but, because the array is so small, the sidelobes are only down −25 dB. There is no error term, because correlated RCS is being calculated. There is an additional 22-dB gain from the coherent summation of the array. Each row of the table was calculated in the same way using the parameters unique to that set of scatterers. The last column calculates the power sum (i.e., uncorrelated) of the random and correlated RCS. The column is then power summed to obtain an overall predicted RCS at the observer look angle. The process is easily automated as shown in an Excel spreadsheet given in the appendices.

Table 6.6 Calculation of Example Antenna RCS

Element	Off Normal/retro RCS							Correlated RCS –Δs				RSS
	G²	λ²/4π	RL	Err	N×m	SL	dBsm	Err	SL	N×M	dBsm	Net
Radiator	4	-43	-20	-17	22	0	= -54	0	-25	22	= -40	-40
Filter and transition	4	-43	-25	-17	22	-4	= -63	0	-25	22	= -49	-49
Elevation phaser	4	-43	-20	-17	33	-30	= -73	0	-30	11	= -75	-71
Vertical feed	4	-43	-26	-28	43	-30	= -80	0	-30	0	= -63	-63
Circulator, transition, and az phaser	4	-43	-20	-17	33	-30	= -74	0	-30	11	= -56	-56
	G²		RL	Err	N×m	SL	dBsm	G²	RL	Circ.	SL	dBsm	Net
Power divider	Not significant < –80 dBsm							47	-20	-20	-10	= -46	
Horizontal feed and switches								47	-20	-20	-10	= -46	-41
Σ/Δ network								47	-20	-20	-10	= -46	

Total net = –38 dBsm

N   M
12 × 12   θ₀ = 15°   φ₀ = 39°

For the example, the RCS was also calculated using Equations (6.63) through (6.67). The antenna was assumed to be installed in a parallelogram test body with edge treatment. The edge treatment used on the test body improves its RCS in the vertical plane and creates some additional lobe structure. Because the element spacing is 1/2 wavelength, there are RCS grating lobes in real space for many scan angles, but the grating lobes point almost straight up or at endfire and would not be exploitable in most real engagement geometries. Figure 6.45 shows a contour plot of the RCS in u, v space. The location of the mainlobe, reflection from the elevation phasors, the principal grating lobe, the test body response, and a 35° cut are shown in the figure.

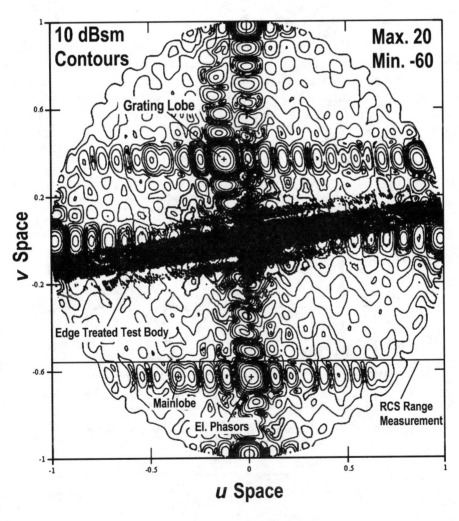

Figure 6.45 Small array in test body—correlated and uncorrelated RCS.

The first-order estimate of the mainlobe RCS from Table 6.6 was – 41 dBsm, and the second-order calculation summarized in Fig. 6.46 above shows –43 dBsm, which is good agreement between the two methods. The particular example used for Figs. 6.45 and 6.46 and Tables 6.5 and 6.6 was realized in hardware. Figure 6.47 shows the example unit in the parallelogram test body and with a measured RCS cut in a 4° elevation window through the -35° elevation plane. It shows good correlation between measurement and first- and second-order prediction. Some of the additional "grass" in the measurement is a result of the RCS range fixtures and other items not modeled, such as external vibration, RCS range walls, and thermal noise.

6.6.3 Estimating Errors Resulting from Circuit Variations

An exploded view of the "stick" element suggested in Fig. 6.41 is shown in Fig. 6.48. The objective of an analysis of this "stick" is an accurate prediction of both the regular and random components of the complex reflection factor, Γ, with frequency and scan angle. A typical phase shifter and other components are shown schematically in Fig. 6.48 and are briefly described in Table 6.7. Each major component in a stick may be made up of a number of minor elements as suggested by the element numbers in Fig. 6.48. For example, the two transitions on each end of the phasor from the waveguide transmission line to the phase shifter itself are made up of elements 24, 23, 22, and 10, 9, 8. To give insight to the level of detail ultimately required for analysis, elements 10 and 22 are the adhesive bond material between the transformers and the ends of the

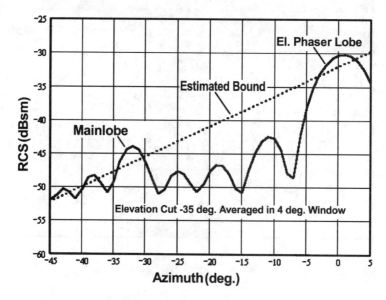

Figure 6.46 Calculated RCS for small antenna example.

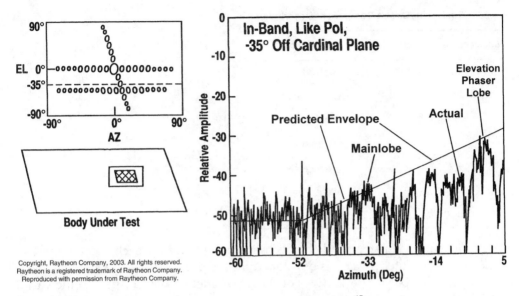

Figure 6.47 Comparison between prediction and measurement (Raytheon[15]).

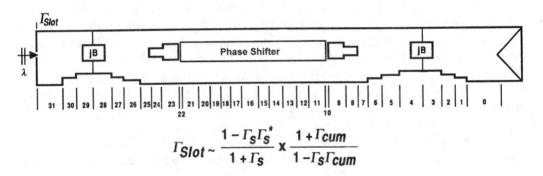

$$\Gamma_{Slot} \sim \frac{1 - \Gamma_s \Gamma_s^*}{1 + \Gamma_s} \times \frac{1 + \Gamma_{cum}}{1 - \Gamma_s \Gamma_{cum}}$$

Figure 6.48 RCS analysis single-element equivalent circuit.

phasor. A stick-by-stick buildup of an entire array will then yield a more accurate antenna system level prediction of the antenna-mode RCS over the field of view. An example analysis is given in this section. What is important here is the method, not the accidentals used in the specific example.[7]

The approach is to calculate the complex reflection factor as a function frequency at the input to the radiator by accumulating the cascade of susceptances from the first termination (usually by an isolator or circulator) all the way to the radiator. The susceptances present a range of values resulting from material variations and manufacturing tolerances, which when summed provide the overall tolerance on the response of the radiator allows accurate calculation of the antenna-mode RCS.

Table 6.7   Equivalent Circuit Component Description

Stick section number	Description
31	Slot radiator
30–26	Filter and transition
25	Waveguide
24–8	Phasor
7	Waveguide
6–4	Vertical feed
3–0	Transition and circulator

The propagation constant and characteristic impedance for each section are determined using computer programs developed for the purpose (e.g., an HFSS model for the element). The partial derivatives of these parameters with respect to the various mechanical dimensions, material constants are also determined numerically using these same programs. For a given set of dimensions, the impedance and propagation constant of each section can be calculated. The phase and amplitude of the total transmission and reflection coefficients can be determined by considering the network as a number of cascaded transmission line sections. The initial dimensions assigned to the various sections are taken to be the nominal dimensions of those components.

Transformer junctions can be represented by an equivalent circuit such as that shown in Fig. 6.49, which consists of a small lumped susceptance between two transmissions lines with different characteristic impedance and propagation constants. Because it is the change in the transmission coefficient rather than the absolute value that is important (constant reflection values are focused), some slowly changing susceptances are dropped. This assumption does not cause a significant error in the computation.

Referring to Fig. 6.50, Γ_n is the reflection coefficient at junction n when transmission line $n-1$ is terminated in a matched load. T_n is the ratio of the voltage propagating to

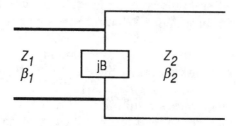

Figure 6.49 Lumped susceptance model for each component.

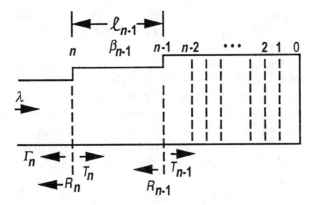

Figure 6.50 nth scattering component analysis.

the right of junction n to the voltage incident from the left of junction n. R_n is the total reflected voltage to the left of junction n. ℓ_{n-1} is the length of section $n-1$ and β_{n-1} the propagation constant associated with section $n-1$. The total transmission coefficient, TT_n, is the ratio of the voltage propagating to the right of junction 0 to the voltage incident from the left of junction n. Junction $n = 0$ is by definition totally absorptive, so $TT_0 \equiv 1$.

Recursion formulas can be developed from Fig. 6.50 as given in Equations (6.68). They can be used to calculate the total reflection and transmission coefficients through the cascaded sections of transmission line to the end point (antenna input terminals or other isolators).

$$R_n = \Gamma_n + \frac{R_{n-1} \cdot (1 - \Gamma_n^2) \cdot \exp(-j \cdot 2 \cdot \beta_{n-1} \cdot \ell_{n-1})}{1 + \Gamma_n \cdot R_{n-1} \cdot \exp(-j \cdot 2 \cdot \beta_{n-1} \cdot \ell_{n-1})}$$

$$T_n = \frac{1 + \Gamma_n}{1 + \Gamma_n \cdot R_{n-1} \cdot \exp(-j \cdot 2 \cdot \beta_{n-1} \cdot \ell_{n-1})}$$

$$TT_n = T_n \cdot TT_{n-1} \cdot \exp(-j \cdot \beta_{n-1} \cdot \ell_{n-1}) \tag{6.68}$$

The cumulative transmission and reflection coefficients occur in the recursion at the two ends of the antenna transmission line path when $n = N$ as given in Equations (6.69).

$$\Gamma_{cum} = R_N \text{ and } T_{cum} = TT_N \tag{6.69}$$

Unfortunately, the cumulative coefficients do not provide insight into the effects of tolerances, as they are made up of many contributors with different absolute sizes. Each element must be varied according to its own tolerances and the cumulative results calculated for each case. The cumulative calculation can use worst-case values (maximum for each) or random values (Monte Carlo) for the set of tolerances for each element. The total number of computations gets large fast, because there will be N elements times Q tolerance sets times F frequencies. The standard deviations of the complex transmission and reflection coefficients are determined using the standard Equations (6.70) and (6.71). This cascade obviously maintains its coherence, and tolerance errors may cancel or reinforce. The best designs actually adjust the transmission line parameters so that the cumulative Γ is optimized at the aperture slot. This often results in a reduction of transmission coefficient and thus antenna gain. As usual, there is a trade-off between passive signature, RCS, and active signature, LPI. For receive-only antennas, large losses are often accepted to improve RCS. For two-way data links and radars, this choice is never best, and losses of 1 to 2 dB are usually the maximum acceptable.

$$\sigma_\Gamma(F) = \sqrt{\frac{\sum\limits_Q \Gamma_{cum}^2(Q,F)}{Q(Q-1)} - \frac{\left(\sum\limits_Q \Gamma_{cum}(Q,F)\right)^2}{Q}} \tag{6.70}$$

$$\sigma_T(F) = \sqrt{\frac{\sum\limits_Q T_{cum}^2(Q,F)}{Q(Q-1)} - \frac{\left(\sum\limits_Q T_{cum}(Q,F)\right)^2}{Q}} \tag{6.71}$$

An example calculation using the method just described is summarized below. This particular example is an all-waveguide stealth array with ferrite phase shifter. Ridged waveguide is used to obtain a tight lattice spacing. The nominal tolerances assigned to the various components are listed in Table 6.8. The transmission and reflection coefficients associated with each of the phase states of a five-bit phase shifter were calculated for 60 sets of tolerance dimensions for the phasor and waveguide.

The dimensions of all the sections shown in Fig. 6.48 and Table 6.7 were allowed to vary, including the phasor position, thickness, width, and length. The phasor was assumed to be perfectly centered in the guide, and this dimension was not allowed to vary. The phasor itself has deviations from commanded phase that vary with frequency as well as dielectric and magnetic constants. Much of these deviations are compensated by a correction table, but there is a residual phase error made up of two terms, quantization and tolerance. For the 5-bit phasor in the example, the rms phase error is 5°. The waveguide cross-sectional dimensions of the phase shifter, Sections 4 through 28 inclusive, were assumed to be independent of those of the aperture, Sections 29 through 31, and the input guide sections 0 to 3. The cross-sectional dimen-

Table 6.8   Nominal Tolerances

Parameter	Dimension	Tolerance
Waveguide	Ridge height	±0.001
	Ridge width	±0.001
	Guide width	±0.001
	Guide height	±0.001
Phase shifter	Width	±0.001
	Thickness	±0.001
	Location	±0.003
	Gap	±0.004
	Magnetic constant	±2.0%
	Dielectric constant	±1.0%

sions within each of those three groups were not allowed to vary from section to section with the exception of the ridge gaps in the matching transformers at the aperture/phasor junction and the phasor/feed junction.

The results of the computations described above are given in Tables 6.9 and 6.10. Calculated errors for the nominal tolerances are given in Table 6.9.

Table 6.9   Variation Resulting from Nominal Tolerance

Dimensions varied	Amplitude var., %	Phase var., deg.	Percent of phase error
Waveguide	1.2	10.8	88
All others	0.3	5.7	46
Total	1.24	12.2	All

It is evident from the results shown in Table 6.9 that variations in the dimensions of the waveguide cross section contribute the bulk of the phase error, and of these the ridge gap, which contributes 48 percent of the total phase error, is the most sensitive dimension. The phase of the transmission coefficient was calculated at the center frequency, f_0 and $f_0 - 0.25$ GHz for the nominal ridge gap dimensions, and for ridge gaps 0.15 percent smaller than the nominal dimensions. The transmission coefficient was computed for each of the 32 phase states of phase shifter. The results of these computations are given in Table 6.10. The average phase error resulting from the change in ridge gap is 16.7° at f_0 and 16.8° at $f_0 - 0.25$ GHz. An additional error of ±5° is incurred at $f_0 - 0.25$ GHz as a result of the frequency change.

Another common engineering technique is calculation using double the target design tolerances as shown in Table 6.11. The errors given in Table 6.11 were computed for dimensional variations of twice the nominal tolerances. This can reveal where manufacturing tolerance sensitivities are highest. Calculations of this type allow mean and variance to be used in RCS prediction of the form given in the last section. Using this technique and the methods previously described, it is possible to calculate RCS within a few decibels of actual measurements.

6.7 Low-RCS Radomes

6.7.1 Introduction

Radomes first were made for weather protection, then for aerodynamics, and recently for stealth. They have been as simple as a fabric or membrane shroud over aperture and feed, such as those used for fixed-point microwave telecommunications. They also have been very complex, simultaneously having weather, aerodynamic, thermal, survivability, and stealth properties. Radome materials have ranged from rubberized cotton to plastic, to structural composites, to ceramic, and to machined metal. They naturally require both mechanical and electrical performance of significant difficulty. All radomes are bandpass in nature, whether intentional or not. The radome design problem is really two separate problems: the in-band transmission performance and the out-of-band reflection performance. Generally, in-band RCS is dominated by the antenna, not the radome. Far out of band, the RCS will be dominated by the shape of the radome. At band edge, the RCS will be a combination of the radome frequency selective surfaces and the antenna.[4,7,16–19]

6.7.2 Antenna and Radome Integration

Any low observables (LO) antenna will require a radome of some type. The combination of a stealth antenna and radome cannot be treated independently. Most commonly, radome designs will require frequency selective surfaces and, as such, will have a significant impact on emitter performance. Bandpass radomes usually require relatively high "Qs" and hence will require multiple frequency selective surfaces or circuits. In addition, such radomes are often polarization selective, and the polarization of the antenna radiation pattern incident on the radome for wide range of scan angles usually causes some polarization sensitive attenuation. Figure 6.51 shows a typical antenna-radome configuration and the dominant sources of scattering.

In addition, LO radomes are often shuttered when the emitter is not in use, and this gives rise to still another set of potential interactions. Bandpass radomes that are shuttered or switchable have been explored for many years. Primarily because of cost and

Table 6.10   Tabulation of Element RCS Simulation (Raytheon)[15]

Nominal phase shift*	Phase shift at f_0*	Phase shift resulting from ridge gap error*	Phase error*	Phase shift at $f_0 - 0.25$ GHz*	Phase error resulting from frequency*	Phase shift resulting from ridge gap at $f_0 - 0.25$ GHz*	Phase error at $f_0 - 0.25$ GHz
0	0	18.24	18.24	0	0	18.30	18.30
11.25	11.20	29.34	18.09	11.52	0.27	29.71	18.46
22.50	22.43	40.48	17.98	23.05	0.55	41.15	18.65
33.75	33.63	51.58	17.83	34.55	0.70	52.55	18.70
45.00	45.05	62.89	17.89	46.30	1.30	64.22	19.22
56.25	56.24	73.98	17.73	57.81	1.56	75.64	19.39
67.50	67.52	85.20	17.70	69.34	1.84	87.07	19.57
78.75	78.71	96.31	17.56	80.58	2.10	98.48	19.73
90.00	90.10	107.55	17.55	92.69	2.69	110.19	20.19
101.25	101.30	118.65	17.40	104.22	2.97	121.60	20.35
112.50	112.52	129.81	17.31	115.77	3.27	133.05	20.55
123.75	123.72	140.89	17.14	127.27	3.52	144.46	20.71
135.00	135.19	152.19	17.19	138.98	3.98	156.10	21.10
146.25	146.39	163.28	17.03	150.49	4.24	167.52	21.27
157.50	157.63	174.50	17.00	162.01	4.51	178.96	21.46
168.75	168.82	185.60	16.85	173.51	4.76	190.37	21.62
180.00	179.99	196.65	16.65	185.13	5.13	201.83	21.83
191.25	191.18	207.75	16.50	196.63	5.38	213.23	21.98
202.50	202.41	218.88	16.38	208.15	5.65	224.65	22.15
213.75	213.61	229.98	16.28	219.65	5.90	236.05	22.30
225.00	225.05	241.33	16.33	231.43	6.43	247.76	22.76
236.25	236.24	252.42	16.17	242.94	6.69	259.19	22.92
247.50	247.50	263.63	16.13	254.46	6.96	270.59	23.09
258.75	258.69	274.73	15.98	266.05	7.30	282.00	23.25
270.00	270.10	285.96	15.96	277.83	7.83	293.73	23.73
281.25	281.28	297.06	15.81	289.34	8.09	305.13	23.88
292.50	292.49	308.21	15.71	300.87	8.37	316.56	24.06
303.75	303.69	319.31	15.56	312.36	8.61	327.96	24.21
315.00	315.18	330.62	15.62	324.10	9.10	339.64	24.64
326.25	326.28	341.72	15.47	335.60	9.35	351.05	24.80
337.50	337.62	352.91	15.41	347.11	9.61	362.48	24.98
Average phase error (degrees)			16.74		4.95		21.73

*All entries in degrees.

Table 6.11   Variation Resulting from Double Nominal Tolerance

Dimension varied	Amplitude var., %	Phase var., deg.	Percent of phase error
Ridge gap	1.9	17.5	48
WG height	1.4	7.0	7.8
WG width	0.6	12.1	23.2
All others	0.6	11.4	21
Total	2.51	25.1	All

power, however, the most common switchable radomes are ones that are mechanically shuttered with either conductive or resistive surfaces. Certain configurations of mechanically shuttered radomes can be switched in 10 to 20 msec. This switching time is often a small fraction of threat surveillance radars' time-on-target. Hence, the radome can be closed as the surveillance system antenna pattern slides up on the low observables vehicle and before exposure to detection.

Behind the radome is an antenna compartment. In some cases, this compartment can be trivially small—perhaps just enough space for vibrational movement between the radome and the antenna. In other cases, the antenna compartment has a significant volume in it relative to the size of the antenna. This gives the low-observables designer more design trade-offs relative to RCS grating lobes and far antenna RCS sidelobes, but it requires the expense of a significantly larger structural design. Antenna compartments of this kind, whether small or large, require radiation-absorbing material (RAM) around the antenna and at the boundary between the radome and vehicle structure. To the extent that any other mechanical surfaces are visible, either di-

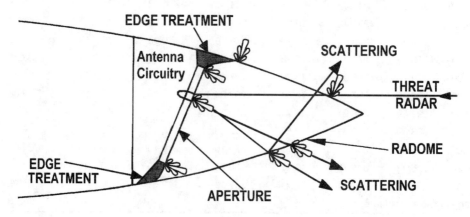

Figure 6.51   Antenna-radome interaction and scattering (adapted from Kay[4]).

rectly or by diffraction or multiple reflective bounces inside the cavity, surfaces must be treated as carefully as the antenna radiating surface itself.

In all cases, however, the radome is in the near field of the antenna and, as it is a frequency selective surface, its interaction with the antenna cannot be ignored. Table 6.12 shows an example radome/antenna interface design goal for antenna-radome interaction. As can be seen from the table, there is often substantial refraction through a bandpass radome. It is not uncommon for as much as a third of a beamwidth of refraction to be experienced over the full range of operating frequency and scan angles in a low-observables radome/antenna combination.

Table 6.12 Example Antenna/Radome Interaction

Radome/antenna interface design goals	Parameter
Frequency	X–band
Transmission (one-way) average complete radome ($f_0 \pm 200$ MHz)	85%
Transmission (one-way) minimum ($f_0 \pm 200$ MHz)	75%
Beam deflection ($f_0 \pm 200$ MHz) maximum	4.0 mils
Repeatability (radome to radome)	± 1.25 mils (max)
Beam deflection, averaged over 5° scan @ (f_0)	0.75 mils per degree
Antenna pattern sidelobe degradation, 1st and 2nd sidelobe ($f_0 \pm$ 200 MHz)	–28 dB or +4 dB increase, whichever is greater
Rms sidelobe level increase	4 dB max

Typically, a bandpass radome will provide an RCS reduction relative to the antenna alone. Out of band with the emitter off, a switched bandpass radome can provide a dramatic reduction in RCS relative to a conventional antenna by itself. Out of band with the emitter on, that reduction is not as great. In band with the emitter on, the RCS is essentially dominated by the antenna RCS. In band with the emitter off, the RCS can be close but poorer than the out-of-band performance with careful design.

The antenna RCS distribution over a hemisphere of arrival angles is usually distributed as follows. Less than 1 percent of the half space will have RCS as high as 15 dB above a square meter. This is, of course, a result of the specular normal to the antenna face. Another 10 percent of the half space will have RCS between 0 and 30 dB below a square meter, and the remainder of the half space can have RCS below –30 dB to –60 dBsm.

Often, the radome normal is far removed from the antenna look direction for obvious reasons. The frequency selective properties of the radome must be optimized for the look direction that requires the best performance (usually, the horizon). Because there is no free lunch, some other directions inevitably will have poorer performance. The reduced performance directions are usually chosen to be at angles where the required range performance is shorter. For example, Fig. 6.52 shows the calculated transmission performance for a single-layer slotted bandpass surface optimized for a successive set of optimum scan angles relative to normal. Typically, the desired optimum scan angle will be between 30° and 60°. In this case, the parameter being optimized is the slot length but, slot spacing also can be used to improve performance, as much of the improvement is the result of mutual coupling.

All radomes have some reflection of the antenna mainlobe, which gives rise to a large sidelobe as suggested in Fig. 6.51. For a radome containing an FSS, the reflection is significant even in band. It is not uncommon for the largest single sidelobe for an antenna-radome combination to be the reflection lobe *and* for it to be the largest single contributor to ISLR. Sometimes, this lobe is retroreflective and is within 20 dB of the mainlobe spike! The sidelobe both active and passive must be carefully analyzed, because the energy involved is relatively large. It can be modeled to a first order using geometrical optics ray tracing.

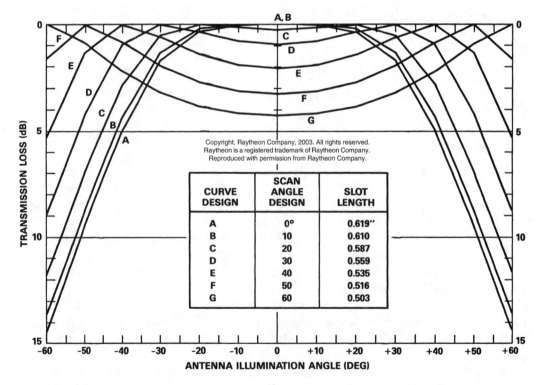

CURVE DESIGN	SCAN ANGLE DESIGN	SLOT LENGTH
A	0°	0.619"
B	10	0.610
C	20	0.587
D	30	0.559
E	40	0.535
F	50	0.516
G	60	0.503

Figure 6.52   Antenna-radome transmission (Raytheon[15]).

6.7.3 General Formulas[4]

The simplest radomes are made of solid sheets of dielectric material. The first-order analysis for this type of radome is given in Equations (6.72). The incident angle and polarization for a given dielectric constant and thickness determine the transmission and reflection coefficients. The apparent dielectric constant of the radome depends on the polarization of the incident wave relative to the local surface normal and, of course, the relative dielectric constant of the radome material. This is usually resolved into two orthogonal components: perpendicular and parallel. The perpendicular component has the incident E field polarization perpendicular to the plane containing the direction of propagation and the local surface normal. That is, the radome shunts the E field at its surface. The parallel component has the incident E field polarization parallel to the plane containing the direction of propagation and the local surface normal. The reflection and transmission properties of these two components is dramatically different. Anyone who has ever used a set of polarized sunglasses has practical experience with this phenomenon. Another important property is the electrical width of the radome, w, which is not a linear function of angle of arrival and dielectric constant. High-strength radomes typically have relative dielectric constants around 2.7 (epoxy-quartz fiberglass), and very high-temperature radomes have dielectric constants around 10 (alumina).

$$|\Gamma_{shell}|^2 = \frac{((1 - \varepsilon_{eff}) \cdot \sin(w))^2}{4 \cdot \varepsilon_{eff} + ((1 - \varepsilon_{eff}) \cdot \sin(w))^2}$$

$$|T_{shell}|^2 = \frac{4 \cdot \varepsilon_{eff}}{4 \cdot \varepsilon_{eff} + ((1 - \varepsilon_{eff}) \cdot \sin(w))^2}$$

$$w = \frac{2 \cdot \pi \cdot d}{\lambda} \sqrt{\varepsilon_r - \sin^2(\theta)}$$

$$\varepsilon_{eff\perp} = \frac{\varepsilon_r - \sin^2(\theta)}{\cos^2(\theta)} \text{ and } \varepsilon_{eff\parallel} = \frac{(\varepsilon_r \cdot \cos(\theta))^2}{\varepsilon_r - \sin^2(\theta)}$$

$$(6.72)$$

where d is the radome thickness and ε_r is the dielectric constant relative to free space (it can be an artificial dielectric as well). θ is the incidence angle relative to the radome surface normal. $\varepsilon_{eff\perp}$ is the effective dielectric constant for incident E field polarization perpendicular to the plane containing the direction of propagation and the surface normal. $\varepsilon_{eff\parallel}$ is the effective dielectric constant for incident E field polarization parallel to the plane containing the direction of propagation and the surface normal.

Figure 6.53 shows the transmission performance as a function of frequency for an epoxy quartz radome optimized for a 60° incidence angle and 9.7 GHz. The thickness is 0.43 inches with dielectric constant of 2.7. The bandwidth is quite narrow for perpendicular polarization and very broadband for parallel polarization.

6.7.4 Composite Radomes[4,7,11,12,16–19]

Often, stealth radomes are a composite of dielectrics and conductive sheets containing slots optimized for the operating band. One example is shown in Fig. 6.54. The conductive sheet is often called a frequency selective surface (FSS). The configuration of Fig. 6.54 provides out-of-band reflective properties, both from the dielectric layers

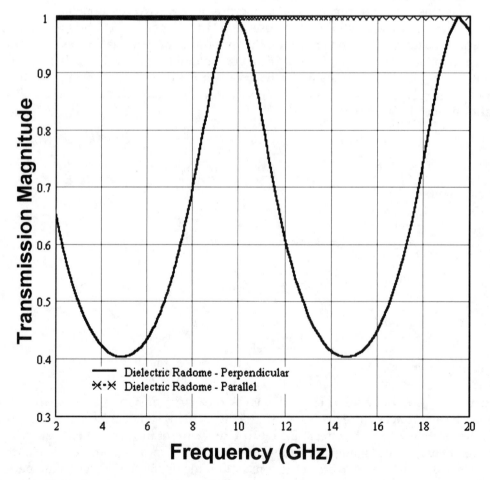

Figure 6.53 Dielectric radome transmission at 60°.[20]

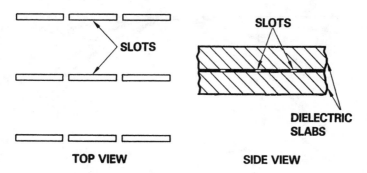

Figure 6.54 Single-layer frequency selective surface (FSS).

and the conductive sheet. The slots are similar to the ones described in Section 6.3. As a result of slot-to-slot mutual coupling, the resonant behavior of the slots in an array are more broadband and broadbeam than a single isolated slot in an infinite conducting plane (Fig. 6.12). Such radomes are surprisingly narrowband, nonetheless. The slot array can be modeled as a single tuned resonant circuit.

Although the structure looks simple, radomes of this type are difficult to make because of dissimilar materials, bonding, and the fact that the optimum slot orientation/polarization varies over the radome surface if it has compound curvature (i.e., spheres, ogives, ellipsoids, and so on). Figures 6.55 and 6.56 show some example composite single layer FSS radome transmission losses for variations in frequency and scan angle for a theoretical and real radome at X-band.

An important point about the transmission performance of Fig. 6.55 is that transmission only means not reflected. At very high incidence angles, transmission is not necessarily into the antenna behind the radome, because the wavefront may be captured on the radome surface or refracted enough to miss or partially miss the aperture. Note also that the radome may have a second passband at roughly double the design frequency as suggested by Figs. 6.53 and 6.56. Because the structure is periodic, there will also be RCS grating lobes at some frequencies. This latter property requires slot spacing as close as possible, typically 1/3 λ. Very tight lattices are a design challenge; sometimes, the slot itself is filled with a very high dielectric constant material to allow it to be made very small. If chosen just right, the second harmonic and grating lobe response can be dramatically lower. Unfortunately, this is at the expense of much greater complexity and manufacturing difficulty.

Figure 6.57 shows the cross section of a typical two-layer composite bandpass radome. Although this example is for rectangular slots, slot shape may be chosen to reduce slot spacing or polarization losses. The thickness of each layer is chosen to optimize the transmission in the maximum look direction similar to the slot trade-offs made in Fig. 6.52. Honeycomb has sometimes been used as the spacer but, unless the cell size is very small, it will have its own set of grating lobes and scattering.

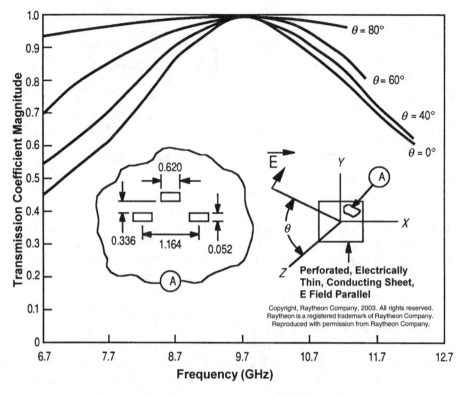

Figure 6.55 Single-layer E-plane transmission versus frequency.[14,15,20,21]

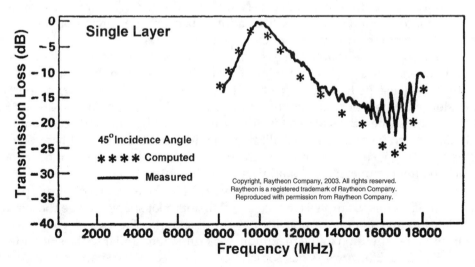

Figure 6.56 Single-layer FSS measured transmission versus frequency (Raytheon[15]).

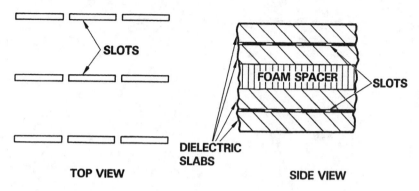

Figure 6.57 Double-layer FSS (adapted from Kay[4]).

Figures 6.58 and 6.59 show some example transmission losses for a two-layer composite radome for variations in frequency and scan angle for a theoretical and real radome at X-band. Of course, as before, the 1/4-wavelength spacing of each layer is chosen to optimize the look direction transmission. Note again that, because the radome causes refraction, transmission through the radome doesn't mean transmission to the antenna. It is possible with layer spacing as close as 1/20 of a wavelength to optimize transmission to the antenna and bandwidth by using resonance and mutual coupling effects. These mutual coupling effects, as well as surface waves, result in out-of-band rejection that is poorer than predicted by design calculations.

However, to place the transmission loss of Fig. 6.59 in perspective, more than 20 dB transmission loss for 8 GHz and below and between 12 to 17 GHz is a 40 dB attenuation of the antenna RCS and a 99 percent reflection of incident power. The radome will make a dramatic difference in the RCS of the antenna/radome combination.

Intimate contact between the dielectric layers and the FSS does not change the qualitative performance of a composite radome, as shown in Fig. 6.60 for calculated and measured transmission. Obviously, the size of the slot is affected by the slower propagation velocity of the radome dielectric.

A single tuned resonant circuit representing each slot is a simple model for a bandpass radome. The slot impedance match to free space determines the complex reflection factor. The radome dielectrics/structural shell can be modeled as discussed in Section 6.7.2. Although these radome elements are clearly not independent, if each element is spaced 1/4 wavelength apart, the reflections cancel, and the composite can be treated as a cascade of independent elements as given in Equations (6.68) and (6.69). Blass[4] suggests a simple model for radiation resistance and admittance for a slot in a ground plane. One form of that model is given in Equation (6.73).

$$Y_{\text{slot}} = g_{loss} + \left(\frac{f \cdot \ell}{13 \cdot c}\right)^2 + j \cdot \left[\frac{f \cdot \ell}{212.5 \cdot c} - \frac{c}{1050 \cdot f \cdot \ell}\right] \qquad (6.73)$$

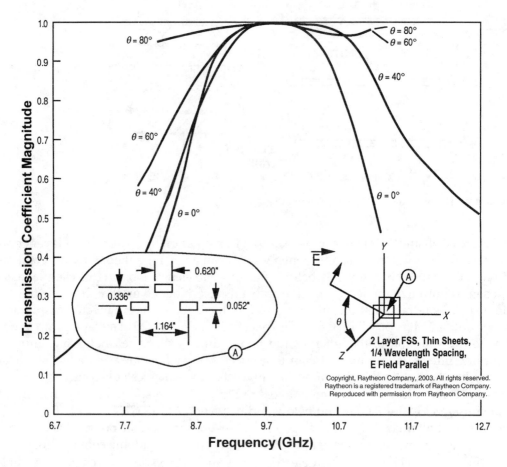

Figure 6.58   Double-layer FSS transmission versus frequency.[14,15,20,21]

where

g_{loss} = slot loss conductance
ℓ = the design length that is close to $1/2$ wavelength (0.48 typical)
f = frequency
c = the velocity of light

Using the above simple slot model, the slot reflection factor can be calculated as a function of frequency as shown in Equation (6.74).

$$\Gamma_{slot}(f) = \frac{Y_0 - Y_{slot}}{Y_0 + Y_{slot}} \tag{6.74}$$

where Y_0 is the admittance of free space 0.00265 mhos.

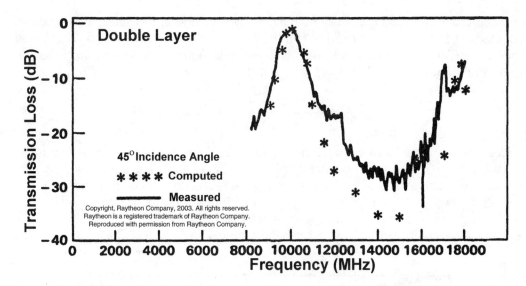

Figure 6.59 Double-layer FSS measured transmission versus frequency (Raytheon[15]).

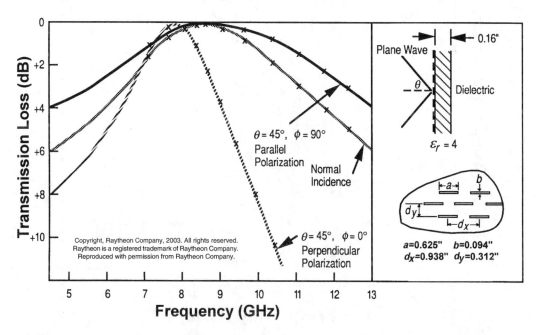

Figure 6.60 Measured and calculated transmission for composite radome (Raytheon[12,15]).

Using the same cascaded dissimilar transmission line approach of Fig. 6.50 and counterpart Equations (6.68) and (6.69), the overall reflection factor for a radome can the be calculated as shown in Equation (6.75). This equation uses the slot reflection factor of Equation (6.74) above and shell reflection factor from Equation (6.72).

$$\Gamma_{radome} = \frac{(1 - \Gamma_{shell}) \cdot \Gamma_{slot}}{1 + \Gamma_{shell} \cdot \Gamma_{slot}} + \Gamma_{shell} \tag{6.75}$$

Depending on the incident polarization, the combination of slotted sheets and a dielectric shell provides reflection factors as shown in Fig. 6.61. The figure uses the shell responses from Fig. 6.53 at 60° incidence angle (assumed to be the threat direction as well as the best performance direction for the emitter) and one or two layers of slots. Depending on polarization, the radome does not control the RCS for as much as ±4 GHz about band center. This is why the antenna RCS performance must be very

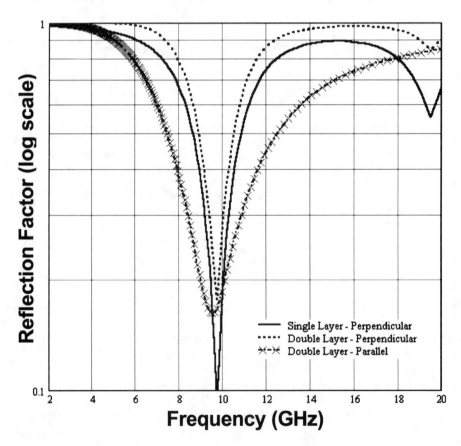

Figure 6.61 Composite radome reflection factor versus frequency.

wideband, and the radome must be complex. It seems that every single slot shape and configuration has been tested and analyzed. Rectangular, round, cross-shaped, H-shaped, Y-shaped, loaded, box-shaped, and mixtures of these shapes have been tried. This is great for graduate students, but there isn't some magic shape any more than there is for the radiators shown in Table 6.2. The biggest challenge is manufacturing complex radome structures. Often, a system designer opts for a conventional radome with an almost flat FSS and edge treatment behind the radome, which is much easier to build and maintain.

6.7.5 Thick Frequency Selective Layers[14]

One possible way to improve the bandpass nature of a frequency selective surface is to make it electrically thick relative to operating and threat wavelengths. The slots or holes then become waveguide transmission lines with exponential cutoff, not slots or dipoles with $1/f$ cutoff. These structures are much harder to analyze and require much more number smashing and experiment to achieve acceptable designs. The analysis usually requires method-of-moments or power series expansion of the possible propagating modes (typically up to 10 modes) in the waveguide-like slots/holes. For normal incidence, the reflection and transmission can be approximated by Equations (6.76) and (6.77).

$$\Gamma_{FSS} = \frac{1}{1 - j(C_{ss} + C_c \cdot \tanh(\beta\ell))} + \frac{1}{1 - j(C_{ss} + C_c \cdot \coth(\beta\ell))} - 1 \tag{6.76}$$

and

$$T_{FSS} = \frac{1}{1 - j(C_{ss} + C_c \cdot \tanh(\beta\ell))} - \frac{1}{1 - j(C_{ss} + C_c \cdot \coth(\beta\ell))} \tag{6.77}$$

where C_{ss} is a function of θ and φ related to the slot shape and spacing far-field pattern, C_c is constant related to slot coupling, and β is the propagation constant in the slot, and ℓ is the slot thickness.[16] There are equations for C_{ss}, C_c (usually called A, B), and β for many slot configurations in the literature, but finding approximations for these coefficients by experiment is usually required, and the slot pattern makes only a tiny difference in the RCS. Because most stealth radomes will be used at angles other than normal incidence, there is an approximation that can be used to estimate transmission and reflection performance at angles away from the normal beginning with Equations (6.76) and (6.77), given below in Equations (6.78).

$$T_{FSS\perp}(\theta) = T_{FSS}(\theta = 0) \cdot \cos^{2(1-p)}(\theta) \quad \text{and}$$

$$T_{FSS\parallel}(\theta) = T_{FSS}(\theta = 0) \cdot \cos^{-1.5(1-p)}(\theta) \tag{6.78}$$

where $T_{FSS\perp}$ is the transmission as a function of θ for perpendicular incidence, $T_{FSS\,\|}$ is the transmission as a function of θ for parallel incidence, and the porosity, p, is the ratio of the area of the holes to the area of the FSS.

Reflection coefficient magnitude as a function of FSS thickness relative to wavelength are plotted in Fig. 6.62 for normal incidence for a rectangular slot array designed for X band. The cutoff frequency of the reflection is periodic in-band as would be expected. The first cycle of this periodicity is approximately equal to the rectangular guide wavelength. Below the cutoff frequency, the reflection coefficient increases with thickness as shown for 4 and 6 GHz. In those cases, the incident wave is totally reflected if the FSS is sufficiently electrically thick.

As shown previously for thin FSSs, the reflection and transmission depend on both the polarization and the angle of incidence. Chen shows examples for three flat brass FSS plates 0, 0.03, and 0.06 inches thick, perforated with circular holes of 0.318 inch diameter. Some results are shown in Figs. 6.63 and 6.64. These holes were arranged in an equilateral triangular lattice and sandwiched between 0.028 inch thick dielectric

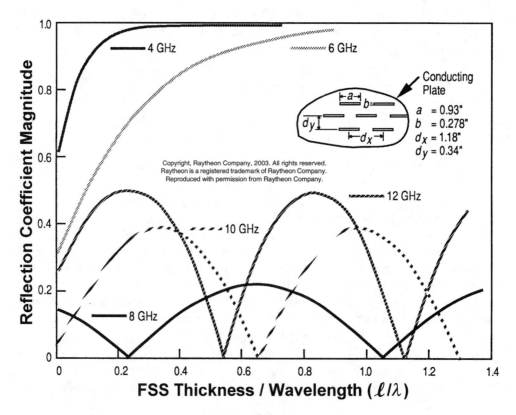

Figure 6.62   Thick FSS reflection coefficient (Raytheon[12,16]).

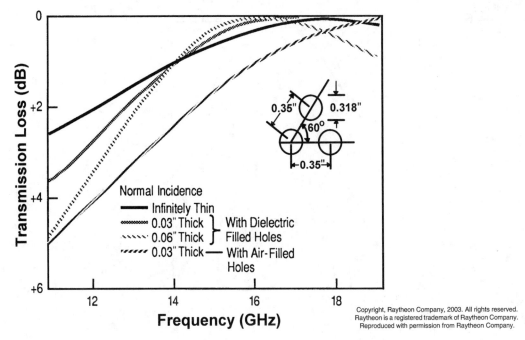

Figure 6.63 Transmission loss versus frequency—circular holes, $\theta = 0°$ (Raytheon[12,15]).

sheets with $\varepsilon_r = 3$ similar to Fig. 6.53. Figure 6.63 shows transmission loss as a function of frequency at normal incidence for four sets of alternate dimensions and dielectric.

Figure 6.64 shows the comparison of insertion loss for perpendicular and parallel incidence at an angle of 60°. As one would expect, when the holes are filled with dielectric, the propagation velocity is lower, and resonance shifts significantly. For perpendicular polarization, as the angle of incidence increases, both resonant frequency and bandwidth decrease relative to normal incidence. The band shape also varies with the plane of incidence in minor ways. However, this variation is small in an equilateral triangular lattice with round holes compared to the rectangular slot configurations shown earlier. It is also possible by selection of just the right thickness and slot size to significantly cancel the second harmonic FSS response with the reflection from the inside surface of the radome FSS.

The round hole type FSS of Figs. 6.63 and 6.64 has a significant advantage for electronic scanning, because off-axis polarization shifts that sometimes occur are not attenuated, and slots do not have to be oriented differently on different parts of a radome. That advantage, of course, means that the cross-polarization RCS of the antenna behind the radome must be much better, as the radome is no help. As hard as stealth radomes are to build, stealthy antennas are harder. Usually, cross polarization is controlled by a combination of antenna crosspol loading and radome rejection.

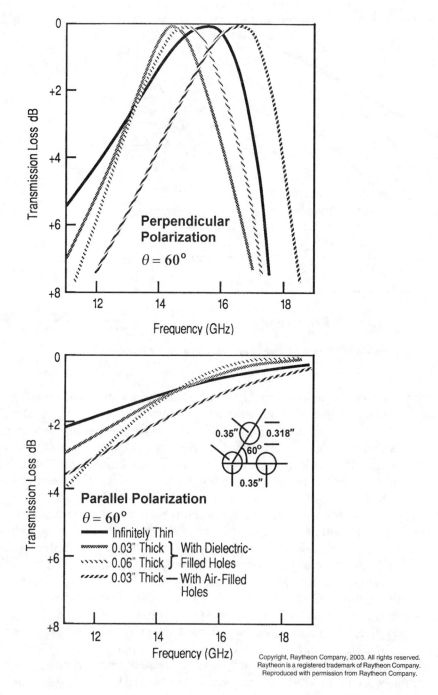

Figure 6.64   Transmission loss versus frequency—circular holes, $\theta = 60°$ (Raytheon[12,15]).

6.7.6 Edge Treatment

Both the antenna aperture and the radome trailing edges require treatment to improve travelling wave and sidelobe RCS. It is possible, with a relatively small amount of lossy material or curvature, to dramatically improve RCS away from the specular (only match to free space helps reduce the specular). For example, consider the RCS test body of Fig. 6.65. Without any edge treatment, it has the classic sins(U,V) shape using the physical optics approximation integrated over the entire body. But where is that RCS really coming from, and what are the major contributors?

The answer to the major RCS contributor question can be found by measuring the RCS of the test body with very high range and cross-range resolution. Figure 6.66 shows an RCS image with 3-inch resolution for the body of Fig. 6.65 from a test range. This image was created with a step chirp range waveform, and cross range was created by Doppler separation by slowly rotating the body through enough wavelengths to allow Doppler filters with the equivalent of 3 inches in cross range. Doppler processing to image a rotating body is usually called *inverse synthetic aperture radar (ISAR)*.

As with all real data, some interpretation is required. As with all filtering, pulse compression, and antenna apertures, there are sidelobes in Fig. 6.66. The most prominent RCS features are the leading and trailing edges of the test body, which peak at –24 dBsm for each resolution cell whose physical area of 3×3 inches is approximately –22.4 dBsm. The other features noted in the image include the test body supporting column, which has been treated but is still not invisible, scattering from a creeping wave returning from the back support brackets on the test body that are –57 dBsm and ISAR filter sidelobes, which are roughly –15 dB down from the edge response. The test body overall dimensions are roughly 3×4 ft with specular approximately 42 dBsm. Virtually all the RCS for the test body is from the edges once the look direction is away from the specular!

Edge treatments can be as simple as lossy metallized mylar film (which you can buy at the hardware store!) placed on the surface of a conductor or dielectric to thick

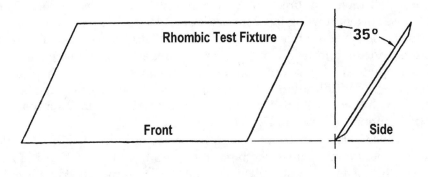

Figure 6.65   Example RCS edge test.

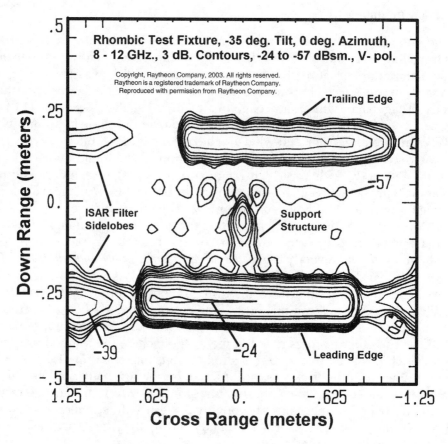

Figure 6.66   Measurement of RCS resulting from edges.

multilayer structures having special electric and magnetic properties with shape optimized for the location and direction of incidence (which are *expensive!*). Surface films
work, because current flowing in a conductor or dielectric has part of the field outside
the conductor or dielectric that is coupled into the film. Rigorously analyzing edge
treatment is difficult, but the phenomena can be modeled easily using convolution in
the physical optics RCS approximation. Convolution of one current shape on another
at the aperture is multiplication of the transforms of those shapes in the far field. For
example, convolution of a 4λ rectangular aperture on a 30λ rectangular plate yields a
current distribution as shown in Fig. 6.67.

This particular current distribution can be achieved with metallized mylar film,
which has a linear ohms-per-square taper that goes from 0 to 1600 Ω in four wavelengths. The RCS of a rectangular plate, Equation (6.56), is modified by the edge treatment multiplier as shown in Equation (6.79).

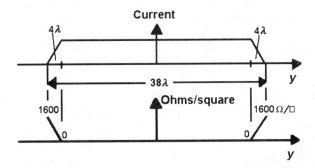

Figure 6.67   Current distribution and edge treatment resistance across rectangular plate.

$$\sigma(\theta,\phi) \cong \frac{4 \cdot \pi \cdot ((30 + 0.5 \cdot (4 + 4)) \cdot \lambda)^2 \cdot a^2 \cdot \Gamma^2 \cdot \cos^2(\theta)}{\lambda^2}$$

$$\cdot \frac{\sin^2(2 \cdot U)}{4 \cdot U^2} \cdot \frac{\sin^2((2 \cdot V))}{4 \cdot V^2} \cdot \frac{\sin^2(2 \cdot V1)}{4 \cdot V1^2} \tag{6.79}$$

where *V1* is given in Equation (6.80), and all the other variables are as defined for Equations (6.52) and (6.56).

$$V1 = \frac{\pi \cdot (4 \cdot \lambda)}{\lambda} \cdot \sin(\theta) \cdot \sin(\phi) = 4 \cdot \pi \cdot \sin(\theta) \cdot \sin(\phi) \tag{6.80}$$

A photo of the test body of Figs. 6.65 and 6.66 with a four-wavelength film edge treatment as modeled in Fig. 6.67 is shown on a test range in Fig. 6.68 (the film is the shiny strip at top and bottom). Surprisingly enough, a four-wavelength edge taper provides more than 20 dB improvement in RCS lobes at 35° off the normal, as calculated in Fig. 6.69. The obvious double-lobe structure for the treated curve of Fig. 6.69 is the result of the width of the edge strip. The actual lobe structure of a treated rhom-

Figure 6.68   Rhombic test body edge treatment (Raytheon[15]).

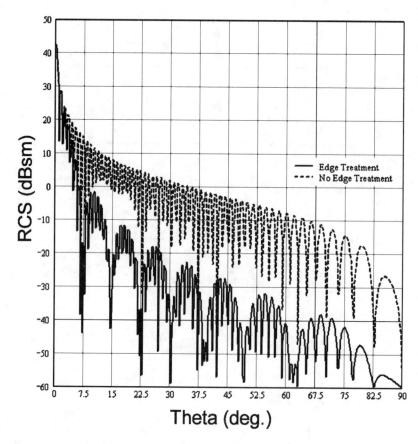

Figure 6.69   Comparison of test body with and without edge treatment.

bus is visible in Fig. 6.47. Calculation and measurement are in reasonable agreement down to about 30 to 40 dB below a square meter.

From an overall stealth point of view, this RCS signature reduction comes at the price of a smaller aperture and hence more transmitted power as well as active interceptability. The right trade-off is based on the postulated threat. A 2-D contour plot in u, v space of the before and after effect of the edge treatment is shown in Fig. 6.70. What is compelling is that almost none of space is below –60 dBsm for the untreated case, whereas more than half of visible space is below –60 dBsm with this simple treatment.

6.7.7   Coordinate Rotations

The equations for active antenna and passive RCS patterns for many basic shapes have been given in the text. As always happens in textbooks, the equations are pro-

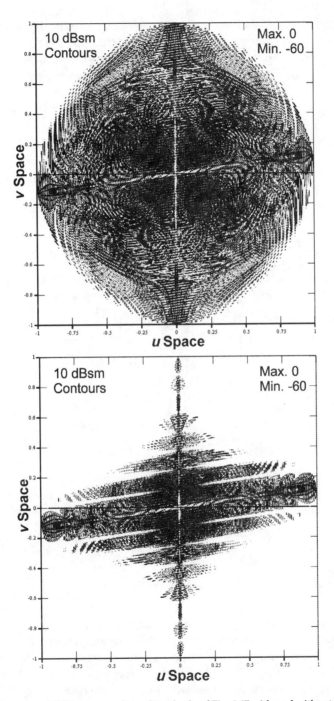

Figure 6.70 Two-dimensional RCS contour plots of test body of Fig. 6.67 with and without edge treatment.

vided in the simplest coordinate space, and transforms to actual locations on a vehicle are left to the student. Unfortunately, the transforms are very messy, but they are straightforward bookkeeping problems. A rotation about all three cartesian coordinate axes is given in Equations (6.81). (I told you they were messy!)

$$x = \rho \cdot \sin(\theta) \cdot \cos(\phi), \; y = \rho \cdot \sin(\theta) \cdot \sin(\phi), \; z = \rho \cdot \cos(\theta)$$

and

$$U_x = \frac{\pi \cdot dx}{\lambda} \sin(\theta) \cdot \cos(\phi), \; V_y = \frac{\pi \cdot dy}{\lambda} \sin(\theta) \cdot \sin(\phi),$$

$$W_z = \frac{\pi \cdot dz}{\lambda} \cos(\theta),$$

α_x, α_y and α_z are rotation angles

about the x, y, and z axis respectively, and

$$\begin{bmatrix} U_x' \\ V_y' \\ W_z' \end{bmatrix} = \begin{bmatrix} a_{11} \; a_{12} \; a_{13} \\ a_{21} \; a_{22} \; a_{23} \\ a_{31} \; a_{32} \; a_{33} \end{bmatrix} \begin{bmatrix} U_x \\ V_y \\ W_z \end{bmatrix}$$

where the a_{ii} are the direction cosines for each axis

$$a_{11} = \cos(\alpha_y)\cos(\alpha_z), \; a_{21} = \cos(\alpha_y)\sin(\alpha_z)$$

$$a_{31} = -\sin(\alpha_y), \; a_{32} = \sin(\alpha_x)\cos(\alpha_y)$$

$$a_{33} = \cos(\alpha_x)\cos(\alpha_y)$$

$$a_{12} = \cos(\alpha_x)\sin(\alpha_y)\cos(\alpha_z) - \cos(\alpha_x)\sin(\alpha_z)$$

$$a_{13} = \cos(\alpha_x)\sin(\alpha_y)\cos(\alpha_z) + \sin(\alpha_x)\sin(\alpha_y)$$

$$a_{22} = \cos(\alpha_x)\sin(\alpha_y)\sin(\alpha_z) + \cos(\alpha_x)\sin(\alpha_z)$$

$$a_{23} = \cos(\alpha_x)\sin(\alpha_y)\sin(\alpha_z) - \cos(\alpha_x)\cos(\alpha_z)$$

$$(6.81)$$

The above rotation equations work equally well for x, y, z or u, v, w space. For example, consider a radome and antenna configuration as shown in Fig. 6.71. It consists of an ogival bandpass radome and a diamond-shaped planar antenna array canted upward at an angle of α_x (Fig. 6.71). The example has edge treatment for both radome and antenna, which also is canted upward. Suppose the cant angle was 25° about the x axis only. The edge can be modeled as a wire loop rotated 25° about the x axis, the

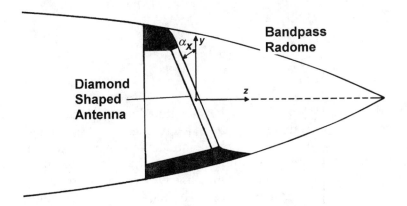

Figure 6.71 Example antenna coordinate transforms.

edge treatment can be modeled as a linear current taper also rotated 25° about the x axis, and the antenna can be modeled as a square array rotated through 45° about the z axis followed by a rotation of 25° about the x axis. The ogive will be assumed to have its axis along the z axis, which has arbitrarily set pointing at the horizon (where the threat usually is). The RCS equations to be used will be from earlier in this chapter and from Chapter 1.

One useful intellectual crutch is to visualize a total rotation as successive rotations about each of the cardinal axes. For the example here, the counterpart Equations (6.82) below show a single z-axis rotation. Each rotation can be cascaded to get the final pointing direction. The 45° rotation about the z axis for an array is then

$$\begin{bmatrix} U_x' \\ V_y' \\ W_z' \end{bmatrix} = \begin{bmatrix} 0.707 & -0.707 & 0 \\ 0.707 & 0.707 & 0 \\ 0 & 0 & 1 \end{bmatrix} \begin{bmatrix} U_x \\ V_y \\ W_z \end{bmatrix}$$

Thus,

$$U_x' = \frac{\pi \cdot dx \cdot 0.707 \cdot \sin(\theta)}{\lambda} (\cos(\phi) - \sin(\phi))$$

$$V_y' = \frac{\pi \cdot dy \cdot 0.707 \cdot \sin(\theta)}{\lambda} (\cos(\phi) - \sin(\phi))$$

$$W_z' = W_z = 0 \tag{6.82}$$

Similarly, the 25° rotation about the old x axis as well as the 45° about the old z axis is given in Equation (6.83).

$$\begin{bmatrix} U_x'' \\ V_y'' \\ W_z'' \end{bmatrix} = \begin{bmatrix} 0.707 & -0.641 & 0.299 \\ 0.707 & 0.641 & -0.299 \\ 0 & 0.423 & 0.906 \end{bmatrix} \begin{bmatrix} U_x \\ V_y \\ W_z \end{bmatrix} \tag{6.83}$$

The desired rotated U_x'' and V_y'' can be substituted into Equation (6.56) to calculate an RCS first scatter from a diamond shaped array tilted up at a 25° angle. For the example here, W_z'' won't be used but, if phase interactions are to be considered between elements, then the z location comes into play. Similarly, if the back edge of the radome is also tilted up at the same angle, then the same type of transformation can be used in the wire loop model of the ogive back edge. The back edge will be roughly elliptical in shape. These elements, when combined with the radome, will provide a first-order estimate of RCS. The resulting U_x'' and V_y'' are given in Equation (6.84).

$$U_x'' = \frac{\pi \cdot dx}{\lambda} \cdot [(0.707 \cdot \cos(\phi) - 0.641 \cdot \sin(\phi)) \cdot \sin(\theta) + 0.299 \cos(\theta)]$$

$$V_y'' = \frac{\pi \cdot dy}{\lambda} [(0.707 \cdot \cos(\phi) + 0.641 \cdot \sin(\phi)) \cdot \sin(\theta) - 0.299 \cos(\theta)]$$

$$\tag{6.84}$$

Figure 6.72 shows the first scatter RCS pattern of a 2500 element symmetrical diamond antenna array at x-band with 0.4 λ spacing with $\Gamma = 1$ and no edge treatment. The figure is based on Equations (6.84) substituted into Equation (6.56). Note that most of space in the absence of manufacturing tolerances is below −50 dBsm. Also note that the cardinal axes appear curved. The peaks in the horizon plane are roughly −10 dBsm (marked by the zero line in the figure).

6.7.8 Radome and Antenna RCS

The antenna and radome are in the near field of one another, and an estimate with an accuracy better than 3 dB requires the phase terms. However, the antenna and radome in the example of Fig. 6.70 are far enough apart that the phase term can be ignored for a first-order estimate of the combined RCS. The first-order model is summarized in Equation (6.85) below.[21]

$$\sigma \approx \left| \sum_{m=1}^{M} \sqrt{\sigma_m} \right|^2 \tag{6.85}$$

For the case here, the combined antenna and radome RCS must take into account the effect of two passes through the radome. Of course, the reflection factor is a func-

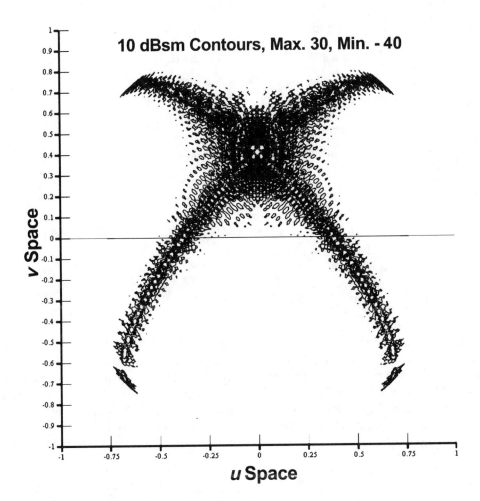

10 dBsm Contours, Max. 30, Min. - 40

v Space

u Space

Figure 6.72   2500-element diamond array, 25° uptilt, $\Gamma = 1$.

tion of frequency as described in Sections 6.7.2 through 6.7.4, and the RCS will be quite different in and out of band. A specific simple model is given in Equation (6.86).

$$\sigma_{ant+\,rad} \approx \left| \sqrt{\sigma_{ant}} \cdot (1 - \Gamma_{rad})^2 + \Gamma_{rad} \cdot \sqrt{\sigma_{rad}} \right|^2 \tag{6.86}$$

where Γ_{rad} is the reflection factor of the radome, which is a function of frequency, polarization, and angle; σ_{ant} is the RCS of the antenna and edge treatment by itself; and σ_{rad} is RCS of the radome by itself. For the extended example here, the antenna will be as described in Section 6.7.6. It will be modeled on Equations (6.56) and (6.79) and the results of Equation (6.84) as tabulated in Equation (6.87).

$$\sqrt{\sigma_{ant}} \approx (2 \cdot \sqrt{\pi}/\lambda) \cdot (M \cdot dy) \cdot (N \cdot dx) \cdot \Gamma \cdot \cos(\Psi)$$
$$\cdot \; \text{sins}(N \cdot 2 \cdot U_x'') \cdot \text{sinc}(12 \cdot \pi \cdot U_x'')$$
$$\cdot \; \text{sins}(M \cdot 2 \cdot V_y'') \cdot \text{sinc}(12 \cdot \pi \cdot V_y'')$$

$$(6.87)$$

where

$$U_x'' = \frac{\pi \cdot dx}{\lambda} \cdot [(0.707 \cdot \cos(\phi) - 0.641 \cdot \sin(\phi)) \cdot \sin(\theta) + 0.299 \cos(\theta)] \quad (6.88)$$

and

$$V_y'' = \frac{\pi \cdot dy}{\lambda} \cdot [(0.707 \cdot \cos(\phi) + 0.641 \cdot \sin(\phi)) \cdot \sin(\theta) - 0.299 \cos(\theta)]$$

and

$$\Psi = (U_x''^2 + V_y''^2)^{0.5}$$

Figure 6.73 shows a contour plot of Equation (6.87) with $M = 50$, $N = 50$, $6 \cdot \lambda$ wide edge treatment, and $\Gamma = 0.1$ for 10-dBsm contours with a maximum value slightly under 10 dBsm and a minimum value arbitrarily limited to –60 dBsm. Because the array is tilted back at an angle of 25°, there is no cross section above –60 at the horizon. Obviously, manufacturing tolerances would cause the realized RCS to be well above –50 dBsm, but that would obscure understanding of the underlying behavior.

Similarly, the ogive radome can be modeled as shown in equation (6.89) and (6.90).

$$\sqrt{\sigma_{rad}} \approx \sqrt{\sigma_{diffraction}} + \sqrt{\sigma_{backend}} + \sqrt{\sigma_{specular}} \quad (6.89)$$

where $\sigma_{diffraction}$ accounts for tip diffraction as well as surface and travelling waves, $\sigma_{backend}$ accounts for the ogive back edge including treatment, and $\sigma_{specular}$ accounts for the ogive specular reflection at large angles as well as the effect of edge treatment. An ogive can be defined by its length from tip to center, L, and its maximum radius, a, which is at the center. This in turn defines the tip angle, α, and the large radius of curvature, r_1. These parameters are mentioned in Table 1.6 and Fig. 1.20. The tip diffraction can be modeled as given in Table 1.6, and the travelling and surface wave can be modeled using a stationary phase physical optics approximation similar to tip diffraction as $a^2 \sin^4(\theta)$. The back end of the ogive of Fig. 6.69, where the edge treatment begins, can be modeled as a lossy elliptical wire loop as in Fig. 1.20, which in turn can be approximated by $\pi \cdot a^2 \cos^2(\Psi 1) \cdot J_0^2(2 \cdot k \cdot \Psi 1)$ where $\Psi 1$ is the locus of the elliptical loop in u, v space. Lastly, the ogive specular can be modeled as given in Table 1.6, but it must take into account the mild attenuating effects of the edge treatment for surface

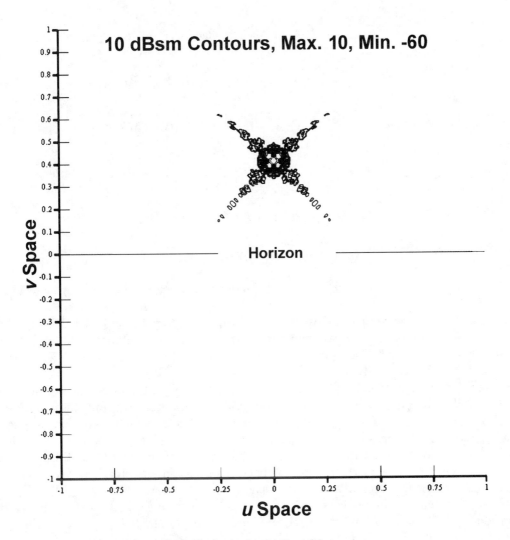

Figure 6.73 2500-element antenna RCS, 6λ edge treatment, $\Gamma_{ant} = 0.1$.

normals near the back of the ogive. The foregoing attenuation can be modeled as the $\cos(\theta)/\sin(\alpha)$. Although this all seems tortured, this is a simple model that provides reasonable agreement below band and in band. What is missing are the grating lobe effects for bandpass slots/holes, which will arrive in real space above band. The slot spacing strategy is to place the slots close enough together so that there is no RCS grating lobe near the horizon ($\pm 10°$) until the slots are well into cutoff. This requires 0.3λ inband spacing, which is difficult. Equations (6.90) summarize the elements of the radome RCS.

$$\sqrt{\sigma_{diffraction}} \approx \frac{\lambda}{\sqrt{2 \cdot \pi}} \cdot (a \cdot \sin^2(\theta) + \tan^2(\alpha))$$

$$\sqrt{\sigma_{backend}} \approx \sqrt{\pi} \cdot \Gamma_{be} \cdot a \cdot \cos(\Psi 1/a) \cdot J_0(2 \cdot k \cdot \Psi 1)$$

$$\sqrt{\sigma_{specular}} \approx \sqrt{\pi} \cdot \Gamma_{sp} \cdot r_1 \cdot \left[\frac{\cos(\theta)}{\sin(\alpha)} - \cot(\alpha) \cdot \cot(\theta)\right]^{0.5}$$

where

$$r_1 = \frac{L^2 + a^2}{2 \cdot a}, \quad \alpha = \mathrm{asin}\left[\frac{2 \cdot a \cdot L}{L^2 + a^2}\right]$$

$$\Psi 1 = \left[\begin{array}{c}(a \cdot \sec(\alpha_x) \cdot \sin(\theta) \cdot \cos(\phi))^2 + \\[6pt] (a \cdot \cos(\alpha_x) \cdot \sin(\theta) \cdot \sin(\phi) - a \cdot \sin(\alpha_x) \cdot \cos(\theta))^2\end{array}\right]^{0.5}$$

$$(6.90)$$

For example, assuming $a = 0.61$ m., $L = 2.4$ m., $\Gamma_{sp} = 0.15$, $\Gamma_{be} = 0.1$, $\alpha_x = 25°$ and $\lambda = 0.0308$ m. in band with vertical polarization, the radome by itself has an RCS pattern as shown in Fig. 6.74. The annulus above 0 dBsm in the figure peaks at 75° of the ogive axis and is typically not exploitable by the threat.

Combining the antenna and radome RCS elements using Equations (6.85) through (6.88) yields the total RCS. Two cases are provided in the figures below. Figure 6.75 is below band at 3 GHz, and Fig. 6.76 is in band at 9.7 GHz. For the below band case, the reflection factor is very close to 1, and the RCS is dominated by the radome. A significant fraction of the forward hemisphere is below −30 dBsm. The RCS still peaks at about 75° at a little over 6 dBsm, and the back edge response is −18 dBsm at 25° uptilt. This performance is readily achievable assuming the radome is carefully made, smooth, and clean. The effects of smashed insects and rain erosion must be removed or repaired. The surface tolerances required are about 1/16 inch. Manufacturing tolerances have not been included in the calculation of Fig. 6.75 for clarity.

The second case, which is in band at 9.7 GHz, is much more complex, because the frequency is higher, and both the antenna and the radome contribute to the RCS response. As previously mentioned, the 2500-element array used in the example is assumed to have the same parameters as Fig. 6.73. For clarity, and because the RCS patterns are symmetrical in the example, the contour plot in Fig. 6.76 is split in half, and the lower threshold is set at −50 dBsm on the left side of the figure and thresholded at −40 dBsm on the right side. The antenna peak RCS is 10 dBsm at 25° uptilt, and the radome peak at 75° is still a little over 6 dBsm. One can see on the right side that most of the half space is below −40 dBsm. A horizon line has been drawn in to provide a maximum threat reference. Again, manufacturing tolerances have not been

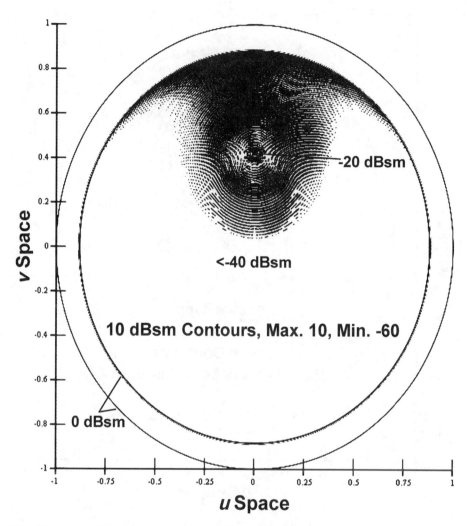

Figure 6.74 Edge-treated bandpass ogive radome RCS, back edge 25° uptilt.

included, for clarity's sake. In this case, however, all tolerances are much harder to meet. The RSS must be nine times better, and there are many more contributors, so individual component tolerances must be very tight indeed.

6.8 EXERCISES

1. Calculate the ideal antenna sidelobes for a 50-element linear array with uniform, Hanning, and Hamming weighting with 0.5λ element spacing at 30° from broadside.

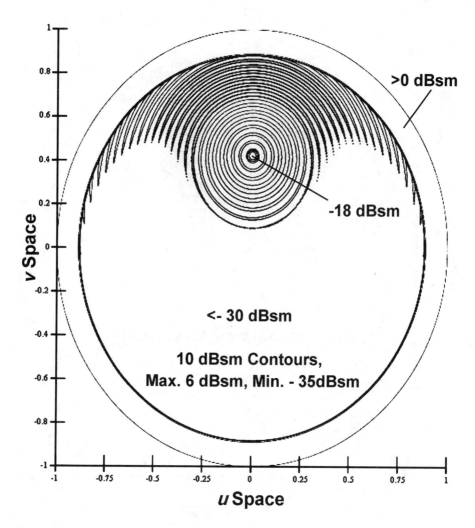

Figure 6.75 Radome and antenna RCS below band.

2. Calculate the ideal antenna sidelobes for a 50-element linear array with raised co-sine paired echo weighting of spacing $1.33\lambda/d \cdot N$ and $\alpha = 0.6$ with 0.5λ element spacing at 30° from broadside.

3. Calculate the antenna sidelobes for a 50-element linear array with Hamming weighting, 0.5λ element spacing, element factor $0.5 + 0.5\cos(1.33\theta)$, and a periodic error of 10 percent every 3 elements.

4. Calculate the antenna sidelobes for a 50-element linear array with Hamming weighting, 0.5λ element spacing, element factor $0.5 + 0.5\cos(1.33\theta)$ and a random error of 10 percent rms.

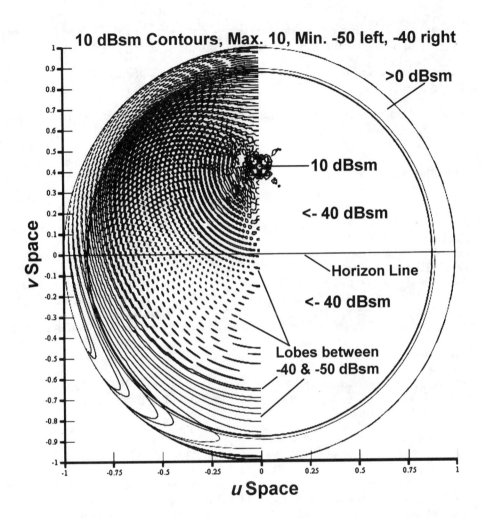

Figure 6.76 Radome and antenna RCS in band.

5. Calculate the RCS of a bandpass radome below band for a radome made up of four right triangles as shown in the figure below.

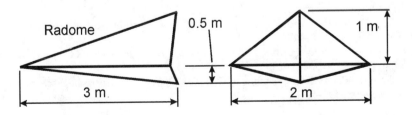

6. Calculate the RCS of a right dihedral where each side is $a \times a$ in size in the plane perpendicular to the dihedral intersection line for $a = 0.5\lambda$, 10λ.

7. Calculate the RCS of a rectangular array at $\theta = 45°$, $\phi = 45°$ for $M = 60$, $N = 40$, $\Gamma = 0.3$, and $E_e = 1$.

8. Calculate the location of the first active and RCS grating lobes for a 100-element linear array for element spacings of 0.3λ, 0.4λ, 0.5λ, 0.8λ.

6.9 REFERENCES

1. Skolnik, M., *Introduction to Radar Systems*, McGraw-Hill, New York, 1962, pp. 260–269, 336–343.
2. Silver, S., Ed., *Microwave Antenna Theory and Design*, MIT Radiation Laboratory Series, vol. 12, McGraw-Hill, New York, 1949, pp. 92–104.
3. Balanis, C., *Antenna Theory*, John Wiley & Sons, New York, 1982, pp. 100–128, 446–522.
4. Jasik, H., Ed., *Antenna Engineering Handbook*, McGraw-Hill, New York, 1961, pp. Jasik 2-30 to 2-41, Blass 8-1 to 8-4, Kay 32-1 to 32-29.
5. Buxton, C., "Design of a Broadband Array Using the Four Square Radiating Element," Doctoral Dissertation, Virginia Polytechnic Institute, Jun. 2001, pp. 6–22.
6. Shelby, S., Ed., *CRC Standard Mathematical Tables*, 16th ed., CRC Press, Boca Raton, FL, 1968, p. 92.
7. Jenn, D., *Radar and Laser Cross Section Engineering*, AIAA, Washington, DC, 1995, pp. 441, 248–306, 313–360.
8. Brookner, E., Ed., *Radar Technology*, Artech House, Dedham, MA, 1977, pp. 289–301.
9. Ridenour, L. N., *Radar System Engineering*, MIT Radiation Laboratory Series, vol. 1, sec. 3.7, McGraw-Hill, New York, 1947.
10. Severin, H., "Nonreflecting Absorbers for Microwave Radiation," *IRE Transactions on Antennas and Propagation*, vol. AP-4, pp. 385–392, July, 1956.
11. Rope, E. R., T. E. Fiscus, and G. Tricoles, "Perforated Metallic Shells for Radome Applications," General Dynamics, Technical Report R-70-032-F, July 1971.
12. C. C. Chen, "Transmission of Microwave through Perforated Flat Plates of Finite Thickness," *IEEE Trans. Microwave Theory Tech.*, vol. MTT-21, Jan. 1973, pp. 1–6.
13. Harrington, R. F., *Field Computation by Moment Method*, New York: Macmillan, 1964.
14. Hughes Aircraft, "LPIR Phase 1 Review," unclassified report, 1977.
15. A few photos, tables, and figures in this intellectual property were made by Hughes Aircraft Company and first appeared in public documents that were not copyrighted. These photos, tables, and figures were acquired by Raytheon Company in the merger of Hughes and Raytheon in December 1997 and are identified as Raytheon photos, tables, or figures. All are published with permission.
16. Mittra, R., C. H. Chan, and T. Cwik, "Techniques for Analyzing Frequency Selective Surfaces – A Review," *IEEE Proceedings*, Vol. 76, December 1988, pp. 1593–1615.
17. Knott, E. F., J. F. Shaeffer, and M. T. Tuley, *Radar Cross Section*, 2nd ed., Artech House, Norwood, MA, 1993, pp. 407–447.
18. Bhattacharyya, A., and Sengupta, D., *Radar Cross Section Analysis and Control*, Artech House, Norwood, MA, 1991, pp. 224–235.
19. Stone, W. R., Ed., *Radar Cross Sections of Complex Objects*, IEEE Press, Piscataway, NJ, 1990, pp. 30–46, 283–298, 496–520.

20. Bargeliotes, P., "Resonant Grid Dichroic Filter for Radome Applications," *Hughes Aircraft IDC*, September 14, 1977, unclassified memo.
21. Pelton, E. and B. Munk, "A Streamlined Metallic Radome," *IEEE Transactions on Antennas & Propagation*, Vol. AP-22 no. 6, Nov. 1974, pp. 799–803.
22. Cheng, D., "Study of Phase Error and Tolerance Effects in Microwave Reflectors," Syracuse University Research Institute Report EE276-H, 1955.
23. Radant, M., D. Lewis, and S. Iglehart, "Radar Sensors," *UCLA Short Course Notes*, July 1973, unclassified, given multiple times in the 1970s.
24. Skolnik, M. I., Ed., *Radar Handbook*, McGraw Hill, 1970, pp. 9-35 to 9-39, 13-1 to 13-27.
25. Mailloux, R., "Phased Array Theory and Technology," *IEEE Proceedings*, Vol. 70, No. 3, March 1982, pp. 246–291.
26. Harrington, R., "Electromagnetic Scattering by Antennas," *IEEE Transactions on Microwave Theory and Techniques*, MTT-11, correspondence.
27. Kahn, W., and H. Kurss, "Minimum Scattering Antennas," *IEEE Transactions on Antennas and Propagation*, Sept. 1965, pp. 671–675.
28. Yaw, D., "Antenna Radar Cross Section," *Microwave Journal*, September 1984.
29. Hanson, R., "Relationships Between Antennas as Scatterers and Radiators, "*IEEE Proceedings*, Vol. 77, May 1989, pp. 659–662.
30. Glaser, J. "Some Results in the Bistatic Radar Cross Section (RCS) of Complex Objects," *IEEE Proceedings*, Vol. 77, no. 5, May 1989, pp. 639–648.
31. Ruze, J., "Antenna Tolerance Theory—A Review, "*IEEE Proceedings*, Vol. 54, no. 4, April 1966, pp. 633–640.

7

Signal Processing

7.1 Introduction to Stealth Signal Processing

Stealth signal processing consists of squeezing the maximum performance out of the signal processed, regardless of the amount of "number smashing" that may be required. This includes minimizing straddling losses, reducing RF interference, and limiting self-noise. It also means choosing a waveform that has lowest inherent losses. Furthermore, the output signal-to-noise ratio must be the minimum acceptable, not the best available. Even so, there are inevitably cost-performance trade-offs that must be made, which revolve around word length and total number of arithmetic operations.

7.1.1 Example Signal Processors

The signal processing in a stealth system can be viewed in two ways: the way the hardware is organized and the way the data flows through the functions performed. Two examples of stealth hardware will be given first. Processing hardware consists of low-level RF, signal conditioning, A/D conversion, one or more programmable digital signal processors (PSP), one or more general purpose processors (GPPs) (e.g., your PC), one or more microprocessor control units (MCUs), one or more shared bulk memories (BMs), one or more data busses, control-interface hardware, recording means, and one or more displays. Each unit is "smart" in the sense that it calculates its detailed parameters from general commands or applications software. The hardware requires software: both a real-time operating system (RTOS) and applications programs unique to the system mode.

The first example is of a system that has both an airborne and a ground segment. Systems of this type must be loosely coupled and semiautonomous, because the location of the airborne segment is not completely controllable. Usually, the signal processing in the airborne segment is just enough to minimize the downlink bandwidth for encryption, LPI, and jam resistance. The uplink bandwidth is usually quite small and can be made quite rugged. Figure 7.1 shows the block diagram of an airborne segment of such a system. There must be motion sensing to position the antenna beams of the radar, the weapon data link, and sensor data link. There is usually one general purpose processor in an airborne segment and several microprocessor control units.[63]

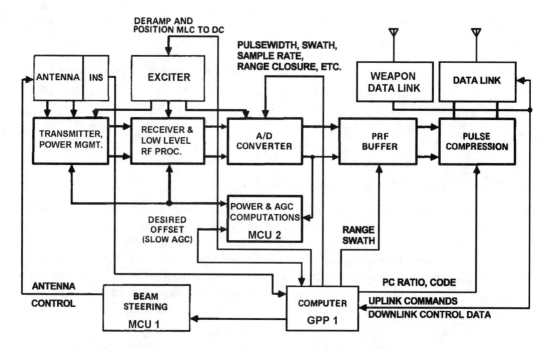

Figure 7.1   Example 1: airborne segment of stealth signal processor.

The ground segment usually has most of the number smashing, mission control, recording, data dissemination, and displays as well as the other end of the data link. The ground segment has a RTOS, applications programs, diagnostic software, and development tools for new modes and debugging. The data link may include a satellite relay. A typical ground segment block diagram is shown in Fig. 7.2. It typically contains all of the functions enumerated in the figure as well as many other minor functions. There are multiple slow-speed busses (10 Mbps) as well as several high-speed busses (10 Gbps).

The second example is made up of two parts: a front-end multichannel receiver and a general purpose signal processor. The multichannel receiver block diagram is shown in Fig. 7.3. The channels may come from segments of the antenna array, the entire antenna at separate frequencies, or separate antennas with different functions (e.g., data link, ELINT, IFF, radar, comm., and so on). Each channel must have several levels of coarse automatic gain control ($CAGC_{1,2}$), a phase modulated first local oscillator (1st LO), bandpass filtering for interference and crosstalk rejection, a second local oscillator (2nd LO), and gain stages with fine automatic gain control (FAGC). Following establishment of the noise figure and selection of *a priori* frequency shifts and of general frequency band, the channels are quadrature detected (I/Q) with a variable frequency oscillator (VFO) reference. The VFO is used to provide some

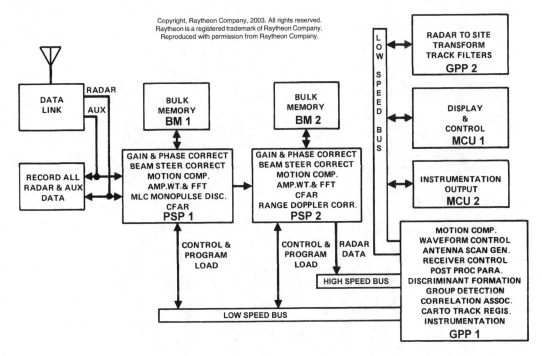

Figure 7.2 Example 1: ground segment of stealth signal processor.[13]

dechirp or to position the largest amplitude portion of the input spectrum to DC to minimize the effects of inphase and quadrature imbalance.[12]

Subsequent to quadrature detection the signal is lowpass filtered (LPF) in an approximation to the matched filter for the chip time. The lowpass filtered inphase and quadrature outputs are converted from analog to digital (A/D) with sample intervals chosen for acceptable performance (not necessarily the chip spacing, as discussed later). There are usually several places in the front-end where built-in test (BIT) signals can be multiplexed into the data stream. Prior to pulse compression, the sampled digital signals are delivered to a digital automatic gain control function that buffers and averages the high-speed samples for further manipulation in the analog microprocessor control unit (MCU). The analog MCU controls all the functions in the front end. The analog MCU delivers AGC and other commands to the receiver MCU (RCVR MCU). Because most RF circuits are very nonlinear, the RCVR MCU uses internal hardware calibration data as well as the analog MCU commands to set the parameters for the low-level RF circuits. Each MCU has a common interrupt driven operating system with watchdog timing designed for an embedded real-time operation system (ERTOS). The ERTOS is usually in every location where an MCU is used, even though the applications might be quite different (e.g., in the power supply).

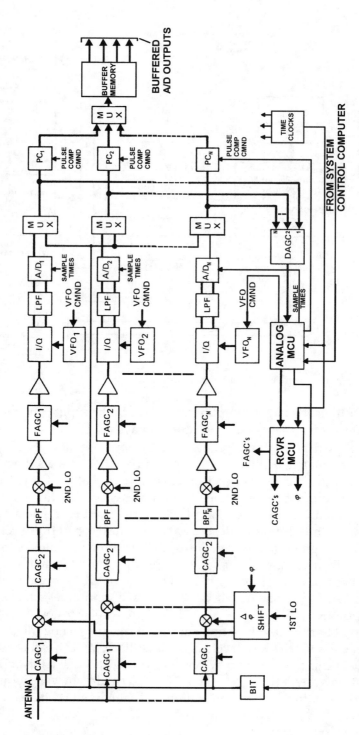

Figure 7.3 Example 2: multichannel receiver.

Pulse compression (PC) is applied to each channel as described in Chapter 5. The PC is not necessarily complete or the same in each channel. Further pulse compression may be performed after narrowband filtering. The pulse compression outputs are multiplexed into a rate change buffer memory and subsequently read out into the general purpose signal processor complex.

The general purpose signal processor portion of Example 2 is shown in Fig. 7.4. The example signal processor architecture has been in continuous use for 30 years[2-4] and has been deployed in many systems including stealth systems. The architecture is known as an heterogeneous multiple instruction multiple data (MIMD) configuration. This architecture is capable of very high throughput, is completely programmable, and has several different types of optimized processing units. It has enough redundancy to allow continuous but reduced performance operation in the presence of failures. It is made up the same building blocks previously mentioned and used in Example 1. The particular example is made up of four clusters, each consisting of an MCU, GPP, BM, I/O, and 3 PSPs. Each cluster can communicate with itself as well as the other clusters through a data network (DN), which is a nonblocking multiway switch.

Surprisingly, most signal processors are memory bandwidth bound, not memory size or arithmetic bound. The solution to this constraint is to cluster a small number of

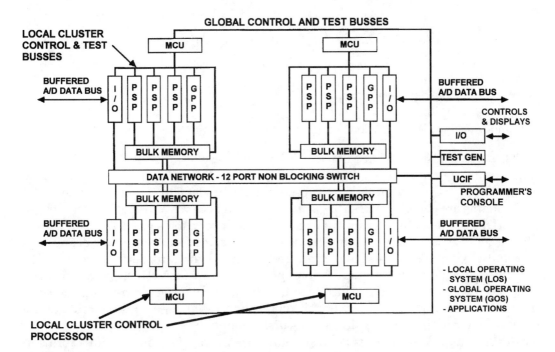

Figure 7.4   Example 2: general purpose signal processor architecture.

PSPs with a GPP and a BM so that data access collisions are minimized. The individual units can talk to each other in one of two ways: direct memory access (DMA) in a shared bulk memory or over a control bus. Unlike conventional general purpose processing, control, test, and data busses are separated, have different word lengths, and have different speeds. Each cluster is controlled by an MCU with an ERTOS. The control unit function is the distribution and initiation of programs in the execution units (BM, I/O, PSP, GPP). The DN can be controlled by a cluster controller or external control. The bulk memory contains an MCU and uses small applications programs to calculate its own addresses.

In addition to the local ERTOS, there is a global ERTOS that may be hosted in one of the cluster MCUs or in an external controller or system computer. Signal processing applications programs are usually short but execute on very large amounts of data. Because signal processing has a very tiny percentage of conditional branches, the units can be deeply pipelined.

7.1.2   Basic Digital Signal Processing

One general equation characterizes most digital signal processing, and it is shown in Equation (7.1). Hints to this equation were given in Chapter 5 in several different forms. This is sometimes called a finite difference equation.[1]

$$E_{out}(N \cdot t_s) = \sum_{n=0}^{N} a_n \cdot E_{in}(n \cdot t_s) + b_0 + \sum_{n=1}^{N} b_n \cdot E_{out}(n \cdot t_s) \tag{7.1}$$

where

t_s = time between sampling instants

E_{in} = digitized input signal vector

a_n and b_n = complex coefficients, which may be dynamic

In its complete form, it is a recursive or infinite impulse response (IIR) filter. When all the b_n terms are zero, this is a finite impulse response (FIR) filter of which digital pulse compression and beamforming are subsets. Similarly, when the b_n terms are zero and the a_n terms are equal to $\exp(-j \cdot 2 \cdot \pi \cdot m \cdot n/N)$, it represents a discrete Fourier transform (DFT). When all b_n terms but b_0 are zero, and only the sign of E_{out} is retained, then it represents an adaptive threshold. Thus, the same form is used for beamforming, pulse compression, filtering, and thresholding. Signal processors must be optimized to execute vector multiplication and addition recursively on large arrays of data. Because each time step is incremented for each new output, most data and coefficients are indirectly addressed, not absolutely addressed. Signal processors must be optimized for indirect addressing and recursive looping.

Most signal processing uses the fast Fourier transform (FFT) in one of its forms to create the bulk of the filters used in beamforming, pulse compression, Doppler filtering, and image compression. In one sense, the DFT is the matched filter for a sampled pulsed monotone of unknown phase. One of the simplest forms of FFT is shown in Fig. 7.5. It is sometimes called a perfect shuffle or "slosh" FFT. There are forms that have half as many arithmetic operations but much more complex addressing and control. It is a designers choice as to which form to use. This form adds and subtracts data values that are always half the record length $(N/2)$ apart. Only the difference term is multiplied by a complex phase shift by the cosine i sine (CIS) generator. The the add and multiplier outputs are successively interleaved (perfect shuffle) and delivered to the register opposite to the source register (ping-ponged). After k passes through the data, the resulting Fourier transform is read out from one side while the input data for the next transform is written into the opposite side. The clocking instant, $C(t)$, occurs at $C(0)$.

Several recursive forms of the FFT are given in Equations (7.2). The perfect shuffle form of Fig. 7.5 is shown at the top of the list of Equations (7.2). Two other recursive FFT forms are also given in Equations (7.2). The perfect shuffle form of the FFT was simultaneously discovered in several places, including by the author in the late 1960s.[8,9] The recursive forms are especially efficient for use in most modern programmable digital signal processors (PSPs or DSPs). Usually, sample records of arbitrary length are filled with zeros to create a convenient, easy-to-calculate FFT input record length. Most modern FFTs are dominated by the number of memory ports and memory port bandwidth, not total quantity of words of memory holding data and coefficients.

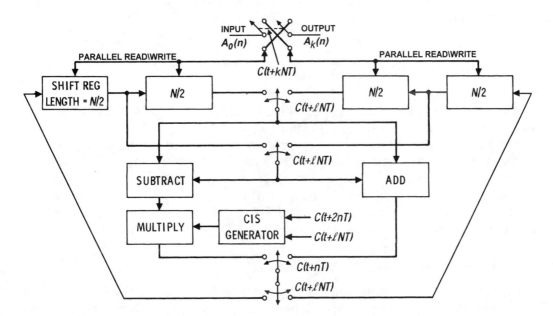

Figure 7.5 Simple FFT processor.

Perfect shuffle bit reversed output order

$$A_\ell(2 \cdot n) = A_{\ell-1}(n) + A_{\ell-1}(n + N/2)$$

$$A_\ell(2 \cdot n + 1) = [A_{\ell-1}(n) - A_{\ell-1}(n + N/2)] \cdot W^{2^{(\ell-1)} \cdot n}$$

Time decimation natural output order

$$A_\ell(n + p) = A_{\ell-1}(n + p) + A_{\ell-1}(n + p + 2^{k-\ell}) \cdot W^{m/2}$$

$$A_\ell(n + p + 2^{\ell-1}) = A_{\ell-1}(n + p) - A_{\ell-1}(n + p + 2^{k-\ell}) \cdot W^{m/2}$$

Frequency decimation bit reversed output order

$$A_\ell(n + m) = A_{\ell-1}(n + m) + A_{\ell-1}(n + m + 2^{k-\ell})$$

$$A_\ell(n + m + 2^{k-\ell}) = [A_{\ell-1}(n + m) - A_{\ell-1}(n + m + 2^{k-\ell})] \cdot W^{2^{\ell-1} \cdot n} \tag{7.2}$$

where

$$k = \log_2(N); \quad W = \exp(-j2\pi/N); \quad \ell = 1,2,\dots k \text{ (index of recursion)};$$

$$p = 0,1,2,\dots 2^{\ell-1}-1 \text{ (index of butterfly sequence)}; \quad m = p \cdot 2^{k-\ell+1};$$

$$n = 0,1,2,\dots 2^{k-\ell}-1 \text{ (index of butterfly)}$$

All of the forms of FFT given in Equations (7.2) are for records whose length is a power of 2. There are also algorithms for base 3, 4, 5, 6, 7, and arbitrary factors.[1,9] There is roughly a 2:1 difference in the arithmetic hardware between the most arcane factored FFT and the base 2 algorithms. The most commonly used FFTs are base 2 and base 4.

Many engineers are more familiar with FFT flow diagrams than recursive equations and a flow diagram for the perfect shuffle FFT algorithm is given in Fig. 7.6. The flow diagram is identical for each recursion except for the complex coefficients.

7.1.3 Matched Filtering and Straddling Losses

Because receive losses require more transmit power, it is essential to minimize losses for stealth systems. Most systems allocate only one sample per resolution cell. As shown in Chapter 6, element spacings of 0.8λ, which are common in many antennas,

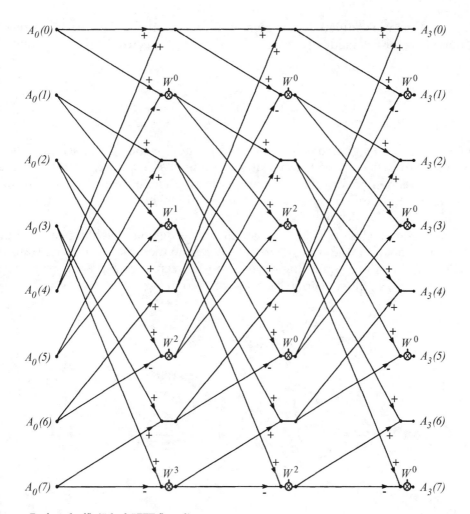

Figure 7.6   Perfect shuffle/"slosh" FFT flow diagram.

may not provide acceptable RCS performance. Similarly, filter bins or range samples must be more closely spaced or there will be straddling losses that are unacceptably high. For example, assuming a perfect matched filter to a single chip width rectangular pulse, the peak straddling loss as a function of sample spacing is given in Equation (7.3).

$$L_s = 0.5 \cdot (t_s/t_c) \tag{7.3}$$

where t_s is the time between sampling instants, and t_c is the chip time width. For the case $t_s = t_c$, the peak loss is –6 dB, and the average loss is –2.5 dB. Similarly, for a filter bank output using modified Taylor weight as shown in Fig. 7.7, the frequency bin straddling loss for filters ideally matched to the input spectrum is given in Equation (7.4).

$$L_{s\max} \approx 0.5 - 0.5 \cdot \cos\left(\frac{\pi \cdot f_s}{2 \cdot B}\right) \tag{7.4}$$

where B is the input signal matched bandwidth, and f_s is the filter spacing. If weighting is as shown in Fig. 7.7, then the peak straddling loss is approximately –6 dB, and the average loss is –1.65 dB, which is probably too high for a stealth system. The filters can be placed closer together by making the record longer either by filling with zeros or dwelling longer (which may not be possible).[5,37,39]

One other notion shown in Fig. 7.7 is that filter sidelobe weighting can be implemented either before or after narrowband filter formation by fast Fourier transform (FFT) or DFT. To reduce straddling losses almost to zero, beams, range, or chip sampling bins and filters must overlap 2:1, as suggested in Fig. 7.8.

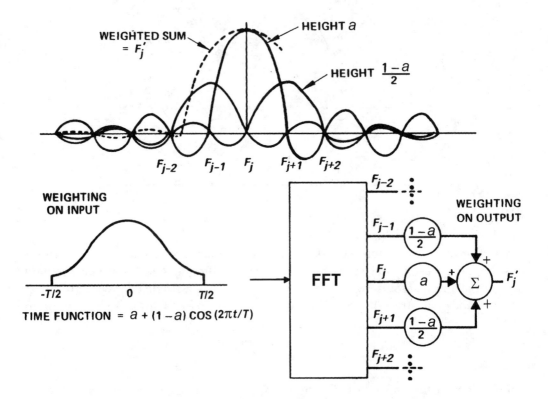

Figure 7.7 Filter sidelobe weighting.

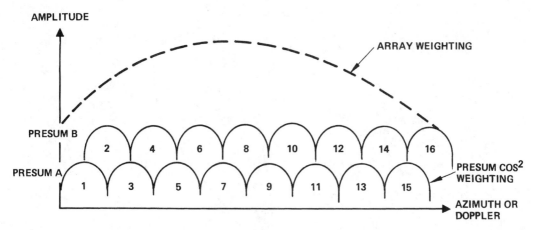

AMPLITUDE

ARRAY WEIGHTING

PRESUM B

2   4   6   8   10   12   14   16

PRESUM A

1   3   5   7   9   11   13   15

PRESUM COS^2
WEIGHTING

AZIMUTH OR
DOPPLER

Figure 7.8 Overlapped range or Doppler bins minimize straddling loss.

7.1.4 Multiple Time around Echoes, Multipath, and Eclipsing

Stealth waveforms require high duty cycles, which naturally give rise to more multiple-time-around-echoes (MTAE) and more multipath interference and greater eclipsing losses. Any range-ambiguous waveform in which there is a significant probability of multiple scattering in the antenna mainlobe will have MTAE. Figure 7.9 shows a simple version of the problem in which two reflectors at different ranges appear to be at the same range (there are actually many cases in range, Doppler, and angle that exhibit the ambiguity). The trick is to find waveform additions and processing strategies that minimize the effect.[12,14,15–19]

Any data link or radar has the MTAE problem but, because low duty ratios or low PRFs significantly increase interceptability, the problem is usually dramatically worse for stealth systems. In fact, the highest duty ratio and PRFs are usually required. Typically, 50 dB of MTAE rejection is necessary in most applications. For example, near-range targets such as over-the-road heavy trucks are 100 m^2, and the minimum conventional target is nominally 1 m^2, for a 20 dB difference. Typical range coverage is 6:1 to 10:1, for an uncompensated dynamic range 31 to 40 dB (see, for example, descriptions of international airborne fire control radars in Streetly.)[64] Couple this range of targets with range coverage, and the result is 51 to 60 dB of required rejection.

PRF stagger can be applied that moves all the ambiguities, and, in a sparse reflector/target space, the ambiguities often can be resolved. PRF stagger does result in range and Doppler blind zones moving to obscure some targets previously detected. These blind zones reduce average detection power on some targets, and so medium PRF systems usually have more average processing losses. This is not quantitatively different from the straddling losses described in Section 7.1.3. PRF stagger must be used judiciously and adaptively, based on the currently detected/tracked target set.

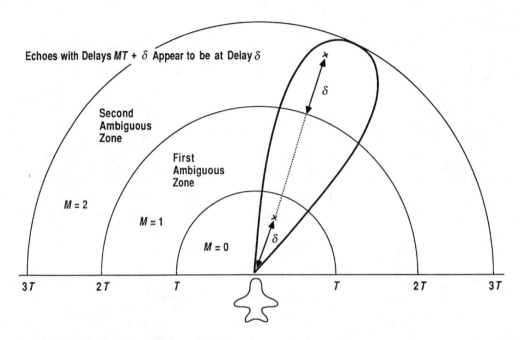

Figure 7.9 Multiple time around echoes (MTAE).

Also, because the scatterers are at different ranges, application of some post-pulse compression as well as post-range resolving sensitivity time control (STC) often improves rejection. Another important technique is to apply pulse-to-pulse phase coding, in essence making the equivalent PRF lower. This has its own set difficulties and ambiguities but, in a dense target/scatterer space such as ground map, it is the only technique that works well. In summary then,

1. Stealth/LPI requires highest possible duty ratio and PRF.

2. Multiple ambiguities are significant.

3. Typically, 50 dB rejection is required.

4. Several MTAE reduction techniques must be used, including STC, PRF stagger, and PRF phase coding.

STC and PRF stagger to improve performance have been used for more than 50 years and are well known. PRF phase coding has been less widely used and is shown in simplified form in Fig. 7.10. The form shown in Fig. 7.10 has been implemented in several systems but is obviously naïve and susceptible to countermeasures. More sophisticated means, based on low sidelobe phase codes typical of those shown in Chapter 5, are more desirable. The concept is most easily explained with Fig. 7.10. Suppose phase shifts of 90° are applied to successive transmitted pulses. On receive,

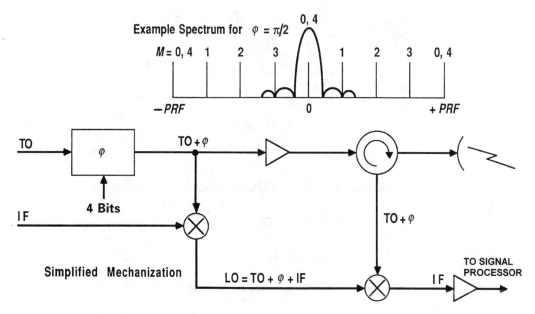

Figure 7.10 Simplified MTAE code/decode example.

that phase shift is applied to the transmit oscillator, which is then multiplied by the intermediate frequency offset to create a receiver reference oscillator. The received signal is then mixed with the reference signal and amplified in the IF amplifier. Any signal with the wrong phase match will not pass through the IF amplifier.

Because the phase is stepwise, and correlation is unweighted, the IF output spectrum is sinc(x) form. Clearly, better ambiguity functions can be achieved but, assuming the ambiguity space is mostly empty in Doppler, as is the case for ground map, the MTAE rejection is effective even for the simple example waveform (see Table 7.1).

Table 7.1   MTAE Summary

MTAE rejection method	Improvement (dB)
Post-PC STC, 0.25 duty	24
PRF stagger 4 or 8:1	6–40
PRF phase coding, 16:1	26
Total rejection	56–59

Another problem related to high duty cycles is eclipsing. The notion of eclipsing is shown in Fig. 7.11. When pulses are long, the probability that a received pulse will arrive during the time that the receiver is blanked by the transmitter is significant, as

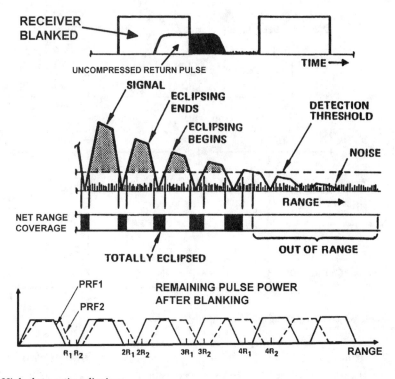

Figure 7.11 High-duty-ratio eclipsing.

shown at the top of Fig. 7.11. This degrades the signal power and resolution after pulse compression and filtering as shown in the middle of Fig. 7.11. Of course, independent of eclipsing, the signal power decreases with range. But, because the threshold is ultimately determined by noise, the net coverage in range decreases as range increases. Obviously, the use of a second PRF can fill in the gaps in range coverage, as shown at the bottom of Fig. 7.11. Unfortunately, this may result in only half of the total transmitted energy returning from some targets at long range, which defeats the purpose of high duty ratio in the first place. A 50 percent duty cycle causes all but the farthest range bin to be eclipsed; similarly, a 25 percent duty cycle has only the last half of the interpulse interval uneclipsed. Accordingly, single-frequency channel duty ratios typically are limited to approximately 25 percent for radars and only a little more for data links.

7.2 Noise in Digital Signal Processing Systems

7.2.1 Introduction

Saturation, roundoff, and filter impulse response must be considered carefully to predict digital filtering performance. In the design of digital signal processing equipment, it is common to base first-order performance predictions on linearized models of the actual digital circuitry. These models provide fairly accurate information on such characteristics as filter bandwidth, passband ripple, sidelobe response, and so on. When more accurate information is desired, however, the processes of signal quantization and saturation must receive more attention because, at each point where these effects occur, they are equivalent to the addition of noise in the linearized model.[1,5]

In the design of any stealth signal processor, the equipment should be capable of producing very low loss performance. The effects of signal quantization, saturation, roundoff, truncation, and dynamic range in the signal processor must be considered. The same analytical techniques are useful in guiding the design of the digital processor hardware and software in such factors as gain, memory size, and ordering of processing operations.

The first element in a digital processor is the A/D converter. Sampling and quantizing are used to convert signals to numerical form. An A/D converter with, say, 10 bits capacity produces numerical output data that may have integer values ranging from –512 to +511. Voltage inputs corresponding to noninteger output values are assigned the nearest integer value in the quantization process, with a corresponding error being introduced. Voltage values corresponding to numerical outputs outside the range –512 to +511 are assigned the values –512 to +511, depending on their polarity, with a corresponding saturation error being introduced. The greater the gain ahead of the A/D converter, the greater the likelihood of saturation; saturation error, therefore, may be minimized by using low gains. Quantization errors, on the other hand, are determined by the quantum level or step size and are relatively independent of signal level. To maximize signal-to-quantization error ratio, it is therefore desirable that gain be high. These conflicting requirements are best met by a compromise that maximizes the signal-to-total distortion power ratio. This compromise, which is analyzed below, dictates that the gain be set so that the rms signal level is approximately 10 dB below the saturation level of the A/D converter.

In a common signal processor configuration, the functional block following the A/D converter is the pulse compression and/or beamforming circuitry. In other arrangements, pulse compression and/or beamforming is performed after filter processing to simplify Doppler compensation. When beamforming and/or pulse compression are performed directly following A/D conversion, the arithmetic procedure consists of adding a number of range and/or sensor samples together as described in Chapter 5, perhaps with complex weights, to form a single compressed pulse or angle output. A very strong point source or reflector would distribute its energy through N range, frequency, and/or angle samples. The largest possible signal that would not saturate the example (10-bit) A/D converter would have amplitude –512 or +511 in each of N adjacent sensor samples. Summing over N sensor samples with uniform weighting would produce an output with amplitude $-512 \times N$ or $+511 \times N$. Having increased the possible amplitude range of the data by filtering, beamforming, or pulse compression, provision must be made for $\log_2(N)$ additional bits of data-handling capacity to accommodate it, or introduce errors from saturation (discarding the most significant bits, MSB), or truncation (discarding the least significant bits, LSB). Saturation and truncation errors introduced in the digital processor are equivalent in effect to saturation and quantization errors introduced at the A/D converter and may be analyzed similarly.

Before digital data are entered into the main memory, they are usually prefiltered into subbands with expected common signal statistics. The prefilter combines many individual samples into a single sequence of data for each subband associated with range, frequency, or angle. As with pulse compression, filtering or beamforming, weighting, and summing several digital words to form one filter output necessitates providing additional bits of data-handling capability to avoid errors of saturation and truncation. Because the word length size is an important cost-determining factor, it is generally cost effective to accept some saturation and truncation errors to reduce word size. The extent of these errors and their effect on overall system performance is an important design consideration and is analyzed below.

Each time saturation and/or truncation occurs, perturbation noise is introduced. The optimal stealth processor, from a minimum perturbation noise standpoint, would provide whatever additional bits of data-handling capacity required by signal buildup. Because large, fast buffer memories and large-word fast arithmetic capacities are expensive, it is desirable to reduce the sample rate to reduce the arithmetic rate. Often, pulse compression or beamforming is located after the main memory buffering and some prefiltering in the signal processing sequence. Furthermore, it is obvious that the sidelobe performance of the linearized error-free processor model establishes a design limit beyond which nothing is gained by reducing the noise contributions of saturation, quantization, and truncation, and at which point the processing configuration can be optimized with respect to other factors.

7.2.2   Noise and Gain in Signal-Selective Networks

When a composite spectrum containing both continuous and discrete signals and noise is filtered, the relationships between peak signal, rms signal, and rms noise are substantially altered. One well established criterion of processor performance may be interpreted in terms of Fig. 7.12. Here, a signal spectrum is depicted with a strong point target at one frequency (angle or range) and a "hole" in the spectrum at another frequency (angle or range), representing a signal of very low SNR. The dynamic range criterion is the measure of how much weaker than the average background a weak target can be without being obscured by noise and/or sidelobes of surrounding signals, and of how much stronger than the surrounding average background a target can be before its sidelobes obscure the background signals.

In a situation where the system performance is not limited by saturation, quantization, or receiver noise, the weakest area signal that the processor could handle would

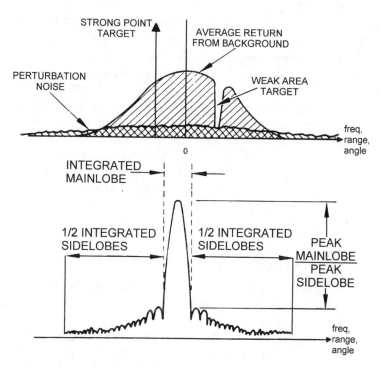

Figure 7.12 Sidelobe and dynamic range considerations.

be determined by the frequency (range or angle) integrated sidelobe ratio (*ISLR*) as mentioned in Chapters 5 and 6. This is a measure of the total sidelobe power of the array or filter and represents the interference to a point signal from all adjacent signals. A typical *ISLR* design goal for high-performance processors is 40 dB; a processor achieving this goal could recognize signals whose power is 40 dB below the surrounding average background power, assuming no receiver, quantization, or saturation noise.

A processor whose design *ISLR* is 40 dB has this fundamental performance limit against weak signals. In the analysis of the effects of saturation and quantization—and in the determination of memory size, scale factors, roundoff procedures, and so on—it should be remembered that the total level of perturbation noise is established by the processor ISLR (i.e., perturbation noise, *ISLR*, and other error sources are combined root-sum-square to form overall performance predictions). Figure 7.12 illustrates the situation in which the perturbation noise spectral density in the region of the weak point target determines the noise interference to that target. Because spectral density is significant and not simply total perturbation noise power, analysis should emphasize the distribution throughout the sampling spectrum of the noise added by saturation, quantization, FFTs, multiplication, magnitude calculation, and so forth.

Signals much greater than the surrounding background level will have range, angle, and frequency sidelobes that compete with the average background level. A target stronger than the average background by the peak sidelobe ratio will produce interference with targets at the location of that peak sidelobe. If the average or root-mean-square sidelobe level of the pulse compressor, beam, or filter sidelobe pattern is taken (rather than the peak sidelobe level), then targets stronger than the background level by the rms sidelobe ratio, on the average, will compete with all targets in the area. Thus, once the processor design is established, for a certain rms sidelobe level, a fundamental performance limit is also established. This limit determines the maximum target strength for which signal handling capacity must be provided. In other words, targets so strong as to generate sidelobe interference at all other angle/frequency locations may be allowed to saturate the processor as well, because the detection performance is degraded in either case.[5,14,34,60]

7.2.3 Effects of Processing Order on Dynamic Range

Pulse compression, Doppler filtering, and beamforming can be done in several locations in the overall signal processing sequence. Because these processes affect dynamic range, they require gain changes, more memory, and perhaps truncation/roundoff. For a 10-bit A/D converter, the value of K_{AD} (see Section 7.2.4) that minimizes the total perturbation power is about 4.5 and, for this value, the ratio of total perturbation power to rms signal is seen to be about 10^{-5} or –50 dB. The largest signal that would not saturate the A/D converter is 4.5 times the rms signal level, or about +13 dB. The largest signal is assumed to be concentrated at a single spectral point; the rms signal has the spectral distribution impressed by the real sensor pattern, and the perturbation noise is partly white and partly colored. The total dynamic range at the A/D for this example is approximately 63 dB. Pulse compression or beamforming will change this dynamic range substantially. Figure 7.13 is a generalized block diagram of a signal processor that will serve as a model for the following discussion.[5]

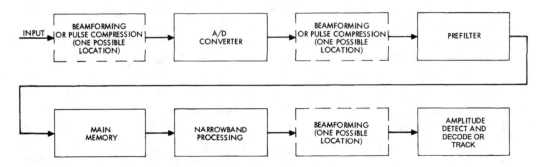

Figure 7.13 Signal processing order considerations.

7.2.3.1 Pulse Compression and Beamforming. Pulse compression or beamforming can be performed at three points in the signal processing sequence: before A/D conversion, immediately following A/D conversion, and following narrowband processing. Because angle or range sidelobes in a beamforming or pulse compression system are a reality whether done by analog or digital methods, and because angle or range and frequency processing are to a first-order orthogonal operations, for this application, only the effect of beamforming or pulse compression on the frequency response of the processor is considered.

It can be shown that, when the A/D converter input signal level is set to minimize the ratio of total perturbation power to average signal power, the ratio does not improve when the pulse compression and/or beamforming circuitry is moved from ahead of the A/D converter to behind the A/D converter in the signal processing sequence. Clearly, an improvement occurs in peak signal-to-rms noise (and, therefore, peak signal-to-rms signal) when such a rearrangement is made. A similar argument applies when the pulse compression and/or beamforming circuitry is relocated to a position following the narrowband processor. It usually turns out that the total number of cells to be processed by subsequent operations is reduced by each processing block. Therefore, it often turns out that there is a processing order that yields the lowest total operations and/or required dynamic range (word length).

The pulse compression or beamforming circuitry, being relatively wideband compared to the ultimate processed bandwidth, does not alter the relative spectral distribution of perturbation noise and average clutter return. The peak compressed pulse or angle response is larger than the peak amplitude of the uncompressed pulse or angle response by the compression ratio or angle gain; a pulse compression or beamform circuit operating on 10-bit input data words could produce output requiring $10 + \log_2 N$ bits to represent them, where N is the compression ratio or array gain. Hence, it is apparent that, for a specified processor peak signal-to-rms noise capability, a larger main memory is required for prestorage pulse compression or beamforming than for the same process performed after narrowband processing. Other factors may make prestorage pulse compression or beamforming more desirable on an overall basis (see Section 5.5.3). On the other hand, pulse compression may require Doppler compensation before compression processing, but pulse compression after narrowband filtering requires none, as the compensation occurs naturally in the filtering.

7.2.3.2 Scale Factor Changes, Saturation, and Truncation. When the digital word size grows as a result of pulse compression and/or beamforming (or other signal processing), extra bits must be provided in the circuitry to handle the larger words. Alternatively, the word size must be reduced by discarding MSBs or LSBs or some combination of the two. Another possibility is the use of floating point to handle a large dynamic range with a fixed word size. Each of these processes adds noise. Discarding LSB is equivalent to dividing the signal by 2^N (where N is the number of

LSBs discarded) and then dropping the fractional part below the least significant bit, i.e., to integer denominate fractional values. The process of truncating or rounding the rescaled data to integer fraction values is almost equivalent to quantizing in the A/D converter. The slight difference arises at the A/D converter where the data before quantization may take on any value between quantum steps, whereas, in the truncation or roundoff process, the fractional part that is dropped has only 2^N discrete values.

Discarding MSBs is equivalent (assuming the use of proper overflow or saturation logic) to allowing the signal to saturate at peak values, and the analysis applied to the A/D converter is usable. It was shown in Chapter 5 that the rms output of a digital pulse compression or beamform circuit is the rms input signal times the square root of the compression or gain ratio; hence, the rms output signal can be computed when the rms input is known.

7.2.4 Saturation, Quantization, and Optimum Signal Level

When analog data are passed through an A/D converter and sent to a digital filter, the proper measure of the effect of the saturation and quantization noise introduced is the proportion of the noise power spectrum within the filter passband. As long as there are signals significantly larger than a quantization step size, quantization noise may be considered, to a very good approximation, to be white or uniformly distributed in the spectral region between sampling spectral lines. Thus, the proportion of quantization noise power that falls within a filter bandwidth is given by the ratio of that bandwidth to the sampling frequency. On the other hand, saturation noise tends to be concentrated in the spectral region of the input signal. The spectral relationships of signals (assumed to have a Gaussian-shaped lowpass spectrum) and saturation noise for four different ratios of sample rate-to-signal bandwidth (f_s/B_s) and for different degrees of saturation are shown in Fig. 7.14. The parameter, K_{AD}, is the ratio of saturation level to rms signal. Note that the case $K_{AD} = 0$ corresponds to hard limiting or 1-bit quantization; in this case, quantization and saturation are one, and the curves represent the total perturbation noise spectrum. If the input spectrum has a lobe structure (bump shape) spectrum and the sampling ambiguity spacing is comparable to the lobe bandwidth (Fig. 7.14a), the saturation noise spectrum is approximately white. In a nonminimum sampling saturation [Fig. 7.14(b), (c), and (d)], the saturation noise density is not constant, and the variations must be considered. It is reasonable to treat saturation noise, in this case, as having a bandwidth approximately equal to the input bandwidth.

The idealized transfer function of an A/D converter is shown in Fig. 7.15, where the saturation level has been set at K_{AD} times the input rms voltage, σ_c. The resulting error signal $e(x)$ is shown to the right in the figure, as well as the assumed input probability density function $\mathcal{P}(x)$. The output noise power resulting from some input signal probability density function can be calculated by integrating the errors for the com-

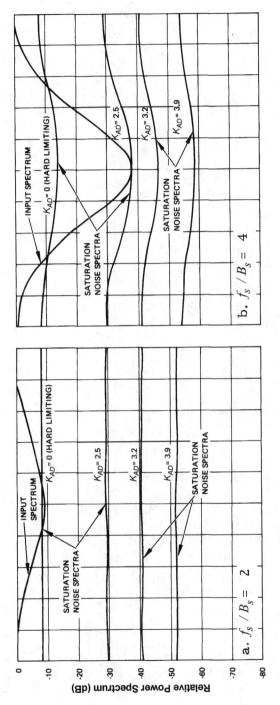

Figure 7.14 Saturation noise spectra (continues).

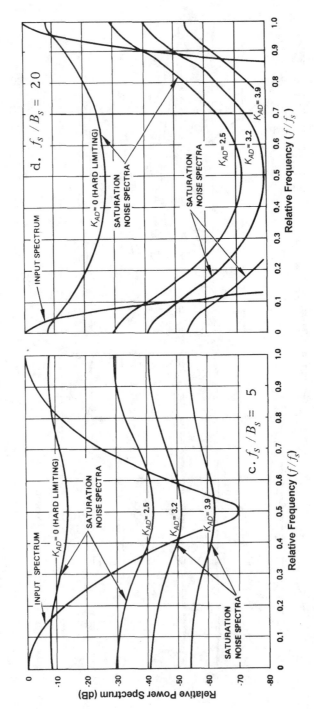

Figure 7.14 continued.

478

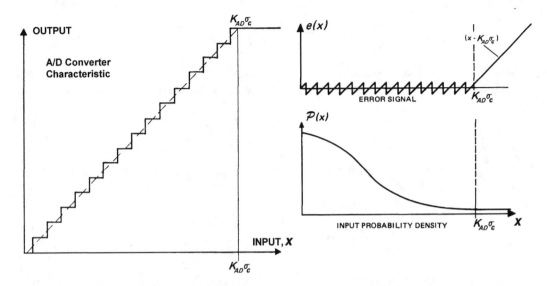

Figure 7.15 A/D converter transfer characteristics.

plete range of inputs. Because the error function is very nonlinear, the integration needs to be performed numerically or by the use of simple approximations.

The total noise power P_n caused by the error term, $e(x)$, can be computed from Equation (7.5).

$$P_n = \int_{-\infty}^{\infty} e^2(x) \cdot P(x)dx = 2 \cdot \int_{0}^{\infty} e^2(x) \cdot P(x)dx \qquad (7.5)$$

This noise power can be broken into two parts, one resulting from quantization (Q_N) and one from saturation (S_n). These terms are given by Equations (7.6).

$$Q_n = 2 \cdot \int_{0}^{K_{AD} \cdot \sigma_c} e^2(x) \cdot P(x)dx$$

$$S_n = 2 \cdot \int_{K_{AD} \cdot \sigma_c}^{\infty} e^2(x) \cdot P(x)dx \qquad (7.6)$$

where $K_{AD} \cdot \sigma_c$ is the saturation level of the A/D. Consider first the integral of Equation (7.6). If x_n represents the value of x at each zero crossing of $e(x)$ in Fig. 7.15 and assuming that over any one interval $-Q/2 < x < Q/2$, that $P(x) \cong P(x_n) = $ constant, then it can be shown that the first part of Equations (7.6) is approximately equal to Equation (7.7).

$$Q_n \cong 2 \cdot \frac{Q^2}{12} \cdot \int_0^{K_{AD} \cdot \sigma_c} P(x)dx \qquad (7.7)$$

The second part of Equations (7.6), the saturation noise term, steadily increases with increasing input signal and can be computed as shown in Equation (7.8).

$$S_n = 2 \cdot \int_{K_{AD} \cdot \sigma_c}^{\infty} (x - K_{AD} \cdot \sigma_c)^2 \cdot P(x)dx \qquad (7.8)$$

For example, the expression for a Gaussian probability density function with zero mean is given in Equation (7.9).

$$P(x) = \frac{1}{\sigma \cdot \sqrt{2 \cdot \pi}} \exp\left(\frac{-x^2}{2 \cdot \sigma^2}\right) \qquad (7.9)$$

The quantization step size Q (from Fig. 7.15) is given by Equation (7.10).

$$Q = \frac{K_{AD} \cdot \sigma_c}{2^{M-1} - 1} \qquad (7.10)$$

where M is the number of bits (including sign) in the A/D converter. Equations (7.7) and (7.8) can be expressed in terms of K_{AD}, $N = 2^{M-1}$ and the tabulated error function Equation (7.11).

$$F(u) = \int_u^{\infty} \frac{\exp(-u^2/2)}{\sqrt{2 \cdot \pi}} du \qquad (7.11)$$

Substituting Equations (7.9), (7.10), and (7.11) into (7.7) and (7.8) yields the normalized quantization and saturation noise given in Equations (7.12) and (7.13).

$$\frac{Q_n}{\sigma_c^2} = 2 \cdot \left(\frac{K_{AD}^2}{12 \cdot (N-1)^2} \cdot (0.5 - F(K_{AD}))\right) \qquad (7.12)$$

$$\frac{S_n}{\sigma_c^2} = 2 \cdot \left((K_{AD}^2 + 1) \cdot F(K_{AD}) - \frac{K_{AD} \cdot \exp(-K_{AD}^2/2)}{\sqrt{2 \cdot \pi}}\right) \qquad (7.13)$$

Similar equations can be generated for other input signal probability density functions (PDFs). For a given value of N, Equations (7.12) and (7.13) have been computed, and their sum, the normalized total noise power, has been plotted versus K_{AD} in Fig. 7.16 for several different PDFs. From these curves, the values of K_{AD} corresponding to minimum total noise power can be determined for different values of M. For all values of M shown, a ±10 percent change in K_{AD} will result in a very small change in total noise power. Hence, the A/D converter input gain need not be closely controlled.

The optimum values of K_{AD}, as well as the corresponding values of Q_n/σ_c^2, S_n/σ_c^2 and P_n/σ_c^2 for a Gaussian PDF for various values of M are given in Table 7.2. The curves from which these tabulated values are derived (Fig. 7.16) show the sensitivity of the total noise power to change in K_{AD}. It is also seen that, at optimum K_{AD}, Q_n is always several times larger than S_n.

It is interesting to compare the total noise power P_n with the value $Q^2/12$, which is often assumed as an approximation. If one writes $P_n = \eta\, Q^2/12$, then the parameter, η, is found to vary between 1.38 and 1.09 for $M = 3$ to 10. Also, it is found that, over this

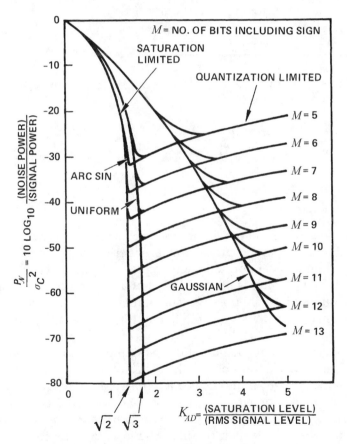

Figure 7.16   A/D conversion noise versus operating level (after Harris[5]).

Table 7.2   Noise at K_{ADopt} for Values of M

M	K_{ADopt}	Q_n/σ_c^2	S_n/σ_c^2	P_n/σ_c^2
3	1.9	3.15×10^{-2}	1.5×10^{-2}	4.65×10^{-2}
4	2.5	1.05×10^{-2}	0.24×10^{-2}	1.29×10^{-2}
5	2.9	3.1×10^{-3}	0.6×10^{-3}	3.7×10^{-3}
6	3.26	0.92×10^{-3}	0.13×10^{-3}	1.05×10^{-3}
7	3.6	2.7×10^{-4}	0.4×10^{-4}	3.1×10^{-4}
8	3.9	7.9×10^{-5}	1.0×10^{-5}	8.9×10^{-5}
9	4.24	2.3×10^{-5}	0.2×10^{-5}	2.5×10^{-5}
10	4.5	6.5×10^{-6}	0.5×10^{-6}	7.0×10^{-6}

range of M values, the normalized total noise can be approximated within about 2 dB by Equation (7.14).

$$\frac{P_n}{\sigma_c^2} \approx 7 - 6 \cdot M \text{ (dB)} \tag{7.14}$$

7.2.5   Noise in Prefiltering

Prefiltering is the process of combining a set of successive samples to form a single data record before narrowband processing. Prefiltering has several effects. Because it is a filtering and amplitude normalizing process, it is equivalent to forming a noise whitening filter. The prefilter is often read out at a fraction of the sample rate; therefore, sampling ambiguities are possible (sometimes called *multirate filtering*). Because the prefilter passband width is some fraction of the total sampling interval, the effect of the prefilter on the signal is different from its effect on perturbation noise. The input signal is concentrated in a region of the sampling interval determined by the real signal space. The perturbation noise is the collective effect of receiver noise, saturation noise, quantization noise, and truncation noise. Receiver, quantization, and truncation noise may be considered white; i.e., the total noise power has an even spectral distribution throughout the sampling interval. Saturation noise is colored with a bandwidth somewhat wider than the input bandwidth. The saturation noise spectrum for an input signal with a Gaussian-shaped input spectrum was shown in Fig.

7.14. When those noise spectra are applied to a digital filter, the output noise can be estimated using the filter noise bandwidth. The frequency response transfer function of a sampled data filter is given in Equation (7.15).

$$H(f) = \sum_{n=0}^{N_p} A_n \cdot \exp\left(\frac{-j \cdot 2 \cdot \pi \cdot n \cdot f}{f_s}\right) \tag{7.15}$$

where
   A_n = impulse response coefficients
   N_p = number of significant terms in the impulse response
   f_s = sampling rate

The meaning of Equation (7.15) can be better understood by considering Fig. 7.17. When an impulse is applied to a digital filter, it sweeps out the time-sampled coefficients of the impulse response of that filter. If the filter is feed-forward only, the impulse response coefficients can usually be written by inspection. Every stable filter has a finite number of significant (greater than the LSB) impulse response coefficients. There are two effects for a set of samples as shown in Fig. 7.17. First, there is a frequency response ambiguity at the sampling rate. Second, because the response is practically limited in the time domain, it has a sinc(x) form convolved on it in the frequency domain, which gives rise to ripples.

When a filter operates on a white noise input, the output may be calculated by multiplying the input noise spectral density by the noise bandwidth of the filter times the

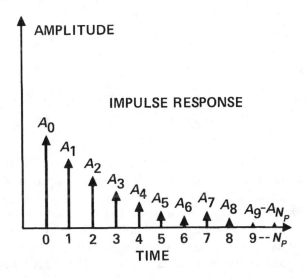

Figure 7.17 Filter impulse response.

square of the filter's center frequency gain. The noise bandwidth of a digital filter of the type used in prefiltering is given by Equation (7.16).

$$
B_n = \frac{\sum\limits_{n=1}^{N_P} A_n^2}{\left(\sum\limits_{n=1}^{N_P} |A_n| \right)^2} \cdot f_s
\tag{7.16}
$$

The filter center frequency maximum gain is given in Equation (7.17).

$$
Gain_{f_0} = \sum_{n=1}^{N_P} |A_n|
\tag{7.17}
$$

When the white noise part of the perturbation power is σ^2, the spectral density is σ^2/f_s. If spectral density, noise bandwidth, and square of center frequency gain are multiplied, then the total output power due to white noise input is given in Equation (7.18). The output noise power is just input noise variance times the sum of the squares of the impulse response coefficients as shown in Equation (7.18).

$$
P_{n\text{out}} = \frac{\sigma^2}{f_s} \cdot \frac{\sum\limits_{n=1}^{N_P} A_n^2}{\left(\sum\limits_{n=1}^{N_P} |A_n| \right)^2} \cdot f_s \cdot \left(\sum_{n=1}^{N_P} |A_n| \right)^2
$$

and

$$
P_{n\text{out}} = \sigma^2 \cdot \sum_{n=1}^{N_P} A_n^2
\tag{7.18}
$$

Colored noise can be handled similarly for most cases of interest. The saturation noise occupies a spectral region slightly wider than the input signal bandwidth and may be regarded to a first approximation as having uniform spectral density over the prefilter bandwidth. Filter output noise resulting from saturation noise input may then be calculated by a formula similar to Equation (7.18) by using an adjusted value of σ^2 to account for the greater spectral density of the input noise in the filter pass-band. If saturation noise power is defined as σ_{sat}^2, and the noise is assumed to be concentrated entirely in the bandwidth of the prefilter, B_s, then the value $\sigma_{sat}^2 \cdot (f_s/B_s)$

should be used in Equation (7.18). The signal can be handled identically when calculating average signal power at the prefilter output; the signal power at the input is treated as if it had uniform spectral density across the prefilter bandwidth and zero density outside.

Sometimes, the above approach to colored noise is unsatisfactory. Under these circumstances, the procedure is to obtain the Fourier transform of the impulse response, find the magnitude squared of this frequency response, and multiply it by the energy density spectrum of the input. The resultant is the output energy density spectrum for subsequent input to the next filter in the system. The Fourier transform of the energy density spectrum is the autocorrelation of the signal whose value at $\tau = 0$ is the variance in each time slot. The DFT frequency response, $H(f_i)$, of a time-sampled filter is given in Equation (7.19).

$$H(f_i) = \sum_{n=1}^{N_P} A_n \cdot \exp\left(\frac{-j \cdot 2 \cdot \pi \cdot n \cdot i}{N_P}\right) \tag{7.19}$$

and the continuous output spectrum is given in Equation (7.20).

$$S_{out}(f) = S_{in}(f) \cdot |H(f)|^2 \tag{7.20}$$

where

$S_{in}(f)$ = input energy density spectrum at frequency f
$S_{out}(f)$ = output energy density spectrum

so that the normalized sampled total output power is given in Equation (7.21).

$$\sigma_{out}^2 = \frac{1}{N_P} \cdot \sum_{i=1}^{N_P} S_{in}(f_i) \cdot |H(f_i)|^2 \tag{7.21}$$

7.2.5.1 Saturation and Truncation Introduced by Prefiltering. Because amplitude weighting is done digitally in the prefiltering, the digital word sizes grow according to the number of bits used to represent a coefficient. If the extra bits so generated are truncated immediately, the error introduced in the overall noise total must be included. An expression similar to Equation (7.18) with $A_n = 1$ describes the output noise of the prefilter contributed by this source.

Because the prefilter has gain, the digital words at its output require more bits to represent them than did those at its input. The prefilter output usually is stored in the main memory before narrowband processing; hence, the word size of the memory determines whether truncation is necessary. If the prefilter output word has more bits than the main memory word, the saturation and truncation noise considerations that prevailed at the pulse compression or beamforming output apply equally to the prefilter output.

7.2.5.2 Coefficient Noise and Stability of Prefilters.

Coefficient approximations caused by limited word length can lead to significant degradation of performance and, in some cases, limit cycle oscillation. The most serious problem associated with digital filters is stability. Because a digital filter can exhibit the same behavior as active analog filters, digital filters can produce ringing and oscillation. Obviously, undesired oscillation destroys performance. The same rules used for analog circuit stability must be applied to digital filters. In addition, there are forms of oscillation that are unique to time and amplitude quantized systems. Limit cycle oscillation is the result of the feedback in IIR filters and amplitude quantization. It can result from saturation as well as quantization. Even low-level oscillation will desensitize a system, thus increasing stealth risk. A necessary but not sufficient condition for stability is that Equation (7.17) is less than infinity as N_p goes to infinity.

One common technique to minimize the effect of coefficient errors in filters is to group coefficients pairwise so that complex conjugate coefficients lie on the same radius circle about the origin.[1] Alternatively, design coefficients can be restricted to exact binary weights and an exhaustive search for the best set about the desired pole locations can be used. The noise power at the output of a filter resulting from coefficient quantization can be estimated as shown in Equation (7.22).[1]

$$\sigma_{coeff}^2 \approx \frac{2 \cdot N_P - 1}{3} \cdot 2^{-2 \cdot M} \tag{7.22}$$

where N_p is the number of significant terms in the impulse response, and M is the number of bits in the coefficient words. Normally, coefficient errors are designed to be a small fraction (<0.1) of the total expected noise at the filter output. For example, suppose the data is quantized to 16 bits, and N_p is 50; then, if coefficient noise is to be 0.1 of quantizing noise, coefficients must be quantized to 19 bits, and no roundoff can occur until after the final sum at the filter output.

7.2.6 Noise and Signal Response in Narrowband Filtering

Often, narrowband filtering is realized with a discrete (often fast) Fourier transform. Its elemental frequency response makes it somewhat different from prefiltering, but the same analysis used earlier still applies. Narrowband processing is performed on the prefilter outputs as stored in the main memory. A sum (complex weighted) of the stored elements is formed to represent a single narrowband filter element. The complex weight is regarded as a phase rotation followed by a (real) amplitude weight. The phase rotation, performed simultaneously with the amplitude weight, is in effect a combination of real and imaginary channel data with no net amplitude change; therefore, no change in signal-to-noise relationship is produced, and no noise is added. The narrowband processor, therefore, may be analyzed in the same light as the prefilter as far as its effect on signals and noise is concerned. The noise bandwidth is calculated and used to compute the power at the output of the narrowband filter processor, after proper account is taken of the differences in noise densities at the input (i.e., that the truncation noise introduced at the previous stage is spread uniformly over the effective sampling bandwidth, whereas the signal and other noise components are more heavily concentrated in that bandwidth about the stage bandwidth in which the narrowband filter is centered).

Narrowband filtering obtained through the use of discrete or fast Fourier transforms has certain distinctive properties that affect the spectral density relationships. Because the output from each discrete Fourier transform filter is an unweighted but phase-rotated sum of input samples, it is not surprising that the frequency response has a $\sin s(x)$ form [see Equations (6.31) and (6.32)]. The frequency response of a fixed-interval discrete Fourier transform, $G_{DFT}(f)$, to a complex sinusoidal input can be found by assuming an input, $g(t)$, as follows in Equation (7.23).

$$g(t) = \exp(j \cdot 2 \cdot \pi \cdot f \cdot t) \tag{7.23}$$

where
   t = time
   f = frequency

The counterpart sampled signal is given in Equation (7.24).

$$g(n \cdot \Delta t) = \exp(j \cdot 2 \cdot \pi \cdot f \cdot n \cdot \Delta t) \tag{7.24}$$

where
   Δt = sample period
   n = index of each sample
   N = maximum number of samples

Now, substituting into the DFT equation yields the frequency response of each of the M DFT outputs as a function of frequency as shown in Equation (7.25).

$$G_{DFT}(f, m) = \sum_{n=0}^{N-1} g(n \cdot \Delta t) \cdot \exp(-j \cdot 2 \cdot \pi \cdot m \cdot n/N) \tag{7.25}$$

and solving (7.25) yields

$$G_{DFT}(f, m) = \sum_{n=0}^{N-1} \exp(-j \cdot 2 \cdot \pi \cdot n \cdot (m/N - f \cdot \Delta t)) \tag{7.26}$$

Each DFT output is a narrowband filter, which is a matched filter for the single-frequency signal of time length $N\Delta t$ at each frequency $m/(N\Delta t)$. But the form of Equation (7.26) was already solved for the linear array in Chapter 6 in Equations (6.34) through (6.37), so the counterpart closed form for the series of Equation (7.26) is given in Equation (7.27).

$$G_{DFT}(f,m) = N \cdot \exp(j \cdot \pi \cdot (N-1) \cdot (m/N - f \cdot \Delta t)) \cdot$$
$$\left(\frac{\sin(N \cdot \pi \cdot (m/N - f \cdot \Delta t))}{N \cdot \sin(\pi \cdot (m/N - f \cdot \Delta t))} \right) \tag{7.27}$$

Let $m = 0$ in Equation (7.27); then it is obvious that each filter is of the $\sin s(x)$ form as previously stated. Also, all of the weighting functions described in Chapter 6 can be used to improve filter sidelobes in the DFT. There is one important weighting function difference, and that is realizability is greatly expanded—limited only by word length (e.g., –80 dB sidelobes *are* achievable).

7.2.7 Roundoff, Saturation, and Truncation

When the digital word size grows as a result of pulse compression (and/or beamforming or other signal processing), extra bits must be provided in the circuitry to handle the larger words, or the result must have the word size reduced by discarding MSB, LSB, or some combination of the two and/or the use of floating point. Discarding LSBs is equivalent to dividing the signal by 2^N (where N is the number of LSBs discarded) and then dropping the factional part (where the data are regarded as having only integer values before being divided), as shown in Fig. 7.18. The process of truncating or rounding the rescaled data to integer values is almost equivalent to quantizing in the A/D converter. The slight difference lies in the fact that, at the A/D converter, the data before quantization may take on any value between quantum steps whereas, in the truncation process, the fractional part that is dropped has only 2^N discrete values.

The noise contributed by truncating or rounding various numbers of LSBs is tabulated in Table 7.3. The noise is of two types, an AC component and a DC bias term. Often, the DC bias term is the most problematic, because subsequent filter integration just makes it bigger, thus diminishing dynamic range. For a stealth system, this added noise must not degrade the system performance in any way that requires more power to be transmitted.

7.2.8 Multiplier Roundoff Noise

Errors generated in multiplier roundoff for equally probable multiplicands and multipliers are not exactly equal to the normal roundoff noise just discussed. The usual as-

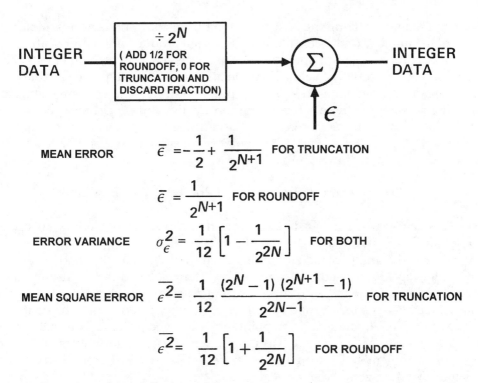

$$\bar{\epsilon} = -\frac{1}{2} + \frac{1}{2^{N+1}} \quad \text{FOR TRUNCATION}$$

MEAN ERROR

$$\bar{\epsilon} = \frac{1}{2^{N+1}} \quad \text{FOR ROUNDOFF}$$

ERROR VARIANCE

$$\sigma_{\epsilon}^2 = \frac{1}{12}\left[1 - \frac{1}{2^{2N}}\right] \quad \text{FOR BOTH}$$

MEAN SQUARE ERROR

$$\overline{\epsilon^2} = \frac{1}{12}\frac{(2^N - 1)(2^{N+1} - 1)}{2^{2N-1}} \quad \text{FOR TRUNCATION}$$

$$\overline{\epsilon^2} = \frac{1}{12}\left[1 + \frac{1}{2^{2N}}\right] \quad \text{FOR ROUNDOFF}$$

Figure 7.18 Truncation and roundoff noise model.

Table 7.3 Truncation and Roundoff Noise

No. of bits discarded	Truncation			Roundoff		
	Mean error	Mean square error	Error variance	Mean error	Mean square error	Error variance
1	−0.25	0.125	0.063	0.25	0.125	0.063
2	−0.375	0.219	0.078	0.125	0.094	0.078
3	−0.438	0.273	0.082	0.063	0.086	0.082
4	−0.469	0.302	0.083	0.031	0.084	0.083
5	−0.484	0.318	0.0832	0.016	0.0835	0.0832
6	−0.492	0.326	0.0833	0.0078	0.0834	0.0833

sumption made in determining errors resulting from truncation and roundoff is that all bits of the number under consideration (e.g., the product of a multiplication) have equal probabilities of being 1 and 0 and are uncorrelated with each other (i.e., that all possible values are equally likely). Therefore, it follows that all possible values of the N least significant bits are also equally probable and that the errors resulting from truncating or rounding these N bits are uniformly distributed. The mean error, mean-square error, and error variance resulting from these assumptions for several values of N are listed in Table 7.3. The binary point is considered to be to the left of the N least significant bits, and the error is defined as

$$\text{Error} = \text{Truncated or Rounded Product} - \text{True Product}$$

The model described above is summarized in Fig. 7.19, including an example roundoff of $M = 5$ bits and a coefficient of $K = 24$.

However, if the multiplicand and multiplier (rather than the product) are assumed to be uniformly distributed and uncorrelated, somewhat different results are obtained. In this case, the values of the N least significant bits (and the errors) are not equally probable. The results of a computer program written to investigate this case are listed in Table 7.4. The significant point to be observed in Table 7.4 is that the mean error does not approach zero or 0.5 as rapidly with increasing numbers of discarded bits. Thus, slightly more DC bias may occur in the filter or FFT output than classically predicted, depending on the signal statistics.

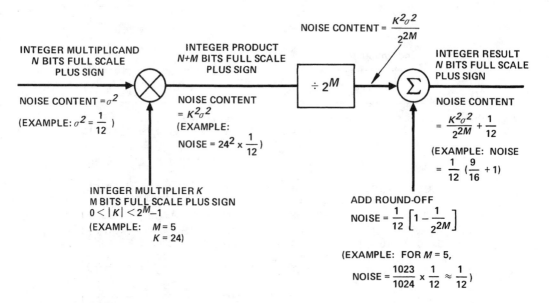

Figure 7.19 Multiplication noise model.

Table 7.4   Multiplication Noise Summary

No. of bits discarded	Truncation			Roundoff		
	Mean error	Mean square error	Error variance	Mean error	Mean square error	Error variance
1	–0.125	0.0625	0.04687	0.125	0.0625	0.04687
2	–0.25	0.14062	0.07812	0.125	0.07812	0.0625
3	–0.34375	0.20703	0.08887	0.09375	0.08203	0.07324
4	–0.40625	0.25488	0.08984	0.0625	0.08301	0.07910
5	–0.44531	0.28638	0.08807	0.03906	0.08325	0.08173
6	–0.46875	0.30597	0.08624	0.02344	0.08331	0.08276

7.2.9   Saturation and Truncation During FFT Processing

The truncation and saturation noise introduced during fast Fourier transform (FFT) processing may be analyzed in identical fashion to those introduced in the proceeding sections. FFT noise is summarized in this section. It has been shown both theoretically and experimentally, with white noise input or sinusoidal input, that fixed point rounded coefficient errors can be calculated as shown in Equation (7.28).[1]

$$\frac{P_n}{\sigma_c^2} = \frac{2}{3} \cdot 2^{-2 \cdot M} \log_2(N) \tag{7.28}$$

where
  M = number of coefficient bits including sign
  N = record length or FFT size
  σ_c^2 = variance of the input signal
  P_n = noise power generated in the FFT operation

Similarly, for the same signal input types, assuming scaling after every transform pass by $1/2$ for a fixed point rounded system, the noise can be calculated as given in Equation (7.29).

$$P_n = \frac{5}{3} \cdot N \cdot 2^{-2 \cdot M} \tag{7.29}$$

where the same variable definitions as in Equation (7.28) apply to Equation (7.29). Obviously, both of these equations are approximate, but their correspondence with actual FFT noise outputs in real signal processors is very good. The agreement between theory and the experimental results is shown in Fig. 7.20.

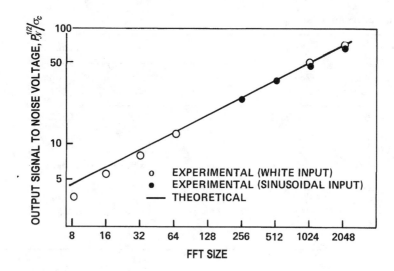

Figure 7.20   FFT noise, scaled by 1/2 after each pass.

Once narrowband processing is completed, it may be desirable to reduce the digital word size if further processing, such as beamforming, pulse compression, post detection integration (PDI), and tracking, is to be done. The same analysis approaches described for pulse compression (Chapter 5), prefiltering, narrowband filtering, and even tracking apply to these cases.

7.3   AIR TARGET—SEARCH, ACQUISITION, TRACK[12]

7.3.1   Air Target Search

The general architecture of the processing in Section 7.3 is shared in common with every modern air-to-air radar, including those in the Swedish (design origin in parentheses) JAS.37 (Hughes), JAS.39 (Hughes), Taiwan Ching-Kuo (G.E.), Japanese F-2 (Westinghouse), French Rafale (appears to be Westinghouse), Eurofighter (Hughes through Marconi to Ferranti, subject of a major international lawsuit), Tornado (T.I.), German F-4G (Hughes), Russian MIG-29 (Phazotron), MIG-31 (Phazotron), MIG-33 (Phazotron), U.S. F-14 (Hughes), F-15 (Hughes), F-16 (Westinghouse), F-18 (Hughes), and perhaps others. The basic approaches have existed since the 1970s and have been steadily refined.[12–14,17–21,22–32]

Like all stealth modes, the object is to squeeze the maximum performance out of the transmitted power. As previously mentioned, this means minimizing straddling losses and matching the waveform as close as feasible to the target characteristics. Range bin sizes should be chosen so that several sample bins can be smoothed into a filter closely matched to target extent; i.e., all the available return from the target but the minimum clutter. For the example here, a minimum size target might have a 40-ft extent and 10-ft resolution, which would require four range bins to be integrated before thresholding. Similarly, the minimum Doppler filter width is determined by target Doppler spread, which is primarily the result of geometrical acceleration. Also, the best available false alarm control and thresholding algorithms are required, because the system must continuously operate with the lowest possible *SNRs*.

Table 7.5 summarizes an example stealth air target search leading parameter set. The waveform chosen in this case is medium PRF (MPRF) to provide all aspect performance. Medium PRF, even with best processing, will always have more losses than, say, a range gated high PRF waveform for closing targets. However, high PRF has poorer performance for opening targets and so, on balance, MPRF might be selected. Eight different PRFs between roughly 8 and 16 kHz are required to provide reasonably complete coverage of the range-Doppler space. Unfortunately, at least some of the PRFs will be blind at most ranges, so the trick is to make sure that the entire PRF set is clear at the maximum design range so that best performance can be obtained there. Note also that the scan rate is slower than conventional air-to-air radars, but the assumption here is that the nine channels are used to form separate beams in elevation so that the total frame time is about normal. This gives adequate update rate but more time to gather energy from a target at the expense of lots of number smashing and receiver complexity. As shown in Chapter 5, channels must be spaced approximately twice the per-channel bandwidth to provide adequate rejection of NEXT. Except for its stealth properties, this parameter set provides standard current generation MPRF air-to-air performance. A mode similar to this easily detected conventional targets at 140 percent of design range.[36]

The processing block diagram for the example mode is shown in Fig. 7.21. In the middle of the figure is a simplified time line. There are 8 phases of 160 pulses each for this set of PRFs. Returns from each transmitted pulse are digitized, and the innermost portion of the pulse compression code is compressed. A/D outputs are used to generate a digital automatic gain control for the receiver as shown in Figure 7.3. The data record is buffered and, when all 160 pulses in one phase are collected, they are amplitude weighted, and the record is filled out with zeros to 256 samples, sometimes called data turning. Each range bin sample record is filtered in a 256-point FFT. The remainder of the pulse compression is completed after filtering, because Doppler phase warp has been removed by the FFT, some filters are discarded, and eclipsing allows only partial compression of the early range bins. One reason for a compound pulse compression code is that it allows relatively graceful handling of eclipsing (the five-chip code is the outer code). After compression, which includes sidelobe suppression, each bin is magnitude detected, filtered to match target extent, and separate frequency channels are summed. Sensitivity time control (STC) is applied after partial pulse compression, prior to first threshold and after range and velocity resolving.[17,18,36,60,62]

Table 7.5   Air Target Search Example Parameters

Parameter	Value
Waveform	Medium, PRF, 8 PRF set
Maximum peak power	16 W
Pulse compression	$16 \times 13 \times 5 = 1040$ Barker and Frank
Range resolution	10–20 ft
Duty ratio	25 percent
Frequency channels	9
Maximum range vs. 1 m^2 target	25 mi
Range bins processed	2600 to 5200
Doppler filters	256
Antenna scan rate	13.3°/sec
Azimuth coverage	60°
Elevation coverage	25°
Frame time	4.5 sec
Total instant bandwidth	1.17 GHz
Operating band	~8–10 GHz

The outputs prior to frequency overlay are used to make an estimate of the center of mainlobe clutter. They are also used to detect countermeasures in an ECCM processing function that discriminates against false targets that have impossible accelerations, sudden increases in RCS, and other inconsistent signature features.

The summed outputs are used in a sliding window ensemble noise estimator. The output of the noise estimate coupled with the threshold crossing count provides an environment adaptive threshold multiplier, which is combined with *a priori* thresholds based on the range and azimuth geometry. The statistical tails of mainlobe and sidelobe clutter are not Gaussian and require a larger number of standard deviations to ensure a low false alarm rate.

Best performance is achieved with a multilevel threshold, sometimes called alert-confirm. A lower threshold allows higher detection probability (alert) with low power, but the higher false alarm rate is suppressed by revisiting a target location with an optimized waveform and a higher threshold (confirm). Very large ground moving targets may still get through all this filtering, and they are eliminated by a combination of amplitude and apparent location.

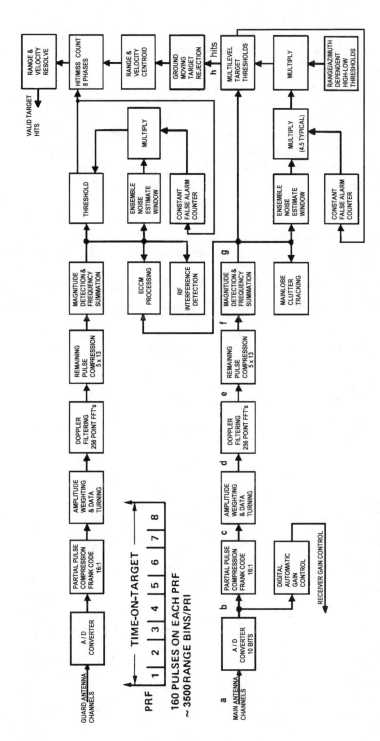

Figure 7.21 Air target search signal processing.

Because target returns will straddle multiple bins in range or velocity because over-sampling to reduce straddling losses has been used, the centroid of each target in range and Doppler is estimated.

The alert-confirm thresholding strategy is shown in Fig. 7.22. An alert threshold is set at 4 dB *SNR*, which has a 50 percent probability of detection and a false alarm probability of 1 percent. Dwell time is set to just detect the minimum target. A table of detections is assembled, and the radar returns to each detection location. The dwell time is set to just detect the minimum target with a high confirm threshold. The confirm threshold set to 13.5 dB *SNR*, which has a probability of detection of 90 percent and false alarm probability of 10^{-7}. Although this scheme is used in conventional radars, its stealth feature is that lower power can be transmitted for a given target detection performance.[61] This multilook strategy works best with an electronically scanned antenna but can also be used with mechanical scanning. It is effective in angle, time, or frequency, but it depends on a sparse target set. The Russian MIG-31 uses alert-confirm.

Another thresholding strategy is track-before-detect, in which many possible target trajectories in range-Doppler space from several antenna looks are speculatively inte-

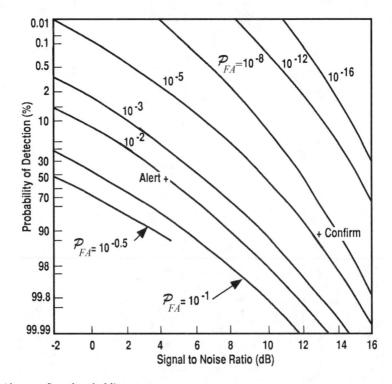

Figure 7.22   Alert-confirm thresholding strategy.

Table 7.6 Signal and Noise Trace Air Target Search

Function	Fig. 7.22 ref.	Min. target	Max. target	RMS clutter	RMS noise	Sum A_N	A_N rms	Formulas	# bits	Added noise	Signal to clutter (dB)	Clutter to noise (dB)	Min. target SNR (dB)
IF	a	0.0144	25.50	114.00	2.41	1	1	Range equation	Inf.	0.000	−78.0	33.5	−44.5
A/D	b	0.0144	25.50	114.00	2.43	1	1	(7.14)	10	0.301	−78.0	33.4	−44.5
Partial PC	c	0.23	408.01	456.00	9.71	16	4	Fig. 5.78	14	0.000	−65.9	33.4	−32.5
Amp wt. and turn	d	0.12	204.02	171.00	3.65	0.50	0.38	Fig. 7.19 and Table 6.3	16	0.289	−63.4	33.4	−30.0
FFT	e	29.49	52229	1.71	62.01	256	16	(7.29) and −40 dB ISLR	16	20.650	24.7	−31.2	−6.5
Remainder PC	f	29.95	53045	0.22	7.82	65/64	8.06/64	Fig. 5.78	16	0.289	42.9	−31.2	11.7
Mag. det. and freq. sum	g	33.70	59675	0.08	2.95	9/8	3/8	(7.18) and (7.20)	16	0.289	52.4	−31.2	21.2
Threshold	h	20.44	59662	0.00	0.00	N/A	N/A	Sig. −4.5 × rms noise	16	0.000	426.2	−260.0	166.2

grated noncoherently and then thresholded. If the target does not move, there is no improvement over conventional detection. If the target moves between looks, there can be an improvement similar to alert confirm.[33] Again, the stealth feature is lower transmitted power. The disadvantage of both schemes is more required number smashing.

The noise performance of air target search can be assessed by performing a signal and noise trace through the block diagram of Fig. 7.21. The inputs and outputs of the processing steps are designated with a letter in Fig. 7.21. For this particular case, the parameters are taken from Table 7.5, except the total power is assumed to be 4 W. The minimum processed range is 3km, antenna gain is 37 dB, average clutter cross section is 0.01 m^2/m^2, the minimum point target is 1 m^2, the maximum point target is 100 m^2, noise figure is 3 dB, and internal losses are 1.5 dB. The altitude line is removed with a crystal filter in the IF. The clutter and large targets are smeared in range ahead of the A/D, and the DAGC applies STC and gain control to keep RMS clutter at the optimum value 0.222 of full scale at the A/D input. The range equation is used to calculate the values of targets, clutter, and noise at the A/D input. It is then a matter of calculating the gain for signal, input noise "plus" roundoff noise and clutter through the processing steps as tabulated in Table 7.6. A word length of 16 bits for I or Q has been assumed as the maximum in the processor in this case, and the LSB is assumed to have the value 1. Once the equivalent "voltages" have been calculated, the signal to clutter and noise ratios are calculated. The maximum target signal must be calculated to prevent overflow. The Table 7.6 numbers are optimistic, because AM, FM, and "birdie" self-noise have not been incorporated into the calculation. Self-noise usually limits performance in high-clutter-to-target environments, but it is beyond the scope of this text.

Returning to the block diagram of Fig. 7.21, guard channels are multiplexed with the main antenna channels at a lower rate and processed in a similar manner to the main channels. The concept of a guard channel is shown in Fig. 7.23. Guard antenna elements have relatively low gain or may be approaching omnidirectional, and thus the received spectrum is the result of very many continuously visible scatterers or interferers and doesn't change very fast. The guard channel outputs are used for near range sidelobe blanking as well as for ECCM and RF interference processing. Any guard channel output that exceeds a predetermined ratio with respect to the main channel represents sidelobe clutter, interference, or jamming. Those hits coming through the sidelobes that seemingly are real targets are blanked in the final hit/miss counter. Alternatively, the guard channel can be used for jammer nulling rather than just hit blanking (this is not shown in Fig. 7.21 or 7.23). The counter outputs above a predetermined hit count (usually three to five) are real targets, and the eight phases are used to resolve the targets in range and velocity to their true location in range-Doppler space.[31,46,60]

Medium PRF surface-to-air search is not fundamentally different from air-to-air search except that a mainlobe clutter map about the radar site is formed instead of

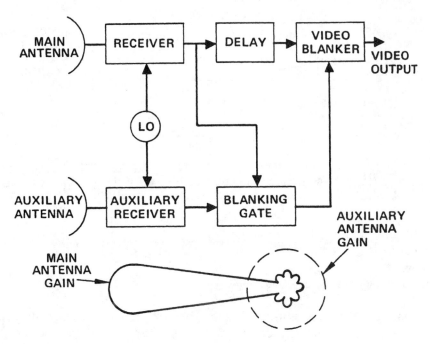

Figure 7.23 Guard channel concept.

mainlobe clutter Doppler tracking, because the radar site isn't moving very fast. The clutter map is used for thresholding and stationary but moving target suppression. Clutter Doppler spread is much smaller for surface systems, and the terrain masks long-range clutter, often reducing dynamic range. Unfortunately, even a very stealthy surface radar is ultimately detectable, as integration times can be very long. (See Chapter 3, especially Figs. 3.45 through 3.47.) The emitter either must move or must have multiple spoofing antenna/transmitters that alternately operate or cover the transmit pulses in the sidelobes. Another operational technique is to create intentional ground bounce, which increases the emitter location uncertainty volume.

7.3.2 False Alarm Control

False alarm control is crucial in any modern radar, but the necessity of working at very low signal levels makes this more important for stealth systems. The basic concept for an adaptive threshold based on the noise environment is shown in Fig. 7.24. An important element of the concept is the assumption that targets are sparse relative to the total number of bins processed in range, angle, or Doppler. The top part of Fig. 7.24 shows a target in the center bin. To prevent the target from self-biasing the threshold, guard bins immediately around the target are excluded from processing. Bins near the bin to be thresholded, early and late noise samples, are summed/aver-

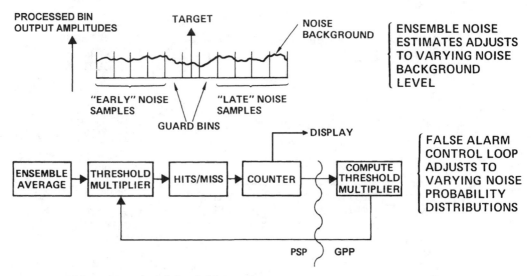

Figure 7.24 Noise-estimate-based threshold control.

aged to create an ensemble noise estimate. That estimated "standard deviation" is multiplied by a factor whose intent is to move the threshold high enough that the probability of noise alone crossing the threshold is less than some predetermined number. The total number of threshold crossings counted, less the number of bona fide target crossings, give an estimate of the false alarm rate. A control loop, usually closed through a general purpose processor, adjusts the standard deviation multiplication factor. There may be other external conditions that suggest that the threshold must be set higher than the minimum set by thermal noise alone (i.e., clutter, jamming, birdies, RFI from other equipment, and so forth). It is known from long and painful experience that some types of clutter and almost all interference have non-Gaussian statistics with large "tails" that lead to dramatically higher false alarms. When these types of signals are recognized, heuristically determined and tabulated threshold multipliers are introduced by the general purpose processor. As will be seen in ground moving target detection, thresholding schemes to control false alarms can be quite elaborate.[17,18,35,60]

One example sliding window noise ensemble estimator is shown in Fig. 7.25. This particular design is widely used in airborne radars. Although simple, it has all the features previously mentioned, early and late windows, center guard notch, threshold circuit, standard deviation multiplication factor of 4, and "tilt logic." The tilt logic is designed to detect if the window is riding up the side of a large clutter or interference lobe. If either early or late noise samples by themselves are substantially larger than the other, then the tilt logic switch directs that side only to the threshold. This scheme can be used in range or Doppler with various numbers (8 to 32, typically) of bins and suitable scaling.

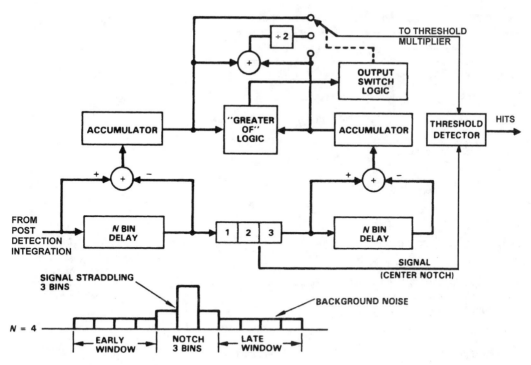

Figure 7.25 Example sliding window ensemble average.

7.3.3 Air Target Tracking

The leading parameter set for tracking a single air target is summarized in Table 7.7. Because the target location and return power are known with some accuracy, the PRF can be chosen to be clear in range and Doppler. The transmitted power can be chosen to be lower than that required for search, thus ensuring that, if intercept did not occur before track initiation, it will never occur. Total bandwidth used is less than for search, but the spectral density is unchanged, because track SNRs can be 10 times less than search. Tracking in angle requires that the receiver channels are used to form azimuth and elevation discriminants, which leaves three sets of three frequency channels for track in the current example. The number of pulses in a coherent array is increased from 160 to 256. The pulse compression waveform is unchanged, but eclipsing is no longer a limitation. A patch of bins around the target location are processed to allow detection of missile launch, countermeasures, and target recognition signatures. Once in track, a target can be followed over a much larger range of angles and distances. Depending on target dynamics or threat level, tracking can be continuous or interrupted to interleave other target tracks or search for new targets. Use of a Kalman filter in the GPP will normally allow a target to be tracked through mainlobe clutter, even if attempting evasive maneuvers.

Table 7.7   Air Track Parameters

Parameter	Value
Waveform	MPRF, optimum for target
Peak power	Power managed—R^3 and 3-dB SNR
Pulse compression	$16 \times 13 \times 5 = 1040$ Barker and Frank
Range resolution	10–20 ft
Duty ratio	25%
Frequency channels	3
Maximum range vs. 1 m^2 target	45 mi
Range bins processed	128
Doppler filters	256
Antenna scan rate	As required—usually low
Azimuth coverage	±60°
Elevation coverage	±45°
Frame time	Continuous
Total instant bandwidth	400 MHz
Operating band	~8–10 GHz

Because most military pilots consider transition to single target track to be a hostile act, tracking must appear to the target as no different in case intercept has occurred. Similarly, missile launch and guidance must also be disguised until the missile is inside the no-escape region.

A simplified block diagram of air target tracking is shown in Fig. 7.26. A notional time line is also shown in Fig. 7.26. The same number of channels are used as in search, but their functions are different, because three channels are required per frequency to form monopulse discriminants. Most of the signal processing is the same as search, but the per-target processor loading is small, because the number of bins required is small. Several features of the track waveform are significant. First, a number of pulses (16 in the example) are usually used to set the automatic gain control/environment parameters. Second, an interval, often called *sniff*, is set aside every few coherent arrays to passively listen for jamming signals and cause movement to frequency channels that are clear.[62]

Track discriminants are formed using both the magnitude and phase of the processed data. Range, velocity, angle, acceleration, and target vertical and horizontal extent changes are estimated for the Kalman filter tracker. The track filter also requires

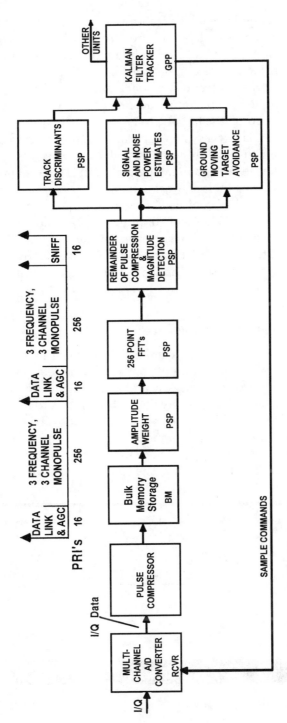

Figure 7.26 Air target track signal processing.

estimates of signal and noise power. Large ground moving targets may still arrive at the filter bank output and must still be censored from the data provided to the tracker. Usually, there are multiple versions of the Kalman filter tracker based on the assumed target maneuvers with logic to select the most likely target behavior at any instant. The track filter outputs are scaled and sent to the antenna, RF front end, timing, transmitter, and exciter. Sample instant commands are sent to the A/D converters to keep the sampling bins centered on the target.

An example of the display of a system similar to that described above is shown in Fig. 7.27. This was an air-to-air flight test in which the target was an F-111. The target was co-altitude with the stealth-radar-carrying aircraft. The total maximum power in this test was 4 W. The three panels show the display during search mode, acquisition, and track mode. The display is completely clean of false alarms. The target was acquired and placed in track at 25 km and continued to track all the way to gimbal lock (very close range). The stealth radar was never detected, even when the two aircraft were visible to one another. The annotation on the display shows the running time, flight number, horizon, threshold multipliers, target range, azimuth, elevation, heading, and velocity.

7.4 TERRAIN FOLLOWING/TERRAIN AVOIDANCE

Even a very low-cross-section vehicle can be detected if it is totally in the clear. As shown in Chapters 1 and 4, operating at low altitude in clutter is best. Additional stealth for aircraft is obtained by flying low and fast with terrain following (TF) or terrain avoidance (TA) systems. The objective of a TF system is to fly over higher terrain, whereas TA systems fly around higher terrain. Such systems have been used for more than 30 years in cruise missiles and manned aircraft. The concept for stealth terrain following or avoidance uses three sets of measurements, real time altitude and terrain as well as a prestored hypsographic map. The concept is shown in Fig. 7.28.

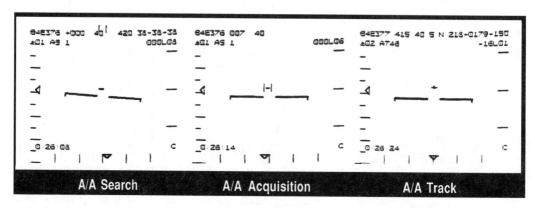

Figure 7.27 Example stealth radar A/A displays.

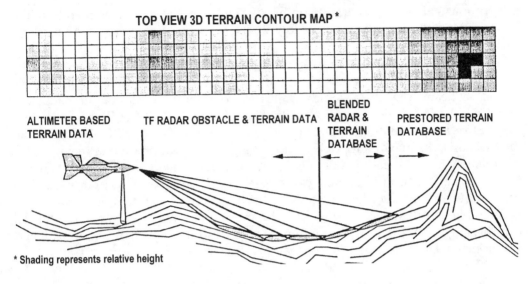

TOP VIEW 3D TERRAIN CONTOUR MAP *

ALTIMETER BASED TERRAIN DATA   TF RADAR OBSTACLE & TERRAIN DATA   BLENDED RADAR & TERRAIN DATABASE   PRESTORED TERRAIN DATABASE

* Shading represents relative height

Figure 7.28 Stealth TF/TA concept.

TF/TA requires a three-dimensional map of the region along the aircraft flight path with accuracy of about 10 percent of the desired flight altitude. The object for low-flying aircraft is to maintain a relatively fixed altitude above the terrain without hitting anything. Unfortunately, the geometry of TF/TA doesn't allow high duty ratios, so best LPI performance must be obtained from the application of other information such as a prestored map database. TF/TA systems must measure some small targets, such as suspended cables, as well as large targets that scintillate. Historically, TF/TA systems were noncoherent, but this leads to unnecessary processing losses for a stealth sensor.

The three sets of measurements used in the concept of Fig. 7.28 must be of comparable scale and accuracy. The first set of measurements is a prestored digital terrain elevation database made from survey data, maps, satellite measurements, and so on. The prestored terrain database provides control points for long range TF or TA manuevers. Because the terrain elevation database is not updated very quickly when new features arise, and it contains some known systematic errors in the distance between measured points as well as height inaccuracies, real-time measurements are also required. The second set of measurements is terrain profile data from a TF/TA radar. The radar must also detect and measure obstacles too small to be in the database but still lethal to aircraft. The third set of measurements is the actual height of the aircraft above the terrain which, when coupled with inertial data to remove flight perturbations, allows matching and registration of the prestored profile with the actual measurements. This is, in essence, a calibration of all three sets of measurements. It also allows blending of the current measurements with the database proportional to relative accuracy/SNR. The advantage of this system is that the power required from the radar and the altimeter can be quite low so, the probability of intercept is extremely small.

An example set of leading parameters for stealthy TF/TA is shown in Table 7.8. The most important features are the duty ratio and the number of frequency looks required per beam. The typical aerodynamic time constant for a subsonic aircraft at low altitude is 2 to 5 sec. For aircraft travelling at a nominal 800 ft/sec, this implies that the control range minimums must be between 1600 and 4000 ft, with 3200 ft typical. On this basis, the maximum pulse width, taking into account partial eclipsing, is approximately 5 µs. Assuming a nominal 0.5-g ride (very rough, but most pilots won't throw up) and flying at El Capitan cliff face (a typical worst case), accurate elevations are required to 3 mi. Very-low-range sidelobes are required, so a complementary code is desirable, which suggests a PRF of twice maximum range (i.e., 6 mi). The typical control range that has maximum influence on aircraft maneuvers is about 2 mi, and power management is usually set so that this range has 20 dB SNR.[12]

Table 7.8   TF/TA Leading Parameters

Parameter	Value
Waveform	15 kHz PRF
Maximum peak power	4 W
Pulse compression	64:1 complementary code
Range resolution	40 ft
Duty ratio	8 percent
Frequency channels	3
Frequency looks/beam	24 ft
Maximum range	6 mi
Range bins processed	512
Doppler filters	1
Antenna scan rate	75°/sec
Azimuth coverage	70°
Elevation coverage	30°
Frame time	TF 0.4 sec, TA 1 sec
Total instant bandwidth	870 MHz
Operating band	~8–10 GHz
Elevation measurement accuracy	20 ft

All TF/TA systems must be designed to detect, recognize, and measure the following obstacles:

1. Towers and new structures

2. Vertical and horizontal cables

3. Terrain

4. Rain and chaff

5. Jamming

These objects may range in reflectivity from 0.001 to 100 m^2/m^2 and may scintillate by 100:1, which requires multiple looks in frequency or angle. The required elevation accuracy is substantially less than a beamwidth, so monopulse off-boresight processing is required. During one beamwidth dwell time, a scatterer will close as much as one-half of a range bin, so range closure compensation is required over the frame to minimize straddling losses.

Figure 7.29 shows an example time line for these modes. During a frame time, the equivalent of 10 major (full beamwidth) and 8 minor (fractional beamwidth) beam steps occur. Each major step will cycle through 24 frequencies of 20 pulses each. Nine receiver channels are allocated as three offset beams of three monopulse channels. Each channel at each frequency has a bandwidth of approximately 12 MHz and is stepped pseudorandomly over a 290 MHz bandwidth. The three beams will cover a total of 870 MHz. Statistics are formed from each pulse across all 480 pulses. Each frequency coherently integrates the 20 transmitted pulses in a single Doppler filter, which also rejects the complementary code 1/2 PRF image and the MTAE returns, if any. The 20-pulse array time is 1.25 msec, which is roughly matched to the Doppler spread of the ground return across the beam.

Atmospheric and ground effects in low level flight often result in substantial deviations in vehicle attitude and heading relative to the desired or ideal flight path. TF/

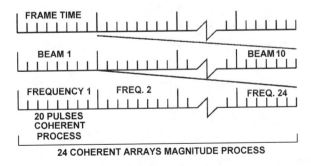

Figure 7.29   Example terrain following (TF)/avoidance (TA) time line.

TA antenna scan patterns must anticipate the potential for these deviations to en-sure proper measurements and safe operation. An idealized antenna scan pattern for the example stealth TF/TA radar is shown in Fig. 7.30. Three separate beams are formed in azimuth to provide more rapid coverage around the ground track for TF and a rapid cut at the flight altitude for TA. The scan in TF must go above the local horizontal to ensure detection of cables. The three beams are separated in frequency by approximately 290 MHz, so NEXT is not a problem. Frequency generation uses the method described in Chapter 5. The actual beam positions in TF are stepped by 1/8th beamwidth every three frequencies to provide a more accurate estimate of terrain height. This stepping requires scan compensation prior to discriminant for-mation.

A block diagram of the processing for the example TF/TA radar is given in Fig. 7.31. Three separate beams of three-channel monopulse are processed. All nine chan-nels, after heterodyning to baseband, are digitized, pulse compressed, and multiplied by the appropriate MTAE phase decode. The early range bins are eclipsed and must be compressed differently as described for air targets. All of the integration for pulse compression sidelobes and MTAE rejection is accomplished in a single Doppler filter for each channel. Most monopulse patterns are not linear, so gain and phase correc-tions from a periodically updated calibration table are applied to each channel. Be-cause the beam is stepped between coherent arrays, scan compensation is applied. The monopulse discriminants are formed and accumulated over the 24 separate fre-quencies in each beam. The discriminants work best when integrated with positive *SNR*. Subsequent to discriminant formation, centroid and extent are calculated and used for target type discrimination and top-of-terrain estimation.

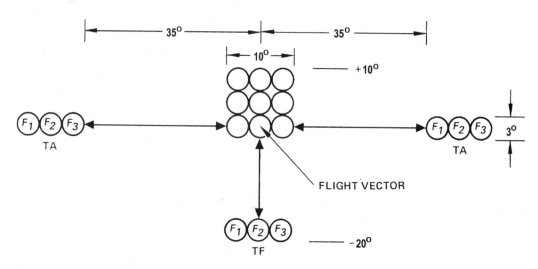

Figure 7.30   Example TF/TA antenna scan pattern.

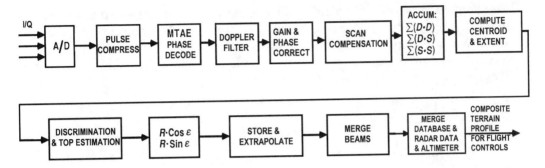

Figure 7.31 TF/TA processing.

The concept for top of terrain estimation is shown in Fig. 7.32. The antenna beam is large compared to the range cell size and the desired height accuracy at the control range. If the *SNR* is high, an angle estimate that is an order of magnitude better than the angle resolution can be obtained with off-boresight processing. The left side of Fig. 7.32 shows an antenna beam intercepting a segment of terrain. The right side of the figure shows a single range bin containing terrain. Somewhere in that beam is a band of terrain perhaps covering the entire azimuth extent of the beam. Estimating the centroid of that terrain return provides the mean elevation of the terrain in that range cell. Estimating the extent of the terrain about that centroid provides information about the height of the terrain above the centroid in that range cell. The one standard deviation estimate of the top of the terrain is $C + 0.5E$. Decades of experiment and deployment have shown that multiple measurements with uncorrelated scintillation and noise provide accuracies of 20 ft.

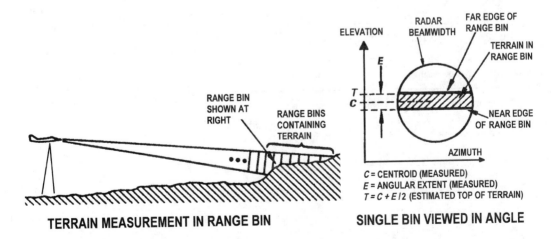

Figure 7.32 Off-boresight monopulse processing.

The relations for centroid, extent, and top of terrain are given in Equations (7.30). Higher-order moments can also be calculated and are useful for detection of jamming. As is well known, all of the experimental statistical moments higher than the mean do not necessarily converge, and so multiple samples of each of these calculations must be averaged to obtain a stable estimate of each.

$$P_r = \sum_{i=1}^{N} |S_i|^2 \quad \text{Power Received}$$

$$C_r = \text{Re}\left\{ \frac{\sum_{i=1}^{N} S_i \bullet D_i}{P_r} \right\} \quad \text{Centroid}$$

$$E_r^2 = \frac{\sum_{i=1}^{N} |D_i|^2}{P_r} - C_r^2 \quad \text{Extent Squared}$$

$$T = C_r + 0.5 \cdot E_r \quad \text{Top}$$

$$(7.30)$$

Obstacle discrimination uses the moments of Equations (7.30) as well as some of the higher-order moments to assist in detection and recognition of TF/TA returns that might be lethal to a low-flying aircraft. Some example recognition criteria are suggested in Table 7.9. The absolute definition of large and small depends on the details of the design and minimum allowable false alarm rate.

Table 7.9 Off-Boresight Obstacle Discrimination

Obstacle	Azimuth extent	Elevation extent	Higher moments
Discrete/power line	Small	Small	Small
Terrain	Large	Small	Small
Tower/balloon cable	Small	Large	Small
Rain/chaff	Large	Large	Small
Jamming	Large	Large	Large

Returning to the processing block diagram, the smoothed angle estimates are converted to lineal dimensions. The lineal dimensions are stored and extrapolated between frames. All the beams are merged into a single terrain profile. That profile is merged with the prestored terrain database and the retrospective altitude data based on measured SNR/known database uncertainty. The result is presented to the flight controls for automatic terrain following and turning flight.

7.5 REAL BEAM GROUND MAP

Most airborne radars have a real beam ground map mode that scans a volume as large as π steradians. It is usually one of the simplest modes in the system. Unfortunately from a stealth point of view, this is one of the most dangerous modes in a radar. Just as much care in design and processing is required as needed for air search. Pulse compression, high duty ratios, MTAE coding, multifrequency, multiple PRFs, coherent processing, and aggressive power management are required. Mapping must go from near range to far range with successively lower PRFs for each scan bar to limit long-range intercept exposure. Because resolution is gross, on-times can be limited to 1 percent or less and should be pseudorandom. Real beam is best used only to fill the blind region at the velocity vector in Doppler beam sharpening. The good news is that resolution cells are large, so transmitted power can be low with a maximum of less than 0.5 W. Example leading parameters for real beam map are summarized in Table 7.10.

Figure 7.33 shows an example of real beam map signal processing. As in the previous example modes, the processing uses pulse compression and Doppler filtering to improve SNR. Because the Doppler spread relative to the PRF changes with angle off the velocity vector, the FFT outputs are *post-detection integrated (PDI'ed)* as a function of angle. Multiple beams and frequencies are also PDI'ed. The output is compressed into eight gray shades for display.

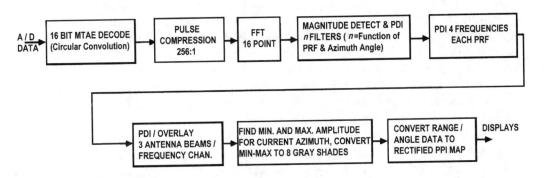

Figure 7.33 Real beam ground map processing.

Table 7.10   Real Beam Map Leading Parameters

Parameter	Value
Waveform	1.6–6.4 kHz PRF
Maximum peak power	<0.5 W
Pulse compression	256:1 complementary code
Range resolution	46 ft
Duty ratio	25%
Frequency channels	3
Frequency looks/beam	4
Maximum range	50 mi
Range bins processed	5100
Doppler filters	16
Antenna scan rate	75°/sec
Azimuth coverage	90°
Elevation coverage	Ground stabilized
Frame time	1.2 sec
Total instant bandwidth	280 MHz
Operating band	~8–10 GHz

7.6   DOPPLER BEAM SHARPENING

Doppler beam sharpening (DBS) is the first of the higher angular resolution ground mapping modes. It uses the Doppler spread of ground return across the beam to split the antenna beam into 10 to 100 segments. For the example here, the beam is split into a constant 16 segments but, because the Doppler spread changes dramatically with off angle, this requires a variable presum filter, FFT size, and post-FFT filter collapsing and PDI. In addition, three simultaneous frequency channels are used in each antenna beam, and three antenna beams are formed. This mode is best done with 2-D electronic scanning but has been implemented with a combination of mechanical and frequency scanning. Example DBS parameters are given in Table 7.11. The range resolution is way out of proportion to azimuth resolution, and so multiple range bins (16 to 32) are integrated after detection. The frame time is long but, fortunately, the required power is usually small. The maximum range is 25 mi, and the PRF ambiguity is chosen just beyond the last range cell, so MTAE rejection is required. Although it is not obvious from the table, the dwell time is 0.25 sec per beam dwell. The use of three

Table 7.11   Doppler Beam Sharpening Leading Parameters

Parameter	Value
Waveform	3.38 kHz PRF
Maximum peak power	<0.1 W
Pulse compression	4096:1 complementary code
Range resolution	9 ft
Duty ratio	25%
Frequency channels	9
Frequency looks/beam	3
Maximum range	25 mi
Range bins processed	11,000
Doppler filters	16–128
Antenna scan rate	Funny
Azimuth coverage	90°
Elevation coverage	Ground stabilized
Frame time	6 2/3 sec
Total instant bandwidth	900 MHz
Operating band	~8–10 GHz

azimuth beams that cover the same ground requires a rapid scan between dwells followed by a very slow scan that just compensates for platform motions over the dwell time.[12,63]

Figure 7.34 shows the signal processing normally associated with a stealth DBS mode. Pulse compression and MTAE rejection are standard (for a stealth mode, anyway). The first DBS unique processing is presummation. The ground return in each frequency channel has been positioned by the VCO before the A/D as close as possible to zero Doppler. There is enough range closure over the presum and FFT process that range closure compensation must be performed before presummation. Each frequency channel is presummed (lowpass filtered) to match the clutter spectrum and resampled at a lower rate for FFT formation. The FFT size ranges from 16 points at 5° off the velocity vector to 128 points at 45° off the velocity vector. The FFT array is filled with zeros to make the transform sizes round numbers at each beam position. The FFT outputs are used to form discriminants used for AGC and clutter Doppler error, because it is not possible to predict Doppler or set the front-end VCOs accurately enough to move the center of the ground return to exactly zero frequency. Each filter

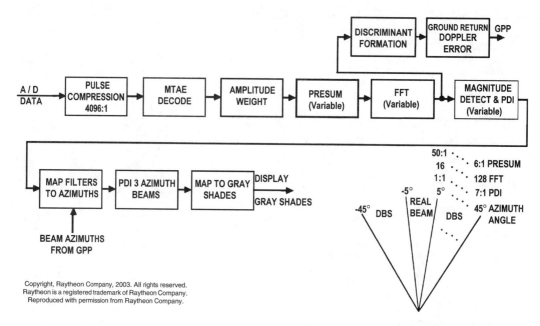

Figure 7.34 Doppler beam sharpening (DBS) processing.[12]

output is magnitude detected, and adjacent filters are collapsed/post detection integrated (PDI) to form 16 outputs across the azimuth beam. With the long frame time, the aircraft and hence the beam pointing direction can easily deviate from predicted position, and so each beam actual angle must be used to integrate (PDI) the outputs from the three separate beams to form the final map.

The dynamic range from any mapping process can easily be 60 dB, whereas the typical display can generate a dynamic range of only 20 dB. Significant dynamic range compression is required. Usually, a nonlinear mapping of the amplitude range in any angle bin is mapped into eight shades of gray (usually yellow-green on the display, where the human eye is most sensitive). Also shown in Fig. 7.34 is the composite processing used in a DBS scan over ±45° about the velocity vector. There are three azimuth sectors: –45° to –5°, –5° to +5°, and 5° to 45°. The central sector about the velocity vector is normally filled with real beam map (see previous section). The two other sectors are DBS and have a different presum ratio, FFT size, and corresponding filter collapsing/PDI ratio for each angle as shown in the figure. The intent of DBS is to provide a PPI sector map with reasonable resolution over 90°.

7.7 SYNTHETIC APERTURE RADAR (SAR) MAPPING

Another commonly used radar mode is synthetic aperture radar (SAR) mapping. Any time two objects have relative rotation to one another, a SAR map can be formed in the plane of rotation. This is the basis of space object imaging, inverse SAR, and SAR imaging. This is another technique that has been around for 40 years. Most SAR is inherently LPI, with very little LPI unique requirements with the exception of staying away from linear FM waveforms and using lower than normal antenna sidelobes. SAR imagery of your house with 3-ft resolution can be purchased over the Internet if you know your latitude and longitude.[6,7,39–45]

The particular example mode described here is sometimes called spotlight or tracking telescope, because the idea is to provide a high-resolution image of a relatively small patch on the ground. The resolution is high enough that recognition of vehicles, most man-made objects, tire tracks, buildings, fences, piping, and so on is quite reliable. Such a mode allows targeting and high-accuracy weapon delivery for virtually all military targets. The principal challenge for a stealth radar is the fact that the cell size is quite small, and hence its radar cross section is tiny. The power required may be the largest or the smallest of any stealth mode, depending on geometry. Large pulse compression ratios are commonly used, but PRFs are restricted, as shown below. Studies have shown that per-pixel SNRs of 4 dB are adequate.[6] As in other modes, maximum ranges are restricted by interceptability, not feasibility. Motion compensation is essential in SAR, and autofocus techniques often are required to obtain crisp maps.

Figure 7.35 shows the processing block diagram for the example SAR mode. Depending on the range to the mapped area, antenna illumination, isoDoppler curvature, Doppler/PRF ratio, phase motion correction, depth of focus, and so on may not be uniform. Although the figure and table show fixed values for many parameters, the actual PRF, presum ratio, phase corrections, pulse compression ratio, FFT size, number of illuminating beams, and multilook overlays are all geometry dependent.

The bottom part of Fig. 7.35 shows representative spectra at points in the block diagram. After each filtering process, the outputs are resampled to reduce throughput demands. As usual, filter aliasing caused by resampling forces higher processing rates than one would hope based on the 3-dB filter width. The dynamic range relative to noise grows dramatically with each step in the process as shown for the three example spectra. Inevitably, filters at band edge, as shown in the lower right portion of Fig. 7.35, will have higher noise levels and thus lower dynamic range. There is one fundamental difference between ISLR in angle or range space and filter processing space: the resample rate is variable all the way to the input sample rate. This allows the designer to be subject to fewer integrated sidelobes by using a higher resample rate and restricting the extent of the processed window. Recall that almost all weightings cited in Table 6.3 had ISLRs of approximately –22 to –25 dB. Those same weighting windows, if close-in sidelobes can be excluded, can produce ISLRs lower than –40 dB when the sidelobes are defined as outside $3 \cdot U_{-3dB}$.[55]

Table 7.12 summarizes the leading parameters for a stealthy SAR mode. This mode requires the highest power of the modes described in this chapter, even though the

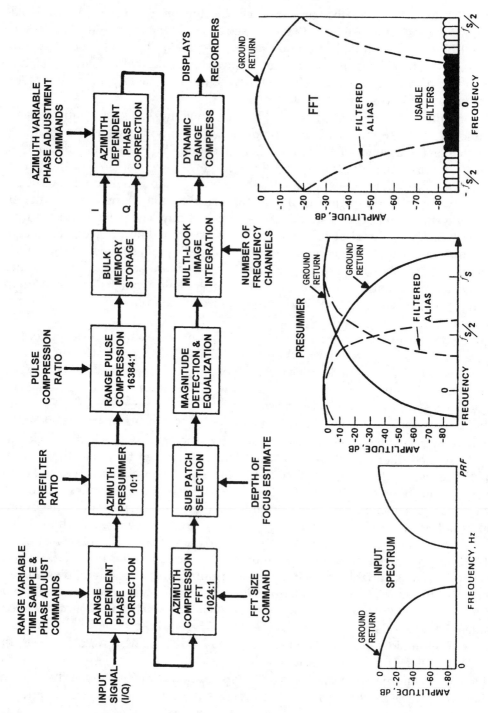

Figure 7.35 Synthetic aperture radar (SAR) signal processing.

Table 7.12 SAR Map Leading Parameters

Parameter	Value
Waveform	2.5–6kHz PRF
Maximum peak power	7 W
Pulse compression	8192–16384:1 complementary code
Range resolution	3 ft
Duty ratio	25%
Frequency channels	9
Frequency looks/beam	5–2
Maximum range	30 mi
Range bins processed	1024
Doppler filters	1024
Antenna scan rate	Ground stabilized
Azimuth coverage	90°
Elevation coverage	Ground stabilized
Frame time	5 sec
Total instant bandwidth	1.5 GHz
Operating band	~8–10 GHz

mode has the highest level of integration, because the cell RCS is so small as a result of grazing angle and absolute size. Multiple frequency looks and output dynamic range compression usually make map outputs look almost photographic, even with low *SNR*.

Stealth systems require high PRFs and duty ratios, but SAR maps require at least an uneclipsed range of the swath distance, so the range of allowable *PRIs* is given in Equation (7.31). The challenge for stealthy SAR is finding a PRF that is high enough.

$$\frac{\lambda}{2 \cdot V \cdot \theta_{az} \cdot \sin(\theta)} \leq PRI \leq 2 \cdot \left[\frac{R_{swath} + R_{min}}{c} + T \right] \tag{7.31}$$

where

λ = transmitted wavelength
c = velocity of light
V = vehicle velocity
θ_{az} = antenna azimuth beamwidth

θ = angle between the velocity vector and the antenna beam center
R_{swath} = swath length
R_{min} = minimum ambiguous range to the swath
T = transmitted pulse length

For example, for a vehicle velocity of 800 ft/sec, beamwidth of 3°, angle off the velocity vector of 45°, wavelength of 0.1 ft, swath of 1000 m, a transmitted pulse of 100 μsec, and an ambiguous minimum range of 10 km (i.e., $n \cdot 40$ km unambiguous), the maximum PRI is 1768 μsec, and the minimum PRI is 274 μsec or 3.65 kHz.

In general, all map targets, whether real beam, DBS, or SAR, are area targets in which multiple cells are involved. The target contrast ratio with respect to the background defines detection and recognition probability. The maximum power from a target can be represented by its autocorrelation function. The maximum detectability relative to its background for a pixelized target can be represented by the sum of the contrast ratios in all the pixels on the target. That sum can then be compared to a conventional threshold normalized to the background power level. Typically, shape and contrast ratio based target detection and recognition is done by a human. Human factors studies done in the 1970s showed that an additional 3 dB or more over that implied by a pure Gaussian statistic detection/false alarm threshold was required for humans to detect area targets in a map background.[7] Thus, the signal-to-noise ratio for some target shape is given in Equation (7.32).

$$SNR_{shape} = \sum_{i=1}^{mm} |CR_i| \tag{7.32}$$

where mm is the number of resolution cells/radar pixels within the target shape, and CR_i is the contrast ratio in each cell/radar pixel. The contrast ratio is the difference between the ground return mean as measured in the radar and the power in the ith cell normalized by the ground return mean. The mean ground return as measured in the radar contains the noise power and the power from true ground return times the composite ISLR, which can be approximated as given in Equation (7.33).

$$CR_i \approx \frac{GR_{cell\ i} + n + ISLR \cdot GR_{mean} - GR_{mean}}{GR_{mean}} \tag{7.33}$$

where

$GR_{cell\ i}$ = ground return power in the ith cell
n = noise power
$ISLR \cdot GR_{mean}$ = integrated sidelobe power
GR_{mean} = mean ground return measured in the radar which contains the noise and sidelobe power
$ISLR$ = the integrated sidelobe ratio for antenna and processing

The contrast ratio can be either positive or negative. A dark spot such as a paved road or lake can be just as useful for recognition as a bright building or stationary vehicle target. For a contrast ratio example, assume a vehicle size 5×10 ft, $\mathcal{P}_D$ of 0.9, $\mathcal{P}_{FA}$ of 0.1, resolution cell size of 1 ft^2; the required SNR_{shape} is $13 + 3 = 16$ dB, which is a factor of 40, then the required contrast ratio per cell is

$$CR_{cell} \geq \frac{5 \cdot 10 - 40}{50} = 0.2/cell \qquad (7.34)$$

In other words, the average contrast ratio per cell must be either 20 percent smaller or 20 percent greater than the surrounding background.

7.8 Ground MTI and MTT

7.8.1 Ground Moving Target Overview

Ground moving target indication (GMTI) and ground moving target track (GMTT) radar modes have a different set of challenges. First, target detection is usually the easy part; the RCS of most man-made and many natural moving targets is large (10 to 1000 m^2). Unfortunately, many moving but stationary objects (e.g., ventilators, fans, water courses, and suspended objects) lead to apparent false alarms. Often, slow-moving vehicles have fast-moving parts (e.g., helicopters and agricultural irrigators). Most areas have very large numbers of vehicles and scatterers that could be vehicles. Processing capacity must be adequate to handle and discriminate thousands of high SNR threshold crossings and hundreds of moving targets of interest. In most cases, all targets must be tracked and then recognized on the basis of Doppler spectrum (helicopters versus wheels versus tracks), rate of measured location change (ventilator locations don't change), and consistent trajectory (60 mph where there are no roads is improbable for a surface vehicle). In addition, vehicles of interest can have relatively low radial velocities requiring endoclutter processing.[20,21]

Table 7.13 summarizes the leading parameters for a spotlight GMTI mode similar to that described in Section 2.3.1. Transmitted power is often less than 100 mW. PRFs are ambiguous in both range and Doppler but unambiguous in the mainbeam and near sidelobes. Complementary codes are often used with the sidelobe ambiguity in an uninteresting part of the return spectrum and rejected by Doppler filtering. A 2048:1 pulse compression code is used, and one-half or less of the bins are retained in an Hudson-Larson compression scheme as described in Section 5.5.3. A 10-ft range cell size is chosen to match the smallest vehicle of interest and break them out of clutter. As usual, high duty ratios and multiple frequency channels are used. A spotlight MTI mode may have 1024×1024 processed bins. Antenna illumination is ground stabilized and can have 2 to 5 beam positions, depending on range. Either a minimum dwell or a minimum power mode will be used, depending on the threat with corresponding frame times from 0.2 to 5 sec. Total bandwidth might be 500 MHz.

Table 7.13   Ground Moving Target Spotlight Example Parameters

Parameter	Value
Waveform	6–12 kHz PRF
Maximum peak power	<0.5 W
Pulse compression	2048:1 complementary code
Range resolution	10 ft
Duty ratio	25%
Frequency channels	3
Maximum range	30 mi
Range bins processed	1024
Doppler filters	1024
Antenna scan rate	Ground stabilized
Azimuth coverage	60°
Elevation coverage	25°
Frame time	0.2–5 sec
Total instant bandwidth	500 MHz
Operating band	~8–10 GHz

A processing block diagram for GMTI/GMTT is shown in Fig. 7.36. Although there are alternative ways to perform endoclutter processing, a monopulse-based processing scheme is given in Fig. 7.36. Sum and difference monopulse channels for multiple frequencies are digitized and pulse compressed. Periodic calibration signals are used to create a gain and phase correction table for all frequencies, antenna beam steering, and monopulse channels, which is then applied to the digitized measurements. Motion compensation to a fraction of a wavelength for platform deviations is applied to the data. High-resolution Doppler filtering is performed in a conventional FFT. Doppler filter outputs are used to form mainlobe clutter discriminants for the final motion compensation. Mainlobe clutter is not in the same frequency location for each range bin, and so filter output indices must be adjusted to present a common input to the threshold detector. The Doppler filter bank outputs are applied to a multilevel threshold detector similar to those previously described in Section 7.3.

The multilevel threshold has several unique features. In addition to the alert-confirm properties, it also uses monopulse discriminants (or multiple phase centers if you prefer) as well as near sidelobe threshold multipliers. Threshold crossings are

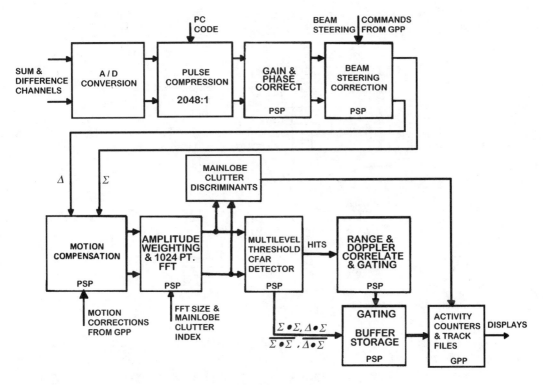

Figure 7.36 Ground moving target detection and track processing.

correlated in range and Doppler and buffered along with corresponding monopulse or phase center discriminants, which are presented to tracking filters or activity counters.

7.8.2 Ground Moving Target Thresholding

The concept for ground moving target thresholding is shown in Fig. 7.37. There are three regions of thresholding: mainlobe, near sidelobe, and Doppler clear. Conventional airborne targets move fast enough that detecting targets in near sidelobes is usually not necessary. Near surface targets of interest will often have radial velocities of a few miles per hour for long periods of time, which forces the detection of ground moving targets well into mainlobe clutter. Monopulse processing allows the first-order cancellation of clutter for many slow-moving targets. Unfortunately, clutter does not always have well behaved statistical tails, and, to maintain a constant false alarm rate, the threshold must be raised for endoclutter targets. The output of the Doppler filter bank might be thought of as a two-dimensional range Doppler image. After index adjustment, the amplitude versus Doppler profile in any range bin is as shown in

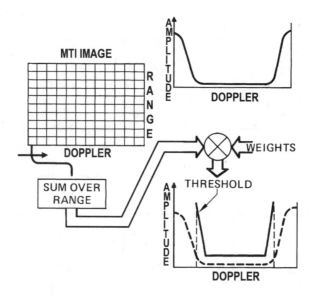

Figure 7.37   Variable threshold concept.

the upper right in Fig. 7.37. By summing over segments of range, an estimate of clutter amplitude can be formed and then weighted by known clutter behavior to form the threshold profile shown in the lower right of the figure. There will still be parts of mainlobe clutter that are completely discarded except for motion compensation.

An actual thresholding scheme based on this concept is shown in Fig. 7.38. The range Doppler space is broken up into a grid. Each cell in the grid is perhaps 64 × 64 bins with 256 grid cells total. The grid locations spanning mainlobe clutter are used for forming mainlobe clutter discriminants only and are otherwise discarded. The bins in each grid cell are ensemble averaged *(EA)* in sum and difference for each frequency channel. The power in each bin in a grid cell in the clear region is compared to a threshold, $P_{TH1}(EA)$, which is a function of the *EA* in that grid cell. In the sidelobe fringe region, a discriminant, C_s, is formed in addition and used to provide monopulse clutter cancellation prior to thresholding. Again, the threshold, $P_{TH2}(EA)$, is a function of the *EA* in that grid cell and *a priori* knowledge of the clutter statistics. Although only one threshold is described, two are actually used before hits and their counterpart monopulse discriminants are passed to the track files. All low-threshold hits are passed to activity counters. As complex as this thresholding scheme seems to be, it is very power efficient, which is exactly what is required for stealthy GMT radar.[63]

7.8.3   GMT Tracking

There are normally so many near ground moving targets that they all must be tracked even in search to select the interesting movers. Figure 7.36 alludes to this by showing

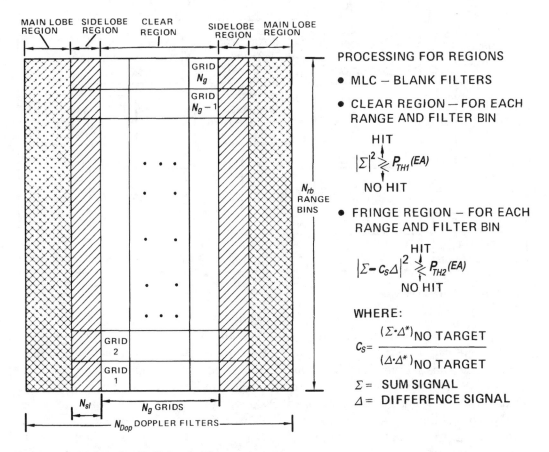

Figure 7.38 Multiregion GMT thresholding.

a track function in the GMT search block diagram. Typically, the movers are sorted out by their signature and movement characteristics. For example, man-made targets are usually moving in a specific direction, whereas birds and insects tend to change directions abruptly and stay in the same area. Ventilators stay in the same location even though they may be moving fast, and helicopters have high subsonic moving parts but move much more slowly. It is not uncommon to have hundreds of moving targets in track simultaneously.

An overview of the major functions GMT tracking is shown in Fig. 7.39. Because tracks may often cross each other, radar observations and correlations with prior tracks are essential. The track gate acceptance functions are conditioned on the target signal-to-noise ratio. Once observations are associated with existing tracks, they are used to update individual Kalman filters. The parameters of high-priority targets are provided to displays as well as used to select the transmitted waveform to maintain

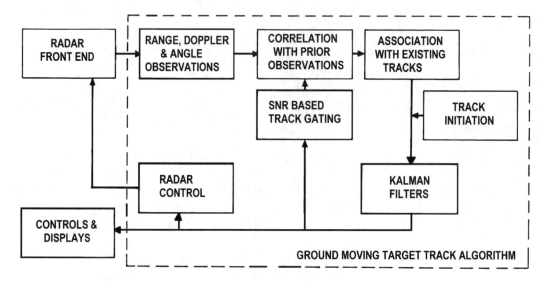

Figure 7.39   GMT track overview.

high-quality tracks on those targets. Because terrain can mask targets, they often will abruptly appear or disappear, which requires a sophisticated track initiation function.

Experience has shown that GMT tracking is best done in a ground-referenced coordinate system. This is true for several reasons. First, most targets are on or so near the ground that they are ground stabilized. Second, hypsographic databases are often used to correct the location of target tracks. Third, target speed relative to ground features are used to recognize target signatures (e.g., surface vehicles cannot travel in roadless country at 70 mph, but helicopters can). A block diagram of all the elements of ground stabilized tracking is shown in Fig. 7.40. First, the observing platform navigation data in all six degrees of freedom must be known and extrapolated to the target measurement time for input to the Kalman filter. Next, track error terms in range, range rate, altitude, azimuth, and elevation as well as noise estimates are input to the Kalman track filters. The track filter output is propagated to the next forecast measurement time. These estimated outputs $(R, \dot{R}, AZ, EL)$ are subtracted from the next measurement and used to form error terms for the track filters. Because most targets are on or near the ground, the known ground altitude (h_G) from the hypsographic database is subtracted from the tracker estimated altitude (h_T) to form an altitude error term. The required antenna beam position in ground stabilized coordinates for the next set of measurements as well as the apparent range, velocity, and acceleration (P_F, V_F, A_F) for a fixed ground location at beam center is calculated (often called the ground stake). The antenna is then pointed in that direction and, because the actual measurements are in radar based angle and range coordinates, they $(R_F, \dot{R}_F, AZ_F, EL_F)$ are added with suitable transformations to the measurements to provide an esti-

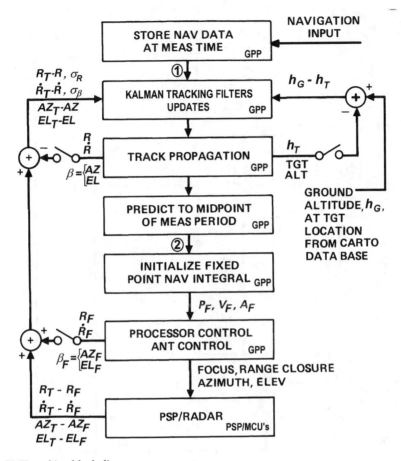

Figure 7.40 GMT tracking block diagram.

mate of target location in Earth stabilized dimensions. The radar and signal processor are provided with calculated parameters to allow proper waveform and processing of the expected MTI image. The processor provides the next set of target measurements (discriminants) in radar-based coordinates to repeat the whole cycle. The normal location of the calculation in the processor complex as described in Fig. 7.4 is given in the lower right in each block.

More details of the Kalman track filters and track propagation described in Figs. 7.39 and 7.40 are given in Fig. 7.41. Figure 7.41 shows the details between point 1 and 2 in Fig. 7.40. The natural tendency is to lump all these filter functions into a single Kalman filter with a large number of degrees of freedom (i.e., everything but tide tables and prayer times). This usually results in filters that are very unstable. Alternatively, more stable tracking is achieved by having dedicated Kalman filters, which track only range or velocity. Often, multiple Kalman filters are used that hypothesize

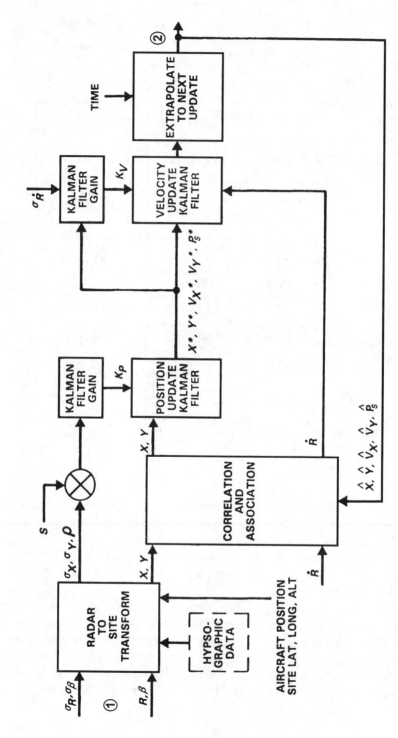

Figure 7.41 GMTT Kalman filter diagram.

a specific class of movement (e.g., around a corner, straight line, sinusoidal in N-E, and so forth), and the best one is selected. The simple filter of Fig. 7.41 is still reasonably complex. Radar range, R, angle, β, and estimated standard deviation, σ_R, σ_β, are transformed to ground stabilized site coordinates, X, Y, conditioned on vehicle position and attitude as well as the hypsographic database. Ground coordinates are provided to a position Kalman filter after being gated by their forecast value. Their counterpart correlation, ρ, and standard deviations, σ_X, σ_Y, conditioned by the signal power are also used to set the Kalman filter gain for the current measurement. Smoothed position velocity and signal power estimates are provided to the velocity Kalman filter. Again, gain is conditioned on velocity standard deviation, range rate, and signal power estimates. The velocity filter outputs as well as propagated positions are provided to an extrapolator based on the expected time of the next radar measurement. The extrapolated position and velocity conditioned on power-based variation is used to form the correlation and association window for the next measurement. Then the cycle begins again for each of several hundred targets (thousands?) that might be in track.

7.9   QUIET WEAPON DELIVERY

Quiet weapon delivery embodies tracking of the form described in Sections 7.3.3 and 7.8.3, with the additional constraint that the missile must also be tracked, and the target being attacked must be range and Doppler clear. This places additional constraints on the PRF and the transmitted pulse width. The waveform changes must be subtle enough that the target is not alerted in the event the transmitted waveform has been intercepted. Recall from Fig. 7.26 that part of the waveform interval has been dedicated to automatic gain control and radar from/to missile data link. Having that interval always in the waveform, whether there is a missile on the fly or not, will allow stealthy target-to-missile updates as well as command inertial guidance until missile acquisition of the target. The missile can also send video or attitude data to the radar for guidance aiding. With precise timing between radar and missile and high update rate tracking, the missile-radar system can have very robust performance in the presence of evasive maneuvers and jamming. Using the radar as a data link has the advantage that data is passing through a very high-gain antenna tracking the missile, and so link margins are excellent. The only time radar as data link is a weakness is if the launcher must immediately flee, requiring "over the shoulder" communications.[47-50]

For ground target weapon delivery, the target may stop or start during the time of flight of the missile, and so both MTI and SAR images of the target must be made simultaneously to ensure continued tracking. As before, the missile must be tracked as well until the endgame acquisition of the target by the missile seeker. Figure 7.42 shows an example of combined waveform processing for SAR target track, missile track, and GMTT. The achievable update rate of roughly 2/sec is too slow unless the weapon has its own inertial guidance.

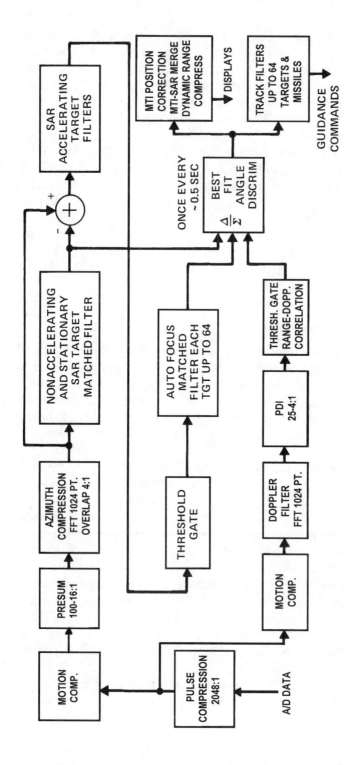

Figure 7.42 Near ground targeting/weapon delivery processing.

Although semiactive missiles have been explored for stealthy weapon delivery, best stealth operation occurs with command inertial mid-course and a separate end game seeker–active radar, ARM, microwave radiometric, imaging EO, acoustic, and so on. An example generic radar missile block diagram with an end game seeker is shown in the next section, in Fig. 7.44.

7.10 DATA LINKS

7.10.1 Introduction

The two main uses for data links associated with stealthy vehicles are high-bandwidth imagery transmission from a weapon or sensor platform to the control platform or ground station, and low-bandwidth transmission of targeting data and guidance/ housekeeping commands. The largest quantity of data links are associated with weapons. The waveform selected must not compromise the signature of the platform at either end of the link. All the stealth constraints on radars apply to data links, especially low-sidelobe antennas, random discontinuous operation, and large transmitted bandwidth. Unfortunately, stealthy imagery transmission already requires high bandwidth, and data compression is an important issue. Low-bandwidth data links can use all the bandwidth to improve encryption and signal-to-jam ratios. However, the data link on a weapon is travelling to the target, which will inevitably attempt to protect itself. When the weapon is near the target, the signal-to-jam ratio can be very unfavorable. Antenna jammer nulling is always required, because transmitting more power to burn through is not stealthy. Clearly, the data from/to a weapon also must be sufficiently encrypted to prevent take-over of the weapon in flight.[47–49]

An example of an interleaved radar and data link waveform is shown in Fig. 7.43. The example LPI system has multiple channels and multiple antenna beams. In this example, some number of beams and frequencies are dedicated to tracking multiple

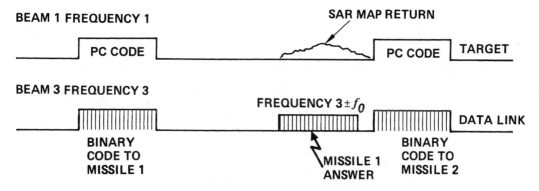

Figure 7.43 Example weapon data link message format.

surface targets as shown in the top part of Fig. 7.43. Time synchronized with that radar transmission on a different set of beams and frequencies, messages are sent to one or more missiles on the fly to the targets. Obviously, all the random frequency diversity, spread spectrum, and encryption necessary for robust communication are incorporated into the message. Each missile may answer back at a known but randomized offset frequency and time with image or housekeeping data. Again, a waveform as robust as possible is used but, because the baseband data and link geometry may be quite different, the compression, diversity, and encryption may be different. The missile waveform must be stealthy and greatly attenuated in the direction of the target, because one countermeasures strategy is a deception repeater jammer at the target. High-accuracy time and frequency synchronization, including range opening and Doppler effects between both ends of the link, can dramatically reduce the effectiveness of jamming by narrowing the susceptibility window. Synchronization also minimizes acquisition/reacquisition time.

A generic radar missile block diagram is given in Fig. 7.44 to assist understanding the missile end of the data link waveform given above. The missile is moving with respect to the other end of the link, so the link geometry is continually changing in time, frequency, aspect, and attitude. The block diagram shows all the elements such a system might include, but not all missiles will have all of the elements shown. There will be some sort of end-game seeker whose aperture is pointed by a servo, because the missile axis is seldom aligned with the target. There may be a transmitter for the seeker and the data link. There will always be a local oscillator (LO), receiver/signal processor, and tracker. The signal processor will generate waveforms for transmission by the seeker or data link. It will also measure target range, angle, Doppler, and so forth and provide those to the tracker. The tracker in turn generates outputs for the seeker servo, for the signal processor for new measurement gate windows, for the au-

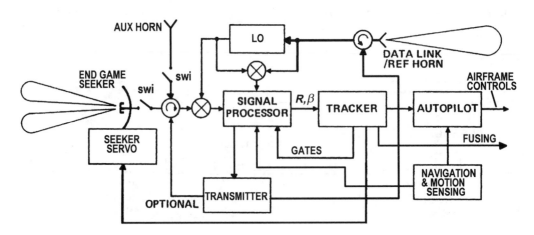

Figure 7.44 Generic radar missile.

topilot to steer the missile to the target, and for fusing when close to the target. The signal processor needs motion sensing and navigation estimates to correct measurements; to track, encode, and decode data link messages; and to perform jammer nulling. And the autopilot needs motion sensing and navigation estimates to maintain stable flight along the chosen trajectory to the target.

The data link serves several functions. First, the data link carrier frequency is used to synchronize the local oscillator and timing to the LPIS including the relative Doppler. Second, the data link from the LPIS to the missile provides target location and flight profile commands to the missile prior to and during target acquisition by the end game seeker as well as other house keeping commands. Third, the data link from the missile to the LPIS provides its version of ownship location and attitude, state of health/house keeping, and imaging from the seeker if any. The imaging may be used to close an outer guidance loop or for probable damage assessment. The data link antenna may be steerable in some simple way.[49,50,53]

7.10.2 Data Compression

Usually, command and housekeeping data bit rates are low enough that only spectral spreading and encryption are required. Imaging data from an end-game seeker or from an imaging LPIS almost always requires compression to get antijam, encryption, and discontinuous operation into the required band. Image compression has been studied since the 1960s for commercial communications reasons. The availability of high-performance image/signal processors has finally enabled commercial image compression to be quite good. Most of us enjoy this image compression through satellite TV, digital cable, DVD movies, and digital cameras. Commercial availability has enabled this same technology to benefit stealth systems. Image data compression can be reasonably expected to reach approximately one bit per pixel with acceptable degradation. Both wavelet and discrete cosine transform compression routinely reach this performance (e.g., MPEG). An example image data compression block diagram is shown in Fig. 7.45.[51–53]

Several methods are used to reduce the total bit rate per pixel in images. Some are lossy but depend on the fact that image features of interest are dominated by low spatial frequencies. Other methods, such as differential pulse code modulation (DPCM), run length coding (RLC), and variable length or Huffman coding (VLC), are loss free but more susceptible to bit errors. Returning to Fig. 7.45, the image input is processed in a discrete cosine transform or wavelet transform in one or two dimensions. The transform outputs usually have many very low-value terms that, after conversion to differential PCM, normalizing by the DC or total image brightness, and scaling according to some weighting function, are truncated to zero. The truncation is obviously a lossy process. After truncation, many terms are zero, and interleaving is used to scatter the nonzero terms throughout the data block representing the image. Run length coding is used to compress the data block so that only the distance between

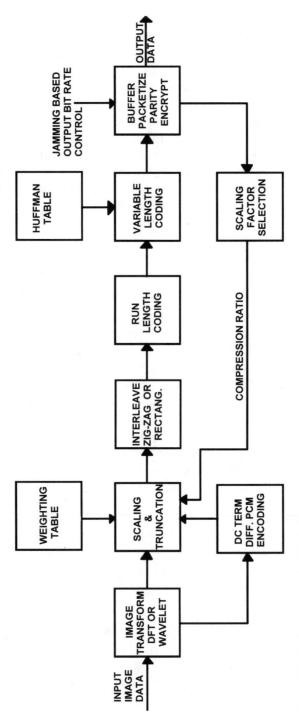

Figure 7.45 Image compression block diagram.

changes from 0 to 1 or 1 to 0 are passed on to the next process function. As mentioned, this is loss free in the absence of noise. Variable length coding, which depends on the probability that certain sequences of 1s and 0s are more probable and are replaced with a simpler bit pattern, is then used to obtain further compression. The most probable sequences and their replacement code are contained in a Huffman table. Subsequent to the compression, the data is placed into blocks or packets, which are typically 250 bits long. The packetizing prevents error propagation from one portion of the image to another. The packet has multiple parity bits added to allow error correction on receive. Approximately 5 to 10 percent of the bits in a packet are parity. The data stream is then encrypted, which randomizes and obscures all structure in the output data. The encryption need not be very robust, because it must resist a decryption attack only for the time of flight of the missile. Unfortunately, the encryption degrades the net compression factor and signal-to-jam ratio by 2 or 3:1. Because the compression is lossy, there usually is a jamming-based output bit rate control, which increases the compression ratio in the presence of jamming at the expense of image quality. As jamming gets larger, the image gets poorer and, ultimately, the image resembles what you would see out your automobile windshield in a heavy rainstorm. In a compression system such as this, accurate time synchronization is essential to avoid resynchronization, which is very time consuming, but also to reduce the effects of jamming.

Most engineers are familiar with various forms of the Fourier transform, but they may be less familiar with what are sometimes called *wavelet transforms*. Although not the only implementation of wavelet transforms, one simple form is a cascade of orthogonal highpass and lowpass filters with rate reducing resampling. This technique is also called *multirate filtering.* Consider the cascade of filter blocks in Fig. 7.46. Each highpass or lowpass filter has an identical impulse response relative to its sample rate. Because the lowpass filter is by definition orthogonal to the highpass filter, the transform is completely reversible up until the time the output of the filter is thresholded. At the end of this cascade of filters, there will still be N samples but, after thresholding, the higher-frequency components will be mostly zeros. The figure shows filtering in only one dimension but, more generally, there is a high- and lowpass filter in both the horizontal and vertical dimensions at each level of the cascade.

The idealized image spatial frequency spectrum is shown in Fig. 7.47 for various designated points in Fig. 7.46. Obviously, the real spatial spectrum is at least two dimensional, but the concept is easier to understand in one dimension. Each filtering stage is carried out in as many dimensions as exist in the image. At the input, the spectrum is shown at **A;** after the first set of filters, the image spectrum is broken into two pieces **B** and **D** by filters $G(f)$ and $H(f)$, which are the frequency response counterparts of the impulse responses $g(t)$ and $h(t)$. (It is four pieces for two dimensions and eight pieces for three dimensions.)

After resampling downward by 2:1, the output spectrum at **C** is the same but is represented by half as many samples. When the entire process is complete, 1/2 of all

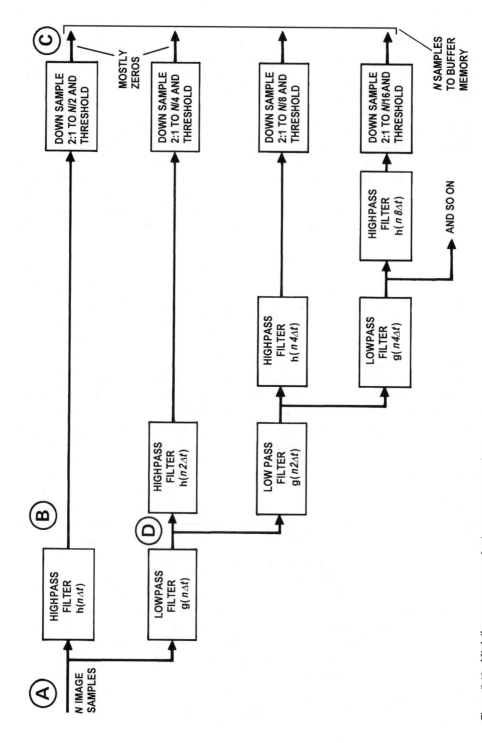

Figure 7.46 High/lowpass or wavelet image compression.

the samples will represent the highest spatial frequencies, the next fourth of the spectrum will be represented by 1/4 of the samples, the next eighth of the spectrum by 1/8 of the samples, and so on as shown in the bottom of Fig. 7.47. The next question is the nature of g and h. The simplest g and h set is called the Haar transform, and g is the sum of two adjacent samples, while h is the difference of the same two adjacent samples. Very elaborate wavelets might have 4 to 6 values of a_n; in other words, these transforms are *very* simple. Obviously, the actual filter responses don't look exactly like Fig. 7.47, but the principle is the same.

7.10.3 Jammer Nulling

The last important topic for stealthy data links is jammer nulling. This technique has been used since the mid-1960s. Because transmitting more power is usually not stealthy, some other methods of improving signal-to-jam ratio (SJR) must be used. The basic concept shown in Fig. 7.48 consists of using a second antenna that has relatively low gain in the direction of the main antenna and relatively higher gain than the main antenna outside the mainlobe. Assuming a jammer is in the main antenna sidelobes, one can subtract the jamming signal received in the auxiliary antenna weighted by the cross correlation of the main antenna signal with the auxiliary antenna signal. This has the effect of nulling a strong jammer signal in the output. If there is no correlation between the two antennas, then the weight is zero, and nothing is subtracted. The integration period, T, must be shorter than the rate of change of the geometry around the data link. Typically, the weight is constrained to stay constant

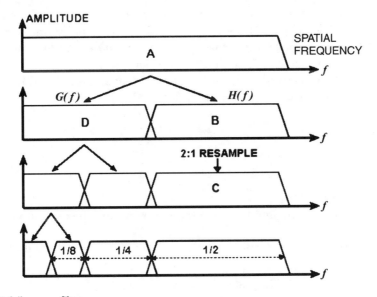

Figure 7.47 High/lowpass filters.

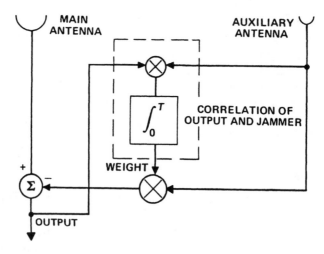

Figure 7.48 Simple jammer nulling concept.

during one message or packet interval. As much as 20 dB improvement in *SJR* can be obtained routinely with the configuration of Fig. 7.48. Higher improvements are often claimed, but only for unrealistically narrow jamming bandwidth.[10,54,57]

The block diagram of the generic radar missile in Fig. 7.44 shows an auxiliary antenna that can be multiplexed with the seeker antenna to provide data link jammer nulling. Similarly, the data link antenna can be used to provide jammer nulling for the seeker antenna. If the jammer is in the mainlobe of the seeker antenna, then it may be on the intended target; missiles are often programmed to destroy the jammer first in this circumstance. The scheme shown in Fig. 7.48 was originally implemented using low-level analog RF circuits; today, this algorithm is usually implemented in the signal processor in a more elaborate digital form described in the next section. It can be shown that the jammer cancellation ratio (*JCR*) for a small signal and in the presence of noise is approximately as provided in Equation (7.35).[10]

$$JCR \cong \frac{1 + JNR}{1 + JNR \cdot (1 - \rho^2)} \tag{7.35}$$

where
JNR = jammer-to-noise ratio
ρ = the correlation coefficient between the main and auxiliary channels

Often, the correlation ratio is assumed to be Gaussian as given in Equation (7.36).

$$\rho = \exp\left(-\left(\frac{d \cdot B \cdot \sin(\theta)}{c}\right)^2\right) \tag{7.36}$$

where
    d = spacing between the main and auxiliary antenna
    B = channel bandwidth
    c = the velocity of light

7.11 ADAPTIVE PROCESSING

In general, an LPIS adaptive nulling antenna has a main beam look direction and some set of auxiliary elements, which may be parts of the main antenna, that are capable of independent measurement and separable steering, which allow jammer cancellation. The hardware block diagram of such a system is shown in Fig. 7.49.

An experimental cross-correlation matrix is calculated for all of the channels: main and auxiliaries. Inevitably, this process results in some loss of performance, and so nulling is used only if there is jamming. The cross-correlation matrix is then used to calculate a set of adaptive weights, which are sent to the beam steering computer to adjust the phases and amplitudes of the corresponding elements so that a nulling signal can be produced, which is subtracted from the main lobe channels. The update rate is synchronous with each coherent array, because updates during an array always lead to bad Doppler filter sidelobes. Because of the latency between experimental cross correlation and the update, an open-loop scanning term is often applied to the cancellation weights to compensate for platform velocity. If the signal to be pro-

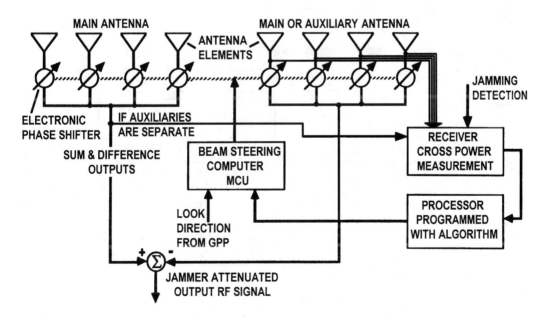

Figure 7.49   LPIS adaptive nulling hardware block diagram.

cessed is clutter that has a high return power, then the auxiliary channels must have a null in the mainlobe. Mainlobe jamming, if it is not right on top of the target, can be attenuated using a combination of the sum and difference channels that places a difference null on the jammer in the mainlobe. This results in a significant reduction of the *SNR* but a substantial improvement in the *SJR*—typically a *JCR* of more than 20 dB for 3 dB degradation in *SNR*.

From a mathematical point of view, one can visualize the adaptive processing as shown in Fig. 7.50. There are two sets of weights (vectors) that are not adaptive: conventional beam steering, $\mathbf{W}_c$, and a matrix of sidelobe canceller weights (a set of vectors), $\mathbf{W}_s$, that are close to orthogonal.

Typically, the sidelobe weights are a series of focused beams that point in every direction but the main beam. For each main beam pointing direction, the sidelobe beam weightings can be precomputed and stored in a table. There is one fewer sidelobe beam than there are inputs. The complex conjugate of the sidelobe beam outputs, $y_i(t)$, are cross-correlated with the processed main beam output, $z(t)$, to form a set of adaptive weights, W_{ai}, which are applied to the sidelobe beams. The equation for W_{ai} is given in the upper right in Fig. 7.50. The particular scheme depicted in Fig. 7.50 is only one of many adaptive array approaches. The most elaborate approaches may do matched filtering in range, Doppler, and angle. These techniques are often called space-time-adaptive-processing (STAP). They all revolve around the idea of calculat-

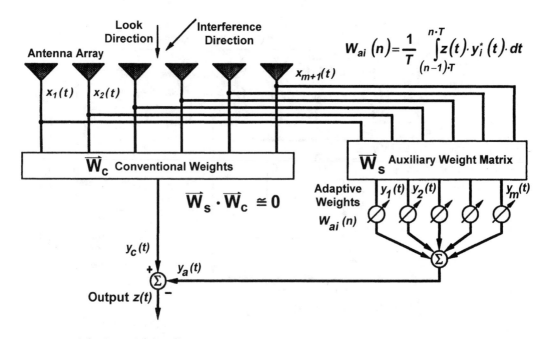

Figure 7.50 Adaptive weighting diagram.

ing a multidimensional covariance function for a set of elemental measurements in range, Doppler, and antenna space. The weighting function is then derived so that all signals are minimized except the desired signal.[10,11,38,54,56–59]

As an example of the type of performance achievable for adaptive arrays, consider Fig. 7.51, which is the gain versus angle cut in the principal horizontal plane of a 50 × 50 element array with one row of 50 elements as depicted in Fig. 7.50 above. The case simulated is a 100-MHz bandwidth spot noise jammer at 22° with a *JNR* of 40 dB and a target at 0° with an *SNR* of 0 dB. The jammer is completely eliminated! Unfortunately, the receive sidelobes are significantly degraded everywhere else, as can be seen in Fig. 7.51. The ISLR goes up, and the SNR goes down, but the jammer goes away. This is often a good trade-off but, obviously, not always.[56]

Figure 7.52 shows the same 50 element adaptive jammer nulling array for patterns with 5 and 40 jammers. As might be expected, the total rejection per jammer is less and, as will be shown subsequently, the *SNR* is significantly degraded. Note that sometimes the jammer configuration improves the adaptive sidelobes, but more often it degrades the receive sidelobes significantly.

In summary, if the signal-to-jam ratio and signal-to-noise ratio are plotted as a function of the number of jammers, the *SNR* and *SJR* get steadily worse as a function of the number of jammers for a fixed number of auxiliary elements. Although the theory says that you can cancel one less jammer than the number of elements, as a practical matter, one can cancel only a fraction (perhaps 25 percent) of the theoretical number of jammers, as suggested in Fig. 7.53. Nonetheless, the investment in hardware is well worth it in most cases.

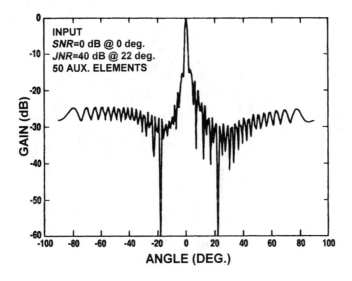

Figure 7.51   Example adaptive arrays, one jammer.

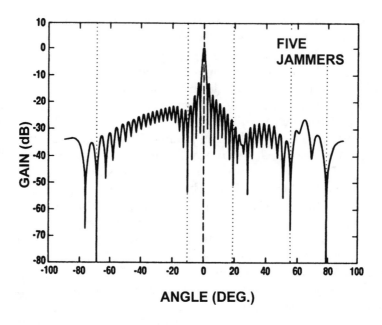

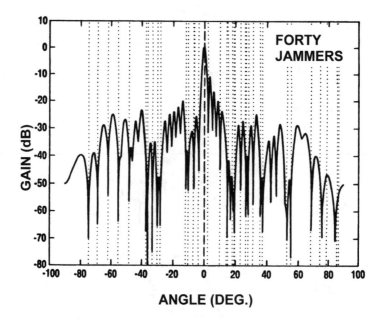

Figure 7.52 Example adaptive array performance, 5 and 40 jammers.

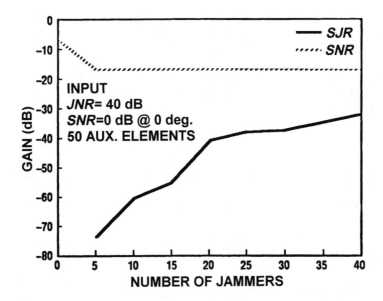

Figure 7.53 Multiple jammer adaptive array performance.

7.12 EXERCISES

1. Estimate the total noise at the output of a 4096-point FFT with I and Q word lengths of 24 bits each and RMS input of 1000.

2. Calculate the signal and noise trace for the GMT mode given in Fig. 7.36 and Table 7.13.

3. Calculate the signal and noise trace for the SAR mode given in Fig. 7.35 and Table 7.12.

4. Estimate the output contrast ratio for Problem 3 above.

5. Calculate the signal-to-jam ratio for a missile data link receiver of 1000 bps, 250-MHz bandwidth, 20-dB receiver antenna gain, single jam canceller, 5 W transmitter with transmit antenna gain 40 dB.

6. For a stacked nine-frequency, 3-m resolution, 10-m^2 target RCS, 25 percent duty waveform, calculate the transmitter power required at 150-mi range and 0.1-sec dwell when a 1000-W, 2-GHz jammer is placed at the target. Assume 38 dB radar antenna gain and 15 dB jammer cancellation. Calculate the power at an intercept receiver in the mainlobe at 200-mi range for a receiver with 50-MHz bandwidth, −60-dBm sensitivity, and 20-dB gain intercept antenna.

7. Calculate the multiple time around echo (MTAE) reduction for a third time around echo with 4:1 PRF stagger, 16:1 PRF phase coding, and a 20 percent duty ratio for a 3-m resolution ground map mode.

8. What is the number of jammers that a 50-element adaptive canceling array can handle that results in only 0 dB output jammer-to-noise ratio (JNR) when the input JNR is 40 dB?

7.13 REFERENCES

1. Eliot, D., Ed., *Handbook of Digital Signal Processing,* Academic Press, 1987, pp. 364–464, 527–589, 590–593, 594–631.
2. Tower, L., and Lynch, D., "Pipeline High Speed Signal Processor," U.S. Patent 4,025,771.
3. Tower, L., and Lynch, D., "System for Addressing and Address Incrementing of Arithmetic Unit Sequence Control System," U.S. Patent 4,075,688.
4. Tower, L., and Lynch, D., "Pipelined Microprogrammable Control of a Real Time Signal Processor," *Micro6 Conference,* June 1973, p. 175.
5. Harris, F., and Lynch, D., "Digital Signal Processing and Digital Filtering with Applications," *ETI Short Course Notes,* 1971–1983, (February 1978), pp. 366, 744–748.
6. Pearson, J., "FLAMR Signal To Noise Experiments," *Hughes Aircraft Report No.* P74-501, December 1974, declassified 12/31/1987.
7. Craig, D., and Hershberger, M., "FLAMR Operator Target/OAP Recognition Study," *Hughes Aircraft Report No.* P74-524, January 1975, declassified 12/31/1987.
8. Lynch, D., "SLOSH Filter Processing," *IEEE AU Symposium on Digital Filters,* Harriman, NY, January 1970.
9. Brigham, E. O., *The Fast Fourier Transform,* Prentice Hall, 1974, pp. 172–197.
10. Schleher, D., *Electronic Warfare in the Information Age,* Artech House, 1999, pp. 279–288.
11. Monzingo, R., and Miller, T., *Introduction to Adaptive Arrays,* John Wiley & Sons, 1980, pp. 78–279.
12. Hughes Aircraft, "LPIR Phase 1 Review," unclassified report, 1977.
13. A few photos, tables, and figures in this intellectual property were made by Hughes Aircraft Company and first appeared in public documents that were not copyrighted. These photos, tables, and figures were acquired by Raytheon Company in the merger of Hughes and Raytheon in December 1997 and are identified as Raytheon photos, tables, or figures. All are published with permission.
14. Stimson, G., *Introduction to Airborne Radar,* 2nd ed., Scitech Publishing, 1998, pp. 355–381, 463-465.
15. Gerlach, K., "Second Time Around Radar Return Suppression Using PRI Modulation," *IEEE Transactions on Aerospace and Electronic Systems,* Vol. AES-25, No. 6, Nov. 1989, pp. 854–860.
16. Rohling, H., and W. Plagge, "Mismatched-Filter Design for Periodical Binary Phased Signals," *IEEE Transactions on Aerospace and Electronic Systems,* Vol. AES-25, No. 6, Nov. 1989, pp. 890–896.
17. Mooney, D., "Post-Detection STC in a Medium PRF Pulse Doppler Radar," U.S. patent 4095222.
18. Frost, E., and L. Lawrence, "Medium PRF Pulse Doppler Radar Processor for Dense Target Environments," U.S. patent 4584579.

19. Durfee, L., and W. Dull, "MPRF Interpulse Phase Modulation for Maximizing Doppler Clear Space," U.S. patent 6518917.

20. Schleher, D., "Low Probability of Intercept Radar" *IEEE International Radar Conference*, 1985, p. 346.

21. Carlson, E., "Low Probability of Intercept Techniques and Implementations," *IEEE National Radar Conference*, 1985, p. 51.

22. Groger, I., "OLPI-LPI Radar Design with High ARM Resistance," *DGON 7th Radar Conference* 1989, p. 627.

23. Schleher, D. C., *Introduction to Electronic Warfare*, Artech House, 1986, pp. 280–283, 347–356.

24. Clarke, J., "Airborne Radar," parts 1 & 2, *Microwave Journal*, January 1986, p. 32, and February 1986, p. 44.

25. Morris, G., *Airborne Pulse Doppler Radar*, Artech House, 1988.

26. Long, W., and K. Harriger, "Medium PRF for the AN/APG-66 Radar," *IEEE Proceedings*, Vol. 73, no. 2, p. 301.

27. Frichel, J., and F. Corey, "AN/APG-67 Multimode Radar Program," *IEEE NAECON 1984*, p. 276.

28. Nevin, R., "AN/APG-67 Multimode Radar Performance Evaluation," *IEEE NAECON 1987*, p. 317.

29. Lynch, D., "Real Time Radar Data Processing," *IEEE Solid State Circuits 4.10 Committee*, Digital Filtering Meeting, New York, October 30, 1968.

30. Klemm, R. "New Airborne MTI Techniques," *International Radar Conference London*, 1987, p. 380.

31. Klemm, R., "Adaptive Airborne MTI: An Auxiliary Channel Approach," *IEE Proceedings*, Vol. 134, part F, no. 3, 1987, p. 269.

32. Klemm, R., "Airborne MTI Via Digital Filtering," *IEE Proceedings*, Vol. 136, part F, no. 1, 1989, p. 22.

33. Harmon, J., "Track Before Detect Performance for a High PRF Search Mode," *National Radar Conference* 1991, pp. 11–15.

34. Aronoff, E., and N. Greenblatt, "Medium PRF Radar Design and Performance," *Hughes Aircraft Report* (National Radar Conference 1975?).

35. Aronoff, E., and D. Kramer, "Recent Developments in Airborne MTI Radars," *Hughes Aircraft Report* (Wescon 1978?).

36. Erhardt, H., "MPRF Processing Functions-Issue 2," *Hughes IDC*, 18 October 1977, unclassified memo.

37. Radant, M., D. Lewis, and S. Iglehart, "Radar Sensors," *UCLA Short Course Notes*, July 1973.

38. Technology Service Corp., "Adaptar Space-Time Processing in Airborne Radars," *TSC-PD-061-2*, 24 February 1971, unclassified report.

39. Lynch, D., "Signal Processor for Synthetic Aperture Radar," *SPIE Technical Symposium East 1979*, paper no. 180-35.

40. Hovanessian, S., *Introduction to Synthetic Array and Imaging Radars*, Artech House, 1980, Chapter 5.

41. Curlander, J., and R. McDonough, *Synthetic Aperture Radar Systems and Signal Processing*, Wiley & Sons, 1991, pp. 99–124, 427–535.

42. Kovaly, J., *Synthetic Aperture Radar*, Artech House, 1976, pp. 72–79, 118–123, 249–271.

43. Lewis, B., *Aspects of Radar Signal Processing*, Artech House, 1986.

44. Schlolter, R., "Digital Realtime SAR Processor for C & X band Applications," *IGARSS 1986*, Zurich, Vol. 3, p. 1419.

45. Fabrizio, R., "A High Speed Digital Processor for Realtime SAR Imaging," *IGARSS 1987*, Ann Arbor, Vol. 2, p. 1323.

46. Hovanessian, S., "An Algorithm for Calculation of Range in Multiple PRF Radar," *IEEE Transactions on Aerospace & Electronic Systems*, Vol. AES-12, no. 2, March 1976, pp. 287–290.

47. Cullen, T., and C. Foss, Eds., *Janes Land-Based Air Defence 2001–2002*, Janes Information Group, 2001, pp. 129–134.

48. *International Defense Review—Air Defense Systems*, Interavia, Geneva, Switzerland, 1976, pp. 61–103.

49. Eichblatt, E., *Test and Evaluation of the Tactical Missile*, AIAA, 1989, pp. 13–39, 52–54.

50. Macfadzean, R., *Surface Based Air Defense System Analysis*, 1992 and 2000, pp. 213–243.

51. Robin, M., and M. Poulin, *Digital Television Fundamentals*, 2nd ed., McGraw-Hill, 2000, pp. 345–425.

52. Pratt, W., *Digital Image Processing*, Wiley & Sons, 1978, pp. 662–707.

53. Simon, M., J. Omura, R. Scholtz, and B. Levitt, *Spread Spectrum Communications Handbook*, McGraw-Hill, 1994, Chapter 4, "Low Probability of Intercept Communications," pp. 1031–1093.

54. Nitzberg, R., *Radar Signal Processing and Adaptive Systems*, Artech House, 1999, pp. 207–236, 267-290.

55. Gabriel, W., "Nonlinear Spectral Analysis and Adaptive Array Superresolution Techniques," *NRL Report* 8345, 1979, approved for unlimited public distribution.

56. Griffiths, L., and C. Tseng, "Adaptive Array Radar Project Review," *Hughes Aircraft IR&D performed at USC*, July 18, 1990.

57. Ko, C., "A Fast Adaptive Null-Steering Algorithm Based on Output Power Measurements," *IEEE Transactions on Aerospace and Electronic Systems*, Vol. 29, no. 3, July 1993, pp. 717–725.

58. Wang, H., H. Park, and M. Wicks, "Recent Results in Space-Time Processing," *IEEE National Radar Conference* 1994, pp. 104–109.

59. Ward, J. "Space-Time Adaptive Processing for Airborne Radar," *MIT Lincoln Laboratory Report* 1015, approved for unlimited public distribution.

60. Hill, R., D. Kramer, and R. Mankino, "Target Detection System in a Radar System Employing Main and Guard Channel Antennas," U.S. patent 3875569.

61. Kramer, D., and G. Lavas, "Radar System with Target Illumination by Different Waveforms," U.S. patent 3866219.

62. Kirk, J., "Target Detection System in a Medium PRF Pulse Doppler Search/Track Radar Receiver," U.S. patent 4079376.

63. Hughes Aircraft, "Pave Mover TAWDS Design Requirements," November 1979, unclassified specification.

64. Streetly, M., Ed., *Janes Radar and Electronic Warfare Systems 1999–2000*, Janes Information Group, 1999, pp. 235–256.

Appendix A

The software provided on the CDROM is for information only. The programs are written in M/S Excel 2000, M/S Quick Basic, and MathCad 2001. They worked on four different computers with both AMD and Intel CPUs in M/S Windows 98, ME, and 2000. *But,* as anyone familiar with these host applications knows, they may not work on your machine and operating environment without some modifications. *There is no warranty!* The following items of software are in Appendix 1 on the CDROM.

1. `Fig1-12.xls`. This Excel program calculates mainlobe intercept versus predetection bandwidth and antenna gain.

2. `INTJAMPWR1.MCD`. This MathCad program calculates the power received at an interceptor from a screening jammer or natural interferer that will protect an LPI radar emitter.

3. `Intvrcs1.mcd`. This MathCAD program calculates the desired available power at a threat for an emitter with a screening decoy.

4. `IRSignature1.mcd`. This MathCAD program calculates the apparent skin temperature due to aerodynamic heating for an aircraft as a function of speed assuming an IR sensor in the 1 to 3 micron band.

5. `IRSignature1a.mcd`. This MathCAD program calculates the apparent skin temperature due to aerodynamic heating for an aircraft as in 4 above, with slightly different graphics and some additional approximations.

6. `ogiveRCS1.mcd`. This MathCAD program calculates ogive geometrical optics RCS plus tip and ogive geometrical diffraction as well as back edge RCS at S band.

7. `Prolate.mcd`. This MathCAD program calculates geometrical optics prolate spheroid RCS.

8. `R90rcs.mcd`. This MathCAD program is an analysis of the R90 acquisition range versus radar cross section for several different radars.

9. `SkinTemp.mcd`. This MathCAD program approximates the apparent skin temperature versus Mach number for an aircraft with an emissivity of 0.8 at sea level.

10. `Triangle.mcd`. This MathCAD program calculates the physical optics RCS of a triangular plate.

11. `Wireloop.mcd`. This MathCAD program calculates the physical optics RCS of a wire loop and simpler approximations.

12. `wireloopL.mcd`. This MathCAD program calculates the physical optics RCS of a wire loop and simpler approximations for a 4-ft loop at L band.

13. `wireloopX.mcd`. This MathCAD program calculates the physical optics RCS of a wire loop and simpler approximations for a 4-ft loop at X band.

APPENDIX A.2

The following items of software are in Appendix 2 on the CDROM. See warranty comments in Appendix 1.

1. `Probdet1a.mcd`. This MathCAD program calculates the probability of detection versus signal-to-noise ratio with probability of false alarm as a parameter using the Rice model for linear detection with steady target in Gaussian noise.

2. `Probdetxa.mcd`. This MathCAD program calculates the probability of detection versus signal-to-noise ratio with probability of false alarm as a parameter using the Rice model for linear detection with steady target in Gaussian noise. It also calculates a power series linear fit approximation for P_{FA} of 10^{-6}.

3. `footprnt81.mcd`. This MathCAD program calculates footprints and power at an intercept receiver using the more accurate model including antenna pattern, weather losses, and flat Earth range geometry.

4. `ftprt3.mcd`. This MathCAD program calculates cookie cutter footprints.

5. `ftprt4.mcd`. This MathCAD program calculates cookie cutter footprints.

6. `Table2-5-6-8.xls`. This Excel program calculates the radar equation using parameters from Table 2.5 and calculates the radar equation results given in Tables 2.7, 2.8, and 2.9 as well as several sensitivity variations.

7. `Table2-5MINDWEL.xls`. This Excel program calculates the radar equation using parameters from Table 2.5 and calculates the radar equation results for the minimum dwell strategy in Table 2.7.

8. `Table2-5MINPOW.xls`. This Excel program calculates the radar equation using parameters from Table 2.5 and calculates the radar equation results for the minimum power strategy in Table 2.7.

9. `Table2-6back.xls`. This Excel program calculates the interceptability equations using parameters from Table 2.5 and the detailed waveform parameters for the minimum dwell and power strategies for Table 2.10.

10. `Table2-6back1.xls`. This Excel program calculates the interceptability equations using parameters from Table 2.5 and the detailed waveform parameters for the minimum dwell and power strategies for Table 2.10 with slightly different waveform assumptions.

11. `Table2-6ext.xls`. This Excel program calculates the interceptability equations using parameters from Table 2.5 and the detailed waveform parameters for the minimum dwell and power strategies with different waveform power and interceptor density assumptions.

12. `Table2-6ext1.xls`. This Excel program calculates the interceptability equations using parameters from Table 2.5 and the detailed waveform parameters for the minimum dwell and power strategies with slightly different waveform and interceptor assumptions.

APPENDIX A.3

The following items of software are in Appendix 3 on the CDROM. See warranty comments in Appendix 1.

1. `PassiveDF6a.mcd`. This MathCAD program calculates the idealized Gaussian patterns for broadband cavity backed spirals used for amplitude comparison passive direction finding (amplitude AOA). The channel match is assumed 1.25 dB. Quadrant beamwidths chosen to limit discriminant range to less than 20 dB. The results are used for Figs. 3.24 through 3.30.

2. `PassiveDF7a.mcd`. This MathCAD program calculates the idealized Gaussian patterns for broadband cavity backed spirals used for amplitude comparison passive direction finding. The SNRs are held constant but noise voltage is assumed to be 10 times lower at the log detectors (It makes a large difference and is one of the problems with log detectors). The channel match is assumed 1.5 dB. The results are used for Figs. 3.24 through 3.30.

3. `PassiveDF8.mcd`. This MathCAD program calculates the total error for phase comparison passive direction finding (phase AOA) as a function of SNR as well as antenna and processing mismatch.

4. `PassiveDF9.mcd`. This MathCAD program calculates the total error for phase comparison passive direction finding (phase AOA) as a function of SNR as well as antenna and processing mismatch with somewhat different parameters than `PassiveDF8.mcd`.

5. `PassiveDF15.mcd`. This MathCAD program calculates the total error for range estimation using ground-based interceptors and phase comparison passive direc-

tion finding (phase AOA) as a function of various ranges and SNR as well as antenna and processing mismatch.

6. `PassiveDF16.mcd`. This MathCAD program calculates the total error for range estimation using ground-based interceptors and phase comparison passive direction finding (phase AOA) as a function of a different set of ranges and SNR as well as antenna and processing mismatch.

7. `PassiveDF17.mcd`. This MathCAD program calculates the total error for range estimation using ground-based interceptors and phase comparison passive direction finding (phase AOA) as a function of a different set of ranges and SNR as well as antenna and processing mismatch.

8. `Radiometer1.MCD`. This MathCAD program calculates the equivalent available noise temperature for a LPI radar and data link for an intercept receiver operating in a radiometer mode.

9. `Radiometer2.MCD`. This MathCAD program calculates the signal-to-noise ratio for an intercept receiver operating in a radiometer mode. The signal is assumed to integrate according to Swerling 1 and grows approximately to the 1.5 power, because the input SNR begins at less than 1. (See pp. 81 and 82 in Nathanson, *Radar Design Principles*, 1st ed., 1969.) It assumes there is no interference.

10. `Radiometer3.MCD`. This MathCAD program calculates the signal-to-noise ratio for an intercept receiver operating in a radiometer mode. The signal is assumed to integrate according to Swerling 1 and grows approximately to the 1.5 power, because the input SNR begins at less than 1. It assumes different thresholding and processing efficiency than `Radiometer2.MCD`. It assumes there is no interference. (See pp. 81, 82 in Nathanson, *Radar Design Principles*, 1st ed., 1969.)

11. `Radiometer4.mcd`. This MathCAD program calculates the signal-to-noise ratio for an intercept receiver operating in a radiometer mode. It assumes different thresholding, and processing efficiency from `Radiometer3.MCD`.

12. `Radiometer5.MCD`. This MathCAD program calculates the signal-to-noise ratio for an intercept receiver operating in a radiometer mode. It assumes different parameters, thresholding, and processing efficiency from `Radiometer4.mcd`.

13. `cornoise3.mcd`. This program calculates simulated experimental statistics for a correlation-type intercept receiver. The typical configuration multiplies the left and right channels of a monopulse antenna-receiver system together and follows this with a square law detector and integrator over the look time. For typical monopulse feed spacing, the two channels will be adequately close in TOA for any emitter in the mainbeam so that only a few decibels will be lost in decorrelation for a broadband binary phase code. There will be recurrent ambiguous sidelobe peaks that will also correlate sidelobe interferers. X and Y are the input

channels, which are assumed Gaussian. The simulation assumes homodyne operation, but the statistics are the same for IF processing. The experimental statistics are plotted along with other density functions to provide an intuition of the relative performance. Moments are calculated. Output mean and variance are estimated. The inputs and comparison statistics have variances of 1.

14. `cornoise2.mcd`, `cornoise4.mcd`, `cornoise5.mcd`. Similar to the above but with different parameters.

15. `TEMPAPPROX1.MCD`. This MathCAD program is a temperature approximation for use with radiometer interceptors.

16. `table3-10.xls`. This Excel program calculates mainlobe and sidelobe available clutter temperature for a radiometer interceptor.

APPENDIX A.4

The following items of software are in Appendix 4 on the CDROM. See warranty comments in appendix 1.

1. `AmbientPWR1.mcd`. This program calculates the ambient power spectrum and pulse density in a 3-MHz band with threshold of –115 dBW for a hypothetical set of emitters distributed over the battle space.

2. `AmbientPWR2.mcd`. This program calculates the ambient power spectrum and pulse density in a 1-GHz band with threshold of –90 dBW for a hypothetical set of emitters distributed over the battle space.

3. `AmbientPWR3.mcd`. This program calculates the ambient power spectrum and pulse density in a 1-GHz band with threshold of –90 dBW for a hypothetical set of emitters distributed over the battle space with emphasis on communications.

4. `AmbientPWR4.mcd`. This program calculates the ambient power spectrum and pulse density in a 1-GHz band with threshold of –90 dBW for a hypothetical set of emitters distributed over the battle space with emphasis on communications and a small battle space. It is set for mainlobe intercepts only.

5. `AmbientPWR5.mcd`. This program calculates the sidelobe ambient power spectrum and pulse density in a 1-GHz band with threshold of –90 dBW for a hypothetical set of emitters distributed over the battle space with emphasis on communications and a small battle space. It is set for sidelobe intercepts only.

6. `AmbientPWR6.mcd`. This program calculates the ambient power spectrum and pulse density in a 30-MHz band with various thresholds for a hypothetical set of emitters distributed over the battle space with emphasis on communications and a small battle space. It is set for both mainlobe and sidelobe intercepts.

7. `grtcirrt.mcd`. This program calculates the slant range, the ground range, and the grazing angle to a penetrator flying a great-circle route at constant altitude from an observer/threat.

8. `Probden1.mcd`. This program calculates one side of several different probability density functions useful in approximating terrain-masking statistics including the Cauchy density function.

9. `Probden2.mcd`. This program calculates one side of several different probability density functions useful in approximating terrain-masking statistics including the gamma and Weibull density functions.

10. `Probden3.mcd`. This program calculates one side of several different probability density functions useful in approximating terrain masking statistics including the Gaussian and chi squared density functions.

11. `Probden4.mcd`. This program calculates one side of several different probability density functions useful in approximating terrain-masking statistics including a Cauchy like density and chi squared density functions.

12. `rdrcls51.mcd`. This is a very simple and approximate program to calculate the amount of the terrain visible to an observer as a function of altitude and ground range to the horizon.

13. `rdrcls52.mcd`. This is a very simple program to calculate the number of emitters visible to an observer in masking terrain as a function of altitude and ground range to the horizon.

14. `Table4-6a.xls`. Tabulation of typical battlefield and civilian emitters supporting Table 4.7 in text, calculation of ambient power and associated plots.

15. `Table4-6c.xls`. Tabulation of battlefield and civilian emitters supporting Table 4.7 in text, calculation of ambient power, calculation of ambient pulse density, and associated plots.

16. `visible.mcd`. This simple and approximate program calculates terrain visibility probability density as a function of apparent altitude and ground range or grazing angle. In addition, it calculates the probability that a visible terrain location will stay visible as a function of azimuth angle or cross range displacement and ground range. It works in MathCAD 2001.

17. `Visible9a.mcd`. This program calculates low-altitude target visibility probability density as a function of surveillor altitude, target/emitter altitude, apparent altitude, and slant range or grazing angle. In addition, it calculates the probability that a visible terrain location will stay visible as a function of azimuth angle or cross range displacement and ground range. As is typical of MathCAD programs, this works in MathCAD 2001 only if BACKWARDS COMPATIBILITY is checked

in the MATH/OPTIONS/CALCULATIONS/pull-down menu. Minor parameter changes from 9.

18. VISIBLE10.MCD. This program calculates low-altitude target visibility probability density as a function of surveillor altitude, target/emitter altitude, apparent altitude, and slant range or grazing angle. In addition, it calculates the probability that a visible terrain location will stay visible as a function of azimuth angle or cross range displacement and ground range. As is typical of MathCAD programs, this works in MathCAD 2001 only if BACKWARDS COMPATIBILITY is checked in the MATH/OPTIONS/CALCULATIONS/pull-down menu. Minor parameter changes from 9a.

19. Visible10a.mcd. This program calculates low-altitude target visibility probability density as a function of surveillor altitude, target/emitter altitude, apparent altitude, and slant range or grazing angle. In addition, it calculates the probability that a visible terrain location will stay visible as a function of azimuth angle or cross range displacement and ground range. As is typical of MathCAD programs, this works in MathCAD 2001 only if BACKWARDS COMPATIBILITY is checked in the MATH/OPTIONS/CALCULATIONS/pull-down menu. Minor parameter changes from 10.

20. VISIBLE11.MCD. This program calculates low-altitude target visibility probability density as a function of surveillor altitude, target/emitter altitude, apparent altitude, and slant range or grazing angle. In addition, it calculates the probability that a visible terrain location will stay visible as a function of azimuth angle or cross range displacement and ground range. As is typical of MathCAD programs, this works in MathCAD 2001 only if BACKWARDS COMPATIBILITY is checked in the MATH/OPTIONS/CALCULATIONS/pull-down menu. Minor parameter changes from 10a.

APPENDIX A.5

The following items of software are in Appendix 5 on the CDROM. See warranty comments in appendix 1.

1. BarkDopp2a.mcd. This analysis is a calculation of the degradation in ISLR as a function of Doppler or phase shift across the uncompressed pulse for a Barker code.

2. ChanSep1a.mcd. This program is the calculation of the output spectrum from a transmitter with a 10-pole filter transmitting a 13-chip Barker code and 169-chip Barker-Barker code with its leakage into an adjacent receiving channel also with a 10-pole filter. The chip width is 50 nsec.

3. `ChanSep2a.mcd`. The following is the calculation of the output spectrum from a transmitter with a 6-pole filter transmitting a 13-chip Barker code and its leakage into an adjacent receiving channel also with a 6-pole filter. The chip width is 50 nsec.

4. `Islr2.mcd`. This program calculates the ISLR for unweighted, triangular weighted, and Parzen weighted FFT filters. This is primarily oriented toward chirp stretch processing. C is the multiplier of the 3 dB down point to be used in the ISLR calculation. N_{rb} is the number of equivalent range bins, T_E is the effective pulse width, and Δf_{IF} is the input bandwidth to the A/D.

5. `LinFMAmb1.mcd`. This MathCAD program calculates the ambiguity function of a linear FM waveform with a 4-μsec pulsewidth and 5-MHz bandwidth. Larger time-bandwidth products are easily performed with this program, but it is more difficult to gain an intuition for waveform behavior.

6. `LinFMAmb2.mcd`. This MathCAD program calculates the matched filter output for a linear FM in the presence of Doppler. It is related to the ambiguity function of the waveform. It doesn't appear to be stable for large compression ratios somewhere between 100 and 1000.

7. `Multifreq1a.mcd`. This MathCAD analysis is for multifrequency waveforms used for radar detection and data link signaling. It assumes propagation through a Rayleigh fading channel with orthogonal transmitted waveforms and square law detection with unknown received phase followed by noncoherent integration. The analysis is based on the *Spread Spectrum Communications Handbook*, p. 605, and the *Radar Handbook*, pp. 15-6 through 15-27.

8. `PCOUT9.xls`. This Excel spreadsheet hosts a visual basic for applications (VBA) program `compcode()` which calculates simple complementary codes and plots the resulting outputs. Be sure to turn on TOOLS/MACRO/ALL OPEN WORKBOOKS.

9. `Compcdz2.bas`. COMPCDZ2 is a QuickBasic program that generates interleaved biphase complementary codes, applies a video filter and timing mismatches, compresses the code, and sums the complements for final output. The generated code is written to a file called PCCODEZ2. The compressed output is written to a file called PCOUTZ2, and the maximum code length supported by the program is L chips.

10. `Compcodg.bas`. This is a QuickBasic program that generates polyphase and biphase complementary codes, applies a series of video filter and phase/timing mismatches, compresses the code, and sums complements for final output. It calculates the peak sidelobe ratio, integrated sidelobe ratio, and the rms sidelobe ratio for a single code and the sum of the complements. The generated code is

written to a file called PCCODEG. The compressed output is written to a file called PCOUTG, and the maximum code length supported by the program is 512 chips.

11. Compcodz.bas. This is a simpler QuickBasic program than Compcodg.bas but still generates interleaved biphase complementary codes, applies a video filter and timing mismatches, compresses the code, and sums complements for final output. It calculates the peak sidelobe ratio, integrated sidelobe ratio, and the rms sidelobe ratio for a single code and the sum of the complements. The generated code is written to a file called PCCODEZ. The compressed output is written to a file called PCOUTZ, and the maximum code length supported by the program is L chips.

12. Polycoda.bas. This is a very simple QuickBasic program that generates type II polyphase complementary codes, applies a video filter and timing mismatches, compresses the code, and sums complements for final output. The generated code is written to a file called Polycode. The compressed output is written to a file called Polyout, and the maximum code length supported by the program is 512 chips.

13. Compcodb.bas. This is a very simple QuickBasic program to generate and compress biphase complementary codes. The code resides in **A** and the complement in **B**. The maximum code length supported is 1024 chips.

APPENDIX A.6

The following items of software are in Appendix 6 on the CDROM. See warranty comments in appendix 1.

1. Islr3.mcd. Antenna sidelobe and ISLR calculation with element pattern and raised cosine weighting, square linear array of 50×50 elements at 1/2-wavelength spacing.

2. Islr4.mcd. Antenna sidelobe and ISLR calculation with element pattern and Hamming cosine weighting, square linear array of 50×50 elements at 1/2-wavelength spacing.

3. Islr5.mcd. This program calculates the ISLR and sidelobes for cosine weighted with edge value 0.25 for a square linear array of $n \times n$ elements at 1/2 wavelength.

4. Islr7.mcd. This program calculates the ISLR and antenna sidelobes with element pattern and segmented parabola weighting, square linear array of 50×50 elements at 1/2-wavelength spacing.

5. Islr8.mcd. This program calculates the ISLR and antenna sidelobe with element pattern and cubic segment—Parzen weighting, square linear array of 50×50 elements at 1/2-wavelength spacing.

6. `Islr9.mcd`. This program calculates the ISLR and antenna sidelobe calculation with element pattern and cosine cubed weighting, square linear array of 50×50 elements at l/2-wavelength spacing.

7. `Radome1.mcd`. This MathCAD program calculates the transmission and reflection performance of single-wall radomes using the lossless case in Jasik pp. 32-4 and 32-5 as a function of angle and frequency for epoxy quartz dielectric.

8. `Radome2.mcd`. This MathCAD program calculates the transmission and reflection performance of several frequency selective radomes using the lossless case as a function of angle and frequency for epoxy glass dielectric.

9. `Radome3.mcd`. This MathCAD program calculates the transmission and reflection performance of several frequency selective radomes using the lossless case in Jasik p. 32-35 combined with slots as a function of angle and frequency for epoxy quartz dielectric.

10. `Radome4.mcd`. This MathCAD program calculates the transmission and reflection performance of several frequency selective radomes using the lossless case in Jasik p. 32-35 combined with slot radiators in Jasik pp. 8-2 through 8-12 as a function of angle and frequency for epoxy quartz dielectric.

11. `rdrclas1a.mcd`. This MathCAD program calculates the antenna sidelobes of a linear array with element pattern and periodic error.

12. `rdrclas7a.mcd`. This MathCAD program calculates antenna sidelobes for a 50-element linear array with raised cosine weighting where the paired echoes are 3 beamwidths apart instead of 4 beamwidths. This is a much easier weighting function to realize in a real antenna, but it has poorer sidelobes.

13. `rdrclas8a.mcd`. This MathCAD program is an antenna sidelobe calculation with element pattern, Hamming cosine weighting for a linear array of N elements of d spacing at 1 wavelength.

14. `rdrcls11.mcd`. This simple MathCAD program is a very approximate antenna RCS calculation for a linear array near broadside.

15. `rdrcls12.mcd`. This simple program is a very approximate antenna RCS calculation for a circular array with elements below cutoff, which works to about 30°.

16. `rdrcls13a.mcd`. This simple program is an antenna sidelobe calculation for Hanning weighting.

17. `pointerr.mcd`. This MathCAD program calculates the allowable rms per element phase error for an array as a function of required angular accuracy or cross-beam dimensional accuracy. The approach is taken from *Antenna Engineering*

Handbook, by H. Jasik, Ed., pp. 2-36 through 2-41 and *Reference Data for Radio Engineers,* 6th ed., pp. 27-34 through 27-36.

18. `Soninel.mcd`. This MathCAD program is the calculation of the Sonine distribution and sidelobes for circular and rectangular antennas. Integrated sidelobe ratios are also calculated.

19. `ArrayRCS1.mcd`. This program is the RCS calculation of a small 2-D antenna array considering only the correlated and uncorrelated first scatter from radiators.

20. `ArrayRCS10.mcd`. This program is the antenna RCS calculation of a large 2-D array first scatter from radiators with correlated and uncorrelated components and 6 λ edge treatment. Both element spacing and number of elements are selectable. Array is rotated 45° around z-axis and tilted back 25°, which are also adjustable.

21. `ArrayRCS2.mcd`. This program is the RCS calculation of a small rectangular linear array antenna with mainbeam pointed off the normal.

22. `ARRAYRCS3.MCD`. This program is the RCS calculation of a small antenna 2-D array with skewed radiating elements embedded in a parallelogram plate with edge treatment. The mainbeam is pointed below and left of the normal.

23. `ARRAYRCS4.MCD`. This program is the RCS calculation of a small antenna 2-D array with skewed radiating elements embedded in a parallelogram plate with edge treatment. The mainbeam is pointed above and right of the normal.

24. `ARRAYRCS5.MCD`. This program is the RCS calculation of a small antenna 2-D array with skewed radiating elements embedded in a parallelogram plate test fixture with edge treatment. The mainbeam is pointed to the same location as a counterpart real test article. The median RCS of the test article is compared with the predicted performance.

25. `ARRAYRCS6.MCD`. This program is the RCS calculation of a small antenna 2-D array with skewed radiating elements embedded in a parallelogram plate test fixture with edge treatment. Uncorrelated errors and first level feeds are included in the calculation. The mainbeam is pointed to the same location as a counterpart real test article. The median RCS of the test article is compared with the predicted performance.

26. `ArrayRCS7.mcd`. This program is the RCS calculation of a large 2-D array antenna first scatter from radiators below cutoff.

27. `ArrayRCS8.mcd`. This program is the RCS calculation of a large 2-D array antenna first scatter from radiators below cutoff. The array is rotated to 45° and canted back to 25°.

28. `ArrayRCS9.mcd`. This program is the RCS calculation of a large 2-D array antenna first scatter from radiators below cutoff with 6 λ edge treatment. Array is rotated 45° around z-axis and tilted back 25°.

29. `EdgeTreat1.MCD`. This MathCAD program is the calculation of the cross-polarization RCS of an antenna embedded in a parallelogram shaped plate with edge treatment.

30. `ogiveRCS2.mcd`. This MathCAD program calculates ogive geometrical optics RCS plus tip and ogive geometrical diffraction as well as back edge RCS including edge treatment and canted trailing edge at X band.

31. `ogiveRCS2s.mcd`. This MathCAD program calculates ogive geometrical optics RCS plus tip and ogive geometrical diffraction as well as back edge RCS including edge treatment and canted trailing edge at S band.

32. `ogiveRCS4.mcd`. Ogive radome RCS using the physical optics approximation in the perpendicular plane and stationary phase in the parallel plane at X band. This program is not very stable; be careful in choosing parameters.

33. `RDRCLAS19.MCD`. This MathCAD program calculates the patterns of multibeam electronically scanned linear array antennas with Hamming weighting and 100 MHz beam step frequency. Array size, element spacing, center frequency, and frequency step size are easily changeable.

APPENDIX A.7

The following items of software are in Appendix 7 on the CDROM. See warranty comments in appendix 1.

1. `Table7-6a.xls`. This Excel program generates Table 7.6 in the text. It is the calculation of the *SNR, CNR,* and dynamic range through the functions of a typical stealth signal processor (Fig. 7.22).

2. `Table7-9MINPOW.XLS`. This Excel program is a tabular calculation of the radar equation for the example modes in Chapter 7 under the minimum power assumption.

Glossary

α Cone angle or secondary phase angle.

AMTI   Airborne moving target indication.

AOA   Angle of arrival.

AOR   Area of regard.

ARM   Antiradiation missiles.

ASIC   Application specific integrated circuit, sometimes called a gate array.

$B_{()}$ Bandwidth for subscripted item.

$\beta_{()}$ Secondary phase angle for subscripted item.

bandpass   Passing only those frequencies in the normal emitter band of operation.

BER Bit error rate.

BM   Bulk memory.

$BW_{()}$ Beamwidth for subscripted item.

c   Velocity of light, 2.9979×10^8 m/sec.

CAGC   Coarse automatic gain control.

CFAR   Constant false alarm rate.

clutter/radar clutter   Unwanted signals, echoes, or images that appear on face of display screen, interfering with observation of desired signals.

crosstalk   Unwanted signals in communication channel caused by transference of energy from another circuit.

CW   Continuous wave.

$d_{()}$ Element spacing or incremental range resolution for subscripted item.

DARPA   Defense Advanced Research Projects Agency.

data link   Radio used to transmit digital or analog information in machine-readable format. Used to transmit radar, sonar, imagery, and other sensor data, as well as weapons guidance command, fire control information, and missile telemetry.

deinterleave   To separate multiple mixed pulse trains from one another to allow measurement of each.

DFT   Discrete Fourier transform.

DN   Data network.

DOA   Direction of arrival systems.

E   Vector electric field.

E Measured voltage.

$E_{()}$ Electric field component for subscripted item.

ε Depression angle, total error standard deviation, permittivity

ε_0 Permittivity of free space, 8.85×10^{-12} F/m

ECCM   Electronic counter-countermeasures.

ECM   Electronic countermeasures equipment.

ELINT   electronic intelligence systems.

emission   Electromagnetic waves radiated by any electronics or physical process; substances (exhausts) discharged into the air.

EOB   Electronic order of battle. Deployment and utilization of communications, radar, electronic countermeasures, and radar intercept equipment in (adversary) military doctrine and tactics. *See also* order of battle.

ERPP   Effective radiated peak power.

ERTOS   Embedded real-time operation system.

ESA   Electronically scanned array.

ESM   Electronic support measures.

$f_{()}$ Frequency for subscripted item.

ϕ Space angle around the z axis measured from the x-z plane.

φ Electrical phase angle.

FAGC   Fine automatic gain control.

FEBA   Forward edge of the battle area (pp. 92, 173, 223).

FEXT   Far-end crosstalk.

FFT   Fast Fourier transform.

FIR   Finite impulse response.

FLAMR   Forward Looking Advanced Multimode Radar, a U.S. Air Force research program.

FSS   Frequency selective surface.

$G_{()}$ Antenna or processing gain for subscripted item.

γ_c Conductivity.

$\Gamma_{()}$ Complex reflection factor for subscripted item.

Γ_{EU} Incomplete Euler gamma function.

GMTD   Ground moving target detection.

GMTI   Ground moving target indication.

GMTT   Ground moving target tracking.

GO   Geometric optics.

GPP   General-purpose processor.

GPS   Global positioning system.

$h_{()}$ Altitude for subscripted item.

h   Circuit impulse response.

H   Vector magnetic field.

$H_{()}$ Magnetic field component for subscripted item.

H   Fourier transform of the impulse response.

i Index.

I_0 Modified Bessel function of the first kind, order 0.

interleave   To mix multiple independent codes or functions into a single time line.

I/O   Input-output circuitry specialized for application.

IR   Infrared.

ISAR   Inverse synthetic aperture radar. Doppler processing to image a rotating body.

ISLR   Integrated sidelobe ratio.

j   Square root of minus one.

j Index.

J_0 Bessel function of the first kind, order 0.

J_1 Bessel function of the first kind, order 1.

JTIDS   Joint Tactical Information Distribution System, sometimes called Link 16.

k Index.

k_B Boltzmann's constant.

KFC   Ku-Band Fire Control radar.

l Incremental length or index.

λ Operating wavelength.

$L_{()}$ Loss.

LAPSIM   Low-altitude penetrator simulation (software program).

LO   Low observable system designed to make object difficult to detect. *See also* signature, passive.

LOS   Line of sight.

lossy   Causing weakening or dispersion of electrical energy.

LPI/LPIS   Low Probability of Intercept System, or system designed to make object (stealth aircraft or weapon) difficult to detect and/or intercept. *See also* signature, active.

LREW   Long-Range Early Warning radar.

m Index.

MCU   Microprocessor control unit.

MDL   Missile data link.

MIMD   Multiple instruction multiple data.

Monte Carlo simulation   Random input variable stimulation to a system or model and output characterization; technique used to obtain approximate solution to mathematical or physical problem.

MPPS   Million pulses per second.

MTAE   Multiple time around echoes.

MTI   Moving target indication.

MTT   Moving target tracking.

n Index.

$N_{()}$ Total number for subscripted item.

$\boldsymbol{n}$ Noise power.

$\mathbf{n}$ Vector surface normal.

μ_0 Permeability of free space, $4\pi \times 10^{-7}$ H/m.

NED   North East Down coordinate system.

NEXT   Near-end crosstalk.

$NF_{()}$ Noise figure for subscripted item.

NRL   Naval Research Laboratories.

ω Angular frequency.

Ω Solid angle.

order of battle   Identification, strength, command structure, and disposition of the personnel, units and equipment of a military force. *See also* electronic order of battle.

$P_{()}$ Power for subscripted item.

p Index or incremental probability.

$\boldsymbol{P}_{()}$ Probability for subscripted item.

$\overline{\mathbf{P}}_{i,s}$ Vector incident or reflected power.

planform Contour of object as seen from above.

platform A weapon system entity such as an aircraft, ship, tank, mobile shelter, spacecraft, etc.

PCR Pulse compression ratio.

PDI Post detection integration.

PUL(t,T) The pulse function, zero everywhere except during a pulse of height 1 and duration associated to T, starting at $t = 0$.

PO Physical optics.

PRF Pulse repetition frequency.

PRI Pulse repetition interval.

ψ Grazing angle.

PSLR Peak sidelobe ratio.

PSP Programmable signal processor.

$r_{()}$ Local distance for subscripted item.

radar (*Radio detecting and ranging*) radio device or system for locating objects by means of ultrahigh-frequency radio waves reflected from object for purpose of determining object location and direction in which traveling.

radiometer Instrument for measuring thermal, electromagnetic, or acoustic total radiated energy.

radome Housing to protect antenna assembly of a communications, intercept receiver, or radar set.

RAM/RAM(s) Radar-absorbing material(s).

range Maximum distance at which a radio transmission can be effectively received.

RCS Radar cross section; (1) ratio of the scattered power to the incident power in the direction of an observer at infinity, (2) image produced by radar signals reflected off target surface.

RDA Research & Development Associates, a systems engineering and analysis company.

RF Radio frequency.

RHAW Radar homing and warning receivers.

rms Root-mean-square.

RMSLR Root-mean-square sidelobe ratio.

RSS Root sum squaring.

RWRs Radar warning receivers.

$s_{()}$ Speed or spectral power for subscripted item.

$\sigma_{()}$ Radar cross section or standard deviation for subscripted item.

S&TA Surveillance and Target Acquisition radar.

signature, active Observable emissions from platform: acoustic, chemical (soot and contrails), communications, radar, IFF, IR, laser and UV.

signature, passive Platform observables, requiring external illumination: magnetic and gravitational anomalies; reflection of sunlight and cold outer space; reflection of acoustic, radar, and laser illumination; reflection of ambient RF (a.k.a. splash track).

sinc( ) Symbol for sin x/x function

sins( ) Symbol for sin(Nx)/(N sin x) function.

SMTD Spotlight moving target detection.

SNR Peak-power-signal-to-mean-noise-ratio.

spoofing Transmitting approximation of enemy waveform interleaved with LPIS waveform.

STAR Simultaneous transmit and receive.

Stealth Popular term for "low observables," i.e., technologies, materials, and designs intended to reduce the radar, infrared, acoustic, and visual signatures of aircraft, missiles and ships. This reduces (ideally to zero) the range at which they can be detected by radar, sonar, and infrared (IR) sensors.

t Time.

t_L Time per look in a single cell.

T_I Interceptor search time.

T_{OT} Emitter illumination time of interceptor.

TA   Terrain avoidance.

TACAN   Tactical air navigation.

TDMA   Time division multiple access, a communications multiplex method.

TF   Terrain following.

θ Space angle measured from z axis.

TWT   Traveling wave tube (type of transmitter).

u $\sin\theta\cos\phi$, x direction cosine.

UCIF   Universal console interface; standard PC keyboard, mouse and display interface.

v $\sin\theta\sin\phi$, y direction cosine.

VCO   Voltage controlled oscillator.

VFO   Variable frequency oscillator.

w $\cos\theta$, z direction cosine.

WAHTD   Wide area hard target detection.

WAMTD   Wide area moving target detection.

XFC   X-Band Fire Control radar.

Index

About the Author

The above photo shows the author in front of the radar and data link on the Tacit Blue aircraft. You can't see anything, which is exactly as it should be on a stealth aircraft. Although there were many more important people on the Tacit Blue program, the author was the first program manager on the Tacit Blue radar avionics. Tacit Blue was extremely successful and reached all its program objectives ahead of schedule. Tacit Blue spent 20 years in the black, and it was unveiled and partially declassified in the late 1990s. The author was involved in many stealth programs including Have Blue, F-117, Sea Shadow, Advanced Cruise Missile, F-22, B-2, and others. The author has been elected a Pioneer of Stealth, a Senior Member of American Institute of Aeronautics and Astronautics, and a Fellow of the Institute of Electrical and Electronic Engineers. The author was an inventor, leader, or contributor to many world firsts, including manned spaceflight, telecommunications, digital signal processing, synthetic aperture radar, and stealth. He was a company officer of General Motors Hughes Electronics and is currently president of DL Sciences, Inc.

The SciTech Radar and Defense Series

How to Speak Radar, by Arnold Acker
This self-teaching CD is the most accessible and least mathematical product available for those needing to know only the radar basics. 2001 / CDROM / ISBN 1-891121-17-0

Radar Principles for the Non-Specialist, 3rd ed., by J.C. Toomay and Paul Hannen
Solid explanation of radar fundamentals and applications that has become a favorite for short courses and quick study. Exercise solutions provided. 2004 / Softcover / 240 pages / ISBN 1-891121-28-6

Introduction to Airborne Radar, 2nd ed., by George Stimson
World's leading radar book—a history, reference, tech manual, and textbook rolled into a beautiful volume, packed with full-color photos and drawings! 1998 / Hardcover / 592 pages / ISBN 1-891121-01-4 or CD 1-891121-14-6

Understanding Radar Systems, by Simon Kingsley and Shaun Quegan
An easy-to-read, wide-ranging guide to the world of modern radar systems. 1999 / Hardcover / 392 pages / ISBN 1-891121-05-7

Air and Spaceborne Radar Systems, by P. Lacomme et al.
Intended for actual users of modern airborne radar, this book examines radar's role within the system when carrying out its assigned missions. 2001 / Hardcover / 496 pages / ISBN 1-891121-13-2

Radar Design Principles, 2nd ed., by Fred E. Nathanson
A classic since 1969—cutting-edge guide to today's radar technology. 1998 / Hardcover / 656 pages / ISBN 1-891121-09-X

Radar Foundations for Imaging and Advanced Concepts, by Roger Sullivan
Designed to lead the reader from basic concepts to a survey of imaging and advanced concepts. For readers with a calculus background seeking a bootstrap to today's most exciting and relevant radar applications and techniques. 2004 / Softcover / 475 pages / ISBN 1-891121-22-7

Understanding Synthetic Aperture Radar Images, by Chris Oliver and Shaun Quegan
Now back in print! This practical reference shows how to produce higher quality SAR images using data-driven algorithms, and how to apply powerful new techniques to measure and analyze SAR image content. 2004 / Hardcover / 512 pages / ISBN 1-891121-31-6

Introduction to RF Stealth, by David Lynch, Jr.
The first book covering the complete aspects of RF Stealth design. Explains first-order methods for the design of active and passive stealth properties. CDROM included with book. 2004 / Hardcover / 570 pages / ISBN 1-891121-21-9

Introduction to Adaptive Arrays, by Robert Monzingo and Thomas Miller
Now back in print! An introduction to adaptive array sensor systems for enhancing the detection and reception of desired signals. 2004 / Softcover / 543 pages / ISBN 1-891121-24-3

Low Angle Radar Land Clutter, by J. Barrie Billingsley
This book shows how to design and predict the performance of radars that operate in situations where land clutter is prevalent. 2002 / Hardcover / 700 pages / ISBN 1-891121-16-2

Airborne Early Warning System Concepts, by Maurice Long
Now in paperback! This comprehensive discussion of airborne early warning (AEW) system concepts encompasses a wide range of issues, including capabilities and limitations, developmental trends, and opportunities for improvement. 2004 / Softcover / 538 pages / ISBN 1-891121-32-4

Microwave Passive Direction Finding, by Stephen Lipsky
Now back in print! This classic text unifies direction-finding (DF) theory and brings together into a single source wide-ranging information on the technology of measuring the direction-of-arrival of microwave signals. 2004 / Softcover / 320 pages / ISBN 1-891121-23-5

Digital Techniques for Wideband Receivers, 2nd ed., by James Tsui
Now in paperback! Shows how to effectively evaluate ADCs, offers insight on building electronic warfare receivers, and describes zero crossing techniques that are critical to new receiver design. 2004 / Softcover / 608 pages / ISBN 1-891121-26-X

Visit the SciTech website for details and pricing
www.scitechpub.com